eurotopia

Directory of Communities and Ecovillages in Europe

The cover cardboard is perforated. If you cut or rip along the perforation, you can see the picture of the party through the window you created. You're doing this entirely at your own risk to damage the book. You can also just turn the page to see the picture.

Imprint

This book is the English version of the sixth edition of the eurotopia European Communities Directory, which has been published every 3-5 years since 1996. It is also available in German language (see bibliography).

eurotopia Directory | Poppau | Sieben Linden Ecovillage | 38489 Beetzendorf | Germany

www.eurotopia.directory info@eurotopia.de

Editor-in-chief:	Michael Würfel (publisher)

We thank all communities that have answered our calls for participation!

IT:	Volker Peters
Data acquisition/input and translations:	Michael Würfel, Rahel Gaedt, Sigrid Künstner, Martin Mütterlein, Mariam Dadgar, Marsilio Passaglia, Jonas Duhme, Anna Kleinert, Peter Reinhold, Lennart Schütz
Additional translations:	Tomas Remiarz, Jenny Lusias, Mathias Lederle, DeepL.com, Google Translator
Layout:	Michael Würfel
Maps:	Maps from the following countries from d-maps.com: Europe-East, Italy, Finland, Poland, Russia, Spain. Large scale map from Russia by Uwe Dedering (Wikipedia) Thank you all for permission to use.
Printing:	Zimmermann Druck, Balve printed on recycled paper

The drawing on the inside cover is taken from the book "Unser Zuhause" by Doro Göbel and Peter Knorr (ISBN 978-3407795984, Beltz-Verlag). We absolutely recommend their picture books for children and parents alike.

Photos have been provided by the participating communities. We thank all photographers, especially the ones who are not credited by name.

eurotopia Directory of Intentional Communities and Ecovillages in Europe
2020 edition, Sieben Linden: Blühende Landschaften e.V.

ISBN 978-3-9816860-5-0

La Città della Luce, Italy, page 289.
"Relationships are at the top of our value system. We share time, property, talents and emotions, together with the common spaces of our home."

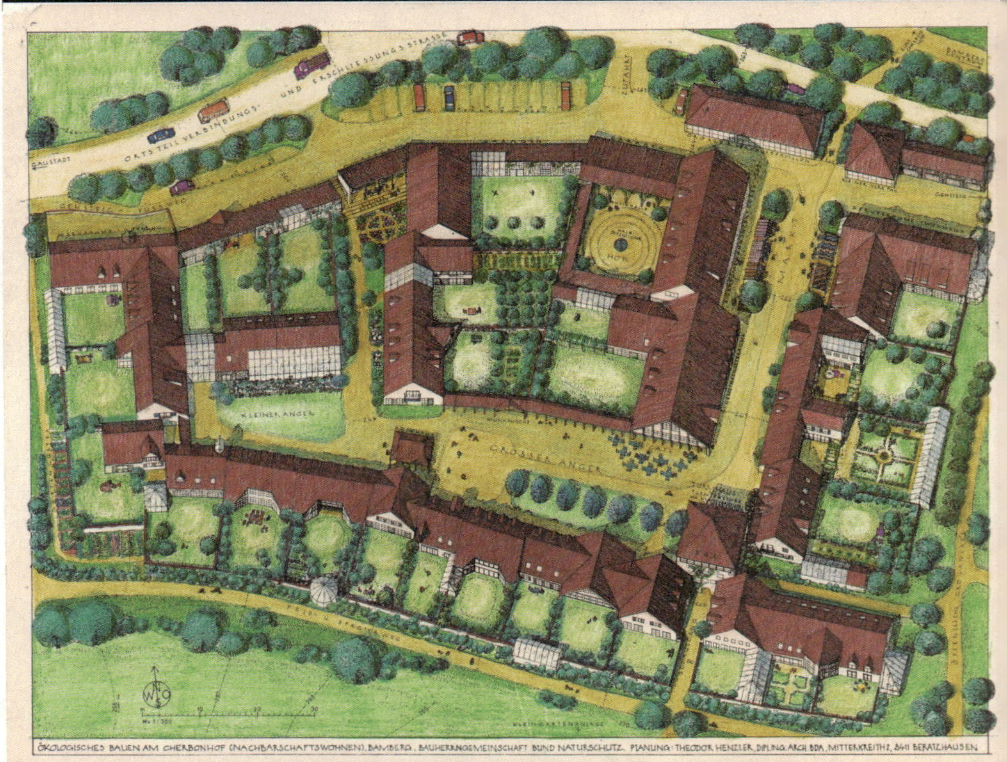
ORTSTEILVERBINDUNGS- UND ERSCHLIESSUNGSSTRASSE
KLEINER ANGER
GROSSER ANGER
M= 1:300
ÖKOLOGISCHES BAUEN AM OHERBONHOF (NACHBARSCHAFTSWOHNEN), BAMBERG, BAUHERRNGEMEINSCHAFT BUND NATURSCHUTZ. PLANUNG: THEODOR HENZLER, DIPL.ING. ARCH. BDA, MITTERKREITH 2, 8411 BERATZHAUSEN

Left, above: Schloss Tempelhof, Germany, page 247)
(shot from exactly the opposite direction than the photo on page 248).

Left, below: Ecological Settlement at Cherbonhof, Germany, page 229.
"The settlement concept is characterised by a varied sequence of streets, alleys and squares. The private interior gardens, which should be protected from view as far as possible ('living rooms in the green'), are contrasted with residential courtyards and street spaces that can be used in a variety of ways and extend right up to the buildings."

Nature Community, Germany, page 231:
"A former holiday village near the Czech border in the middle of pristine nature offers living space to our ideas: We live vegan, ecologically and sustainably, connected to humans and animals, honestly and consciously."

Above: Sunflower Cohousing, France, page 131
Below: Tamera, Portugal, page 327

Above: Aerium, France, page 98: "Our collective offer lots of freedom and therefore requires high personal motivation, spirit of initiative and responsibility."

Middle: Keveral Farm, UK, page 393: "Some people call this place paradise."

Below:
Giesserei, Switzerland, page 372: Ca. 310 residents from 12 nations, 156 living units, 14 commercial operations.
Photo: Kurt Lampart

Above: "The Udgården is a small ecological community consisting of a farm and 16 apartments just outside of a small village in Central Jutland." Denmark, page 92
Below: "AzGrad is not to produce ready-made citizens for a system that can no longer find solutions for the problems it has created." Russia, page 330

ELTRIRIA 2009 Thierry Oger

Above: Interneland/Innerland, page 349:
"We have international communities in Spain, Brazil and Argentina as well as centers in 30 countries and we also offer virtual communities for online meetings and meditations on our websites."

Below (not a computer generated image):
Damanhur, Italy, page 284: "Our respect for the environment goes beyond an ecological vision - it recognizes the existence of a conscious and sensitive life in the plant and animal worlds, and researches contact with intelligences that inhabit this universe."

SMALL FOOTPRINT

Left: Small Footprint, Estonia. "A drawing of our premises by Paulinha Soares Junqueira", page 417

Above: Cambium, Austria. "We wish for mindfulness and respect in our personal relationships as well as an open attitude towards the other and her/his potential", page 57

Middle left: Cohousing Pomali, Austria, page 58

Middle right: Osho Shangri La, Czech Republic, page 415

Right: Keuruu Ecovillage, Finland, page 94

Arche im Nauen, Switzerland, page 367: "As a community we would like to live with each other on the basis of the Gospel with our abilities and limits."

Trelay Cohousing Community, UK, page 410: "We work hard but we also have fun."

Youth of Kirschblüte Community, Switzerland, page 370: "We reject as hostile to community all forms of authority and hierarchy that do not result as a matter of course from natural power relationships."

Above: Wilde Gärtnerei, Germany, page 267: "We open our door for all kinds of guests who either are looking for first experiences in agriculture or want to extend their limits in life."

MasCobado, France, page 127: "We have co-designed our residence and apartments with the architects and we manage our living space collectively."

Krisna Völgy, Hungary, page 422: "Europe's largest and most well organized ecological community."

Left page, above: Lilleoru, Estonia, page 416
Below: Bijgaardehof (planned, Belgium) consists of three co-housing-projects: Wijgaard, De Spore & Biotope. Page 442

Right page, above left: Sannerz community is a christian community and collective which belongs to the Bruderhof movement. Page 245
Above right: Ecovillage "Ecohameau Les Trois sources", France, page 105

2nd from above: Sambruket i Sösdalabygden, Sweden, page 562: "Sambrukets aim is to extend eco village ideas to the large number of people who cannot or do not want to join something as tight and definitively located as a formal community."

3rd from above: InGe Wohnprojekt, Germany (planned), page 456

Below: Tinkers Bubble, UK, page 408: "Life is lived mostly outdoors, so it's cold in the winter, but we live on the top of a steep hill, so there's plenty of chances to get warm!"
(photo by David Spero)

Contents

Cloughjordan Ecovillage, Ireland, page 277 - Foto von Eoin Campbell

This eurotopia Directory...

Congratulations. You found this book! It shouldn't be so difficult, since eurotopia is the most comprehensive and, at least within the first years of its publication, most up-to-date account on intentional community life in Europe – but it is. Distributing this international edition of our directory is difficult. Of the last eurotopia Directory, published 2014, we sold more than 7000 German books, but less than 1400 English copies. Although it was light and we could offer wordwide shipping for just 2 Euros, it was difficult to spead the world. Even in English, even with the internet helping people to find it. Now, this edition is even more packed with data and wisdom, which is great, but it weighs beyond the magic metric pound which makes it a seven-Euro-parcel to ship. How should we get it into the world?
This is why I'm considering to strike a deal with – well, I don't want to use fairy tale terms here, but if you read the following article, you might see why we aren't too fond of global commerce. Anyway, you might have obtained this book from an online sales giant – and it's ok. We are behind this. It's a way to get the book to you. And the only job that was sacrificed was myself standing in our tiny packing booth listening to the tunes that the old stereo from the free shop is still willing to play, shedding tears about the German postage that leaves us on these eurotopia envelopes much too fast.

That said, I can cheer up because now, with the distribution taken care of, there's hope to actually find readers for this preface of mine. And again, that's a good thing because this book is a treasure – again. It continues to present as many communities as possible in Europe. It shows that community life is a living, functioning model of life. This is not supposed to be a secret, and most people have already experienced and enjoyed the state of "community" outside of intentional communities, ecovillages, communes and co-housing projects. But even this edition of the eurotopia directory will not impress mainstream perception. It stroke me that for the last edition of eurotopia, not a single commercial newspaper or magazine has reacted to the review copies I sent out. And I sent them out to pretty much everyone. I thought that the eurotopia Directory could inspire exciting articles: that there are hundreds of real-life examples in Europe of people living together in a completely different way than letting themselves be carried around by economic, social and political tides. Well, it didn't happen. So we have come to terms with the insider tip existence and let it flow into the title design: The adventure of community life is well hidden when you're dwelling in mainstream culture. You have to do the first step yourself: even if it means tearing up a book cover (I'd recommend just turning the page).

That community life remains largely unknown, however, is one of many small indications that our society is not as permeable between the mainstream and all other lifestyles as it likes to present itself. This is what my long article about the waterfall is about, which, I admit, is a bit dystopian, although I am a declared opponent of conspiracy theories. Probably a bad sign, that even I speak of the mainstream going over the edge. But on the other hand, there are plenty of exciting opportunities to work on a different future. This is what the article "Becoming, being and remaining a community" is about: That in long-standing communities there might be even more challenges to be found than in community founding. And challenges are a good thing, of course, because you grow with them. You can call them adventures.

To write two long articles and print them because I am the editor and nobody keeps me from doing it is, of course, bold. I only took the liberty after inviting comments from other communities, like in a blog. An exciting editorial experiment that reframes the author's perspective (mine) and brings in completely different opinions from completely different communities. It emphasizes the main asset of this directory: The breathtaking variety of attitudes that can be found within the community movement.
The articles that appeared in the last editions of eurotopia are still highly recommended, especially for eurotopia newcomers, and most of them can be found on our website.

Making Of
It is curious and perhaps hard to imagine how difficult it is to communicate constructively with the communities we know. The communities published in the book have been contacted three times and more by e-mail and reminded to contribute to this book (in English). Some were also contacted by letter post or even by phone. Most recently, there were e-mails in the respective national languages, probably translated rather brutally by big brother Google, which speeds up the waterfall considerably (see article), but provides amazing tools for global digital communication. We are very pleased with the response - in this issue sensational 604 communities are described (107 of them house projects and 65 settlements each with "distinct community character"); compared to 430 in the last issue. Nevertheless, we have received no response at all from about 1800 communities (or rather "from 1800 mail recipients of whom we have good reason to believe that they belong to communities in Europe"), although the emails did not bounce back to us. So for every community in the book there might be two to three that are not in the book.

The book was compiled again in Sieben Linden Ecovillage, Germany, with numerous volunteers who first "searched" for communities and then transferred their answers to our database. Apart from the great work they did, these were wonderful times: With Rahel, Sigrid, Martin, Mariam, Jonas, Marsilio, Anna, Peter, Lennart... And then on with Jonas and again Rahel. The English editing was done by Eliška. Most of the collaborators mentioned have responded to my call in the eurotopia newsletter. The indispensable and patient Volker from Schloss Glarisegg community in Switzerland was in charge of IT. May these people be welcome in all communities, thank you!
One of the helpers researched the Anastasia movement on his own initiative and wrote the article on pages 49ff. In doing so, we are clearly positioning ourselves for a cosmopolitan society without excluding people who like family estates. We don't question the notion to live a healthy and self-determined life with a peaceful mindset, but put fingers on some of the statements in the Anastasia books. This has already led to interesting discussions and feedback that has confirmed our standpoint.

The biggest thanks goes to the communities that have patiently endured our technical problems and our constant e-mailing and agreed to be presented. There are several lists of communities on the web, but they are usually not as up-to-date and comprehensive as one would expect from the internet. We try to compensate the disadvantages of the printed book again with our book assisted database search on eurotopia.directory/booksearch, and there is also a secret idea how we can perhaps make a digital offer despite our reservations about that. Visit our website from time to time or sign up for the newsletter if you are interested. We send it so rarely that it won't bother you. And the reverse flow of information is also welcome: send us feedback! Sometimes a good argument is enough to do something different for the next issue. This time, for example, the tagging of communities that were not included in the last issue: Those who know the last book and want to know which new projects have been listed can now check for the laughing star - this is the result of a reader's suggestion. Thanks also for that!

Planned!

I must still mention a quite important change. The planned communities in this issue are no longer listed between the existing ones, but in a separate part. That's probably not so great for these groups, but it allows readers to distinctively search for existing or planned groups. A planned community is just more of an idea, much less concrete than an existing community (see article "Becoming, being and remaining a community"), and the number of planned communities has almost tripled since the last issue. The step to list them separately seemed appropriate, and I apologize for any inconvenience, especially to the planned groups, who may feel like they had to move to a back row seat.
As far as eurotopia is concerned, there are a number of plans. First of all, the directory must be completed and brought into the world - then we will pursue an idea to perhaps make the directory digitally accessible, along with other new features. Why we haven't done this so far has been described in detail in the last issue and it can also be found in the "FAQ" on our website. Also, the world needs a good film about community life. Since no one else delivers it, we probably have to do it ourselves. 18 years after "Living under Palm Trees" (see bibliography, page 570 top) I am quite determined to give it a well aimed new shot again. Whoever wants to support this, for example financially, simply writes to eurotopia.

I now wish you an exciting and inspiring time in the world of communities, starting on page 52. If you read the following articles: The comments below are really good and important and put everything in a different light!

With community greetings:

Micha Würfel

Collective exercise, from left to right: Rahel, Jonas, Micha. A eurotopia story: Jonas was going to just take a look at Sieben Linden and visit many other communities after - and then just stayed. In 2019, he answers phone calls and emails at the Sieben Linden reception...

The Waterfall Ahead

Intentional Communities and Mainstream Culture

by Michael Würfel | translated by Tomas Remiarz

With all the freedom we enjoy in Europe, our lives are very much embedded within a cultural mainstream. This stream determines how we talk and the topics we choose, what we strive for in our lives, what makes us laugh or cringe. How we dress, how we treat each other, what we find acceptable or not. As soon as many or even most people like, do or buy something - like smartphones, takeaway coffee, tattoos, super hero movies or the evening news - it becomes mainstream.

Mainstream can be what is tried and tested, or simply convenient. Sometimes it's just what's always been there. Some suspect that the mainstream is a political construct – with secret intentions beyond democracy. Certainly, and not secret at all, the mainstream consists of what is easily marketable: the market economy is mainstream. Whatever is popular can and must be supplied, because that is where profit can be made, and a lot of effort goes into making things popular.

The existence of a cultural, commercial or social mainstream is inevitable. We would have to be very isolated from each other to avoid developing such a shared culture. To speak English in the UK, or some form of Chinese in China – that, too, is mainstream. Nevertheless we cling to our individuality. We make a lot of effort to stand out from the mainstream. Witty T-shirts, wacky haircuts and our very own ambitions make sure that we are a bit different to everyone else.

The reason I am writing about "mainstream" is an image that has lodged itself inside me: the mainstream as a large river heading for a waterfall. The current is the time we live in.
I have known this mainstream for as long as I have lived. And I have been swimming with the current, generally moving in the same flow as everybody else while imagining myself to be a creature of the margins – as almost everybody else, as well.

Ever since I've been able to think I have had the feeling that the speed of this big river is increasing. And as I learned more and imagined I understood more of this world, I began to question its direction. Of course these feelings are not new, it might be called "mainstream" in itself to wail about the general direction the world is going. Two centuries ago many people were horrified by the idea of travelling by train, doubting that the human body was "made" for such hair raising velocity.

Still, when it came to having to make a choice of profession I chose carpentry over media studies because I thought I could already hear the approaching rapids down which the big river (i.e. the state, the financial system, the "mainstream") would inevitably tumble. I figured that even if utter chaos lay beyond the waterfall, houses would be in far greater demand than

TV programs. Maybe I should have become a farmer, as food is probably even more essential than shelter, but as a carpenter I could at least trade house repairs for farm produce.

My apprenticeship passed and still the waterfall itself hadn't come into view, despite the overwhelming amount of bad news about species extinction, water scarcity, poverty, exploitation and an epidemic of depression among young people. After having learned to build houses (and therefore being prepared for systemic breakdown) I figured I might as well enjoy the fruits of the mainstream while they are in vast supply, and for a few years I lived a life of simple jobs that did not take much commiting, cheap accomodation in East Berlin, a comfortable student status at a film school. As well as the cheap food, clothing and consumer goods that come as general perks of life in Western Europe. I claimed to be leading an "alternative" lifestyle, but the only things that weren't mainstream about me were that I didn't own a car and never held down a 9 to 5 job.

Eventually I moved to Sieben Linden Ecovillage in 2007, and with hindsight I can say that even then I didn't stop being part of the mainstream, at least not right away. Personally, because it took time to grasp what intentional community actually means. And as a community, because economically we are very much part of the main economic flow of society, be it through our professions, as tax payers, recipients of health insurance and so on. Not that there is much wrong with that. After all, most our friends, families and even readers of this article are moving down the same river.

From a different perspective, though, one could argue that we at Sieben Linden – and others that got together in intentional communities – have exited the mainstream and are now moving down a smaller, slower (and safer?) branch of the water. Or maybe that the point where our branch departs is at least coming into view. I would like to take on that perspective: That by living in a 150-strong intentional community I have placed myself at a certain distance from what I have just described as mainstream.
The boundaries between our communal life and mainstream culture may be very open and diffuse, but they are palpable. When I (as I rarely do) watch the news or visit a city I often feel like an external observer of the mainstream. And when I do, my unchanged impression is that "progress" is accelerating beyond our control. The noise of the waterfall is growing.

Only yesterday I was flabbergasted by mobile phones. Today I am no longer surprised about the existence of Google glasses. 30 years ago, I looked up to people boycotting the census – today many of the same people contact me on Facebook, although everybody knows that our data is harvested and used. Donald Trump used Facebook data for his campaign. The next big thing, apparently, will be self optimisation using artificial intelligence: computer-aided thought, perception and memory, available via chips implanted into the body. By contrast, there are not even mobile phones in our village. And here I am so rooted, occupied and challenged that I forget how unusual our life is these days: To head out barefoot to a meeting which is based on consensus and starts with a feelings round, wearing clothes from our free shop, and not for a moment finding anything unusual about it - I think I indeed may call this a different culture.

For the sake of argument, let's call these cultures (mainstream vs. community) divergent evolutionary paths. Where it becomes the norm in the mainstream to merge humans and technology or to genetically "optimise" embryos before birth, and we eco-villagers decide not to do any of these things. Is it conceivable that eventually we may reach a point where we are no longer "compatible", socially if not physically?

Back in the day when I knew nothing about GM, Google or brain optimisation, I imagined that "alternative" people who are open to their senses and feelings, holding on to ethics and ideals, would be unable to cope in a world where people know what one shouldn't do and do it anyway.
We are not quite there yet. Most "greens" are still physically and psychologically able to visit motor shows and Volkswagen managers can come on a tour of our ecovillage without actual panic attacks. I imagine it would be much harder to spend a few weeks in each other's shoes (I used to have an annual job at a 10 day film market. I found it kind of fascinating at the beginning and was desperatele longing for my community life near the end of the event).

There are still plenty of people leaving the mainstream to join communities with very different social norms. And there are many others who don't live in a "commune" but don't surrender to the mainstream either. Many of them are active in the projects, groups and organisations listed under "useful contacts" in the eurotopia Directory. It is even likely that these initiatives and individuals are the most important sources of ideas and solutions for a liveable future, rather than intentional communities. Even the mainstream itself has a role to play; after all we didn't build the solar modules or high tech furnances used in Sieben Linden all by ourselves.
Some commune members quite happily mix with the mainstream, myself included. I enjoy watching films and reading books. I support TV documentaries being done about our community. I even have ambitions to make films and write books that are successful in the mainstream, and of course a major point of the eurotopia Directory is to reach people in the mainstream and maybe even entice some people to join a more communal stream.

Still, when I read that right now companies are developing "smart" contact lenses with display,

camera and Wifi functions, and that in lab tests mice have been successfully "fed" false memories, I once more see the majority of consumer society blissfully drifting towards the waterfall. I see how the stream accelerates, the noise rises to a roar. It is getting increasingly hard to reach the shore, until finally the current may become too strong to get out.

Will we eventually reach a point where old fashioned, fully organic humans are no longer compatible with the continually self-upgrading cyborg mainstream, unable to earn their living and fulfil their basic needs within the mainstream? That would give the term "wild swimming" a whole new meaning. Wild swimmers would be effectively excluded from the mainstream, not because of active persecution by the system (which would be unnecessary, with most citizens willingly installing each new upgrade) but because there would be no jobs, no homes and no mates for the wild ones. In which case it would make more sense than ever to veer off in the direction of like-minded souls and collectively invent a culture that suits biological humans better than the one currently posing as "progress". To the mainstream itself, the very notion would probably appear absurd - but aren't alternatives already discredited by mainstream often enough?

With its culture of agreements, arrangements and compromise, living in community is by no means a guaranteed success for everyone. That is the downside. Even if I establish that there's a problem with the mainstream, I am not to say that there aren't many challenges to a life in community. It especially comes as a disappointment for people accustomed to the illusion of limitless individuality, ever present and noisily touted in mainstream culture. I would just carefully point out that this "individuality" isn't as real as it is sold to be. The mantra may be "Just do it", "Be yourself", "Think different", but you can't 'do it' without the right shoes, can't 'be yourself' with the wrong drink, can't 'think different' without the correct gadget.
If you are "yourself" (yeah!) outside the mainstream (without Twitter account, YouTube channel and smartphone), who will know and who will care?

In my very subjective view as an active participant rather than an objective observer of this process, consumerist individuality is a lie that is easy to see through. A witty T-shirt slogan is no proof of individuality. In fact there is no need to prove anything – individuality comes into being through exchange with others, by taking a stand among those around you. Only your actual fellow human beings can witness your individuality. It may seem like a paradox: individuality only shows up in groups (however large these may be). And if we live together I can tell when you are genuine, when there is a real bond between us. We can learn to cooperate, to trust and understand each other.

To live this way is undoubtedly a tall order. One might say, "I have my doubts about all this doom mongering but anyway, I'll at least listen out for the sound of the rapids. And if it ever comes down to a choice between chip implant and communal living, I will move into a commune." There is a chance that by waiting for change to happen to you, you'll end up missing the crucial moment of leaving the mainstream and end up being caught between two worlds. Have you heard about senior citizens moving together in shared flats to avoid being alone and uncared for at old age? This might be a good approach in principle, but it has its pitfalls in practice. If as an 80-year old you aren't used to negotiating between your personal ideals of a tidy kitchen or cosy bathroom and those of your house mates, you will bitterly struggle. Compromise is tough enough for a 40-year old like me. Communal living is a challenge, and it takes all your wits to make it work.

That is a good reason to look into communities now, while it is still possible to work "outside". While mortgages, health insurance and broadband are still available for people who want to live outside the mainstream. My doom-and-gloom scenarios aside, there is an undeniable consistency in the idea of coming together in order to sort out the fundamental aspects of human existence communally, to value each other and not be divided by superficial attributes or market mechanisms.

We are talking about the real-life enactment of values like belonging and togetherness, in contrast to mainstream culture where the same values are much touted, but only actually tolerated within carefully set boundaries. "You can be part of the Apple-culture, just as long as long as you buy the latest i-gadget." "Share your news, collect friends, join in on social media – and give us and our advertising sponsors your data in the process." In an actual, offline community it is in very much your own hands how much you really are part of something larger.

The mainstream on its hazardous path to the waterfall is no place where a resilient culture can thrive. Well-meaning efforts of right living in the wrong context will always be an uphill struggle. And if you are one of those people who want to leave the mainstream but not run away to the woods by themselves, you will need people who offer cultural alternatives. In other words, communities.

This text has been presented to the communities that are participating in this directory. Some accepted our offer to supplement it with their own views:

Daniel, Zonca (page 299):

I am very happy to have found a community in which there is no single leitmotif other than a shared appreciation of our natural surroundings and of each other. This, to me, represents the maximum of maturity available to us as individuals when we no longer need any specific dogma to hold our communities together but rely instead on our ability to 'create' solutions out of the box, through communication based on love, respect and connection.

I am not anti-technology, however I recognise that technology can become an impediment or distraction from the very elementary human need to physically engage with the world and others. It does not have to be this way though; the digital illiteracy of many within the eco-movement is also an obstacle to the forming of bonds of solidarity across the globe. The rejection of technology is, from a psychological and historical perspective, a form of fundamentalism that sprouts from a very dubious puritanism that I hope we can avoid.

Ecovillages will often experiment with technologies like wind and solar power. But why accept only those technologies that leave us in an insular, isolated situation?

My point it that we need to evaluate the impact of technology in terms of what we ultimately desire — comfort versus connection for example. My hope is that eco-communities will continue to promote this message, not as a dogma but as a choice we can make whatever our situation in life. These values will then obviously also become embedded in some of the technologies we create.

Raphael, Nachbarschaftlich Wohnen in Dormagen (page 229):

If the mainstream will really end up in a waterfall, my wish is that we will be able to develop alternatives together. It should be an alternative that enables most people to row from the current onto dry land. When the first ones gained ground, they can reach out and offer the still floating an open hand and help them on land.
Your description of alternative communities sounds very exclusive to me, and quite rejecting towards the mainstream. I wish for openness towards the mainstream and honest interest in (technological) development, even though it might irritate at the beginning. I can personally understand your rejection of new inventions, I just fear that it will not help to defend against real dangers (e.g. data protection violation).
I am afraid that it is only marginally better to look at the waterfall from a quiet arm of the river. I am afraid that the sight of others standing close by in the vortex will inwardly tear you apart, especially while knowing that you could have reached out to them if you would not have turned away from the people of the mainstream.

Ad Vlems, Ecovillage Boekel (page 476):

Ecovillage Boekel is an ecovillage, one leg knee deep into the mainstream. We hear the waterfall as loud as the writer of this article. We want to inspire mainstream to divert as a whole before reaching the waterfall. We want the waterfall to dry out, because the water has found a new, more sustainable river bed. We want to be the first in that river bed to inspire the mainstream as a pilot project, a best practice, a living lab for innovations.
We talk with ministries and our province about all the problems that lie ahead. One director of one of the ministries had a meeting on the terrain where we are going to build our houses. After we talked, he said 'If you are restricted by legislation, let me know, and I will put someone on it, because this ecovillage needs to succeed!' Afterwards, he gave a presentation about Ecovillage Boekel to all eighty directors and said to them: 'You all need to visit these small, local projects, because out there the real change is being made!'
Ecovillage Boekel never has had any problems with legislation. By law we should have been connected to the sewer system, but the municipality has decided we can filter our own waste water and let the clean water infiltrate in the soil. We are mentioned in a law for innovative, sustainable projects where we can test a different way of building. And when it goes well with our project, the general laws change! That is unique in Europe!
Let the whole river flow down the gentle slope, instead of over the waterfall. Save as much as we can...

Volker Thielmann, Landhüter Housing Co-op (page 209):

My vision is as follows: To build or co-work at a place where I can live tomorrow or even immediately. Where I have the possibility to spend time in this chosen family/community. Even if it is at the beginning only for some days, on the weekends or during the holidays. Through that I have the possibility to step out of the mainstream, to consciously enter an environment of community and to co-create it from now on and to be influenced by it. Just making this decision will change something within me and new developments and experiences will show up. This first step paired with the courage to face the challenges of communities is the pathway towards personal development. This development can lead in the best case to the place for living that some of us are dreaming of. But it involves work and engagement, it does not happen by itself. You are the beginning of this process. You do not need to wait; you can start immediately to make use of your situation and your possibilities. You can start your search now. You do not need to be different. Out of the mainstream, out of your everyday life you can find your new path. The existing communities are one way for that. No matter if you will dwell and live there today, tomorrow or the day after tomorrow.

Neus Buira, Q - Center of Experimentation (page 482):

I've been excited several times reading the article, it transmits many of the feelings that we have - those who have made the decision and live away from the mainstream in the periphery, creating other possibilities for the water of life to flow.

“There is no outside”, we are all in the same place, the planet Earth; isolation does not make sense, and also not to follow the inertia of a materialism that is already dehumanized today. The paradigm shift, the social, cultural and civilizatory transformation is inevitable. The process is slow because it depends exclusively on the fact that more and more people develop, open their consciousness, and act accordingly.

To dare to feel, to dream, to imagine another possible world, without wars, without social or environmental injustices, etc.; it helps to nourish the collective imagination of luminous ideas - not obscure ones such as those that govern the collective unconscious today, based on the dependence on money, machines and hierarchies.
Recognize the old world, as such, old, outdated: and free the imagination, dare to dream free, without fear: How would you like to live? It is a first step.

Another important step is to become aware of the fear that dwells in you, of your own, family, social and cultural conditioning that governs you. Seeing it, accepting it, understanding it and releasing it is another process that leads us to personal empowerment, through which one chooses the path of life and has the strength to walk it.

There are many steps, continuous tests of physical, moral, ethical, spiritual strength. Getting out of the inertia of the mainstream is like being born again! Thousands of small decisions let you tend to do the same every day or let you, little by little, transit to another reality, another way of being here in this life. It is a creative act and depends only on yourself.

“We are alone but not isolated“, with time you realize that there are more people who have done the same process, some have even achieved to create islands and archipelagos of alternative life on this planet.

Life is movement, change is inevitable. Consciousness is the vehicle and each individual is the protagonist. Today, progress is spiritual or is not progress, only decadence and barbarism of a very primitive egocentricity, submerged in fear and existential ignorance.

Holger, Kommune Karmitz (page 194):

Well, you want the comment of such a long and many aspects jolting text to be short. Ok. I find your text to be somewhat dished up, especially at the end, where you paint the “communities“ kind of as the only alternative to mainstream (next to hermitage). For me they are only one of many alternatives that exist in the same way as other “cultural specialities“ within society(ies). Mainstream, for me, has no clear borders or “banks“ that separate it from the remains of the whole of society. My people or my “gang“, who partly used to be punks, were mainly focused on gaining as big as possible of a distance to this zone of greatest conformity – the mainstream. “Outside of society I want to be“ sang Patti Smith. By now I have been part of the commune for a long time, where we, by the way, partially distance ourselves from the term “community“. We have confederates in single persons and nuclear families as well as in “communities“, are part of communities and sometime call ourselves

clans. But which of the different cells that represent the whole society are the "better" ones for us, shows in acting together and in the ability to communicate well. Right-wing extremists, for example, are always striving to offer the best possible distance towards "mainstream" and to create their own living environments. And for me, first of all, there is no reason not to recognize and to respect the individuality of different people who are part of the whole of society, in spite of their smart phone apps, Facebook and fashion conformity. For many of them it is their financial, psychological and social possibilities which have brought them in the place in society where they are living. Most people who are in communities, communes and suchlike, as far as I know, are first of all children of the middle class. On the other side there are the ones which drop out very far from the societal middle field, like homeless people, street children or refugees. It is very nice that the children of the middle class try to create a more positive form of existence through the vision of community for their kind. Nevertheless I question if it is not equally important to fight more and more radically against the inequalities between poor and rich and if the strength for that can really be found in the niches, or rather in the mainstream itself, with the people who create or would create their "communities" differently to us.
This is not meant to be an argument against the existing "culture of community" of the eco leftists, but merely a critic and addition...

Falk Zinke, ARNIA e.V. (page 143):

There is yet another possibility: to live in a village, which organizes itself like Heckenbeck (page 173) or Flegessen - thus providing the ecological, sustainable framework, but practically setting no limits to individuality. The framework is as little rigid as individuality is, so development is possible on both sides, anytime and slowly or faster...

The children, who now strike every Friday all over the world, also mean us, not only politics! We simply don't have time for mainstream anymore, the days count – while politicians think and act in election periods. We need actions like the ones Greta from Sweden shows us: intelligent, effective, sustainable and fast!

Katja Dern, Hofgemeinschaft Kotthausen (page 186):

After two years in the Hofgemeinschaft Kotthausen and many years of searching and founding communities before, I notice an increasing alienation of myself from the mainstream outlined by you. I deliberately rarely move away from our "island Kotthausen", because I am quickly overwhelmed by the dynamics of the city, the people, the mainstream. I only consume media in very small doses; ecology, reduction of consumption and minimalisation are themes that drive me and that I try to bring into the community.
And yet I know that I am part of the system, that I live and act in it, and that I also profit from it (through health insurance, parent/child benefit etc.).
Our community does the same. We consciously seek an open exchange with the outside world and our economic survival is also linked to this exchange.

I think it is good and important to question the mainstream, to separate ourselves from it in the form of a life in community and yet to acknowledge that we are part of the system. I think that both the technical innovations you mentioned and the human alienation that goes with them, as well as the alternative life in a community, are currents of our time.
For even though the majority of society swims in the "mainstream", we in the Hofgemeinschaft Kotthausen have registered an increasing interest in our community in recent years. People from near and far visit us, ask questions, want to start projects themselves, join us. The mass of those who question the mainstream, even want to turn their backs on it, is increasing. And in addition to the suspicion that we communities are sometimes confronted with, in my eyes there is also growing recognition in some places for the courage and willingness to shape social togetherness and coexistence differently and more sustainably.

James Dennis, Editorial team, Diggers and Dreamers (page 568):

I know this waterfall idea and think it is a powerful one.

I feel fear and confusion when I imagine myself there and I guess I want to understand why. So that I can better quiet the fear and enable myself to make decisions.

Maybe it only looks big from our perspective and actually the river is 10cm deep on the edge and we can walk there and hop down into a nice, warm, gentle bath?

Maybe when the mainstream becomes aware of it, it will easily find the resources to overcome it and that is exactly the impetus we needed to unlock our creativity. 'Necessity is the mother of invention'.

Perhaps there is a flying saucer waiting just below and out of sight to catch me and go on a space adventure?

Heinz Weinhausen, SSM (page 253):

Sooner or later everyone notices, even in privileged societies like Germany, that workplace, career, business success; that the possible consumerism is not the whole life. In these mainstream patterns, individual sense, caring for others, trying out new things, leisure, participation in shaping society, etc. can only be realized to a limited extent. The longing to get out of the hamster wheel, the longing to no longer have to function comprehensively and not to sacrifice one's lifetime to the requirements of the market economy is widespread.
At least to some extent many people want to live differently and meaningful.
The step into an intentional community is a quantum leap far too great for many. For most people, it doesn't go beyond dreaming about such a big step. But a considerable part tries to integrate the idea to be something other than a market object into their individual lives, by means of care for children as parents or grandparents, as helping friends, as voluntary or political commitment, as "hobby". And that is like a pile of sparks to me, which can also change from glowing to burning. In intentional communities, the fire is constantly burning and it shines for others.

Another picture is that there are cracks in capitalism that can be widened. The reading of John Holloway's "Crack Capitalism" was very, very enlightening and stimulating.
Ultimately, there is some mainstream and some "beyond mainstream" in every individual soul, whether it lives in communities or not. The emphasis can shift for anyone. What remains for the individual is to live authentically for oneself, to go one's own way and to shine a little contradictorily. At least one's own soul can find peace in this way, having contributed and contributing its part to the change of society.
For the salvation of mankind, however, a revolution with the motto of the 1918 rebels is needed: all to all. This revolution is now maturing because capitalism cannot meet the requirements of non-growth with regard to the climate catastrophe. The worse the consequences of climate warming will be felt, the more the market economy will fall into crisis. Nobody knows whether the revolution will mature, however.

Photographs in this chapter are all from communities that are presented in this book. Unfortunately, for technical reasons, we are unable to say which photo is from which community.

Becoming, Being and Remaining a Community

by Michael Würfel | translated by Tomas Remiarz

I try to warn anyone interested in living in community that it is not an easy task. In a best case scenario, the community members of the future, who will have grown up in the intentional communities of today, will be more collectively minded than we are today, who were usually socialised in mainstream culture and are often still very much part of it. Those future beings may find it easier to negotiate, compromise and come to agreements. They may be even more open hearted and less dependent on perceptions of "success" and "performance" than those attempting to live communally today.

I have been living at Sieben Linden Ecovillage for twelve years now, and I am lucky to be able to say that communal culture feels more "normal" to me than mainstream culture - friendly glances, human contact and goodwill as opposed to a perfumed hygiene and fashion-conscious society and attention-sapping apps. Even so, I keep catching myself as I chase after mainstream ideals. Sometimes, I would love to be that guy in the adverts who everybody is mad for. Sometimes, I wish one of my entrepreneurial schemes would turn into the next viral consumer craze and I'd only have to decide whether or not to sell up to Google for $3bn at the end.

Over the decade-plus that I have lived in this ecovillage I have learned to distinguish between my dreams and the reality of communal life, which can be hard but is ultimately rewarding. It is a privilege of sorts. Not many people are given this experience, as they don't stay in communities as long as I have done, for a variety of reasons. Some move into the community of their dreams, only to find it isn't for them after all. More often, people get together to start a community that never takes off.

A little survey among German speaking communities showed that of the 27 groups listed as "planned communities" in the previous edition of Eurotopia, ten have disbanded and seven didn't respond to my letter of enquiry for this edition.
In many cases groups loose their energy or group members fall out with each other, and only a fraction of planned communities actually manage to get something off the ground. Still, the idea of setting up a new community appears to hold far more appeal than moving into an existing one and having to adapt to its established culture. I can plan a community exactly how I want it to be. Someone who lived at Sieben Linden for six months explained to me that "the most important criterion of the community I want to live in is that I was one of its founders." Well fair enough – you can't get that in an existing community!

Accordingly, there are always a lot of listings of "planned communities" in our directory, and there are always loads of people visiting communities in order to set up their own. And that is, of course, great! Still it makes me sad to see people burn themselves out in the process and loose all enthusiasm for communal living. We could use a lot of this energy in existing communities to tackle the slightly different challenges we face. How can communal life be simplified and made more appealing to more people? How can we keep things from going stale? How much difference do we really make in the world? These and other questions come to the fore once basic necessities such as building repairs, group structure and financial stability are seen to.

Six of the 27 "planned communities" in 2013 are still considering themselves as "planned", and one of them has managed to find land for their project. Finally, four others have turned into actual communities. Successful projects do exist – and they can be sources of great inspiration to others. In Sieben Linden we were amazed at the successful start of the Schloss Tempelhof community in Southern Germany, which generated much support and received enthusiastic media reports. We were surprised to hear that almost everybody there comes to meetings and that people don't have a problem with paying an even higher joining fee than ours at Sieben Linden, where we keep discussing the affordability of our model for people on low incomes. Without wanting to spoil the party we had a feeling that this initial enthusiasm might not last at its current level. (They thrive, but we didn't get that article we were hoping for from them about the state of their progress.)

Communities, just like the people living in them, change over time as they go through different phases in their lives. So it can happen that the initial pioneer spirit gives way to a pragmatism that can make things at times feel boring, regimented, small minded, or just more "normal". In the early days of Sieben Linden, almost everybody living here was young. Those who weren't had always seen themselves as pioneers and therefore felt exactly right, placed among other pioneers. Their initial enthusiasm acted like a shock absorber that made individual bumps along the rough track almost unnoticeable. When in 2001 I made a documentary film about our ecovillage I only had to turn on the microphone to capture self-confident statements about the exemplary nature of the ecovillage model. Today, that would be impossible. Those of us who have been here for a while feel doubt, torment ourselves with comparisons with other communities and occasionally bemoan the death of enthusiasm. Newer arrivals are more reserved and cautious, not knowing if and how they can bring their own ecovillage dreams into the mix. Our community has entered a different stage in its life, and for us it is incredibly important to focus on what is important at this point of our evolution: being and staying a community, to carry on living in the knowledge we are doing the right thing, and to enjoy what we do.

Twenty years on from the beginning, Sieben Linden's population has become much more diverse. Ageing founder members live beside newcomers of similar age who until recently were moving in the mainstream. Old and new inhabitants who still want to be pioneers ask where there is space left for creativity in our half-finished village (everywhere of course, but equally there are signs of previous efforts everywhere, and people who remember those efforts). There are far fewer builders and technicians on site than at the start, in proportion to the overall number of villagers. We have families coming here who are looking for a good place for their children to grow up. There are clever people who want to improve everything and an often silent majority of residents who like the idea of "improvements" in principle, without wanting to go through the reasons why many things are how they are now. There is a risk of throwing out the baby with the bathwater. There are many calls for "new!" and "better!" solutions, and I'm pretty sure that not everyone agrees with me: that most of all we have to work on creating a culture that helps us to live together. And that the place we have to work hardest on is ourselves, individually. As ever it was.

At Sieben Linden we have passionate debates about our decision making structure, and there is a certain self-destructive willingness to repeatedly discard our attempts and reinvent the wheel. We are less dedicated to consistently look at our feelings and resentments. The regular forum-group (ZEGG style forum) has long been abandoned – although we know that we are instantly much more understanding and friendly towards each other after having spent a lot of time sitting together in a circle during one of our intensive periods.

What if we could manage to interact differently in our everyday lives, to trust and appreciate each other more, and to recognise when we're obeying our selfish impulses instead of thinking of the community? In that case we could probably do without the expensive organisational development process we are now getting into (2019). This, I think, is the true challenge of "staying a community" or rather "staying alive as a community". I only have to look at myself – now with three children and several work projects. In the evening I am with the children, or tired, or abandon myself to entertainment in the form of films, books or TV series. I don't often go to communal events and therefore often walk into the trap of intellectual discussions that don't meet the need of a collectively shared experience.

Does this mean that we follow the trend of the 1968-generation from fury to moderation, from adventure to comfort, from the wooden bench to the TV-sofa? Or am I merely projecting my own ageing process onto that of my community?

Despite all these niggles, our ecovillage remains a beacon of social and ecological change compared to mainstream society. The progressive funding body Heinrich-Böll-Stiftung uses us on one of their posters, where we sit as a "pearl of energy transition" together with other projects on a map of Germany. A new study confirms that over the last fifteen years our ecological footprint has shrunk even further. And while I wonder if Sieben Linden has lost some of its edge, the mainstream hasn't stood still either: while we might be slowly shuffling towards the sofa of convenience, it has picked up even more speed on its way to the waterfall. Therefore, relatively speaking, we are avantgarde alternative, more than ever.
We have also changed by learning from some of the mistakes we have made. We may not build houses completely by hand any more, which was an amazing and impressive thing that has happened here from 2001 until 2003. But that also means that we have managed to organise their construction in a way that doesn't burn out those doing the work. And we manage to bring up children while we build our houses. That's something that the group that built that house without any machinery didn't achieve.

Despite all fascination with ideas for setting up new communities I would like to speak up for an awareness of the value of communal structures that have grown over many years.

I believe that as an established community we can be a model to society, even more than an enthusiastic group that is just starting. We can make space to innovate and welcome fresh ideas, without disregarding all the arrangements made over years as fossilised bureaucracy. The crux is, nobody else can help us where we are. Here we stand, as a functioning ecovillage, and nobody really knows where we go from here. Among us we have people who consult on setting up ecovillages, but who do we ask to find out how we can get even better? Who, if not ourselves, could work out how to shape communal living as a functioning alternative to an existence as consumers? We just have to realize that and appreciate that whatever we do here, we are still working on an amazing project and are still in the process of meeting a challenge: The challenge of being and remaining a community, which might just not be much easier than to become one.

Again, this text has been presented to the participating communities of this book. Here a small collection of the comments we received:

Eva, Sieben Linden Ecovillage (founded 1997, page 249):

In many cases you speak to me from my soul, as a co-founder of Sieben Linden, but there is something I want to correct. I wouldn't say that we are at a point where no one can help us any further or give us advice. In our history we have learnt so much from so many different sources, and that we can still do. No school, no adviser has a patent remedy for us – but also not for communities in the making. We can learn a lot from each other and from other people, who stand up for change: in regards of the organisational culture, we can learn from sociocracy and Frederik Laloux (Reinventing Organisations). In regards to consciousness in group processes from ZEGG's Forum, from NVC, from Deep Democracy of the process oriented psychology and many other sources of inspiration. In regards of dealing with information management from many alternative platforms... And so there are still many sources that we can learn from.
And at the end each pathway is unique and each community has to find their own way. The "community compass" that we developed out of our experiences in the Ecovillage Sieben Linden (www.gemeinschaftskompass.de, German language only) is therefore not a "step for step plan", but it shows which areas need attention – that applies for emerging communities as well as for long existing communities like ours.

Gerheart von Bültzingsloewen, Nature Community (founded 2014, page 231):

I also do not experience life as easy and comfortable in the Nature Community in the Bavarian Upper Palatinate. But instead as tremendously exciting, manifold and challenging – and feel like doing something more meaningful as in the previous 20 years of full-time work in the mainstream as a project manager in Munich, being "alternative" only during free time.
The first three years in our former holiday site in Schönsee are lying behind us – it still feels like a part of the pioneer's phase, but at the same time we have already learnt a lot from our inherent errors and I think we are "out of the woods", that means we are in the position to visualize the project in 2025 and 2050 – with very positive scenarios, that we could surprisingly consistently develop in our last intensive time.

One inherent error was an unfiltered welcome culture in combination with the ideal "everyone contributes what she can, and gets what she needs". And added to it the vision of an internal money free system... That attracted people where the balance was not matching up: on one side "mother community please provide for me" and on the other side "overstrain through providing for the ones that are not able to care for themselves – financially, work wise and emotionally". With hindsight I see it even as positive that in the middle of 2018 we threatened to slide into a economic distress – that helped us to go through important processes and set us on course through a minimum financial contribution of each inhabitant and the professionalising of the event operation in regards to sustainability. An important key is also our willingness to look at inner topics consistently, especially the taboo topics sex, money and power, which at the end are all forms of energy. Admittedly, some members, who did not want to confront this adequately left us. The ones that remained are now stronger and more in balance connected with each other. As a motley crew of all ages and professions we are very diverse, and we have gone together on an exploration of how trust and peace can grow and flourish in a sustainable environment.

Martin von Oadien, Jakobgut (founded 1997, page 190):

"Still it makes me sad to see people burn themselves out in the process and loose all enthusiasm for communal living."

Maybe for many it is about direct self-realization? Maybe it is because many seekers hardly know who they are? And can you really live together with others, when deep inside yourself you feel that you do not know that? So they first search for ideas and their actualization, and then for people as a means to that end – instead of actually searching for community.

We try to face this by explicitly inviting co-creation, when the elemental basics from us possibly match with an interested person. But hardly anyone is searching for such possibilities, many also hardly know places, where their thoughts could be realized (therefore an online directory for this – "where are your skills to co-create needed and appreciated?" – would be pretty reasonable).

“The initial enthusiasm acted like a shock absorber that made individual bumps along the rough track almost unnoticeable.”

As I experience it, this is, after more than twenty years, still the case for us of the “hard core”. I hardly experience a rapprochement towards “the mainstream”, like elsewhere, which I consider as great luck. We have rather become more “crazy”.

I find the encounter with interested persons more and more difficult. People who are often highly enthusiastic and who want to convey themselves intensely, hoping that I will be equally enthusiastic. Sharing passionate thoughts, which have sometimes been conveyed to me a hundred times in a similar way.

I do not want to become an actor, not a quasi-commercial sales person, I want to take the people that I encounter seriously, showing myself how I look like within. But many interested persons cannot cope with that, they want to live enthusiasm and don’t suspect how often I have been told something comparable, where more or less nothing came out of it. At least nothing that I have heard of.

Such kinds of people are not aware of how much attention they receive from us, what remains undone because of that. Do they mainly see their own egoistical satisfaction of needs? How many of them have noticeably been interested in how I am feeling? What they could do for us? To really get to know people (us)(in our case first in written form)? To perceive them? And the other way around: Do I still succeed with it towards the “always the same” interested persons, who in my sensation increasingly threaten to turn into one bulk, even though I try to resist it? Who are far too much aiming to please us, so much so that the individual becomes unrecognizable. What a pity.

“At Sieben Linden we have passionate debates about our decision making structure.”

I hardly dare to write the letters down: for us it has lately been rather like we hardly think about that. It seems we can do without it at the moment (it does not have to stay so).

P.S.: If you look at our reported numbers, one could think, we are simply a quite manageable group (in this edition, Jakobgut is listed with only three members). In reality we had many ups and downs, several times we were full and therefore organised more space. At the moment our pork cycle arrived rather at the bottom.

A Few Words About

Decisions in Communities

by Michael Würfel

As long as I don't have concrete examples in mind, it doesn't seem difficult to make decisions in communities. If you look at the descriptions of the communities in this book, you can see quite clearly what the individual communities want: for example, to live together, sustainably, healthily. Obviously everyone agrees. So it is no wonder that every so often, "consensus" is listed as the way decisions are made.

As a long-standing member of an intentional community, I'm actually surprised. I have almost never experienced decisions that are supported by everyone in such a way that they cannot imagine anything better. While consensus does not mean that everyone wants the same thing, it does mean that everyone supports a decision. And that is less easy in practice than you might think. Here are a few examples from Sieben Linden (the reference community for eurotopia, so to speak):

- In the first years of the community it became clear that mobile phones in are undesirable in Sieben Linden. A sign with a crossed out mobile phone was nailed to the welcome sign at the village entrance and we wrote and told about ourselves that we are a mobile phone free village. This was not a documented decision. In 2018 we started to acknowledge that many people were using cordless DECT phones, which radiated the same or more than mobile phones, and that Wi-Fi networks could also be found in the village. For some it was clear that this should be just as banished as mobile phones, others said that the decision (which, as I said, wasn't even formally one) would only apply to mobile phones. In addition, despite the "ban", some switched on their mobile phones from time to time to send or receive text messages or to check their messengers. It was time for a new agreement.
- Some think that a central courtyard in our village should be redesigned. Others don't think so at all. Some want to move an old compost toilet, which occupies a central space and is no longer used much, others see no necessity, find the old structure beautiful and worthy of protection and would rather see the time and money for the redesign campaign invested somewhere else.
- There is an offer to buy farmland nearby. Because we are in a land reorganization process, we could exchange land we would buy now for land in the immediate vicinity of our existing settlement area - a historically unique opportunity to add agricultural land directly to our village and create a buffer between our settlement area and the conventionally farmed fields around it (pesticides!). However, the land is more expensive than average and the purchase would possibly prevent other investments, such as the urgently needed expansion of our community house. Buy or not?

There must be an exchange about all these things, in which the different arguments are put on the table - but even after that not everyone will agree, because people assess the situations and developments differently and set different priorities. It may be possible to make pure consensus decisions (i.e. decisions supported by many and not prevented by the others), but they will certainly not come about easily.

The idea of not being able to do "what I want" is quite daunting for many people who do not live in communities. At least, these people have realized what a central role decision making has in community life - perhaps they have even understood the core problem faster than some community dreamers who are only planning a community, but have already written down exactly how it is supposed to work.

We in Sieben Linden have modified the consensus principle for us in such a way that we implement a decision when two thirds of the votes say "yes" to it and nobody calls "veto". Abstentions and even no votes do not stop such a decision (as long as the make up less than one third of the voters). A veto only has a suspensory effect. This means that the decision will not be taken now and work can be done on the proposed resolution again. This then demands the full commitment of the veto bearer and a willingness on the part of the rest of the group to respond to the veto bearer. If no agreement can be reached, a ⅔ majority can do the job.

Consent is another interesting possibility of decision-making within the framework of sociocracy, which focuses on maximum participation of the participants and promotes decisions that are "good enough for now and safe enough to try" - but it requires a lot of time and the physical presence of the whole group.

Systemic consensus as a further example is one way of not asking for support for different options, but for resistance to them. This is where what hurts the least is implemented - and also often leads to interesting results.

Whether it is a principle of consensus or a way of voting, for decisions to be taken, the community must work well together. Meeting for this is only the very first step (and probably there are decision models that can do without physical meeting). In any case, it is important that the factual arguments on the respective topics are collected without already being mixed up with personal opinions, priorities and emotions. It is certainly also important that the different points of view (opinions) are collected, but this is where it gets difficult:

- People speak in very different qualities. Some can make flamboyant pleas and others hardly dare to say anything at all in front of others, apart from the different language skills or the level of education. This is also a problem in written exchange.
- Some are not afraid (and don't even notice it) to manipulate by underpinning their opinions with concern or polemics. Sometimes facts are cited that would not stand up to scrutiny, but are not scrutinised in the heat of the moment ("recent studies have proven...").

- Some take more time than they are entitled to (actually it would be possible for anybody to divide the available time by the number of participants and time their own contribution).
- A conversation leader should be considerate, trustworthy, tough and 100% present at the same time. But people are not like that.
- Before opinions can be exchanged, personal disturbances must be eliminated, factual questions asked and answered and, at best, appreciation expressed for the proposals under discussion. This can be done in a structured way with the so-called coloured card system. Meetings sometimes lose some of their dynamics - but this is often a good thing. Everyone who wants to say something comes forward with coloured cards. The red ones mean "I have a disturbance" and are allowed to speak first. The orange ones mean appreciation, they come next. Yellow: I have a question. Green: I can answer questions. Then blue: I have an opinion (more information on the internet, search for "coloured cards decisions"). Because we are lazy or naive or simply constantly overestimate our group competence, we (Sieben Linden) usually do it without coloured cards.

I think that we "normal" people with normal "schooling", who have not lived in well-functioning communities from an early age, should actually learn a lot before we try to make decisions together. There is a vast amount of tools that might help to enter into a more connecting and understanding exchange: emotional and physical exchange spaces (ZEGG style forum, sharing, perhaps also playing, cuddling, playfights), communication practices like community building according to Scott Peck or non-violent communication, and then a lot of skills or communication goals that community members – or, at best, all people – should acquire themselves:

- Self-awareness: Listening to self
- Emotional Intelligence: Recognizing feelings and identifying needs
- Personal Expression: Speaking for self
- Listening Accurately to others
- Empathizing with others
- Equity and justice: Commitment to meet everyone's needs as much as possible
- Research: Finding the relevant facts, examples, and knowledge to guide us
- Observation: Clear and accurate perception of what's happening
- Verification: Testing truth claims, evaluating arguments and evidence

- Creativity: Using imagination to open new possibilities
- Comfort with uncertainty and paradox
- Sense of the common good: Serving the well-being of the whole
- Deciding together using shared group values and agreements
- Commitment to growth, learning, healing
- Openness to receiving feedback and working with it

In: Joe Cole, "Skill Building for a Culture of Collaboration", Communities Magazine 181, winter 2018, page 44 (see bibliography)

We should enter the decision making process with the noble air of deciding for our common cause and giving the best we can for it. Apart from the skills we need, it takes time. An extremely scarce commodity for most. But if we take our personal responsibility for good decision making seriously (see list above) and shape the decision-making situations well and furthermore have confidence in each other and in the committees that may exist, then hopefully at some point joint decisions and communities themselves can work so well that the idea of having to struggle to get together and live with unsatisfactory compromises proves unfounded.

Photo on page 40 by Hubert Perschke:
"SSM", see article on next page and description on page 253.
Above: "Stamm" (meaning "tribe"), a planned community, page 464. The first picture in this article shares the same fate as described on page 31: We can't say who those nice people are, but their community can be found somewhere in this book.

Commune of Outcasts

SSM - We learn while moving forward

Photo: Hubert Perschke

by Heinz Weinhausen
translated by Mathias Lederle

The Self Help Group SSM in Mülheim/Cologne has been living new lifes within the old world for almost 40 years. Oddballs, unwilling to build a civic career, have teamed up with "losers" and outcasts of our competitive society to do their own thing. Equality instead of exploitation, togetherness and cooperation instead of elbow mentality, inclusion instead of exclusion, rising up against injustice instead of being powerless are their socialistic principles. Their radical basic understanding already shows: SSM, the Socialistic Self Help Cologne-Mülheim, is a political commune and a practical critique of the capitalistic system. Its goal is nothing less than to change the current system, not tomorrow, but right here and now with the current people. Although change may only be achieved bit by bit amidst an ocean of unthinking economical business as usual, every progress nevertheless proves to the participants of a solidary community that another world is possible.

It all began quite spartan. When Tom Koch cut the gate to the factory premises at Düsseldorfer Straße 74 with his bolt clipper on November 3rd 1979, none of ten occupiers could imagine that this more than a thousand hectare big property with its four buildings waiting for demolition would become their new home. In the past, liquor or schnapps had been distilled here. Now the city of Cologne had bought the premises to demolish everything and build residential houses in this fancied proximity to the river Rhine. The group hoped that the administration would at least compensate them with backup premises for them to start their socio-economic project. Wondrously, it really worked. After fourteen years of occupancy – temporarily without electricity and water – the occupiers received a long-dated, affordable lease agreement. Gradually the living conditions of the occupiers improved. Today we are 25 participants.

One-pot-economy

Since the beginning, everything surrounding the SSM has been very simple and transparent. All the money from the habitation closures which the group undertakes and their second-hand shop goes to the group's financial account. From there, the house and living costs are paid, the trucks maintained, political actions disbursed, tools and machines acquired, social insurances settled et cetera. In the end, there remains an allowance that is paid to every participant of the self-help group in equal amounts for personal use. Every member additionally has a basic housing right.

Living better with less money

Relying on one's own work and self-sufficiency make life at SSM truly sustainable. Lots of used things from tables to books stem from habitation closures. We heat with no-cost wood, collected and chopped by ourselves. We live in gradually self-built and renovated rooms and apartments. Our food comes from our own pots and pans. Our individual know how is available for all of us, which really pays off. That way, much less "gainful employment" is needed to earn our livelihood and subsequently, time for social aspects is freed. People with disabilities have their place in our community like the fitter ones have. We don't need the work rush of a factory. That is "New Work" SSM style. Because of that, there's time to have a plenary assembly on one morning each week. During session, we plan work for the upcoming week and there's time to deal with a lot of problems that we're able to solve in most cases.

Politics is work, too

When you encounter injustice or mischief, resist. According to this credo, SSM always took part in the development of its neighbourhood. During the sanitations of the 1980s it rescued, aided by others, some buildings from demolition that are now collaboratively being used in hereditary lease. A bunker from World War II was transformed into a "Culture Bunker". There were also lost struggles and unsuccessful house occupations that taught some valuable lessons to the participants. Today, the main focus is on taking part in the initiative "Fresh Air" with the goal of driving motorised traffic from the neighbourhood. On Monday nights, the self-help group, aided by supporters, counsels help seekers having trouble with authorities or landlords for free.

A commune for "anyhuman"

You don't need A-levels or money to take part in our community. You just have to be willing to seriously contribute to it. After a week of getting to know each other during shared work, the preliminary membership is usually decided upon, including allowance money und provisional housing. Many people approached the SSM after encountering an emergency situation, e.g. becoming homeless. They usually first have to learn what community spirit means. What does it e.g. mean to take responsibility for oneself and for the entire group?

Many people get to know our "self-help without a boss" as customers during relocation, a habitation closure, in the second-hand shop, during a concert, in a seminar or while practicing yoga. Within our neighbourhood, we cooperate with others to create a better quality of living. We stay in close contact to many initiatives and key figures and in so doing became an important switch point within a local network.

Many "outsiders" are fascinated by our unreserved reception of marginalized people into our midst to give them a chance for a self-determined life independent from social welfare. We even have financial backers supporting us when it comes to making investments. Our SSM secret is a good mixture: "New Work" combines the strength of individuals and levels their weaknesses. For us, not only the small steps count, but also the bigger ones. The biggest task at hand right now is the new construction of our second-hand furniture depository including housing space. Day by day we learn many new things. It keeps us vivid and lets us grow.

Camphill: Communities of Extraordinary People

by Stefan Siegel-Holz | translated by Tomas Remiarz

Until far into the 20th century people with severe cognitive limitations were widely seen as ineducable. Developing in response to this, Camphill Communities created a ground-breaking way of working with disabled people that has become widely respected since its inception by doctor and remedial teacher Karl Koenig in the 1940. The name Camphill refers to the place of the original community in Scotland where Koenig, together with a group of young Jewish refugees, set up a community for people with and without cognitive disabilities.

Koenig's idea was as simple as it was inspired: given the right conditions it should be possible to give people with special needs opportunities to take part in and even shape the life of a community. A communal social context could become a naturally therapeutic environment, as long as it fulfils five fundamental needs: social integration, feeling valued, self-determination, challenge and development. Koening was sure that adults dependent on other people's help need opportunities to contribute to the communal life. He held that they should have fulfilling work that benefits others and becomes a source of recognition by others. Also that they should see themselves as participants in a local community with distinct voices, rights and responsibilities. Of course this has to be balanced with appropriate protection against the challenges of our modern competitive capitalist and intellectualised society.

Koenig highlighted three aspects of shared living: housing, work and culture. These became the basic tenets of his concept of comprehensive inclusion of disabled people which was put into practice by Koenig and his friends from the 1950s onwards. Even he was astonished by the success of the first "Camphill Village Community" in Botton (Yorkshire, England):
"From day one it became clear that village life was neither unusual nor insufficient for the "villagers". It simply was their life, and the villagers were determined to build it from the ground up. In residential buildings, stables and garden, in the field and workshops, in the chapel and hospital they carried this life with a growing feeling of responsibility and awareness. Against all expectations, this sense of responsibility grew almost daily. They naturally helped their weaker brethren and carried one another in a morally admirable manner" (Koenig in: Hans Müller-Wiedemann, "Karl König", p. 300; translated from German).
While Koenig saw the traditional village as the most appropriate form of living together, he didn't see it as the only possibility. For him agriculture and care for the landscape were important occupations, as were the more straightforward physical and social infrastructure in the village. The need for external help tends to go up as the living environment becomes more complex and anonymous.
For Koening it was obvious that men and women, as well as different generations should live in the same building. He wanted to avoid creating mental institutions in disguise, where residents would be seen and treated as patients. Generally, Koenig regarded disabled people not as "ill" but as people with particular life circumstances.

Such a designed village should finance itself as much as possible through income from the work done by its inhabitants. Equally, the residents both with and without disabilities, are meant to complement each other in a continual process of give and take. Rather than existing for disabled people, volunteers live with them and everybody experiences the community as a place of interdependence. Such a social organism depends on farmers and growers, craftspeople, trade and spiritual guidance, rather than on care specialists in a medicalised sense.Koenig denied a clear division between "able" and "disabled" people, between helpers and those in need of help. In his view, new dimensions become visible as soon as one starts to look beyond superficial appearance towards the fundamental character of a person, and the idea of "disability" begins to dissolve when the focus shifts from a person's possible deficits to their competences. In daily collaboration and coexistence capacities and talents become as visible as apparent disabilities. Koenig believed that people with disabilities also had a particular purpose for humanity in general, as the social conscience of a society. They can teach us a lot about how to deal with the frequently hidden struggles of illness and adversity, pain and suffering. They can teach us about interdependence and the relief that can be found through mutual support. For Koening, "disabled people exist not just for us to help them, but so that we can be helped. It is a mutual give and take" (Koenig in: Müller-Wiedemann, p. 478, translated from German). Meeting disabled people allows us to adjust current notions of humanity that idolise youth, beauty, achievement and material success. In Koenig's view, disabled people are also har-bingers of a new, humane society based on solidarity; it is their special mission to bring people into greater awareness.

Koenig expected a lot from the members of these communities: the willingness to share their lives with fellow villagers; readiness to keep learning; the humility to embark on shared processes without arrogance towards those perceived to be "weaker"; the ability to give other people space, and finally the willingness to fundamentally rethink categories like work, pay, leisure, privacy, family and public life. Under these conditions a community can come into being between unusually diverse peo-ple with differing abilities and disabilities, between women and men, families and singles, young and old, people of different nationalities, different cultures and religions. Camphill was an international movement from the very beginning. The core of the living community was the home, the shared roof over people's heads and shared daily activities such as preparing meals, cleaning, gardening and so on.
While Koenig's work has inspired and influenced many others, he did not leave behind a systematic body of work for others to build on. It was left to his successors to develop his approach further and continually adapt it to the evolving environment. Many Camphill communities have been set up over the decades following his death, building on and being inspired by his ideas. Especially in the early days these new endeavours were fantastic achievements of their founders, often against strong resistance, without viable economic foundations or contemporary understanding of special needs pedagogy. Anthroposophy had a strong influence on the movement, embedded in artistic, architectural, medicinal, therapeutic and many other practices. At the same time, life in each individual community has developed its own dynamics. Over time, a tendency emerged towards an internal focus on life in the community, and (against Koenig's intentions) towards withdrawal from wider society.

The social therapy concept stimulated a wide range of creative approaches that built on Koenig's ideas in different ways. Emerging urban communities had the advantage that it was simpler and more straightforward to take part in general social life. Smaller farm com-munities based on agricultural production were much simpler, navigable and embedded in natural cycles. Work communities had a much clearer separation between the spheres of work and home life.

Camphill communities attempt to create a living environment appropriate to the respective needs, possibilities, capacities and challenges of their inhabitants, walking the middle ground between lack of stimulation caused by the repetitive routines of care, and stressful overload through confusing social demands and anonymity. Each community has to be able to create such balance in order to offer a space for disabled people that is both socially inclusive and protective; where participation and withdrawal are possible choices; holding people safely while enabling their autonomy.

We need communities where it is as normal to receive help as it is help others. Communities that are reliable fields of caring relationships and facilities, of valued routines and rituals that profoundly anchor each individual life and give it meaning. Communities, finally, where life is experienced as comprehensible, manageable and meaningful, which according to research makes people stronger and more resilient. At best, this process can lead to a community that has changed from a place to live into an actual home.

Camphill in England - by Tomas Remiarz:

While never quite entering the mainstream, Camphill communities have affected current ideas around "care in the community" which, at its best, might be seen as the fulfilment of Koenig's dream of social integration. The reality sadly lags behind such lofty ambitions. Care regulations in England are a hindrance to integration as often as they help it. In fact draconian regulations regarding the assessment of and support for "the disabled" only serve to deepen the divide the Camphill model attempts to dissolve.

Camphill villages themselves have come under pressure from the profit-driven neoliberal approach to care, with an increasing load of paperwork and regulations being placed on them. Partly this has happened as a backlash from the dysfunctional mainstream care system for the elderly and disabled, where neglect and even abuse of people "in care" are common place. Another factor is the general drive towards a more individualised society, which means Camphill is once again swimming against the stream.

Unfortunately these developments have also affected the shared living model and the trust-based relationships between people that are at the core of the Camphill approach. Some Camphill charities in England try hard to adhere to the government line on this, which has at times led to tensions with long-standing residents. Meanwhile some residents of the original Botton village in Yorkshire have set up their own independent community within the village that continues to uphold the original care model.

People and organisations within the Camphill movement have teamed up with many other social care providers to fight the government's austerity programme and its effect on their work. All this shows that intentional communities (of any kind) cannot pretend to live in a zone unaffected by politics and wider social and cultural shifts, and that active engagement with this context is unavoidable. Retreat is not an option.

N.B. Camphill Scotland and Northern Ireland are legally separate organisations, as those parts of the UK have different care legislation to that of England and Wales.

Addresses and a link to a map of European Camphill Communities: page 510

No room? No way!

by Sonja Beutelmann (Text and Photos)

Arrived, parked, ready to leave at any time - theoretically. But whoever forms a trailer community has come to stay. In Germany there are over 150 trailer community spaces, some of which have existed for decades. Again and again, they are a thorn in the side of urban politics - because trailer life is in a legal vacuum. A city often regrets that there is no free space for community life in trailers and trucks. A community often has no choice but to occupy one of the many urban wastelands. Sometimes the city then offers a contract, rarely the commune is simply tolerated, more and more often it comes to expulsion. Why is it so difficult for people to live together in mobile vehicles?

Trailer places are very different. Punk lives on some, order on the fewest. Probably all have the social thought of life in a community. Some groups find a private area to lease and form a trailer community far away from the public. But behind most of the

trailer communities there is a political intention such as protesting against excessively high rents or for legalizing life in the trailer as a self-chosen form of housing. Occupations also plead for the right to claim a piece of land somewhere as a human being born in this world. Is it not a human right to be allowed to live anywhere, even without paying money for it?

There's no commercial reward for the city out of trailer communities. But the alternative life concepts provide for urban diversity and, with donation-based cultural events, offer a meeting place for people of the most diverse origins and social strata. Many of the trailer communities are involved in anti-sexist, anti-capitalist and anti-racist actions. Some create regular public "Küfa" events (Küfa meaning "kitchen for all"), information events, workshops and offer a place to sleep for travellers or interested people who would like to "try it out". Trailer communities are popular in society, but politics sees things differently.

Established trailer communities with public events can often be found with their own website on the Internet. The autonomous centre of the respective city ("Autonomes Zentrum") can also be a good place to find out about a local trailer community. Since the legal situation is difficult for a trailer life and some communities want to remain undiscovered, there is no comprehensive directory of trailer communities.

The creation of new communities with trailors and trucks is extremely complicated in urban areas. The housing offensive in many cities is displacing existing trailer spaces. Society is threatened with the loss of open spaces, recreational facilities and alternative forms of housing. Politicians are pursuing their commercial interests. But caravans are resisting, because their form of housing also has a right to exist. Or to whom does the world belong?

left: "The city wants to sell their oldest trailer lot - we are supposed to go - are you out of your minds?!"

The Anastasia Movement by Jonas Duhme

In the course of working on this edition of the eurotopia book, we have increasingly come across settlement projects and networks of the so-called Anastasia movement, which originated in Russia in the mid-1990s. By now, there are more than 400 so called "kin domain"-projects, many in the founding phase, some already populated by more than one hundred families. More precise figures are difficult to obtain, as most settlers remain very much "under the radar", which is, however, nothing unusual for eco-settlements and community projects.

In recent years, the Anastasia movement, its underlying books and its concept of family homes have also become more popular and obviously more widespread in German-language environmental, ecovillage and community "scenes" - which our editorial staff can follow most closely. Both independent experts and operators of the German Anastasia websites speak of up to 15 notable Anastasia settlements or serious settlement initiatives in Germany. In addition, in almost all neighbouring countries there are a number of kin domains or projects that consider themselves inspired by Anastasia.

Due to some Anastasia projects submitting their community information to our directory, we came to the point of asking ourselves: What exactly is this Anastasia movement and what does it stand for? Who are its protagonists, what explains this increase in popularity? And: Should we have an attitude? If so, which one?

The Anastasia movement is based on the ten-volume work "The Ringing Cedars of Russia" by the Russian entrepreneur and author Vladimir Nikolaevich Megre (born 1950). In this, the fictional narrator (who is strongly based on Megre himself) describes in a fantastic travelogue the encounter with the noble and pure legendary figure Anastasia, who lives exclusively in and from nature. Anastasia grants him spiritual and profound insights into questions of life, the world and spirituality. From these and further explanations of Anastasia's ancestors Megre develops in the course of his books a very clear conception and idea of a lifestyle, according to which the

imminent end of the world and future wars prophesied by him can be prevented and finally peace would be created in the whole universe. The basis of this life is the pursuit of health and purity, which can only be achieved by the complete self-sufficiency of a family estate. To this end, one hectare of land per family would have to be transformed into their respective basis of life, and they should live there as isolated as possible from the rest of humanity, only with Anastasia neighbours who would populate the surrounding properties.

For Russia, it seems that many people are attracted to the Anastasia myth, after the rapid dissolution of the Soviet Union and the collapse of socialism was followed by an unbridled capitalism and an accompanying overpowering oligarchy principle, which left many people dependent and behind. Loneliness, environmental degradation, globalization, lack of confidence in politics or society, etc., may be factors that have played a decisive role in the upswing of the Anastasia movement (the books were published in 1996) and other groups, whose worldview is also based on a strict exit from society (such as Vissarionites in Sibiria, see "Ekopolis Tiberkul", p. 332). Discussions with more progressive members of the self-proclaimed "post-Anastasia" movement showed that there is a major trend towards radical escapism in parts of Russian society, for which the Anastasia movement is a strong symptom: The dream of turning away from the "evil" society and all its negative consequences towards self-empowerment and the creation of a paradisiacal island of bliss is anything but unknown to us. So why this info-text?

Megre presents absolutely precise ideas of a good, correct lifestyle for every person who strives to reach the completed stage of Anastasia. Anything else, any other way, any other lifestyle is categorically excluded and described as "sick" or "unclean". Megre sees the basic problem of the impure society above all in an alleged control of the world by a secret caste of Jewish priests (Levites), who have controlled the fate of the societies for thousands of years. The Jewish people represented a kind of extended arm and, like "programmed robots", intervened subversively and conspiratorially in societies, peoples and cultures in order to control them in the sense of the Levites. In this programming Megre sees the predominant reason for the alleged greed for money and power of Jewish people and states that a partial guilt for pogroms is to be found among the Jewish people themselves. He also sees Adolf Hitler here merely in line with many rulers who saw themselves virtually "forced" to such acts*. In addition, around the person of Vladimir Megre, his books and the "only truth" proclaimed therein, a commercially very successful cult has emerged, which in many respects can be described as a sectarian cult of leadership and through which Megre also tries to gain political influence.

In addition to totalitarian ideas about lifestyles, an intolerant good-evil dualism towards everything else and open anti-Semitism, Megre's books also include further conspiracy theories, comprehensive historical revisionism and discrimination, such as homophobia (in the "good" world, only "natural" small families are envisaged, hence father-mother-child) or racism. This can be found, for example, in an extreme ancestor cult based on a theory of several original races of individual "pure", homogeneous peoples and their traditional habitats. The aim of the movement as a whole must be to revive or restore them today. In 2016, the Swiss office for sectarian issues "infoSekta" classified the movement as sectarian and problematic.

In the German-speaking countries there are increasing reports that the Anastasia cult has led to great interest on the part of right-wing and anti-Semitic circles as well as the radical part of the so-called Reich Citizens' Movement. It can no longer be argued that there are only a few unfortunate individual cases in which borders are blurred. Rather, on closer inspection it becomes apparent that some central actors of the German-speaking Anastasia movement cultivate relationships with the corresponding scenes, are active in them, often originate from them or open their sites to right-wing extremists and conspiratorial circles and organizations - or even offer events, meeting places and networking sites for these target groups themselves. Topics such as ecology, sustainability and nature conservation are obviously not necessarily apolitical or bound to a politically left worldview, as is often assumed today.

It is far from us to accuse every fan of the books or member of the movement per se of the attitudes or belief patterns addressed. But on the one hand we would like to point out that both the understanding of community as well as the concrete idea of the world of Anastasia-followers are not really in line with our idea of community. Much more problematic are our assessments of the ideological foundations outlined above, as well as of the concrete interdependencies in right-wing and ethnic scenes and circles, whose attitudes and worldviews we firmly reject.

A comment by Dmitry Solovyev of Kind Land (p. 336) and GEN-Russia (p. 535):
For us, Russian ecovillagers, it was really strange to discover that a neo-Nazi movement had adapted Anastasia ideas to their ideology. I discussed the article with different researches of Anastasia movement and their position also differs from the stated one. My opinion, both as a person, who has been living in one of the biggest Anastasia communities (Rodnoe, Vladimir region) for the last 15 years, and as a researcher (my Bachelor's thesis was "Self-conscience of Russian ecovillagers", 2015), quite differs, too.
We didn't find any direct mentions in Megre's books about anti-semitism or racism, but more importantly, I've never met any Anastasia followers who shared these ideas.
About Megre leadership and political ambition - I'd like to say that he is my neighbor in Rodnoe, he is an elderly man and with poor health, so he is not able to rule any societies. In Rodnoe, there is an anarchic style of management. Generally, Anastasia movement is not about sects, there are no administrations or anything similar. Vice versa, Anastasia villages are based on grassroot activity.
As for criticism, I find important to mention quite a romantic attitude to the important sides of life (community management, private and common earnings, waste management and etc.), ideology of escapism (but this is far from asociality), low social activity, interpretation the books as sacred and etc. But, once again, these features of some individuals or groups cannot be extrapolated to the whole ideology in its entirety.

*see Megre 2002/2007: Anastasia Book VI: The Book of Kin, Ringing Cedars Press, Chapter 6 (p. 158-162, depending on edition).

Pictures: all by Wikimedia. P. 49 top: by Димон Рышков, CC BY-SA 3.0-License. P. 49 bottom: an Anastasia settlement in Drabeši, Latvia, by Santa Zembaha, CC BY-SA 4.0-License. P. 50: by Колпаковы, CC BY-SA 3.0-License

At a Glance...

The following table shows at a glance the numbers of existing/planned communities presented on the following pages. In total, we publish 605 detailed descriptions of intentional communities. 85 of those are still in a planning stage (from p. 438), 520 are realized physically (see following pages). We have contacted more than 2500 addresses (successfuly, according to our email logs), so there's still a lot of communities that did not respond - and therefore the following table, inviting as it may seem to community researchers, is not a valid source to assess European intentional community numbers. You can find the respective self-assessment of the communities as a cohousing or settlement project (in any case with a strong focus on community) - but this is also vague. In the UK, for example, many groups that even call themselves 'co-housing' have not ticked the co-housing box.

"Additional Camphill" are the 87 Campill communities from the respective countries that are not listed with the detailed descriptions, but in our comprehensive list starting on page 510. "Total Camphill" are the communities from that list plus the ones with detailed descriptions. Furthermore, we list the number of communities that were part of the last (2014) edition of eurotopia (either with detailed descriptions or as members of community networks), but are not listed in this edition ("additional from 2014", see page 492)(Not in the table below: one each from Moldavia and Belarus).

The following numbers are entirely, but maybe not quite completely, accurate. And more community addresses yet can be found as members of "Networks of Communities" starting on page 508.

	total	planned	existing	co-housing (self-assessment)	settlements (self-assessment)	additional Camphill	total Camphill	additional from 2014	useful addresses
all countries	604	85	519	107	65	87	92	200	96
Austria	22	7	15	7	1	1	1	6	2
Belgium	15	4	11	10				4	3
Bosn.-Herzg.	1		1						
Bulgaria	2		2						
Croatia	1		1					1	1
Cyprus								1	
Czech Rep.	3	1	2		1	1	1		
Denmark	30	2	28	8	5			5	5
Estonia	2		2			1	1	1	
Finland	4		4	1		4	4	2	1

	total	planned	existing	co-housing (self-assessment)	settlements (self-assessment)	additional Camphill	total Camphill	additional from 2014	useful addresses
France	60	5	55	16	4			12	3
Germany	223	27	196	31	19	11	11	70	42
Greece	4		4	1				3	1
Hungary	4	1	3	1				2	
Ireland	2	1	1			17	17	1	
Iceland								1	
Israel	3		3					1	1
Italy	37	9	28	8	3			7	2
Latvia								2	
Lithuania	1		1			1	1	2	
Luxembourg	1	1			1				1
Netherlands	36	7	29	9	8	2	2	15	2
Norway	5	1	4		2	6	8	1	1
Palestine									1
Poland	2	1	1			1	1	4	
Portugal	9		9	1				1	2
Rumania	1		1					3	
Russia	13	2	11		5	1	2	4	
Serbia									1
Slovakia	2		2					1	
Slovenia	1		1	1				1	
Spain	26	5	21	1	6			8	5
Sweden	18	2	16	5	4	3	3	5	3
Switzerland	20	2	18	2	2		1	4	5
Turkey	4	1	3	2				1	1
UK	51	6	45	7		38	39	28	6
Ukraine	2		2					2	

IMPOR

In this eurotopia Communities Directory, intentional communities present themselves with descriptions and addresses. What a gift! We, the editors of this book, have asked them for it and therefore bear part of the responsibility for our readers to regard this generous gesture with care and respect. It might come as a surprise, but not all communities presented seek new members or appreciate visitors. The one reason all theses communities are in this book is to share the idea of a community lifestyle. Of course, another important reason is to enable contact to them. We just ask you sincerely to respect when a community does not provide a way of contacting or visiting them or even states in their description that they can't accomodate any guests at this point.

If an intentional community has not explicitly said otherwise, then it's not cool at all to just show up with (or without) your tent and sleeping bag. That might lead to very uncomfortable situations for community members: They might not want to send you away, even though no one is really ready to host you and show you around at that day. If you then say that you have found that place in the eurotopia Directory - well, it might well be that this community will not let us present them ever again. Please, avoid that! Even if you find that this community isn't any good anyway (since they sent you away - how dare they?), please try (hard) to imagine that for another person it might mean a lot to get to know it.

The good news is, intentional community members are often wonderful people, and they will often try to accommodate and to integrate you. Or they offer seminars to get to know their community. It's just important to say (experience shows that it is not obvious) that community members also need privacy and chances to retreat. THANK YOU!

TANT

Self-Descriptions from Intentional Communities, Communes, Co-Housing Projects and Ecovillages of Europe

How to Use this Directory

All the following information has been put together and sent to us by the communities themselves. Data might be missing or be contradictory.

The following data is from existing communities. Find planned communities in the chapter starting on page 438. Within the chapters, we sort the communities by country and, within the countries, in alphabetical order. Some European countries with few communities are presented at the end of each chapter because they share one map (p. 411). The maps at the beginning of each country roughly show the location of each existing community: The numbers on the map are the page numbers in this directory. A number with a "B" behind it means that you're looking at the *second* description on that page. We have demonstrated this on page 64, where we printed that "community number" on the top right of each community description. Again, you don't need it, cause it's the page number.
Planned communities with no location aren't on the map, obviously.
To help you find the location of a community in countries with a lot of numbers on the map, like France, UK and Germany, use the coordinates on the top right corner of community descriptions in these countries.

The symbols in the descriptions are explained on the next and on the last page of this book.

Phone numbers: Replace the "+" with your country's code for international calling (often "00") and do not dial a "0" later in if there is one in parenthesis ().
If you're *in* the respective country, don't use the first digits lead by the "+"-sign, but do dial the "0" if there is one in parenthesis () - the number to dial will then start with that "0".
If there's more than one phone number, we have often listed the additional number(s) without repeating the country or area code. If you read "+31-(0)123-45678, -87654", then the second number shares the numbers of the first one up to the last hyphen (+31-(0)123-87654).

You're looking for a specific community? In the **index** at the end of the book you can find the communities not only with their complete name, but also with other names they might be known with ("ECOlonie - Centre Ecologique International" can be found as well as "Centre Ecologique International, ECOlonie~"). If the respective community appears at several places in the book, the **bold** page number is the place where the detailed description is printed.

A powerful tool to search the book is the search engine on eurotopia.directory/booksearch - you can insert search terms, search for keywords as well as community size or year of founding. You will also get page numbers as search results (see above), so the book is necessary to use that feature.

Legend

- planned community
- this project was not described in the last edition of eurotopia (2014)
- co-housing (with emphasis on community living)
- settlement (with emphasis on community living)
- ecovillage (self-assessment)
- A1 coordinate on the respective map (countries with many communities)
- telephone number, email address, website
- year of founding
- urban/small town or suburban/in the country (behind: size of premises))
- (...) behind that, if applicable: name of bigger cities close by.
- ♀/♂/ number of women/men/children (if this information is given)
- number of members (inconsistencies between this and the answers above have simply been adopted)
- degree of self-sufficiency with food
- languages spoken in the community (other than language of that country) e=English, f=French, g=German, i=Italian, sp=Spanish
- how many of you are working on the community site? (i.e. how many of you are usually on location?)
- how many of you are working within community-owned enterprises?
- who decides important issues? (in parenthesis: how are these decisions made?
- handling of finances
- ownership of premises and buildings
- ideological, spiritual or religious focus
- main fields of work
- keywords
- open to new members - behind: requirements for new members
- working guests welcome after prior arrangement (free board and lodging)
- ( ) in parenthesis: as above, though not necessarily with free board and lodging
- member of the following community networks

Austria

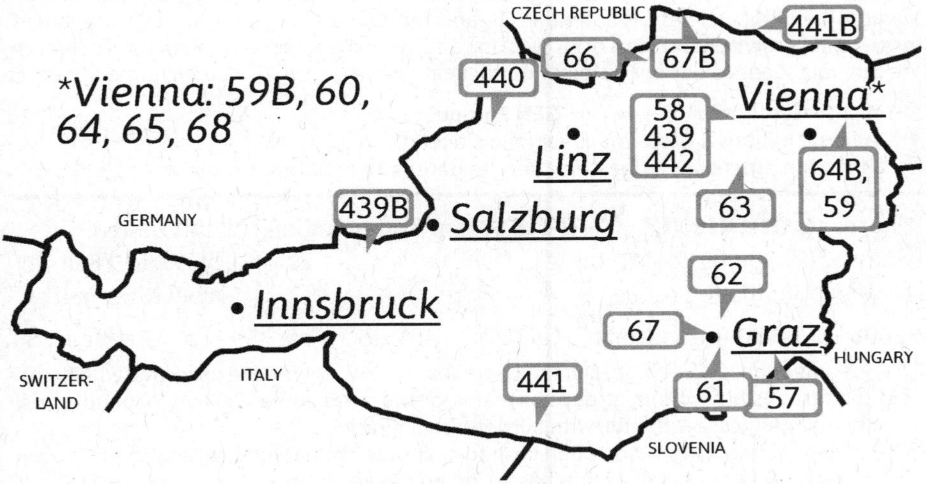

In the last edition of eurotopia, we published 10 communities from Austria - this time it's 22: 15 existing ones and 7 in planning stage (see page 438). Find one Camphill community on page 510. Two existing communities from last time are not published anymore, you can find their adresses on page 42, though. GEN Austria is an active community network (page 522).

Cambium Living in Community

Kasernenstraße 2 | 8350 Fehring ☎+43-(0)3155-285 01
info@cambium.at www.cambium.at

2017 | 16 ha (near Graz) | 15-20 % | ♀26 ♂24 25 75

- g, e, i, sp, Bosnian — planned — 6-8 permanently, 10-15 various times
- sociocratic Circle-Organization Method (SCM) including a management circle and a plenum (sociocratic consent or systemic consensus principle)
- partial collective financing (e.g. household)
- our association bought the property in 2019
- as a community, we don't follow any specific ideological, spiritual or religious path.
- community and personal development, agriculture, renovating and building, household, cooking, childcare, public relations, networking
- ecological, self-managed, non-violent, inner growth, organic diet, adult education, seminar-house, peace-work, environmental protection, permaculture, organic agriculture, organic gardening, animal friendly husbandry, ecological construction, humour, handicrafts, carpentry, clay building, media, culture, body work, music

☺

We strive to create a living space for about 100-150 people – a place for residency and work where people of all ages, different occupations and interests feel at home. We want to preserve traditional skills as well as apply meaningful sustainable technologies.
Living and working in ongoing dialogue with other locals and groups, we aim on building a solidarity economy and support interpersonal relationships on a deep personal level. We

wish for mindfulness and respect in our personal relationships as well as an open attitude towards the other and her/his potential. Organic farming, kitchen and catering, an academy for personal growth, new forms of co-working and other meaningful small business ventures are tangible ideas to put our vision into practical life. *See picture on page 11*

→See also description of network "**GEN Europe**"
regional initiative "Steiermark gemeinsam jetzt"
☺ admission process of at least one year to determine if both sides see this as a match

Cohousing Pomali Verein "Miteinander Zukunft Bauen"

Landersdorf 108 | 3124 Wölbling ☎+43-(0)699 190 866 71
info@pomali.at www.pomali.at

2008 1.3 ha (near Krems) 10 % ♀ca. 25 ♂ca. 25 ca. 25 ca. 75

g, Hungarian, e, Czech, sp, Dutch we share the work we have to do (and with help)
the responsible working groups or teams, or the management group (depending on the defined responsibility), with consent (sociocratic)
individual financing; association cash to cover the annual joint expenses and property management as well as purchases and process support
rented with the goal of being taken over by the association
no, it is important to us that different spiritual/ideological/political approaches can have a place in our community
garden and outdoor space; property management; foodcoop; car sharing; kitchen
ecological, spiritual, Christian, ecumenical, Buddhist, Taoist, esoteric/New Age, inner growth, meditation, yoga, tantra, organic diet, vegetarian diet, vegan diet, permaculture, organic gardening, humour, accessible to the disabled, circle culture, supportive listening, social tools, constellation work, birth work, partly ecological, cross-generational, various occupational groups, sociocratically organised, some self-employed

☺ ()

Since 2013 we have been living in our intergenerational housing project. We have been sociocratically organised since 2012 and each of us works in at least one circle or team. This means that everyone is involved in the ongoing maintenance and further development of our project. We do not have a plenum for decision making, but we use the large group as a space for connection among each other and to better prepare some decisions. Due to this clear organisational structures, we have relatively few conflicts. However, if there is a crunch, then we can trust on extensive expertise for conflict resolution in the project, as several of us are therapists, consultants, process facilitators, mediators or similar. In addition, we also have crafts(wo)men, artists and other professional groups.
The topic of ecology occupies many of us in very different ways, but we have no guidelines to which everyone must adhere, although we make use of some synergy effects that are only made possible by the size of the group (car sharing, foodcoop, eating together; cooperation with local farmers...). As far as child-rearing, political attitudes and spiritual orientation are concerned, it is also more important to us that different approaches have a place next to each other and complement each other as a community as a whole to have a valid guideline. We have been particularly successful in embedding ourselves in the neighbourhood and the village, in which we enjoy a high level of acceptance.

→See also description of network "**GEN Europe**"
☺ participating in the community life and accepting our agreements. Further trainings towards sociocracy; financial independence; cooperative shares

Der Lebensraum Verein zur Förderung Ökosozialer Wohnkultur

Tannengasse 1 | 2230 Gänserndorf

kommunikation@derlebensraum.com www.derlebensraum.com

2005 1.7 ha (near Wien, Bratislava) 2 % ♀27 ♂25 33 85

- g, e, f, sp, Hungarian
- we decide as a group (consensus or majority decisions)
- individual financing — owned
- maintaining our premises, sharing everyday issues, celebrating life
- ecological, spiritual, Christian, ecumenical, esoteric/New Age, inner growth, meditation, yoga, organic gardening, alternative energy, ecological construction, accessible to the disabled, flexitarian

As the first cohousing community in Austria of the new generation, we have been living together for 13 years now. So far, we have rented, which again and again caused a lot of living fluctuation. Now we are in the process of buying the entire settlement, which has brought new people to us.
We are a colorful group of people of all ages, with diverse interests. Exchange and sharing, support in everyday life, cooking together, eating and celebrating are cornerstones of our community. We love our large grassland with tepee, fire pit, playgrounds, vegetable gardens and beach volley and football court. Once a month, we meet in plenary to discuss upcoming issues and make decisions.

Frauenwohnprojekt [ro*sa] Donaustadt

Anton-Sattler-Gasse 100 Top 22a | 1220 Wien

kontakt@frauenwohnprojekt.org frauenwohnprojekt.org

2003 4852 m² 0.04 % ♀21 ♂0 25

- g, e, f, i, sp
- general assembly (majority vote, consensus if possible)
- independent households — cooperative
- feminism
- ecological, self-managed, feminist, non-violent

☺

The house was inhabited at the end of 2009. When an apartment becomes available, we are happy about new women who want to live with us in a neighbourhood of solidarity, where community is possible but everyone has her own retreat.
Available apartments will be published on the homepage. In 2 of 3 cases the association has the right of proposal.

- loose contact with other women's housing projects in Vienna
- ☺ women, alone/with partners, with/without children, migration background, special needs

Gemeinschaft B.R.O.T. Kalksburg

Verein für gemeinschaftliches Wohnen
Promenadeweg 5/1/40 | 1230 Wien
office@brot-kalksburg.at www.brot-kalksburg.at

2009 10 ha ♀ca. 45 ♂ca. 35 ca. 30 ca. 110

g

plenary (everyone in common), management team (aiming at consensus, otherwise majority vote with power of veto)

property development contract

spiritually open

ecological, self-managed, spiritual, seminar-house, accessible to the disabled, intergenerational, historical site (former English park), close to nature, forest, sheep, bees, chickens

Since 2010 we live in parts of the former Jesuit monastery and in newly built low-energy and passive houses on the grounds of the Kollegium Kalksburg.
We are a large community that unites people of all ages. We understand the diversity of individual life plans, life phases and the personal spirituality of all residents as an enrichment for our common life. Which principles and tools help us to grow together again and again in our everyday life as a community can be read on our website: http://brot-kalksburg.at/ueber-uns
B.R.O.T. stands for pray/meet, talk, be open, share. Originally, our project originated from a former B.R.O.T. project with Christian roots.
The members of our community come from different cultures and religions. A 'spiritual consciousness' is important to us, even if it is not bound to any denomination or church. Because we have 'praying' in our name, it always leads to questions about what we are and what we are not... Anyway, we are a very heterogeneous community and we are open to different spiritual backgrounds, which are lived and practiced here individually and collectively.
Altogether there are four B.R.O.T. communities (Hernals, Kalksburg, Aspern, Pressbaum) in and around Vienna with a common umbrella organization:
https://brot-verband.at (B.R.O.T. Network, founded 2008)

association of B.R.O.T., Initiative Gemeinsam Bauen und Wohnen

Haus der Stille

Friedensplatz 1 | 8081 Heiligenkreuz am Waasen ☎+43-(0)3135-826-25
info@haus-der-stille.at www.haus-der-stille.at

1979 (near Graz) 5 % ♀9 ♂6 0 15
10

g, e, i
leadership team (consensus)
financing through guesthouse and seminars, selling of training materials, donations
association "Haus der Stille" (House of Silence)
Franciscan, ecumenical, interreligious
looking after guests, education, meditation
ecological, non-violent, spiritual, Christian, ecumenical, meditation, adult education, peace-work, accessible to the disabled, peace signs of the religions, paths for reflections and meditation, labyrinth, garden of silence

☺

The calm setting 15 km southeast of Graz in the southeast Styrian hilly landscape and the atmosphere of the house invite to reflection, silence, prayer and meditation - healing silence for people who are looking to find inner and outer peace.
Since 1979 the "House of Silence" has been led by a Franciscan-oriented community of women and men and offers space for about 60 guests who are looking for a few days of silence and distance from everyday life, who need a longer break to recharge their body and soul, who want to participate in various courses for deepening their life and faith (focus: meditation, Bible, Franciscan spirituality, peace).
People between the ages of 18 and 35 have the opportunity to spend a year and participate in the community.
Cornerstones of the daily routine and benefits for the guests: simple life, distance from everyday life, prayer and meal times with the community, possibility to withdraw completely into silence (after consultation also into one's own poustinia/retreat), daily hour of silence, Bible discussion and posture exercises, long walks in the surroundings.
The willingness and ability to engage in longer periods of silence and being alone are important requirements to make most of your time with us. Pastoral care and social and life counselling can be offered if needed.

INFAG Austria
☺ good mental health and resilience, 2 week probationary period to live here for a year

KooWo Kooperatives Wohnen Volkersdorf

Rabnitzweg 18 | 8063 Eggersdorf bei Graz
koowo.volkersdorf@gmail.com koowo.at

2015 3.6 ha (near Graz) 70 % ♀14 ♂14 20 48

g, e, i, sp 0 1.5

the working teams and the management circle (consent)

individual financing living space, community financing community space, club contributions, asset pool

in cooperative property

attachment to nature, open to all religions, sociocracy

agriculture, administration of the housing complex, co-working

ecological, self-managed, non-violent, organic diet, environmental protection, self-sufficiency, permaculture, organic agriculture, organic gardening, organic farming, alternative energy, new technologies, ecological construction, accessible to the disabled, sociocracy

Cooperative Living (KooWo) is a lively and sustainable co-living project.
We are a colorful bunch of adults, kids, and animals with different characters, interests, jobs, and ages. Mixed as we are, still we do share something:
We share a common vision. We want to live, work and organize a farm together. Because we believe we can learn from each other and grow as a community, by developing and unfolding ourselves. Because we want to share resources, want to know where our food comes from, want to live in an ecological and energy efficient manner and want to help each other. Because together we are more productive and it's also more fun!
We set up three houses for 28 families. Moreover, we also have an old courtyard and an old house that we will turn into a common area. Specifically, we plan co-working spaces, therapy rooms, a music room, a workshop, and other working areas.
We take decisions following the consensus principle and organize ourselves in thematic working groups. Leaders and a delegate of the working groups make up our higher committee and are elected through open elections. Our building promoter is Die WoGen, who is committed to realize community living projects.

WoGen Wohnprojekte Genossenschaft

to reach our cross-generational vision, we are looking specifically for men, families with teenagers and persons older than 40

Lebensgut Miteinander

Prünst 12 | 3163 Rohrbach an der Gölsen ☎+43-(0)664-885 439 83
info@lebensgutmiteinander.com www.lebensgutmiteinander.com

2014 16 ha (8 ha forest, 7 ha meadow) (near Wien) 50 %
♀7 ♂7 3 17
11 11

- g, e
- Decisions are taken by the authorized committee for the concerned topic (sociocratic consent-principle).
- Individual finances - with shared financial responsibility for refunding the estate and for costs of the association.
- owned by the association "LebensGut Miteinander"
- shared values (see below)
- all-generations-community, children daycare, ecological vegetable agriculture, seminar center, therapeutic group-practice
- ecological, self-managed, non-violent, hierarchical, organic diet, adult education, pedagogical, seminar-house, environmental protection, permaculture, organic agriculture, organic gardening, construction, humour, accessible to the disabled, sociocracy, day care for the elderly

☺ ( )

On "LebensGut Miteinander", we are organizing a agricultural estate as a space for development and experience of full-filling, all-generational living and working together on a foundation of values. We connect agriculture, the companionship of children and elderly people, learning and cultural offers, and sustainable healthcare to a new form of co-operation.
The estate comprises 8 ha of woods, 7 ha of grassland and 1 ha with buildings.
It is seated in the hilly alpine upland - 20 km south of St. Pölten and 50 km to the southwest from Vienna (1 h by car, 1,5 h by train). We have our all-generations-cohousing (we plan to become 25 adults + many kids), ecological and CSA vegetable and herb-agriculture, children's daycare, a seminar center and a therapeutic group-practice (3 rooms). Apart from that we have 5 donkeys, 3 cats, 8 running ducks and 10 chickens. The values we strive for are: ecology, health, security, joy, responsibility and loving attitude.
We cook collaboratively vegetarian once a day - favorably with the delicious seasonal vegetables from our garden. We still accept people who want to realize with us our vision of sustainable living and a cooperation of all generations.

- diverse networks of the region
- ☺ we have a comprehensive process with financial, time and content criteria

Mauerseglerei Wohnen und Leben in Gemeinschaft

64

Endresstr. 59c | 1230 Wien
+43-(0)680-225 76 72
info@mauerseglerei.at
mauerseglerei.wixsite.com/gennesaret

2016 | 3000 m² | ♀20 ♂17 32 69

g, e, f, (i)

mainly working groups, plenary sessions on fundamental issues (we use the sociocratic method)

property of our association

ecological, self-managed, left-wing political understanding, alternative energy, new technologies, ecological construction, humour, accessible to the disabled, non-smoking, elementary school, therapy-center

Our house was built as a communal housing project. This means that we, as a community, have been responsible for the development, planning and construction process. The administration and maintenance of our house is also jointly organised. We adhere to common rules for our discussions and interaction in order to be able to treat each other respectfully, attentively and appreciatively.
There is always something going on in our common rooms. If you are looking for company or need support, you will quickly find open ears or helping hands. We live in a pleasant variety of private space (27 households) and community. We make sure that all residents are well integrated socially and emotionally. We are interested in our flatmates, curious to get to know other models of life and experience our differences as enrichment despite all possible conflicts. We solve conflicts with openness, understanding and honesty and get support from outside if necessary. In financial emergencies, we support residents with means from our Solidarity Fund. Evenmore, we have a common food co-op.
The four pillars of our project reflect the fact that sharing (solidarity), respect for nature and life (ecology), mutual support and tolerance in living together (community) and authentic encounter on a spiritual level (spirituality) are important to us and that we also want to implement them.

☺ see category "mitmachen" (=participating) on our website

Ökosiedlung Gärtnerhof

64B

Hochwaldstr. 37/5A | 2230 Gänserndorf
+43-(0)2282-702 89-0
www.atelierdeubner.at

1988 | 2 ha (near Wien) | ♀22 ♂19 15 56

e, Swedish | 4

everyone in common (consensus)

limited common financing & indiv. financing

partially property of individual members, partially common ownership

- none
- architecture, organic farming, therapy for people with disabilities
- ecological, self-managed, non-violent, esoteric/New Age, chaotic, inner growth, organic diet, vegetarian diet, pedagogical, mentally/psychologically handicapped persons, physically handicapped persons, environmental protection, self-sufficiency, organic gardening, organic farming, alternative energy, new technologies, ecological construction, integration of the handicapped, carpentry, art, media

The settlement is existing for 31 years now. In the meantime, some of the children of the original residents have already taken over the apartments or houses. There are community meetings every 2 months where important decisions are made. 4-6 times a year there are working days where the facility is cleaned, the grassland is maintained. Many points of the plant care such as maintenance of the bathing pond, the plant purification system, the lifts, the sauna and the open spaces are carried out independently. There is also a range of social assistance for needy and elderly residents.

☺ an expansion is planned

Ökotopisches Zentrum c/o Fritz Balatka

Achauerstr. 21 | 2326 Maria Lanzendorf ☎+43-(0)2235-444 26
fritz.balatka@gmx.at

1979 4000 m² (near Wien) ♀16 ♂15 8 39

- g, e, f, i, sp, Portuguese
- all members of the association in monthly meetings (discussions reaching majorities and consent)
- indiv. financing & common financing of maintenance, running costs and investments
- joint ownership of all adult persons who are members of the association
- no, but a general ecological orientation
- cohousing, sharing daily activities, once in a while joint projects
- ecological, self-managed, non-violent, inner growth, environmental protection, alternative energy, social work, therapy, handicraft, art

Founded in 1979, our project is among the pioneers in Austria. The project aimed at creating a group oriented alternative between "individual flat in town" and "idyllic rural family". We founded an association which bought the premises of a former textile factory near Vienna. According to individual and group interests, flats and common rooms were built. Our members earn their living in Vienna and shared activities of building and living together at home. Important for us was to create a child-friendly environment, ecologically oriented and self-administered in a pluralistic and democratic atmosphere. As we do not share a common ideology we try to cultivate our core values of democracy, tolerance, the will to share and to participate in our community. Defined monthly contributions of the members secure the financial base of the association.

PAN Schule für gemeinschaftliches Leben

Harmannstein 2 | 3922 Großschönau ☎+43-(0)2815-6651-0
portal@pan.at www.pan.at

1995 60 ha (near Gmünd in Niederösterreich) 85 %

♀12 ♂11 3 26

g all all

plenary assembly of the community, in specific sections those who are legally responsible (CEO, headmaster) (mostly unanimously)

extended family community, everyone shares everything

property of a non-profit private foundation

original Christian

education, agriculture, kitchen, food processing, workshops

ecological, self-managed, non-violent, spiritual, Christian, inner growth, organic diet, adult education, pedagogical, free school, peace-work, one-world development, animal protection, environmental protection, self-sufficiency, organic agriculture, organic gardening, organic farming, animal friendly husbandry, forestry, alternative energy, new technologies, ecological construction, humour, accessible to the disabled, non-smoking, workshops: metal, wood, textile, graphic, media

☺

"Health for soil, plants, animals and humans" is the PAN guiding principle at which we have aimed our lives in a common and practical way.
Since the foundation almost 25 years ago we have been trying to take Jesus seriously and see ourselves as spiritual beings who make human experiences. This attitude of consciousness leads us more and more deeply into a way of life of community with everything that surrounds us and makes our existence a constant process of learning and development.
Today we are a multiple-generation community of about 30 people with a majority of "young adults" between 20 and 30 years who want to live and work together in dignity.
Therefore we cultivate our 60 hectares of land and forests in a biological way, we keep our animals (sheep, cattle, outdoor pigs, bees, chickens, cats, dogs) appropriately and in a healthy manner and together we care for our vegetable and herb garden (about 1 ha). We refine the obtained raw products in our kitchen as well as in our milk and meat processing facilities. We sell the surplus in the WIR-Selbermacher shop and in self-supplying shops in the region.
Main branches of the PAN project are also a private school for 6- to 15-year-olds, several workshops (wood, metal, textiles) and last, but not least, a business unit focusing on research and development (software solutions, regional network), services (graphic design, accounting, IT) and soil health (regenerative microorganisms).

☺ willingness to develop

Projekt Kooperatives Wohnen

Am Silberberg 4 | 8074 Raaba ☎ +43-(0)316-401 363
gert.civegna@aon.at

1978 1 ha (near Graz) ♀28 ♂18 6 52

g
property of community and private property
self-managed

Our housing project will celebrate its 40th birthday in autumn 2019. We moved in in February 1979 and developed this in the course of the 1968 movement. In contrast to the current housing projects, however, we are a Wohnungs-Eigentums-Gemeinschaft (WEG), i.e. each apartment (house) is privately owned and can be sold or rented without the consent of the other inhabitants. What unites us is the structure of an association, the complete self-administration and mutual social commitment (also care of refugees), but there is no ideological orientation. As everywhere, there are, of course, some who are strongly committed and others who almost only enjoy the low operating costs due to self-administration. The "old residents" are now at 70+, and their children are now scattered elsewhere. Fortunately, there are a few new families with children who will hopefully take over our care at some point.

Seedcamp Gemeinschaft St. Hildegard Stiftung

Hauptplatz 7 | 3851 Kautzen ☎ +43-(0)2864-2573
edu@seedcamp.at www.seedcamp.at

2000 7 ha (near Waidhofen/Thaya) ♀9 ♂9 19 37

g, e, f, i 7
all that are concerned (through conversation) household till
trust administered by association
free spirit
house & gardening, school & kindergarten, animals
ecological, self-managed, anarchistic, left-wing political understanding, non-violent, spiritual, Christian, ecumenical, Buddhist, Taoist, esoteric/New Age, chaotic, anthroposophic, inner growth, meditation, yoga, tantra, sexuality, free love, organic diet, vegetarian diet, vegan diet, macrobiotic diet, adult education, pedagogical, free school, peace-work, one-world development, mentally/psychologically handicapped persons, physically handicapped persons, environmental protection, permaculture, organic gardening, forestry, alternative energy, new technologies, ecological construction, humour, accessible to the disabled, free spirit, art, nature, human, landscape, human dignity and health

☺

We see the "seedcamp" as a creative meeting place, where many different kinds of workshops and seminars are held and interactive events are organised. The goal of this very organic project is the natural development of a regular programme of events with periodically repeated projects, in order to provide a continuum for the development of sustainable models of thinking. In order to fill this extended "permaculture" with a feeling of natural security, we call all interested parties to contribute, either by organising and holding a variety of workshops and seminars in order to share valuable knowledge, or by breathing life into this place and its growing community with full devotion, all the while being sensitive to the concept of freedom. There exists a large, loose community distributed across Austria and beyond. We act together in order to co-evolve this place and we come together for festivals. The community that is actually present on site is still in its infancy, but the information contained within its cells is holistic.

- woodcult
- we want the place and the community to stay open-minded

Wohnen im grünen Markt

Maria-Lassnig-Straße 32-34 | 1100 Wien
wohnen@gruenermarkt.at | www.wohnen.gruenermarkt.at

2014 | 6000 m² | ♀31 ♂24 24 79 | 5

- d, e, Farsi, Korean
- the entire group (consensus)
- indiv. financing with limited common financing (shared spaces)
- registered association, rented by members
- none
- self-managed, meditation, yoga, accessible to the disabled

Welcome to "Wohnen im Grünen Markt" (Living in Green Market). Our house will have 45 flats and several common areas where we regularly want to spend time together by cooking, being creative, playing, tending our gardens or to simply hang out and relax.
We are a colorful community, consisting of families and individuals with lots of different interests and talents. We want to shape parts of our life and our environment together.
Our house is being built in a new, exciting area of Vienna, close to the central railway station. There will be a park, a boulevard with cafés, offices and cultural places for everyone and some other co-housing communities.
"Wohnen im Grünen Markt" stands for sustainability, tolerance, creativity and openness for new ideas. Our house is a lively space and supports diverse concepts of living.

- Initiative Gemeinsam Bauen & Wohnen, Foodcoop
- interest in co-housing projects, sharing our vision

Belgium

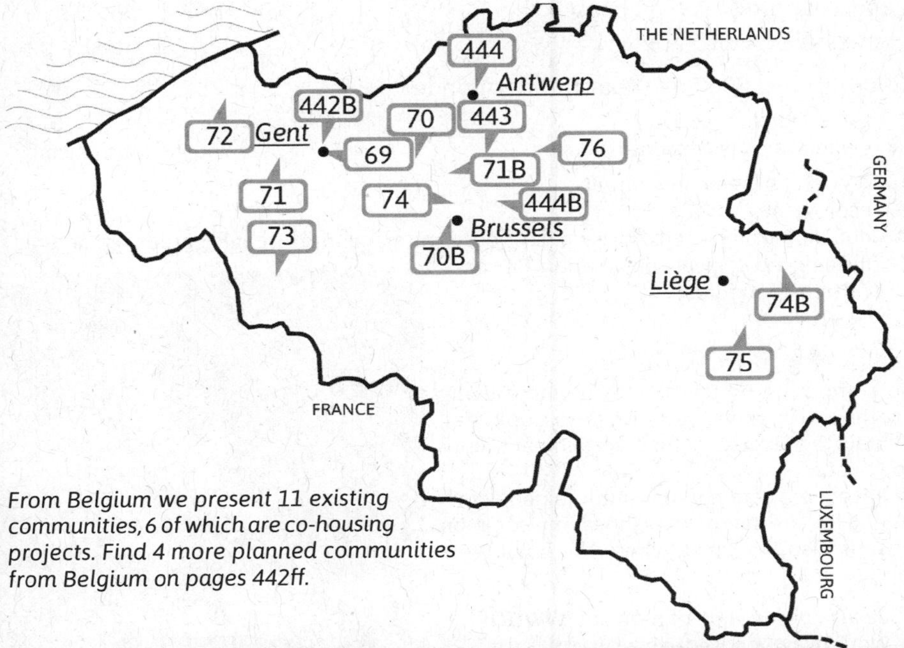

From Belgium we present 11 existing communities, 6 of which are co-housing projects. Find 4 more planned communities from Belgium on pages 442ff.

Cohousing De Schilders

Schildersstraat 5 | 9040 Sint-Amandsberg ☎+32-(0)48-607 26 06
steven@lowimpactman.be

2016 (near Gent) ♀9 ♂3 7 19

- Dutch, e, f
- the whole group (consensus)
- individual financing building the project, limited common financing common purse
- most of us are owners, one unit is rented
- ecological, social, feminist
- living together
- ecological, self-managed, feminist, non-violent, organic diet, seminar-house, environmental protection, alternative energy, ecological construction, humour, non-smoking, regulary art events at our home

We are a small urban cohousing project based on ecological and social values. We live in small private units and share a lot of space and also cars, stuff, time, a grand piano, tools, books and so on. Our new building has a low ecological footprint and is mostly run by renewable energy. We are engaged in local groups and organisations on car-sharing, coop housing, feminism, art and activism and local food. We work together with other local groups and try to inspire people to social and ecological living.

Samenhuizen, active in several local groups and coops

Co-Housing HuisElf Bruno Verschaeve

Kakemanstraat 11 | 9280 Lebbeke ☎+31-(0)49-469 50 15
bewoners@huiself.be HuisElf.be

2006 2000 m² (near Dendermonde) 5 % ♀3 ♂5 1 9

g, f, e, Dutch 2
owners (mostly consensus) individual financing
property of some inhabitants
ecological
non-violent, inner growth, seminar-house, permaculture, organic gardening, organic farming, ecological construction, Perma-culture, food forest

☺

The co-housing project is mostly about living together with some gatherings to cook, eat, maintain the garden... or for entertainment, inner growth etc.
We have a big common house for joined activities and for all kinds of workshops. The common house is also the place where the volunteers stay.

→See also description of network "**WWOOF**"
WorkAway, voedselteams

Communauté de la Poudrière

60 rue de la Poudrière | 1000 Bruxelles ☎+32-(0)2-512 90 22
contact@lapoudriere.be www.lapoudriere.org

1958 50 % ♀14 ♂38 8 60

f all of us
all of us (consensus) shared incomes
recycling furniture and second hand items - farming
one-world development, self-sufficiency, learn to live together; multi-religious

☺

The community "La Poudrière" is a fifty year old experience which shows the possibility to live and work together, even if we are different (religions, social background, age...) and even if we don't receive any financial support from the government. Everybody is asked to give the best he/she can. We share all our incomes, and we keep the private property to a very low level. We eat together 3 times a day and have several meetings. The use of the French language is therefore very important. Our main activity is recycling items or furniture, but we do also run a farm which produces fruits, vegetables and meat. Everybody has his/her own room, or flat (families): privacy is respected, even if we share a lot of other things.

→See also description of network "**Emmaus**"
☺ to speak a little French

De Groene Doening

Beentjesstraat 28 | 9850 Poesele   +32-(0)486 853 907
ruth@veerenvlam.be   veerenvlam.be/DeGroeneDoening/

2017   5250 m² (near Gent)   30 %   ♀7 ♂2 1 10

Dutch, e, f, sp, g
all who are concerned (consensus)   limited common financing
property of two members
ecological, self-managed, non-violent, permaculture, organic gardening

De Groene Doening is a co-housing community in Poesele, a tiny village near Gent.
Our community is based around 10 enthusiastic people between 14 and 61 who have a passion for creating an environment where we take care of ourselves, the others and all that surounds us. We get all excited about connection, DIY-ing, sustainability, permaculture, self-sufficiency, renewable energy, ethical banks, local and organic food, baking bread, composting toilet,... Being aware of our choices is important to us.
The domain is 5250 m² and consists of a main house, 2 smaller units, a workshop, barn, sheep stable, studio, an orchard, an amphibian pond and lots of space for gardening and creating a permaculture oasis. Our green paradise gives abundantly, all of which we process into delicious food. Preserving, air drying, fermenting, we love to experiment!
In the main house there are around 5 people. They share bathroom, kitchen, dining area and living room. The smaller houses, where there's space for a couple or small family, have their own facilities. Some spaces, like the workshop, are shared by everyone.
Once a month we roll up our sleeves alongside volunteers. On this workday we also organize a guided tour for potential co-housers. Welcome!

Samenhuizen
conscious about sustainability

De Okelaar

Oppemstraat 8 bus 13 | 1861 Wolvertem   +32-(0)479 297 126
mia_vr@yahoo.com   www.okelaar.be

2012   3 buildings on 32 ha (near Brussels)   ♀18 ♂9 11 38

Dutch, (f, e)   none (yet)   only volunteers work on the community site
the resident group (consensus)
bio-organic building and living, focus on social life and care for each other and persons in need, spiritually open to all kinds of religious and non religious visions
community centre: eating together and with neighbours, hiring rooms and kitchen, laundry place for neighbours and café, small fair trade and bio-shop, guestrooms, co-working place
ecological, self-managed, left-wing political understanding, non-violent, spiritual, ecumenical, inner growth, organic diet, vegetarian diet, adult education, seminar-house, organic gardening, alternative energy, ecological construction, accessible to the disabled, making community with the locals

The Okelaar:
- cohousing for 23 families
- with a large community centre, open to the neighbourhood, to organisations close by and further away
- with three pillars: ecological - social - spiritual
- very mixed audience in terms of age, intergenerational
- varied socio-economic group
- community centre with: banquet hall, professional kitchen, multipurpose room, laundry bar, organic and fair trade shop, co-working place, 2 guest rooms, 4 youth rooms
- open to the neighbourhood: eating together with residents and with neighbours
- one mini-community house for three very old people
- one care home
- privately owned properties, homes for rent and welfare housing
- all work for the community is voluntary work
- legal form: CVBA-SO = cooperative company with limited liability and social purpose.
- as a cooperative, we pursue the 7 ICA principles
- collaboration with a CSA self-harvest farm
- we share cars, bicycles, the community garden, washing machines, work equipment, garden tools

CSA
engagement in our 3 pillars (ecological, social, spiritual)

Eikenberg Cohousing Eikenberg

Zandtiende 33 | 8310 Brügge ☎+32-(0)470 306 786
cohousingeikenberg@gmail.com cohousingeikenberg.weebly.com

2016 4100 m² (near Antwerp) ♀12 ♂15 9 27

Dutch, e, f, sp
all who are concerned (consensus) individual financing + cooperative
property of all and rented
living together in an ecological way, inclusion
ecological, self-managed, left-wing political understanding, non-violent, inner growth, vegetarian diet, physically handicapped persons, environmental protection, permaculture, organic gardening, alternative energy, new technologies, ecological construction, accessible to the disabled, passive construction, no gas, own electricity

☺

Eikenberg is a community in planning based on the philosophy and foundations of communal living together:

- Passive buildings: 18 residential units and a community house.
- Energy neutral: No fossil fuels.
- Promoting biodiversity: Compensation for cultivated land through green roofs and walls.
- Gentle mobility: As few (shared) cars as possible.
- Ecological life & sharing: Reduce your own ecological footprint and support each other in living ecologically.
- Inclusion: Cohousing Eikenberg aims to form a socially inclusive community. Everyone is involved in common activities as far as possible, according to their abilities.
- Self-government: We assume the self-responsibility of the nursing recipients and the commitment and concern of the residents. We create an environment in which self-government is promoted.
- Participation: In order to involve everyone in the process, the infrastructure will be adapted, co-determination procedures will be established, and a coordinator can be appointed to represent the care recipients and give them a voice.
- Living together: Variety in the group, people of all ages (families / singles with children, couples, individuals). The community is networked and openly embedded in the neighbourhood and is an active part of the local and wider society. Privacy is fundamental.

Cohousing Eikenberg is founded, inhabited and managed by people who support these goals and are committed to implementing our vision. Decision-making is democratic and consensus is sought.

Carsharing, Wooncoop, Samenhuizen

☺ subscribe to our vision

Fondation "Mais Oui!"

Place de Velaines | 7760 Velaines ☎+33-(0)646 006 446

celine.courdain@hotmail.fr

2018 10 ha (near Tournai/Lille) at the moment 25 %

♀6 ♂8 8 22

f, e, sp, Dutch

a special comitee (consensus)

all for free

both individual & common

property of a private social & cultural foundation

no

permaculture, ecology, well-being

ecological, self-managed, inner growth, meditation, yoga, pedagogical, seminar-house, peace-work, environmental protection, self-sufficiency, permaculture, forestry, alternative energy, ecological construction, humour, social tools, community building, therapy, arts

☺

Nine families gathered a few years ago to buy an old castle that was transformed into a monastery in order to make it alive again around shared goals & values. We have to restore it with both ideas of a multigenerational co-housing on one side & on the other with the creation of multiple projects that will allow anyone to reconnect his/her inner nature to the nature around us. We have welcomed two further new families recently and still look forward to find some more. We also welcome woofers & couchsurfers.

☺ questionnaire & time

L'Échappée

Drootbeek 30/05 | 1020 Brussels ☎+32-(0)473 128 810
contact@echappee.collectifs.net www.echappee.collectifs.net

2016 ♀15 ♂15 15 45

f, e, sp, i, Dutch

ecological, left-wing political understanding, feminist, non-violent, seminar-house, peace-work, alternative energy, humour, migrants protections, collaborative economy

Our project was created in order to build a little paradise in the heart of a city like Brussels. Mission complete :). We live peacefully with respect of the environment and of human beings. We often welcome migrants for long periods of time in order to ease their journey within Europe. In addition, we organise an academy in order to help new groups to avoid obstacles and traps. Feel free to contact us!

Photo: children's meeting

La Communauté de la Verte Voie

Verte Voie, 13 | 4890 Thimister ☎+32-(0)497 176 990
maisonvertevoie@protonmail.com dbao.be/communautedelave

1973 0.5 ha (near Verviers, Liège, Aachen) ca. 10

f, e, g

individual financing property of individual community members

managing a property, communal life, education

ecological, left-wing political understanding, non-violent, organic diet, adult education, pedagogical, seminar-house, peace-work, one-world development, mentally/psychologically handicapped persons, physically handicapped persons, environmental protection, organic gardening, social work

☺

The community Verte Voie ("green way") in Thimister, founded in 1973, is already 46 years old today (2019). From the very beginning, it was a concern of the residents to combine the idea of community life with a place of socio-cultural education. Today Verte Voie is the seat of various initiatives and associations for lifelong learning that are very active in the region, especially in the areas of nutrition and recycling, as well as for people with disabilities. The community Verte Voie is:

- A place where community life makes real savings possible and thus provides a response to economic and environmental crises
- A house that promotes encounters, exchange, mutual help and conviviality between the residents and offers an alternative to the surrounding individualism
- A home that offers the opportunity to become an associate in a community project

- An old building with character that offers a pleasant rural living environment and a very dynamic socio-cultural environment.

"De bouche à oreille" (part of "Culture et developpement")
autonomy on an emotional, material and financial level; interest in social issues, politics and culture

L'Arche Communauté d'Aywaille

Rue Mathieu Carpentier, 80 | 4920 Aywaille +32 (0)4384 3143
secretariat.aywaille@larche.be www.larche.be/index.php/aywaille

1975 (near Liège)

f, e, sp, (g) around 15
part of the group (consensus if possible, otherwise administratives)
individual financing and public funds
living and working with (mentally) handicapped people
spiritual, Christian, social work, adult education, pedagogical, mentally/psychologically handicapped persons

L'Arche Aywaille accommodates and cares for 40 adults with mental disabilities in a total of five households. Each household has its own pedagogical focus, which is based on the needs of the respective residents: Severe mental disabilities, autism or behavioural disorders, moderate mental disabilities, care for the elderly and physically disabled people.
During the day, most residents visit the day centre, which offers a wide variety of work and leisure activities. Some residents also work in local employment or are active in voluntary social projects: community crèches, old people's homes, solidarity shops, etc.

→See also description of network **"L'Arche (Jean Vanier)"**

Plukrijp wzv

Trommelstraat 24 | 2223 Schriek ☎+32-(0)15-233 560, +32-(0)487-107 270
frankruymen@gmail.com plukrijp.be

2007 1.5 ha (near Antwerpen) 90 % ♀5 ♂5 2 12

e, f, i, Flemish all of those present
all who are concerned (consensus via sharing circles)
we live from the free donations we receive from our network, paying the basic costs
property of one family
gardening, seedsaving, working with wood/metal/..., preparing meals together, daily course in permaculture
ecological, self-managed, left-wing political understanding, feminist, non-violent, spiritual, Taoist, esoteric/New Age, inner growth, meditation, yoga, tantra, organic diet, vegetarian diet, adult education, adult education, peace-work, one-world development, physically handicapped persons, environmental protection, self-sufficiency, permaculture, organic gardening, alternative energy, new technologies, ecological construction, humour, accessible to the disabled, non-smoking, social tools, community building, crafts, woodwork, therapy, art

☺

Plukrijp is a permaculture farm since 1982. We host six hundred visitors a year as well as people taking courses, interns, students and people from all walks of life. Feel free to join us if you're interested in an alternative lifestyle with a focus on permaculture gardening. Over the years we have become an internationally known place for young people wanting to practice & learn no-till permaculture as a life style.
Work = love in action; we expect all our volunteers to perform valuable help, please no mental or physical slacking. Anyone willing to volunteer 5 hours/day is welcome to join us.
Help consists of: carpentry, building, gardening obviously; scratching, seeding, watering, weeding, making jam & curry, cleaning, etc. a plethora of activities mostly outdoors.
We have an ongoing year-round permaculture theory & practice course. Come and join, we work by the principle of contribution = you give what it's worth to you.

LETS Begijnendijk, LETS Heist op den Berg
☺ willingness to learn new crafts, real interest in permaculturre and to grow through true sharing. No alcohol, no cannabis, no sexual harassment, no addictions, if possible (nicotine, coffee)

Bosnia and Herzegovina, Bulgaria, Croatia, Czech Republik see pages 411ff

Denmark

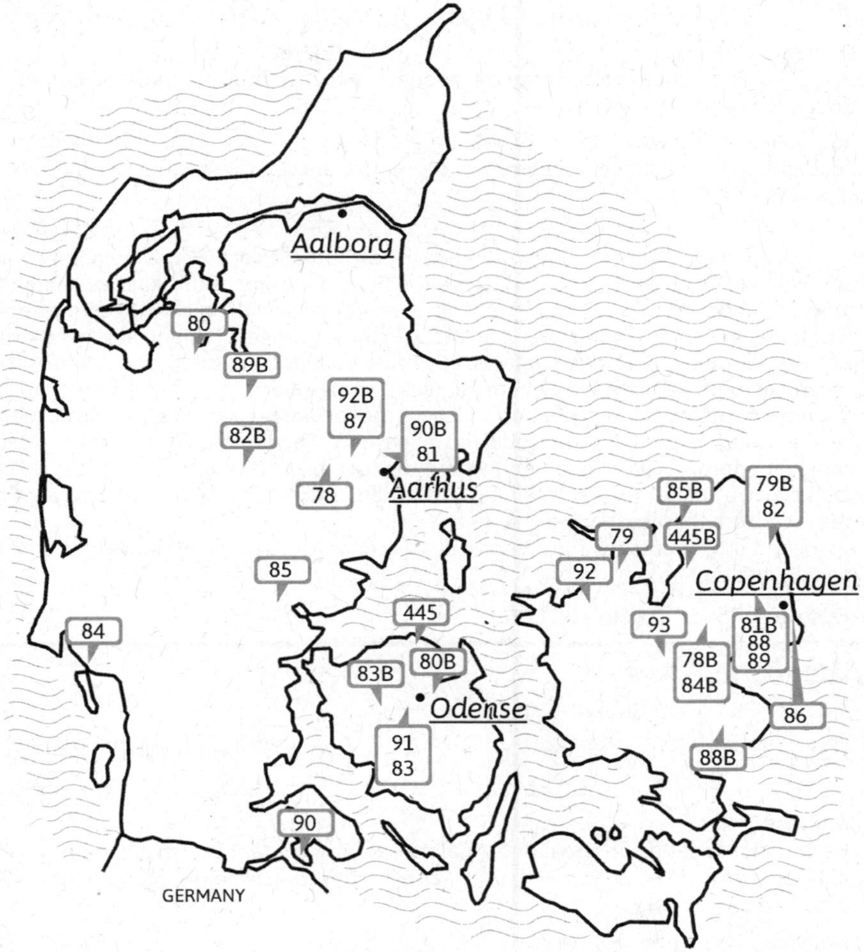

The last eurotopia edition included a chapter called "fun with statistics" in which we concluded that Denmark has the highest ratio of communities per inhabitant. The book presented 17 communities then. The following chapter lists even 28 existing communities in Denmark! There's five communities that were published in 2014 but didn't respond this time (page 493f), plus there's two planned ones on pages 445 - so Denmark is still leading the stats (even though we spared us the exact mathematics). Fantastik!
Find two networks of Danish communities in the networks chapter: On pages 520f there's Foreningen Bofællesskab.dk ("Bofællesskabet" meaning shared flat) and on page 532f there's LØS - Landforeningen Økologisk Samfund.

A-70 Toustrup Mark

Lykkegaardsvej 352-400 | 8472 Sporup
+45-8696-8361
hej@toustrupmark.dk
www.toustrupmark.dk

1971 | 5 ha (near Aarhus, Silkeborg) | 1 % | ♀19 | ♂16 | 28 | 63

- Danish, e
- all adults in the community (common meetings with proposals for major decisions, discussions and decisions)
- limited common financing - food, heating, electricity
- property of all
- ecological, self-managed, organic diet, alternative energy, music, good neighbourship, democracy

Today, Toustrup Mark is a community residence in East Jutland, with 26 private homes with bath and kitchen. As far as we know, we were the first cohousing of this kind in Denmark, the property was bought in 1971 by a group after a member saw the ad "Village for sale" in a local newspaper. Many experiments took place in the early years (among others - shared economy and clothes). This led to conflicts, but Toustrup Mark endured and nowadays it is a strong and colourful community, where all ages live together:
A community with a lot of common areas, which we use for dining room, kitchen, music room, laundry room, billiard room, car workshop and much more. The community of Toustrup Mark is spread over approx. 50,000 m² of land where we keep pigs, have built a football field, a volleyball court, a small "lake" with island, a forest and of course we have planted a lot of fruit trees and berry bushes all over the area.
We enjoy good neighbourship and the practical, economical and environmental benefits of living at Toustrup Mark.

→See also description of network "**LØS**"

Absalons Have

Tønsbergvej 70 | 4000 Roskilde
bestyrelsen@absalonshave.dk
www.absalonshave.dk

2005 | 1 ha | ♀28 | ♂28 | 35 | 90

- Danish, e, g
- the common meeting once a month (majority vote, important issues: 2/3 majority)
- all houses pay ca 180 EUR/month to the community, food is paid separately per meal/head
- property of all, but inhabitants own their own house
- share to make life easy for everybody
- external jobs
- self-managed, humour, easy living, make daily life easier

Absalons Have ("Absalon's garden") is a housing community located in Trekroner, just outside Roskilde. The community consists of 23 families and was established in June 2005. Approximately 90 people, adults and children live here.
The activities of the community consist of: Community dining (Mondays to Thursdays), shared responsibility for recreational areas, maintenance and social activities, such as Christmas

traditions, summer parties, jazz concerts and many more.
There are monthly meetings, where information is shared and major and minor decisions are made by democratic voting.

☺ own a house in the community

☆ Ananda Gaorii Ashram Farm and Learning Center

Holbaekvej 56 | 4560 Vig ☎+45-718 812 73, +45-315 030 80, +45-264 535 21
gaoriiconnections@gmail.com www.anandagaorii.dk

1988 13 ha (1 h from Copenhagen) 30

- we want to provide an immersive experience in spiritual, communal and sustainable living
- yoga, meditation, organic farm, organic bakery and seminar facility

☺ ( )

Ananda Gaorii Ashram is a community of monks and volunteers. We live, work, eat and meditate together. Our daily routine includes yoga and meditation and our activities include growing our own organic food, producing our own organic bread, selling our products at local markets and our own shops, hosting retreats and workshops and doing community activities such as food collection and distribution. We are also very much into chanting and making music. We started in 1988 and we are registered as an organic farm, organic bakery and seminar facility. We have thirteen hectares of land and we are located an hour west of Copenhagen in the Danish countryside. The purpose of our community is to provide an immersive experience in spiritual, communal and sustainable living for young volunteers who live with us for 10 months or longer. We also offer retreats and workshops on topics related to yoga, meditation and holistic lifestyle for short term guests and volunteers. Three monks are living here permanently and we normally have twenty to thirty volunteers and six bakery staff.
Opportunities exist for ten month volunteering through the European Solidarity Corps but we are also open to long or short term volunteers of any age or nationality providing they have an interest in yoga and meditation. Paid positions as bakery workers are also available.

☆ Andedammen

Andedammen | 3460 Birkerød
www.andedammen.net

1984 ♀16 ♂10 14 40

Andedammen ("The Duck Pond") is a living community established in 1984 in the town of Birkerød, Denmark, and still going strong. We are organised as a housing cooperative comprised of 17 units, with approximately 40 inhabitants of all ages.
Each household has a private housing unit, but we share a significant portion of our daily lives and have extensive common facilities to support this. We typically have dinner together multiple times a week, taking turns cooking for the wider group in our common kitchen. All aspects of our common life are governed democratically, though with a high degree of delegation to subgroups with different responsibilities, such as maintenance, social activities, recycling and gardening. A governing board, whose memberships rotate periodically across all inhabitants, ensures coordination and deals with external issues.

Andelsboligforeningen Asgård

Asgårdvej 19 | 7800 Skive
mariannewarburg@gmail.com www.ab-asgaard.dk

1988 1 ha ♀19 ♂12 13 44

e, Danish
annual general meeting (majority vote) individual financing
each householder owns part of the value of the whole community according to the size of their living quarters
common grounds, maintenance and cleaning of buildings, garbage sorting, cooking and eating together
accessible to the disabled, non-smoking

☺

The community was established in the 80es like a great number of similar co-housing projects in Denmark. The main purpose was to create a community where we shared the many daily chores of the families: daily meals, taking care of the houses and gardens, looking after the kids etc., as well as leisure activities.

Bjergager Solby

Aasum Bygade 18 | 5240 Odense NØ ☎+45-702 23380
post@bjergager.dk www.bjergager.dk

2005 4 ha ♀2 ♂2 3 7

e, g, f, Danish 3 3
those involved and financially responsible (consensus or else majority vote)
property of one member
spiritual, political, artistic
arts, festival, teaching, politics, leadership and media
ecological, self-managed, non-violent, spiritual, Christian, esoteric/New Age, anthroposophic, inner growth, meditation, yoga, sexuality, vegetarian diet, adult education, pedagogical, free school, seminar-house, one-world development, new technologies, non-smoking, arts, media, film-making, TV station, esoteric, therapy

☺ ()

Uniting love and power through wisdom and creating a framework for inner growth and initiation into the esoteric mysteries is the key endeavour at Bjergager Solby.

Home to the Excalibur project (www.ringoflife.org) and the largest stone circle built in Scandinavia for over a thousand years, Bjergager has a deep esoteric vein which is reflected in the seven places of power on our land of which the "Sword of Truth" and the "Grail Garden" are the most important ones.

→See also description of networks **"LØS, GEN Europe"**

esoteric networks, Transition Towns

Blåhøjen

Skelagervej 415a | 8200 Århus N ☎+45-315 025 77
gunner.byskov@stofanet.dk www.blaahoejen.dk

1983 | 2500 m² | ♀18 ♂19 17 54

e, g, f
the community (on meetings on the first Sunday every month)
common financing | shared property
no
to maintain the community!
self-managed, humour, accessible to the disabled, non-smoking

☺

The community Blåhøjen is based on managing everything ourselves, and the main point is that we eat together every day. We decide who is going to live here, and we decide how everything is to be made. It has been working perfectly for 34 years, and we expect it to be working perfectly for the next 34 years as well!

☺ acceptance of our values

Bofællesskabet Æblevangen

Æblevangen 41 | 2765 Smørum
johan.jacobsen@aeblevangen.dk www.aeblevangen.dk

1979 | 2 ha (near Måløv, Copenhagen) | 0 % | ♀37 ♂38 24 99

Danish, e | none | none
the monthly community meeting (we talk about it and sometimes we vote)
limited common financing - we pay a certain amount | owners
no
living and eating together

☺ ()

36 houses, half of the houses are inhabited by families (2-3 children), the other half is inhabited by adults, whose children have grown up and left home. A few members live outside the fence but nearby (divorces) - and then the common house with 600 m².
There are groups who eat together every weeknight - everybody finds their own ways - some are eating together twice a week, some less and some more. We have traditions - special days - where we work together on our territory or at the common house. We have special parties a few times a year. There are no written rules, and then on the other hand of course there are things we do in special ways. E.g. in our common laundry, we wash the clothes that are next in line or, when invited to a private party, we help out as a part of the deal. We have a lot of small groups that do yoga and sports and we have groups who deal with common issues, e.g. which trees to plant, which benches to paint next time – or how to renovate our common paths.

☺ buy a house which is for sale

Bofællesskabet Bakken The Co-House Community Bakken

Bakken 23 | 3050 Humlebæk ☎+45-267 445 13
info@bf-bakken.dk www.bf-bakken.dk

1980 <0.1 % ♀27 ♂29 14 71

Danish 0 few
the community (majority) individual financing
ownership
architects, journalists, teachers and much more
self-managed

☺

The community is for families who prioritize a binding community. Each family has their own house. We share 5,5 ha of land, outbuildings on the old farm area, sheep, horses, a small lake and a lovely large common house with common music room, play room for children and other activities.

☺ you have to own one of the 27 houses

Bofællesskabet Bøllingsø

Peter Svinths Vej 100 | 7442 Engesvang ☎+45-615 086 01
info@boellingsoe.org www.boellingsoe.org

1989 5000 m² 2 % ♀9 ♂6 3 18

e, Danish
everybody in a meeting (consensus) individual financing
common property (andelsboligforening)
education, therapy, marketing, childcare, retired
self-managed, organic gardening, alternative energy, humour

Bøllingsø is a project based on self-owned and self-managed cohabitation. The houses are built and based on the idea of friendly neighbourhood. We have a common house with room for about 40-50 people. We eat together every evening from Monday to Thursday. Our common social life is centered around the 4 weekly meals, the different tasks for keeping the hens, gardening and outdoor maintenance of the buildings.
In addition we have some rituals around Christmas, the New Year's celebration, our annual weekend tour and an occasional one week trip by bus to another European country.

Bofællesskabet Bondebjerget

Sædekildegårdvej Bellinge | 5250 Odense SV ☎+45-289 199 01
rudy@bondebjerget.dk www.bondebjerget.dk

1982 58-112 m² per dwelling ♀70 ♂70 60 200

g, e

a grand meeting twice a year (majority vote) individual financing

rented alternative energy ☺

The community "Bondebjerget" was built in 1982 when a group of people was seeking for an alternative life form. Bondebjerget consists of 4 groups, each containing 20 households and a common house. There are different types of housings ranging from 58 m2 to 112 m2 - of which some are suitable for elderly and disabled.
Bondebjerget is a non-profit department under the housing association CIVICA, which means that the department is subject to the law on non-residential housing.
A living community is a different way of living where we know each other better than you might be used to. In addition to the many and great benefits to live this way, we have many practical things to do and we expect you as a resident to participate actively.
If you want to stay at Bondebjerget, you must attend an information meeting before you are on the waiting list. In addition, you must participate in a joint dining in one of the groups where you see and hear about how a common day of life works in a group.

→See also description of network **"Foreningen Bofællesskab.dk"**

Bofællesskabet Drejerbanken

Drejerbanken 21 | 5492 Vissenbjerg ☎+45-275 783 96
lvalentiner@hotmail.com www.drejerbanken.dk

1978 (near Odense) 10 % ♀19 ♂11 17 47

Danish,g,e

all persons above 18 years (consensus on common meeting) ☺

limited common financing property of all

teacher, psychologist, engineer, doctor, dentist, nurse

ecological, non-violent, animal protection, environmental protection, organic gardening, alternative energy, humour

Our buildings are built as a community with 20 homes and a large common house with shared laundry and workshop etc. We eat together almost every day.
We live in a beautiful area in the countryside. We have a large kitchen garden, where we grow part of the vegetables (of course organic) that we eat in the common house.
Half of our homes are rental houses, and the other half is privately owned.
The Drejerbanken was established on the basis of:

- stable and social relations between residents with mutual responsibility and equality
- creating safe and binding childhood conditions for the children, who are able to get close relationships with adults other than their parents
- creating a common housing form that the majority of the population can afford to live in (owner-and-tenant-based residence)

☺ only a wish to live in a community

Bofællesskabet Fladengrund

554 | 6710 Esbjerg
bofaellesskabetfladengrund@gmail.com www.fladengrund.dk

1980 ♀12 ♂12 3 27

Danish, e
monthly meetings (consensus/majority votes)
monthly payment by each house to a common budget
houses individually owned

Fladengrund is a "bofællesskab", a cohousing scheme in Esbjerg, on the west coast of Jutland. There are twelve houses, each with its own garden, grouped around a 'common house', used for communal eating but also containing a launderette, a fitness room, a workshop with tools, and four guest rooms. Behind the common house is a field with benches and tables, swings and a trampoline for the kids, and some fruit trees.
We typically eat together twice a week, which means you get to cook for about 20 people every four weeks. Other opportunities for common meals and festivities are regularly taken, such as Christmas, New Year, mid-summer, and important birthdays. Participation is entirely voluntary. There are also ad hoc activities – gardening, day trips, making beer, and so on. Shared facilities mean that individuals don't need to buy items like washing machines, tumble driers, lawnmowers, or large workshop tools.
It isn't a 'commune' – each house is privately owned, and you still have your own private life behind your own front door. But you know that all the other front doors will be open to you if you need someone. For us, the system strikes the right balance between individual privacy and sociability with your neighbours.

→See also description of network "**Foreningen Bofællesskab.dk**"

Bofællesskabet Ibsgården

Ibsgaarden 62 | 4000 Roskilde +45-294 676 52
ib@ibsgaarden.dk www.ibsgaarden.dk

1983 0.4 ha 2 % ♀18 ♂14 14 46

e, Danish none none
general assemblies, an elected board (we strive for consensus, but sometimes vote)
individual financing property of all
self-managed, left-wing political understanding

☺

Our community has no firmly defined ideological direction. Its foundation is more a mixture of experiences from traditional family life and a more or less experimental communal lifestyle. It consists of a community of normal households combined with several communal features. This means that each family has its private life, but at the same time it shares the dining room, the library, the laundry room and the workshops (such as a photo lab) with the others. Each individual has certain duties, such as cleaning the common rooms, cooking three times a month etc. But apart from the few duties we enjoy the more pleasant sides of life: we play music together, put on plays with the children, celebrate New Year together and cook haute cuisine - usually - only for the adults.

☺ new members must buy a share in the community

Bofællesskabet Lille Grundet Lille Grundet Cohousing

Grundet Bakke 18 | 7100 Vejle ☎+45-219 797 52
interesse@lille-grundet.dk www.lille-grundet.dk

1988 2 ha 5 % ♀22 ♂16 12 50

Danish, e, (g), (f) 0 1 (+10 retirees)
all - joint meetings (consensus or vote)
limited common financing/individual financing share funded
none
cohousing, (vegetables, animals, gardening)
organic gardening, diversity, socializing

☺

We are a co-housing project where we share a lot of common activities: joint dinner (on weekdays), organic garden, activities for children, workshops and social events. The architecture is unique by using a lot of colors on the facades - a happy place!

→See also description of network **"Foreningen Bofællesskab.dk"**

☺ co-living and able to finance a share

Dyssekilde Ecovillage

Regnbuen 11, Torup | 3390 Hundested ☎+45-279 393 53
matawanb@gmail.com, eksternerelationerdyssekilde@gmail.com
www.dyssekilde.dk

1989 13 ha (near Hundested) ♀85 ♂51 60 196

Danish, e, i, f 20
the Common Meeting (4 times per year + eventual extra meetings) (majority, after discussion in groups and with all present ones)
individual financing
common grounds owned by our association, single houses are private
ecology and social engagement
local sustainability and participation to networks of ecovillages
ecological, self-managed, environmental protection, organic gardening, alternative energy, ecological construction, accessible to the disabled, yoga, music festival, health, inclusiveness, tolerance

☺ (✋✋)

Dyssekilde Ecovillage is a vibrant community where we live (and some work) focusing on social interaction, tolerance, respect, sustainability, and ecology. It is made of 7 housing groups, each with their own budget, that are responsible for the area between and around their houses.
For us, sustainability is a major focus in practically everything we do and it shows itself in many different ways and on many different levels. There are no direct rules and no sanctions, but it is expected that everybody does their best in whatever way possible.

In addition to the personal initiative, we have circa 30 working groups of volunteers (taking care of the various common tasks) and a number of communal initiatives that help to ensure a high level of sustainability such as: a windmill, joint laundry, shared geothermal heating system, joint vegetable gardens, shared waste management, waste-water cleaning, car-sharing, low-emission common house etc.
Since the start of the ecovillage a kindergarten, a school, an organic shop, a meeting center and other socially relevant working activities have started in the local mainstream village (Torup, about 350 inhab.) of which we are part of. We have become an example of sustainable growth.

→See also description of networks "**GEN Europe, LØS**"
☺ financial accountability, willingness to volunteer work, attitude to ecology and sociality

Freetown Christiania Information office: Nyt Forum

Bådsmandsstræde 43 | 1407 København K ☎+45-329 565 07
✉nytforum@christiania.org 💻www.christiania.org

1971 | 36 ha | 800-900

- Danish, e, Swahili, morse code
- common meetings and area meetings (consensus)
- individual financing and communal financing
- first as a squat, then legalized since 1989 and now owned by the community
- ideological and ecological sustainability (anarchistic)
- ecological sustainability and autonomy, culture and art
- (editor's note: all possible keywords on our list have been checked for Christania...)

☺

Since its birth, Christiania has been associated with rebelling against the prevailing system, and for just as long, the Freetown has experimented with new terms of democracy and ways of self-government. Based on an idea of as much freedom and as much power of the individual as possible, Christiania has organised itself into various councils: The Common Meeting, The Economy Meeting, The Area Meeting, The "Business" Council, The Co-operative Workers Meeting, The House Meeting - and all that in good, old fashioned neighbourly spirit.
Decisions are not made by majority vote but are made in consensus. On very few occasions, not binding tendency voting is made, to get a clear idea of how much the opinions differ. Everybody has the obligation to live by the decision taken at the meeting and all Christianites have access and the right to speak at all meetings. It can be a long tough process, but

everybody, also the thousands of ex-Christianites around the world, have learned a lot about self-government in this way. In the way that Christiania organises itself lies the real socio-political experiment, which we wouldn't like to live without. People can manage themselves by a direct peoples democracy based on common sense rather than laws.
Christiania has no laws, but we have put up a series of bans which are in force in the Freetown: No hard drugs, no weapons, no violence, no trading with buildings or residential areas.

→See also description of networks "**LØS, GEN Europe**"
☺ participation

Hertha Levefællesskab

Landsbyvænget 14 A, Herskind | 8464 Galten ☎+45-384 034 01
hertha@hertha.dk www.hertha.dk

1996 20 ha (near Århus, Herskind) 30 % ♀60 ♂60 32 152

Danish, e about 40

home owners association, Village Foundation and institution jointly

individual financing village foundation & private ownership

acceptance of the presence of developmentally retarded people; acceptance of decision making on an anthroposophic basis

reverse integration with young developmentally retarded people

anthroposophic, organic diet, pedagogical, mentally/psychologically handicapped persons, organic agriculture, organic gardening, organic farming

☺

Hertha Levefællesskab ("Hertha Living Community") was a social policy initiative with the goal to integrate a group of individuals with special needs within a normal community. The community is based on two ideals: The first is to realize inclusion between the groups. The second ideal is the ideological inspiration from Rudolf Steiner's anthroposophy. Today 137 people are living in the community, of whom 26 have special needs. The estimated value of the buildings is 13.5 Mio. Euros. Half of the value is common ownership or rather "nobody's-ownership", as they are owned by the Village Foundation. The rest are privately owned homes.
Our belief that such a community life situation would bring about dynamics and synergies has proven to be true. Therefore, the project as a whole has great societal relevance and value as:

- a unique example of cooperation between three actors in society: the official (municipality), the private society (Hertha Shared Housing & Workshops, the Village Foundation and other donating foundations and donators) and the civil society (the other residents of Hertha and Herskind)
- an example of a new type of non-institutionalized institution with homely touches and integrated sheltered workshops that provide meaningful work for the benefit of the entire local population as the project is an alternative to the general trend towards more centralized and large institutions
- a project bringing back life to the countryside, life that is about giving care and creating meaning for a marginalized group
- a place with a social and environmental profile that inspires and creates new knowledge.

→See also description of networks "**LØS, GEN Europe**"
☺ none, but we have a long waiting list

Holdepunktet

Fanevaenget | 3520 Farum

peter@rentzen.dk www.holdepunktet.dk

2014 ♀32 ♂26 30 88

Danish, e

common meeting (consensus) individual financing

ownership

academics

self-managed, non-violent, alternative energy, new technologies, ecological construction, humour, non-smoking

We are founded by a group of people with experience from a previous co-housing project. The vision behind Holdepunktet is a community for all generations. Young families with children and seniors, half and half. We want to live in a community where we can help, support and inspire each other. We want to see the children grow up, playing across ages, boys and girls, and to see them develop good social competencies.

→See also description of network **"Foreningen Bofællesskab.dk"**

Karise Permatopia

Køgevej 15 | 4653 Karise

info@permatopia.dk www.permatopia.dk

2018 29 ha (near Kopenhagen) 50 % ♀70 ♂70 60 200

e, Danish 1

in working groups (sociocracy)

individual financing, limited common financing

property of all; some individual member property

permacultural, trying to create a self-sufficient system

farming, renewable energy, waste water treatment, permaculture

ecological, environmental protection, self-sufficiency, permaculture, organic agriculture, organic gardening, organic farming, new technologies, humour

Karise Permatopia is a new, modern ecological community not far from Copenhagen (65 km). It is situated in a lovely natural setting of 29 hectares, with woods, fields, meadows and small ponds, within walking distance to Karise Station. Karise Permatopia comprises 90 ecological and sustainable modular row houses; 44 are rentals under the provisions of the Danish social housing law; 23 are cooperatives and 23 condominiums. The original farm buildings in Karise Permatopia are being renovated to house the many facilities that will be used in common.

We want to establish a self-sufficient, meaningful and modern community that ensures both lower living costs and environmental sustainability based on the principles of permaculture. This ecological community is being realized by developing mutually enriching interaction between habitats and agriculture, through the involvement of all residents in the work of cultivating their own food and administrating the daily life of the community, by supplying heating and recycled water through the use of green energy and recycling waste products into resources.

It is our hope that Karise Permatopia will therefore provide a framework for a strong and meaningful community, since it rests on a foundation of shared work and social interaction. A living community demands sharing among all its members – children and adolescents, young families, mature adults and old folks.

☺ financial, willingness and ability to work in the shared systems

Mageløse

Tårnet 6 | 3500 Værløse ☎+45-267 822 26
info@magelose.dk magelose.dk

2019 (near Jonstrup) ♀22 ♂22 24 54

e, Danish
individual financing

☺

Mageløse is a newly started cohousing community consisting of 29 houses. The location at a former Royal Danish Air Force air base, 20 km from Copenhagen, provides a scenic area. The first residents moved in during February 2019. The community has a common house, where the residents have dinner together during the week. The common house is the physical center and the natural meeting place for the residents. Mageløse also owns the old control tower, which the community will open to the public. There are many ideas for the everyday use of the control tower.
In order to create a pleasant space, the community is car free. There are only paths for walking and biking in the housing area. Cars are parked in the surroundings of the housing area.
All the residents take part in the comprehensive work to establish the new cohousing by being part of our many working groups and participating in our every second week common meetings. Mageløse has a 50/50 split between families with children and seniors. A diversified composition of residents has been a priority from the start.
Visitors are welcome. Google Map will guide you if you search "Mageløse".

Nonbo Hede Poul Birk

Nonbo Hede 28 | 8800 Hald Ege, Viborg ☎+45-202 810 87
poul@pbirk.dk www.pbirk.dk/Birks_hjemmeside/Nonbo_Hede.html

1976 3 32

joint meeting (consensus) property

☺

Nonbo Heath is a commune that was based on the idea that the design of the development should give rise to a social community. There was no programme that clearly stated the community's goals, but the main ideas of living community were:

- that neighbours work together and help each other and thereby achieve a higher quality of life
- a community center with weekly communal meals
- to keep cars outside the village and to have communal parking at the common house
- to have no fences on the property
- ecological approaches

Økosamfundet Soleng

Skeldevej 23 | 6310 Broager ☎+45-298 049 19
info@okosamfundet-soleng.dk okosamfundet-soleng.dk

2015 11 ha (near Sønderborg) 10 % ♀6 ♂7 10 23

g, e, f, sp, Danish 3 none
the board (sociocratic, by consent) individual financing
cooperative
we are based on fellowship and permaculture
gardening, building and hosting a café
ecological, spiritual, inner growth, vegetarian diet, environmental protection, permaculture, construction, sociocracy

☺ ()

At the moment we are 6 families living on the land and in the buildings of two farms. Right now we are waiting for the permission to build houses on the ground. The goal is to grow up to 18 families in 5 years. On the farms live also two pigs, twelve hens and seven cats.
Our intention is to build this community according to the permaculture principles: care for nature, care for people and fair share. We want to be an ecological example for the surrounding area, building ecofriendly houses either by using recycled or sustainable materials for the houses.
The community spirit and self-sufficiency on vegetables and eggs are important to us. Other shared values are resilience, working with nature, respect and inclusion, esthetics and self-development. We have not yet reached our goal, but step by step we are getting there.

→See also description of network **"LØS"**
☺ interested in community and permaculture, being open minded

Overdrevet

Overdrevet 14 | 8382 Hinnerup
www.overdrev.dk

1980 4000 m² (near Aarhus) 25 % ♀30 ♂30 15 75

e, g, f none none
all (discussions) private ownership
no
dwelling
ecological, self-managed, environmental protection, organic agriculture, organic gardening, organic farming, alternative energy, non-smoking

Overdreven is an organic community in Hinnerup, East Jutland, founded in 1980 as Denmarks first low energy community. It comprises 2.5 hectares of 25 homes, which houses around 70 people.
The 25 private homes are divided into two farms and one street with a common house in the middle. The housing community also belongs to a closed farm property. Part of the area is used for organic vegetable garden, which makes the community partially self-sufficient.

☺ buy a house

Solens Hjerte ("The Heart of the Sun")

Planteskolevej 51 | 5250 Odense SV ☎+45-659 634 65
post@solenshjerte.dk www.solenshjerte.dk

1992 | 6 ha (near Odense) | ♀22 ♂19 2 43

- g, e, f, sp, Danish, Swedish, i
- all involved (we talk with each other and find the best solution)
- individual financing
- focus on personal growth, meditation, friendship and love
- cooking, crafts, building, garden work and daily tasks
- non-violent, spiritual, inner growth, meditation, vegetarian diet, seminar-house, peace-work, conventional agriculture, organic gardening, conventional gardening, construction, humour, apiculture, therapy, creativity, music and singing

☺ ()

We live in four large buildings surrounded by very beautiful parks and gardens with a swimming pool. We own 14 acres of land with two horses and a yard of beehives. We have a music studio where we play music every day (drumming, choir, singing and playing, creative workshops), a large meditation hall and various workshops (garage, carpentry).
Our home is an oasis of peace, acceptance and loving life energy. Here hearts can open. Everybody is working on finding his or her own way to clarity and responsibility. We are well grounded on the Earth, active with our hands in daily life matters and anchored in unconditional love.
Of course living together and working in teams brings up emotional stuff. We consider this as a good opportunity to become more aware of our psychological patterns. We can then dissolve them in love and acceptance and create a life in meditation and friendship.
We use various therapeutic and spiritual techniques to let go of old limitations and be able to live more consciously in the present.
We have many guests. Some come here to live and work on topics in their inner life, joining

the work in the community for a limited period of time. Some come to relax and take a holiday in these marvellous surroundings and others come to recover and regenerate after times of illness or crisis.

☺ openness and the desire for inner growth, sharing of our vision of love and of peace on earth

Tranehøj

Skippingevej 25 | 4460 Snertinge ☎+45-209 300 36 ,+45-212 661 33
beboer@tranehoej.dk www.tranehoej.dk

1984 3.5 ha (near Kalundborg) 50 % ♀4 ♂3 2 9

- g, e, f, sp
- all (consensus) — all expenses are shared but not shared income
- property of the association
- no
- jobs outside, animals, vegetable garden, orchard, cultural activities outside
- ecological, self-managed, left-wing political understanding, feminist, one-world development, environmental protection, organic gardening, animal friendly husbandry, alternative energy, humour, non-smoking

☺

We try to base our everyday life on the principles of sustainability, ecology and fair trade. We put a lot of effort into gardening, animals, maintenance and projects which make our place more sustainable (solar panels, windmill etc.).
At the same time we strive to make Tranehøj a resource base for activities related to the world around us - locally as well as globally. We have a fruitful relationship to the local folk high school, and once a year some of us organize cultural visits of young artists from countries in Africa, Asia or Latin America to schools - locally and in other parts of Denmark.

→See also description of network "**LØS**"
☺ social skills, practical skills, political and cultural outlook, enthusiasm

Udgården - Lading Andelsboligforening

Udgården 24, Lading | 8471 Sabro ☎+45-869 496 18
alle@udgaarden.net www.udgaarden.net

1992 17 ha (near Âarhus) vegetables 25 %, beef 100 %
♀16 ♂12 8 38

- Danish, e, g, f, sp
- everyone in common (majority vote/consensus)
- common household budget and individual financing — community
- ecological, nothing more
- usually work and cows, sheep, chickens, vegetable garden
- ecological, self-managed, non-violent, organic diet, environmental protection, organic agriculture, organic gardening, organic farming, animal friendly husbandry, alternative energy, new technologies, ecological construction

☺ ✋✋

The Udgården is a small ecological community consisting of a farm and 16 apartments just outside of a small village in Central Jutland. We are approx. 30 adults and 30 children, 20 cows, 10 sheep, 20 chickens and some dogs.
Each family has its own apartment, but we dine together on weekday evenings. These meals are mainly ecological and consist of products from our own farm (beef, mutton, vegetables etc). The farm is run by all of us together and all work is done voluntarily in our leisure time.
All adults have full-time jobs outside of Udgården.
The apartments are of a low-energy construction: double insulation, super-low-energy windows, and floor heating. Heating is based on a combination of solar panels and a wood-pellet stoker. Rainwater is collected from all roofs and used in our toilets and common laundry.
We have no common religious or ideological reference, but what we have in common is the desire to lead a less polluting and polluted lifestyle. *See photo on page 8*

→See also description of network "**LØS**"
☺ financial and ideological

Valsølillegård

Knud Lavard Vej 94 | 4174 Jystrup/Valsølille ☎+45-61 27 06 58
✉dorte@valsoelillegaard.dk

1983 | 3 ha (near Ringsted) | <40 % | ♀9 ♂9 4 | 23

g, e, f, sp, Danish, Swedish, Norwegian | 2
all shareholders (consensus, sometimes (rarely) majority)
individual financing, maintenance budget
owned in 15 shares, one house for 2 families rented out externally
self-managed, left-wing political understanding, non-violent, adult education, peace-work, environmental protection, organic gardening, construction, some crafts, international work, voluntary work, amateur theater, product development, grandchildren, travelling all over the world

In 1983 some hippies were travelling through the countryside to find a cheap big old house to live in as a community. The dream was to build the place up, live, work and raise children together. That's what we did. 30 years later more than 25 children had a wonderful childhood at "Valsølillegård" and are carrying on the spirit to do something to change the world for the better. 25 years of handicraft work made the place extraordinarily beautiful and well established.
The average age of the inhabitants is now about 65, but we and our dreams are not dead yet. With lots of voluntary external work and long distance travelling, we still spread the idea of living tight together and sharing a dayliday.
One (limited) chance to visit would be to contact Heiner Gringmuth and his wife Dorte Gringmuth Aagaard via "CouchSurfing", "WarmShower/Dachgeber" or "Servas".

Estonia see page 416ff

Finland

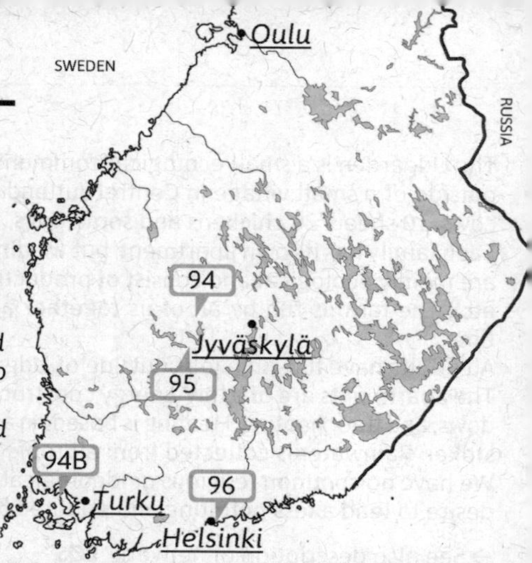

From Finland, we can only present four communities. Four more are listed with the Camphill communities on page 511f. Our database claims to know 32 communities in Finland, but they just did not respond. Would have been better to meet in the sauna then to write emails, for sure.
There is a finnish community network named GEN Finnland/Suomen kestävän elämäntavan yhteisötry, see description on pages 526f.
Onnea (good luck)!

Keuruun Ekokylä Keuruu Ecovillage

Kivijärventie 300 | 42700 Keuruu
info@keuruunekokyla.fi www.keuruunekokyla.fi

1998 | 50 ha (near Keuruu) | 60 % | ♀12 ♂13 | 11 | 36

e, Swedish, Finnish
all adult members (voluntary work)
all members (consensus and majority vote)
individual financing
owned by Keuruu Ecovillage Association
ecological, but we are politically and religiously unattached
work related to self-sufficiency and maintenance of the community
ecological, self-managed, non-violent, inner growth, vegetarian diet, environmental protection, self-sufficiency, permaculture, organic agriculture, organic gardening, animal friendly husbandry, alternative energy, ecological construction, arts and music, environmental art, sociocracy

☺ ( )

Keuruu Ecovillage is situated near to the small town of Keuruu in the middle of the beautiful Finnish countryside. The aim is to promote an ecologically and socially sustainable lifestyle in all areas of life. The ecovillage is mainly sustained by voluntary work of its residents. If you want to visit us, please send e-mail first. There are times when the ecovillage is closed for visitors. *See picture on page 11*

→See also description of networks "**GEN Europe, GEN Finland**"
☺ interest towards ecological lifestyle in a community, will to participate in the daily life of the community.

Livonsaaren yhteisökylä Livonsaari community

Pohjanpääntie 67 | 21180 Livonsaari
www.yhteisokyla.net

2005 | 60 ha (near Turku) | 15 | 55

Finnish, e
commonly owned

Livonsaari Community was founded in 2005. It has become a village within a village, with sustainability and communal living as its foundation.
There are about 40 adults and 15 children permanently living in the community, plus animals - horses, donkeys, chickens, cats and dogs.
Our buildings are as diverse as their inhabitants. For many, the first dwelling has been a sauna, a sacred place for Finns. There are some traditional Finnish log houses and more adventurous techniques, combining clay, straw and reed. Some alternative energy solutions are in use, such as a small windmill and some solar panels.
In the village there are several small businesses, which tie together the old days and the present. There is an organic farm, log house builder, saddle maker, traditional Finnish healer, book publisher and a coopwerative garden producing organic vegetables.

Väinölä Väinölä yhteisö (Väinölä Community)

Väinöläntie 15 | 35700 Vilppula ☎+358-(0)34-713 390
posti@ihmisyydentunnustajat.fi www.ihmisyydentunnustajat.fi

1978 3 ha (near Mänttä, Tampere, Jyväskylä) 25 % ♀20 ♂9 0
29

Finnish, e, f, Swedish, (g) 18 10
all who are concerned (together) shared economy
property of Ihmisyyden Tunnustajat
theosophical
arts & culture, gardening, construction work, cleaning
self-managed, non-violent, spiritual, inner growth, peace-work, organic gardening, organic farming, humour, educational and cultural work, music

We are an independent group, not belonging to any registered society. Within the Finnish theosophical movement, our group was founded by Martta Horjander in 1978. Ethics and human values are the basis. The main objective is to promote and try to put into practice the ideals of the Finnish author, Pekka Ervast. These are peace, brotherhood of all men and a non-violent way of life.
Theosophy emphasizes in Christianity the meaning of the Sermon on the Mount, but also points out the unprejudiced study of all religions. Life is a cycle of development. The immortal spirit of every human being strives, through a series of reincarnations, for the human perfection or for the goal of unselfish love. Karma, the law of cause and effect, ensures that every effort towards the goodness is worthwhile. We are responsible for our deeds, with which we also create our future.
The alternative lifestyle movement can have a greater positive effect on society than voting or doing politics. Individual choices and decisions on the basis of self-improvement can form an important counterbalance to materialistic and militaristic directions.
Brotherhood, warm-heartedness and caring for each other in small occasions of everyday life is something worth practising. It is a matter of making an effort and learning in life.

→See also description of network **"GEN Finland"**
Ihmisyyden Tunnustajat

Vanhankaupungin kulttuuriekologinen klubi

(Vanhakaupunkis cultural-ecological club)

Viides Linja 7 C 59 | 00560 Helsinki
+358-(0)407 098 597
eero.haapanen@gmail.com
www.kuusiluoto.info

1990 | ca. 1.5 ha | 50 % | ♀39 ♂40 | viele | 79

- Finnish, Swedish, e
- annually elected board (4-8 members) (majority vote, consensus)
- membership fees, grants
- rented from the City of Helsinki
- support equal, ecologically sustainable lifestyle and to maintain cultural heritage
- nature conservancy, fishing, gardening, sheep keeping
- ecological, self-managed, feminist, non-violent, inner growth, seminar-house, animal protection, environmental protection, self-sufficiency, permaculture, organic agriculture, organic gardening, forestry, alternative energy, new technologies, ecological construction, construction, humour, woodwork, art, music festivals, Midsummer fest, sauna culture

☺

Vanhankaupungin kulttuuri-ekologinen klubi (Vanhakaupunki's cultural-ecological club), VKEK, was established in 1990. A group of students found an old wooden villa that was left in deteriorating state. The group started repairing the house and surrounding property and was later able to rent it from the city of Helsinki. The property entails half of the island of Kuusiluoto ("spruce islet"), located in the geographical centre of Helsinki. The club has free grazing sheep restoring the meadows and maintaining traditional landscape and nature. There is a duckboards connection to the mainland, therefore Kuusiluoto is a popular recreational target for the public. It is located next to a prominent wetland nature reserve.
The club is politically and religiously nonaligned. Members endeavour to support an equal, ecologically sustainable lifestyle and to maintain cultural heritage. VKEK has about 40 active and 40 supporting members, with an annually elected board.
The members gather monthly to do volunteering work in the property. Club activities include e.g. repairing and upkeep of the villa, gardening, sheep keeping, making nest boxes for birds, nature conservancy, sauna culture, boating, fishing in the summertime, ice fishing and ice boat sailing during winter. Most of the activities are done by traditional methods as much as possible. There are also many theme events, for example Midsummer festival (Juhannus) and music festivals. The villa is mostly used as a leisure time home for the members.
Although the members of the club are people of different ages and different social classes, all of them take an interest in nature and rich cultural environments.

☺ new candidates must take part in the volunteering work a few times, membership fees

photo: Janne Käpylehto

France

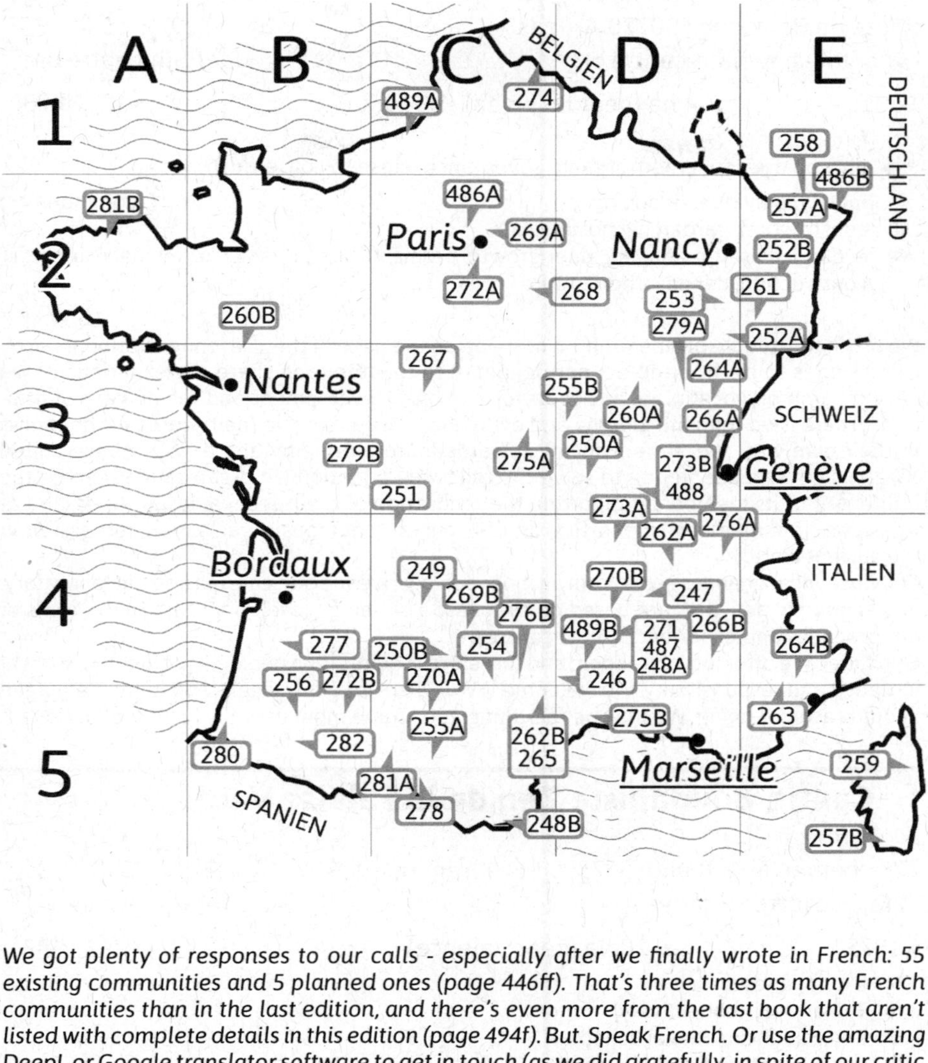

We got plenty of responses to our calls - especially after we finally wrote in French: 55 existing communities and 5 planned ones (page 446ff). That's three times as many French communities than in the last edition, and there's even more from the last book that aren't listed with complete details in this edition (page 494f). But. Speak French. Or use the amazing DeepL or Google translator software to get in touch (as we did gratefully, in spite of our critic of Google in the waterfall article). There's also a few German speaking communities in this chapter; some French regions close to the German border are actually German speaking and are listed with German names (e.g. "Gemeinschaft Loutzweiler Mühle", meaning "Community at the Mill of Loutzviller"). Some German named communities in France are nowhere near Germany ("Lebensgemeinschaft Bazian"), however... Well...
There's two community networks in France: Colibris (formerly known as Mouvement des Oasis en Tous Lieux), page 519, and Réseau Français des Eco-Villages (page 536).

☆ Aerium

D4

Route de Peyraube | 30770 Arrigas ☎+33-(0)499 644 791

contact@aerium-centre.org aerium-centre.org

2010 4 ha (near Le Vigan) 10 % ♀5 ♂8 10 23

- f, (e), (g)
- all who are concerned (organic governance (close to consensus))
- mostly individual financing — community
- construction, gardening, hosting, DIY
- ecological, self-managed, inner growth, permaculture, organic gardening, ecological construction, dance, arboriculture

☺

We are in the middle of a beautiful landscape on the edge of the «Cevennes National Parc». Our vision is to be, live and act our deep nature together and share our experiences. We have a common vocabulary and tools for personal development and taking care of relations. There is no founding dogma and everyone is invited to live their search for deepness in their own way. We are advancing towards more autonomy and a governance called "Organic governance" trying to solve the following equation: an organisation helping the fulfillment of its members, supporting the realisation of their deepest inspirations, and at the same time members supporting an efficient, autonomous organisation, integrated in the current world.

Convinced that changes come from an inner ecology, we see ourselves as a social laboratory. The economic activities are based on the reason of being: develop and share the fruits of our practices and experiences, host courses.

Our collective offer lots of freedom and therefore requires high personal motivation, spirit of initiative and responsibility. We welcome WWOOFers for a few weeks and more if engaged in our reason of being. We may be recruiting, too. Come meet us! *Photo on page 7*

☆ Centre Bouddhiste Zen de la Falaise Verte

D4

Ecole Rinzai

278 chemin de la Riaille | 07800 St Laurent du Pape ☎+33-(0)475 851 039

falaiseverte@orange.fr falaiseverte.org

1975 20 ha (near Valence) ♀1 ♂2 3

- f, e, Japanese — 3
- Taïkan Jyoji (collectively)
- Zen Buddhism, Rinzai School
- sutra, silent meals, samu, zazen
- self-managed, feminist, non-violent, hierarchical, Buddhist, meditation, vegetarian diet

☺

The Falaise Verte Zen Centre follows the tradition of Rinzai Zen Buddhism and is affiliated to its head monastery, Myoshin-ji in Kyoto. It is run by Master Taïkan Jyoji, who founded it after spending seven years in monastic training in Japan.

As well as sesshins, daily zazens and zen retreats, the Centre offers kyudo practice (traditional Japanese archery) in a kyudojo affiliated to the French Kyudo Federation (FKT). Many seminars as well as weekly practice sessions are held in this beautiful setting. Finally, the centre runs zazenkais (zazen sessions) in major French cities.

☺ attend at least one session at the Centre

Collectif Creatif du Castellas

D4

3 impasse du Castellas, Bessege | 30160 Languedoc Roussillon

☎+33-(0)466 603 157

contact@collectifcreatifcastellas.org collectifcreatifcastellas.org/#/

2012 (near Alès) 10 % ♀2 ♂2 4

f, e, g 4

permanent members (consensus) individual, association activities

association

threefold social order

eco-renovation, bio gardening, artist residency, cooperative restaurant, crafts

ecological, self-managed, non-violent, inner growth, organic diet, vegetarian diet, adult education, pedagogical, seminar-house, environmental protection, self-sufficiency, permaculture, organic agriculture, organic gardening, alternative energy, new technologies, ecological construction, humour, threefold social order

☺ ()

The Collectif Créatif du Castellas ("Castellas Creative Collective") is an ambitious experience of a participatory business, a place of life, of activities and experiments, an educational centre. We believe in the undeniable advantages of community life, if it is appropriately led. We learn that the implementation of our goals is dependent on the clarity of our personal and communal vision. We strive for balance and implementation on all dimensions of the project, culturally, legally, and economically. We do this by applying sensibly chosen values for each dimension: personal freedom, legal equality and material generosity. We share our cultural activities and participate in weekly council meetings. During building and further education events we renovate the large 19th Century building, which the association bought for the purposes of testing this approach. The property is self-financed and in communal ownership, it secures utilisation rights for the resident members. The goal is to establish living quarters, communal facilities, a canteen, a theatre and workshops. We finance the project with our monthly contributions, the income from shared activities, perhaps public or private subsidies, your contributions, gifts and donations. In order to organise and coordinate our actions, we use the website and in the second week of each month we welcome visitors, apprentices, as well as people who wish to participate with the project, to discover a communal lifestyle or let themselves in for an adventure...

☺ members give a loan to the association and get it back upon leaving

Douceur et Harmonie Domaine Maman Terre

C5

Domaine de maman terre | 66230 Serralongue ☎+33-(0)468 396 256
enthousiasme@wanadoo.fr www.douceurharmonie.wordpress.com

1964 | 21 ha (near Perpignan, Girona) | 75 % | ♀3 ♂1 0 | 4

f, e | 3

everyone in common (consensus) | shared income

common property

vegan, gentleness ("douceur")

gardening, household, guest house, book

ecological, non-violent, spiritual, esoteric/New Age, inner growth, meditation, yoga, organic diet, vegan diet, peace-work, animal protection, environmental protection, self-sufficiency, permaculture, organic agriculture, organic gardening, organic farming, alternative energy, new technologies, ecological construction, no cell phones/wireless, non-smoking, vegan, harmony, gentleness (douceur), self-sufficiency, simplicity

☺

Les Amis de la Douceur (= "the friends of gentleness"), and that is quite a programme! Here every kind of violence and cruelty is excluded. The most beautiful virtues are united in the harmony that is being promoted here. We let ourselves be inspired by the stars in the sky, by the Earth, the rivers, trees, flowers and birds. They teach us to help one another and grow. Our projects include growing lots of organic fruits, vegetables and flowers...

Le domaine de Maman Terre (="estate of Mother Earth") is an old farmstead situated at an altitude of 850 m in the forested mountains of the Haut Vallespir in the Eastern Pyrenees, it is isolated and very peaceful. We can offer our vegan/vegetarian guests simple accommodation in our gite (a cottage) in summer, and invite volunteers to help us in our daily tasks. Our diet is exclusively vegan and drug free, including such harmful products as alcohol and tobacco.

We prefer visitors to our community to contact us in advance preferably by post or e-mail telling us a little about themselves.

☺ group harmony, commitment to a vegan, ecological and organic lifestyle

Eco-hameau d'Andral

C4

Mas d'Andral | 46300 Le Vigan
ecohameaudandral@live.fr andral.org

2008 | 8 ha (near Gourdon) | 10 % | ♀12 ♂12 7 | 31

f, e

the whole group (consent) | limited common financing

property of all

no

taking care of the land: gardening, permaculture, biodiversity

ecological, self-managed, non-violent, environmental protection, permaculture, organic gardening, animal friendly husbandry, alternative energy, ecological construction, shared governing

☺

Co-housing project for 25 houses, 17 are finished in 2019. They must be built in an ecological way: houses with wooden frame, a renewable energy producer, limited electrical power, no-water toilets and a common sewage system through plants.
Newcomers must spend some time with us in the project before applying and becoming associates.
Self-governing is through four commissions and a plenary meeting organised every month.
A common budget is planned every year for common investments and functioning expenses.
We are widely invested in local associations, like local money, local sustainable agriculture, cultural associations.

→See also description of network **"Colibris"**
AlterHabitat Toulouse, Coordination de l'Habitat Participatif
☺ cooptation

Eco-hameau de La Bénisson-Dieu

D3

5 place des Abbesses | 42720 La Bénisson-Dieu
francois_nolle@hotmail.com

2016 | 5 ha (near Roanne, Lyon) | 50 % | ♀8 ♂8 22 38

2

g, e, f
all who are concerned (consensus)
limited common financing
properties of the members and rented
ecological, hierarchical, Christian, adult education, pedagogical, seminar-house, mentally/psychologically handicapped persons, physically handicapped persons, environmental protection, permaculture, organic agriculture, organic gardening, organic farming, forestry, ecological construction, humour

☺

We are an eco-hamlet composed of several families wishing to put into practice integral ecology. We are planning to build up a primary and secondary school for children, and a pedagogical farm and a reintegration project. We are also welcoming pilgrims on the road to Santiago de Compostela.

Ecohameau de Verfeil

C4

Lieu-dit Négo Soume | 82330 Verfeil sur Seye
+33-(0)648-017 151
ecohameauverfeil@gmail.com
verfeil-eco.over-blog.org

2013 3 ha (near Montauban, Toulouse) 25 % ♀7 ♂6 8 21

f, e, sp, Flemish, Portuguese, g 5

all residents (by consent in important decisions)

monthly payment for collective running costs and purchases, houses individually financed

shared ownership of community areas and individual ownership of residential plots

ecological

eco-construction, permaculture

ecological, environmental protection, permaculture, ecological construction, forest garden, choir, open days

☺ ()

Our aim is to create together a place to live and share our time, knowledge and resources, striving for a balance between people and nature, as well as between individual and collective priorities.

Our group has a wide range of ages, backgrounds and nationalities, but we all share a common desire to live more moderately and reduce our impact on the environment. With this in mind, our houses are all constructed by using ecological principles and materials; we have a permaculture vegetable garden and a forest-garden project; we share a community building with a workshop and kitchen; we have a collective plant-based water treatment system, etc.

Our group life moves along to the rhythm of shared meals, common tasks and projects, shared decision-making, choir rehearsals and other festive events and celebrations. Our project is continually evolving as we are not yet at full capacity and new members always add their ideas and skills to our melting pot.

Our current visitor facilities are weather-dependent so visiting is fairly limited! However, we hope to develop on-site activities in the future, as well as offering workshops and volunteering opportunities.

→See also description of network "**Colibris**"

☺ going through our inclusion process, sharing our values, buying either a plot to build on or an existing house

Eco-hameau du Moulin de Busseix

C4

Route de Busseix | 87500 Ladignac-le-Long ☎+33-(0)555 509 39 81
contact@moulindebusseix.org www.moulindebusseix.org

2004 — 10 ha (near Limoges) — ♀7 ♂7 5 19

f, g, sp, e — 1

meeting of the residents every two weeks; general assembly once a year (consent: if there is a strong opposition of a minority the decision must be reexamined)

no

gardening, building, common life, breeding

ecological, self-managed, non-violent, self-sufficiency, organic agriculture, alternative energy

Our main principles are:

- Reduction of our ecological footprint (e.g. through solar energy, solar water heaters and photovoltaics, dry toilets, phytoremediation of waste water, house insulation and composting, etc.).
- A collective and solidary life (communal food, common grocery shopping/food-coop, common laundry and workshops, a community vehicle, collective gardens, etc.).
- Opening to the outside world: maintaining close contact with the region and its towns, welcoming many visitors, internships and WWOOFers.

The community owns a common house, nine premises for lodging, several barns, a workshop and around 10 ha of land.

☺ to live with us several months before engagement

éco-lieu Thélème

D2

Impasse du Brin de paille | 70160 Breurey-lès-Faverney ☎+33-(0)663 669 565
asso.theleme@gmail.com assotheleme.wixsite.com/theleme

2012 — 6 ha (near Vesoul) — ♀3 ♂3 4 10

f, e, sp — 1 2

all who are concerned (consensus) — individual financing

property of all

medicine, carpenter, plasterer

ecological, non-violent, environmental protection, self-sufficiency, permaculture, organic agriculture, organic gardening, organic farming, animal friendly husbandry, ecological construction, construction

Our group exists since October 2012. The association was created on December 10, 2013 around common values:

- Experimentation of "living together" and pooling our resources by privileging autonomy
- Respect for the human being and his environment
- Reducing our ecological footprint
- Priority is given to being, and not to having.

Our desire to promote the collective interest instead of individualism is fundamental. It is through the realization of this project that we have the ambition to live this ideal family, whose dynamism will benefit, we hope to all citizens of our municipality of residence. Baptized Thélème, named after the abbey of Rabelais describing a place of personal and collective fulfillment, we would like to point out that our project is not related to any religion or political party, nor to an ideology.
Thélème illustrates a way of conceiving the world, placing human beings and their environment at the heart of our concerns.

☺ to be in agreement with the charter, the various "rules" of the place, to participate financially in the collective project, to bring ist know-how and know-how to advance the project

Ecolline éco-lieu de vie

E2

| Saint-Dié-Des-Vosges ☎ +33-(0)329 410 909; +33 (0)768 557 320
palatof@free.fr, flocondesalpes@free.fr www.ecolline.over-blog.com

2008 (near Strasbourg, Freiburg i.B.) 7 24

Ecolline is a community that started 2008 in an already very ecological place in the Vosges mountains, France. Ecolline comprises ten self-built houses from local and biological materials such as wood, straw and clay.
We are 17 adults and seven children who live in close contact with their environment and in particular with each other - striving to respect and accept our differences. Through our shared values on topics as environment, ecology and humanism, we tend to live a frugal life. Strictly speaking, we are not a community, but a collective that shares many things: a large plot of land, most of the premises, equipment, tools, energy, work...
Each household consists of a house built on a small plot of land. The rest of the land (1 ha meadows, gardens, paths and 1.2 ha forest) is jointly managed by the members.
All our bigger decisions are made together. For this purpose we meet monthly in the plenum to discuss the proposals of the small committees, and then carry out a sociocratic consensus process.
We have a open visitors day every year on the last Sunday in May (2pm) and give the possibility to meet us (by working) on the 4th Saturday of each month (from 9am to 6pm) during our collective work day. We also regularly need help with work sites all year round (accommodation and meals in exchange).
Please, inform yourself online and contact us in advance before coming!

ECOlonie - Centre Ecologique International

D2

1 Thietry | 88260 Hennezel ☎ +33-(0)329 070 027
info@ecolonie.eu www.ecolonie.eu

1989 40 ha (near Epinal) ♀7 ♂7 3 18

f, e, g, Dutch
part of the group (consensus)
common property
everyone
all & volunteers
shared income

vision containing ecological and spiritual elements (non-dogmatic)
guest house and camp site, shop, restaurant, organic garden, goats and dairy farm, art studio, renovation, bakery, creative courses and yoga
ecological, spiritual, anthroposophic, yoga, organic diet, vegetarian diet, adult education, pedagogical, seminar-house, environmental protection, organic agriculture, organic gardening, organic farming, alternative energy, new technologies, ecological construction, no cell phones/wireless, non-smoking, artistic activities: workshops

☺ ()

Ecology (the thesis of ecological thinking and acting) at ECOlonie goes beyond healthy and organic gardening and building. Ecology is a philosophy that includes all aspects of life - both physical and spiritual. Eco is derived from the Greek word oikos, meaning to feel at home. To feel at home somewhere simply means to feel at ease, to be able to be yourself.

At ECOlonie, creativity and spirituality inspire the development of this vision to include environmental sustainability and the daily experience of nature. Each individual can define their own spirituality, as we focus on being undogmatic, down to earth and practical.
We aspire to make viable the utopic dream of a durable, multi-faceted, humane society through the atmosphere and activities available at ECOlonie. We really appreciate people coming to help at ECOlonie. All these people, offering their help voluntarily, contribute to the continuing existence of ECOlonie.

→See also description of network "**GEN Europe**"
☺ three months of work with us, during which we can discuss all the little preconditions

Ecovillage "Ecohameau Les Trois Sources" C4

Le Soulayrol | 12550 Plaisance ☎+33-(0)607-252 454
troissources@gmail.com www.ecovillage-3sources.eu

2011 52 ha (near Albi) 30 % ♀3 ♂4 2 9
4

g, e, fr, sp, Dutch
for the association : all together (meetings)
a civil real estate company open to subscribers
we offer open meditation times open to all beliefs
seminars , guesthouse, therapies, open to all projects
ecological, self-managed, non-violent, ecumenical, inner growth, meditation, organic diet, pedagogical, free school, seminar-house, peace-work, environmental protection, permaculture, organic agriculture, organic farming, forestry, alternative energy, new technologies, ecological construction, humour, accessible to the disabled, We love the differences but in a the respect of a frame

☺

We are an intercultural and inter-confessional ecovillage project in the South of France. We are not a commune, but a housing and living community with a civil real estate association. Our common facilities like garden, laundry, sauna, café/cafeteria/restaurant are run by the association.
You can buy shares, rent a place or build a house... We want to scale up our community and welcome families with children as well as individuals. We like people with a project and former community experience. We have a multi-activity room, sauna, spa, shared therapy room, offices, shop, laundry, workshop, meditation room and plenty of space for your activities and skills.
There are still a lot of things to improve and develop but the place is beautiful in genuine real nature (even a bit wild), 2 hours from the Mediterranean Sea (Montpellier) and 1h30 from Toulouse. We have three sources, land, trees and no industry or nuclear power station nearby. A democratic school project for next year is starting. We like freedom and differences but it is necessary to have a clear frame. We like beauty, as well as clean and orderly premises (we French are not specialists, please Germans, help us ;-) !!).
Please do bring your own ideas, you can visit us in holidays, active vacations or as participating stay (séjour participatif). We believe the future will be about loving differences, because they help us growing, not in the imposing of ideas, even the most beautiful ones.

→See also description of network "**Colibris**"
☺ have a real project and have worked on yourself *Photo on page 15*

☆ Eco-village Sainte Camelle

C5

09 100 St Victor Rouzaud ☎+33-(0) 688 967 446 (SMS)
www.ecovillagestecamelle.fr

2011 25

Happiness in action!
We have joined together as a group to live a conscious life for ourselves individually, for other humans and for the planet. The ecology in all its forms and shapes influences all decisions and considerations in our eco-village.
The communal life is inspired from the shared vision of a way of living together that is seeking for mutual interests, lived love, authenticity, sincerity, confidence, relationship, happiness and simplicity. Each member remains fully responsible for her/his inner development and yet is also member of communal life.
In Sainte Camelle we are cultivating vegetables, have permaculture, a forest garden, an orchard and we offer various workshops and an open introductory café. We work on our projects, organize concerts, offer self-experience courses and spaces for internal development - and also rent our premises for events and courses... Almost all of our members work in our project most of their time.

Eotopia

D3

lieu-dit le Jardin Jolivet | 71140 Cronat
+33-(0)667 023 894
contact@eotopia.org
www.eotopia.org

2016 | 3 ha (near Bourbon-Lancy, Moulins) | 30 % | ♀3 ♂3 | 2 | 8

e, f, sp, Danish

everyone (consent reunion/consensus) | limited common financing

property of the association

veganism, gift economy, unschooling, non-violent communication

permacultural garden, clay and straw housing, non-violent way of life, vegan cooking, land art, translations

ecological, self-managed, feminist, non-violent, inner growth, vegan diet, free school, animal protection, environmental protection, self-sufficiency, permaculture, alternative energy, ecological construction, humour, non-smoking, sociocracy, no alcohol, handicraft

☺

Eotopia is an experimental space where we are looking for an harmonic way of life with all beings.

- Gift Economy: Everyone can give and receive freely. We wish to create a place where faith is more important than a bank account. On a practical level it means we produce and organise workshops and free events, without asking for something in return.
- Vegan: We try to not consume any product connected to animal exploitation. We want each sentient being, whatever species it belongs to, to live its life and to pass away the way it wants to.
- Unschooling: We are definitely egalitarian, which means in opposition to racism, sexism, speciesism, "moneyism", "ageism"... and hierarchism. Education is the foundation of our social system. We want to explore if it is possible to establish free education, free from dogmas, borders and limits. We also want the inhabitants and visitors of the ecovillage to be teachers as well as students.
- Togetherness: We feel it is time for humanity to challenge obsolete systems and define new ways of life that are respecting the environment and all the beings within. We will try to let the youngest learn by themselves, encouraging their creativity without putting knowledge in their hands but rather giving them the tools they need to discover it by themselves.
- Permaculture
- Low carbon village

We are on our way towards it... but still pretty far from it... we are going at a slow pace, taking our time! We are very open, to visitors, but just during visiting periods!

→See also description of network "**GEN Europe**"

Seed Networks

☺ ideological, financial, and most important feeling wise

Ferme Légère

B5

Chemin de Pilate | 64410 Meracq ☎+33-(0)559 350 047
fermelegere@greli.net fermelegere.greli.net/doku.php

2015 11 ha (near Arzacq Arraziguet, Pau) 70 % ♀4 ♂5 1 10

e, f, sp, Esperanto 2 everybody (2 employee status)

everyone. People owing shares of the property have certain powers according to the status (consensus)

common purse for food, all the rest is individual. "Rents" to pay back the investments and the everyday costs (telephone, internet, taxes,...)

collective property of all

no

vegetable growing, honey, bread baking, (re)construction work

ecological, self-managed, left-wing political understanding, feminist, non-violent, organic diet, vegetarian diet, seminar-house, environmental protection, self-sufficiency, permaculture, organic gardening, alternative energy, ecological construction, construction, humour, collapse anticipation

☺

Ferme légère ("light farm") aims at experiencing an alternative lifestyle, a practical ecology, having the smallest possible impact on its environment. The project seeks autonomy regarding energy, food and all domains of life. By necessity and by pleasure we are open to the outside. We have recently started to anticipate a global collapse and try to think accordingly. The key of our project is the collective and multigenerational dimension.

Friends of the Earth international, Amis de la Terre

☺ share our degrowth values and experience living with us for a time before any decision of integration is taken

Fontaine Aux Abeilles

E2

9 rue du Luxembourg | 67700 Saverne ☎+33-(0)388 899 708
#7908 at eurotopia.directory/booklinks

(near Karlsruhe, Nancy) 20

Fontaine Aux Abeilles is a participative habitat built around a major idea: to share a good life in a widely shared space where everyone and his/her visions have been involved since the concept was created.

Two thirds of the built area consist of private dwellings of eight families with all in all twenty residents. The ground floor offers three common entrances with disabled access, large meeting room, small kitchen, bathroom, guest room, rented offices contributing to the management costs of the collective, common workshop, bicycle parking, storage, honey shop. Outside, 2/3 of the spaces are very green and shared, with vegetable garden, orchard and a view of the beautiful landscape. The rest are terraces with small private gardens.

We have made architectural decisions, considering aspects of bioclimate, integrating the assets of sun, wind, rain and earth. The building is organic, with a choice of natural and healthy materials, economical and non-polluting technologies such as bioelectricity, ventilation through ground heat exchangers, rainwater recovery for watering, cleaning, toilets or collective laundry etc.

Francischu

E5

Francischu, Route de Bonifacio | 20131 Pianottoli-Caldarello, Corse du sud

☎+33-(0)647 46 23 99

1993 8 ha (near Bonifacio) ♀3 ♂2 2 7

g, f, e, i best: all

everyone in common (consensus) limited community financing and individual

rented/leased

faith in love (Agape, Philos and Eros)

garden, firewood, bread-baking, canning, cheese-making, hide-tanning, sausage-making, animal care, sailing

ecological, self-managed, anarchistic, non-violent, spiritual, Christian, ecumenical, inner growth, organic diet, environmental protection, self-sufficiency, permaculture, organic agriculture, organic gardening, animal friendly husbandry, ecological construction

☺

Our previously almost completely German enclave has turned into a respectable international community. We are more mindful of one another because we do not all speak the same language. We still welcome all new arrivals as members of our community. Experience, knowledge and know-how are exchanged. We are gardeners, farmers, shepherds, craftspeople, artists, musicians, seafarers, adventurers, jesters and wise (wo)men. We are open to new ideas and rejoice in deeds. We want a healthy Earth for ourselves and our children. We want healthy children for our Earth.
Outside the summer season, unannounced visitors are almost always welcome, we would ask for a small donation (10 EUR).

☺ self-development (dealing with childhood problems), openness, enjoying activity and movement (as far as possible)

Gemeinschaft Loutzweiler Mühle

E2

Moulin de Loutzviller | 57720 Loutzviller ☎+33-(0)387 967 524

2000 40 ha (near Saarguemines (F), Zweibrücken (D)) 60 %

♀2 ♂1 1 4

g 2

all who are concerned in common (consensus) limited community financing

property of individual community members

undogmatic

farming, horses, renovating, farm shop, housekeeping

ecological, self-managed, non-violent, spiritual, esoteric/New Age, chaotic, inner growth, organic diet, vegetarian diet, adult education, animal protection, environmental protection, self-sufficiency, organic agriculture, organic gardening, organic farming, animal friendly husbandry, forestry, alternative energy, construction, no cell phones/wireless

☺

You there! Yes, you :-) ! Do you want to lead a simple and down-to-earth life with many animals, plants and some familiar people? Would you like to spend your life in an atmosphere

that is open and benevolent, friendly, familiar and honest? Do you like to work "hand in hand" and enjoy it? Can you drop whatever you're doing because the horses are roaming in the neighbor's fields or unexpected rain threatens and the harvest has to be collected? Can you regard physical work as a challenge? Do you want to help instead of chatting? Can you manage not to turn a mosquito into an elephant?
If you still feel addressed now, then we are really looking forward to getting to know you! The mill people.

☺ empathy, pioneer spirit, imagination, energy, love towards people, animals and plants. Clear head. Interns welcome.

☆ **Gemeinschaft Vitulettu** Vitulettu Community (E5)

Lieu dit Vitulettu | 20234 Pietricaggio, Corsica ☎+33-(0)495-359 331
info@vitulettu.org
www.vitulettu.org, www.workaway.info/775729454116-en.html

1997 | 4.5 ha (near Cervione, Bastia) | >50 % | ♀3 ♂1 1 5

g, f, e — 2, 3

everyone in common (consensus) — limited community financing

property of individual members, rented/leased

living acceptance and attendance

gardening, forestry, apiary, seminars

ecological, self-managed, spiritual, inner growth, meditation, organic diet, vegetarian diet, vegan diet, self-sufficiency, permaculture, organic agriculture, organic gardening, forestry, alternative energy, construction, humour, music, piano, lovely places, inspired by Eckhart Tolle, Byron Katie, "A Course In Miracles" (book), reflected shamanism

☺

At the edge of the wilderness, 800 m above sea level, we live in a former abandoned Corsican mountain village. Our everyday life consists of diverse and hard work but also of nature's gifts all-year-round. Furthermore we focus on our inner growing and changing, alone and together.
The challenges of the present give us opportunities to remember what is most important: creating bonds, helping each other, listening, questioning thoughts, practicing music; listening to the inner glowing and following happiness.
We own five houses/ruins of this village, which we have secured or restored. The community house contains the plant nursery and various cellars for self-sufficiency. For ten years we organized workshops, the shell of the seminar-house is completed. A guesthouse is available for the sometimes numerous international guests. We had to work out most of the villages' infrastructure by ourselves. The water pipes; a solar-system; fences; vegetable gardens and orchards and many ongoing projects. Everything else has become routine, such as the extensive all-year-round vegetable and fruit growing.

The market and work in the neighborhood keep us in touch with people from the valley and other part of Corsica. The 'beekeeper house' is waiting for a beekeeper, workshops are emerging and there is room for new inhabitants and for new ideas, yet unknown... Until you come!

Workaway

☺openness to self-realisation, acquaintance period

Goshen Ferme de la Chaux

D3

Ferme de la Chaux | 21360 La Bussière-sur-Ouche
soko@jesusfreaks.fr
www.goshen.fr

2009 | 13 ha (near Dijon) | ♀5 ♂3 8 16

f

2

the inhabitants (unanimity, consensus, vote) | limited common financing

association

anti-globalization, Christian

hospitality, Seminar House

ecological, self-managed, anarchistic, left-wing political understanding, non-violent, spiritual, Christian, ecumenical, adult education, seminar-house, peace-work, environmental protection, humour, voluntary simplicity, friendliness

In a time of individualisation and the dictatorship of money, we are working towards a world of solidarity, fairness and ecology together. This is what we are creatively trying to implement in our eco-hamlet. Here a group of families and friends live in intentional simplicity, whereby Christian values, participation, common celebrations, as well as hospitality, everyday environmental protection and an alternative economy are important components of our life.

→See also description of networks **"L'Arche (Jean Vanier), Colibris"**

☆ Hameau du Ruisseau Eco-centre du pays Nantais

B3

Route de Louisfert | 44520 Moisdon la Rivière (Loire Atlantique)
☎+33-(0)781 542 455
maison.autonome@orange.fr
www.heol2.org

1984 2 14

f

Between Nantes and Rennes, Vannes and Angers, 80 km from the sea, the Ecohameau is located in the Middle Country, the "Land of the Mée".
Balance, common sense, reality, far from excess - it tries to share its testimony and inspire everyone to begin their personal journey towards their own autonomy and path of life.
After 30 years of experimenting with a new way of life from the smallest social unit - a family - we wanted to continue the adventure on a hamlet scale. As the result of assiduous determination and a clear and global vision, the sale of adjacent land was offered to meet our "demand".
Beyond the high-performance technical achievements of our Ecohameau, we are experimenting on a daily basis with the delicate challenge of "living together" in order to co-create a new life mode over time that inspires a new economic organization:

- Produce by yourself what is possible in a given territory
- Exchange goods and services locally
- Aim for financial autonomy based on the capacities of each individual
- Practice indoor ecology and conscious communication essential for conviviality
- Educate, transmit, welcome
- Celebrate our successes as much as our difficulties

Our activities:
Visits, reception of schoolchildren, trainees, WWOOFers
Seminars about autonomy/ecology
Manufacturing internship: Pigott wind turbine, straw/earth construction...

→See also description of network "**Colibris**"
☺ ee are happy to receive guests but have no more space

☆ Jean le Moine

E2

Jean le Moine 1 | 70310 Amont et Effreney
☎+33-(0)384 944 512
steinmann.mathias@orange.fr

1993 5 ha (near Luxeuil les Bains) ♀2 ♂4 2 8

e, g, f
consensus
barter trade, independence, meditation, free spirituality
care and support for people with handicaps, livestock farming, cheese, bread baking, renovating, gardening, firewood, healing care, creating space for self-discovery
Ecology, self-realization, spirituality, inner growth, meditation, dance, singing, education, working with mentally handicapped people, opening up to arts and culture, bodywork and natural healing

☺

Jean le Moine is a 400-year-old farm in a secluded location on the western slope of the Vosges, on the edge of the "Thousand-Lakes-Region". The farm was acquired by a family with four children 25 years ago and since then inhabited by many people and generations. For nearly 20 years, WWOOFers, Helpx-ers and Workaway-ers have been reviving the farm and took care of goats, sheep, horses, chickens, ducks, peacocks, cats and dogs.
We live in the trailers that we brought with us, also in a hut, a tree house, a renovated bakery and in the old farmhouse, where currently even more living space is being built.
A young mentally handicapped man has lived with us for five years now and currently, five adults and two children live part-time with us.
With respect for our different origins and characteristics, we try to live towards inner growth and creativity, also supported by daily and shared rituals.
We are happy about new people of different ages and origins and there is also room for more animals, teepees or circus carriages, etc.!

Workaway, Helpx
☺ new members are welcome without conditions or requirements (families of all generations or nationalities)

L'Arche de Saint Antoine Communauté de l'Arche

D4

Place de l'Abbaye | 38160 Saint-Antoine-L'Abbaye ☎+33-(0)476 364 597
accueil@arche-sta.com
www.arche-de-st-antoine.com/communaute/stagecommunautaire/

1987 (near St-Antoine-L'Abbaye) ♀18 ♂18 14 37

f, e, sp
permanent residents (halocracy/sociocraty)

all all
shared income

property of all (through an association)
non-violence, Christian with an openness toward other religions
seminar-house, training in and practice of non-violent actions
ecological, left-wing political understanding, non-violent, spiritual, Christian, ecumenical, inner growth, meditation, organic diet, vegetarian diet, adult education, seminar-house, peace-work, environmental protection, organic gardening, humour, accessible to the disabled, singing, dancing, manual work

☺

The Communauté de l'Arche de Saint-Antoine gathers people (families as well as singles) around their commitment for non violence, regarded as a tool to transform oneself and therefore to transform society. Christians with no church affiliation founded the community of Saint-Antoine in 1987 and since then, they welcome people with various beliefs, with respect for each person's own spiritual path.
Around 40 people live in this house (children included). They chose a simple living revolving around sharing and spiritual growth. Some people are permanent residents, some are here for a shorter period.
Our life is made of work, trying to minimise our needs, to reconcile with each other, prayers, meditation, chanting, dancing and partying. We also take part in non-violent actions.
In a "community stay" we welcome people wishing to share our way of life and to experience our principles of "living together". See also: "F.E.V.E." under "useful addresses" in France.

→See also description of network **"L'Arche de Lanza del Vasto"**
☺ mutual acquaintance during an internship

La Borie Noble Regain Association

C5

La Borie Noble | 34650 Roqueredonde ☎+33-(0)4-674 409 89

laborieregain@gmail.com www.association-regain.com

1962 450 ha (near Lodève, Montpellier) 75 % ♀4 ♂4 1 9

f, e, sp, Portuguese all

the residing members (consensus) shared economy for residents

property of SCI (société civile immobilière) "Paix et travail"

non-violent communication in the spirit of Gandhi, manual activities

farming, gardening, bakery, woodwork

ecological, non-violent, spiritual, ecumenical, inner growth, meditation, yoga, vegetarian diet, self-sufficiency, permaculture, organic agriculture, organic gardening, conventional farming, alternative energy, NVC, non-violent activism, Erasmus+

☺

Located in the south of France, we are a small group belonging to the network of L' Arche communities. With several different buildings, some dating back to medieval times, we live simply in a secluded valley high in the hills of the Languedoc region. The unique layout of La Borie allows each person or family the privacy of his own residence whilst enjoying the fullness of life in community.

Our intention is to live in harmony with others, ourselves and nature. Daily life includes working with our hands, learning how to produce what we consume and creating an environment for artistic expression. We share vegetarian meals at least once a day, and yes, we have a few animal friends. We have been producing bread (from farm to table) for 50 years, traditionally, i.e.: with our hands!

With 450ha we are surrounded by nature, fruit trees, rivers and lakes. Our neighbouring villages are some of the most beautiful in the south of France, and we border the traditional French route for the Camino de Compostela.

We value artistic expression and many of our residents are highly creative. We enjoy singing, theatre, musical expression, crafts, and every week we go crazy with a Saturday night of traditional world dance.

→See also description of networks **"L'Arche de Lanza del Vasto, GEN Europe"**

☺ willingness to speak/learn French, one-year trial period, commitment to live ecologically and mindfully

La Collective de Chalvagne

E4

200 route du Col du Buis | 04320 Val de Chalvagne ☎+33-(0)489 779 034

chalvagne@lesgouttesdo.net chalvagne.lesgouttesdo.net

2009 10 ha (near Entrevaux) 15 % ♀3 ♂2 4 9

f, e

all who are concerned (unanimity without decisional meetings) pooling

rented, lent, property of the NGO

support for self-empowerment/self-responsibility

relations, agriculture, guests, building

ecological, self-managed, inner growth, free love, adult education, free school, self-sufficiency, permaculture, organic farming

☺ (✋✋)

The Collective de Chalvagne is located in the middle of nature in a valley in the Southern Alps, in which useful techniques for an independent world and for solidarity can be explored and developed. Since we founded our association, self-help and self-responsibility has been the foundation of all our actions and functions in order to create the conditions in which the living together can blossom. The Collective de Chalvagne is open to all. It is self-governed by the people and the association in an honorary capacity, with the goal of combining the means of communal production in order to fulfil the basic needs of each individual.

☺ integrate into the NGO of the place and integration process to become a residing member

☆ La ferme de Chenèvre

D3

11 chemin de Chenèvre | 39110 La-Chapelle-sur-Furieuse

+33-(0)672 573 406 contact@chenevre.fr

colibris-lafabrique.org/les-projets/la-ferme-de-chenevre-graine-de-liens

2017 25 ha (near Besançon) 20 % ♀5 ♂5 6 16

f, e, i

limited common financing collective ownership

eco-construction, organic agriculture, ceramic, bakery, goats, arts

ecological, self-managed, environmental protection, organic agriculture, organic gardening

(✋✋)

La Ferme de Chenèvre is a French eco-place located in the department Jura. Open to experiments, our local-global transition project is including co-housing, co-learning, co-working and common good economy.

We have been starting a 10-year eco-renovation program in 2017 to re-shape a traditional isolated farm dating back to the Middle Ages. We are developing a range of activities from eco-construction to biological agriculture and bakery, ceramics, arts, non-violent relationships and sociocracy.

We are striving to create social bonds on an open and contributive basis as well as to develop our autonomy together and to be able to live our ideas in harmony with nature and humans.

Our project is part of Oasis network (Colibris) and has been supported by a loan from the cooperative bank La Nef.

→See also description of networks **"Colibris, WWOOF"**

La Ferme du Collet

E4

06260 La Penne ☎+33-(0)49305-8450
arias@no-log.org

2001 (near Nizza) 6

The eco-hameau "La Ferme du Collet" is an intentional community of 6 people, including 5 partners (3 households). At the moment (2019) we cannot welcome new members. Each household is autonomous in its professional activities, we share the management of land, forests and other collective spaces.
Since 2001, the date of the creation, we have been experimenting together with permaculture, dry toilets, phyto-purification of water, eco construction, photovoltaic and passive solar panels, wood heating...
We have also experimented with animal traction. We sometimes welcome people in exchange for help and also welcome WWOOFers.

→See also description of network **"WWOOF"**

La Fleyssière Communauté de l'Arche

C5

Lieu dit La Flayssière | 34650 Joncels ☎+33-(0)467 444 090
accueil.flayssiere@gmail.com www.arche-de-la-flayssiere.fr

1978 15 ha (near Montpellier, Béziers) 60 % ♀3 ♂5 1 9

f, g, sp, e — all
everyone in common (consensus) — shared economy (shared income and capital)
common property
non-violence, spiritual, inter-religious dialogue
farm work, garden, cheese making, pottery, seminars on non-violence, kitchen
ecological, self-managed, non-violent, spiritual, ecumenical, inner growth, meditation, yoga, organic diet, vegetarian diet, adult education, peace-work, organic agriculture, organic gardening, organic farming, alternative energy, ecological construction, pottery

☺

We are a small community in the mountains of southern France, not far from Lodève.
In response to the problems of our time, we strive for a simple life in which there is space for manual and mental work, inner growth and common activities. Interpersonal relationships are at the forefront of our community life; we strive for mindful interaction and group processes; decisions are made by consensus. La Fleyssière has a large self-catering garden, organic farming with dairy cows and cheese, a pottery and from May to September a seminar business.
In our daily routine there is space for silence,

meditation and common rituals, whereby we pay attention to openness for all spiritual traditions. After the morning round to organize the day, we will work in the various areas and then meet again for a joint vegetarian lunch. In the afternoon we take time for meetings, yoga or singing; once or twice a week we dance in the evening. The organization of different celebrations opens space for creativity, artistic representation and sociability.
Despite our remoteness we are well integrated into our social environment thanks to various weekly and handicraft markets, yoga classes and our active commitment to peace and justice. Our own children have grown up in the meantime, but we are happy to affiliate families with children again.

→See also description of network **"L'Arche de Lanza del Vasto"**
Community of Arche, nonviolence and spirituality
☺ agreement with the principles of the Arche

La Lune Nette Les Lunettes

D3

Les Lunettes | 39170 St. Lupicin
☎+33-(0)384 421 765
lalunenette@yahoo.fr

1996 2.75 ha (near St. Claude) 20 % ♀5 ♂6 3 14
5 5

g, e, i, f, sp
everyone in common (consensus) limited community financing and individual
property of individual members and of persons not living on-site
more Amerindian than anything else, e.g. the 4 elements as a guideline, vegetarian if possible
garden, plant and fruit harvest, renovation, teepee, wood, handicraft, honey, culture
ecological, self-managed, non-violent, spiritual, inner growth, meditation, yoga, organic diet, vegetarian diet, vegan diet, seminar-house, peace-work, environmental protection, self-sufficiency, permaculture, organic agriculture, organic gardening, organic farming, forestry, alternative energy, new technologies, ecological construction, construction, humour, baking bread, seminars (wild plants, nutrition, culture), self-determined learning, crafts: Indian jewelry, talking sticks, basket weaving, dream catcher, felting, carpentry, sports: hiking, swimming, canoeing, canyoning etc., music

We are a small community of several tenants in an old beautiful eyewear factory, located in a untouched nature in the French Jura right at a river canyon, with waterfalls and pools, surrounded by woods and meadows. A village is nearby with everything you need. We cherish and maintain this place as a group, discussing everything in regular meetings. We take care of the house and the surrounding nature, for a good social coexistence with the animals and plants. We have deep respect for nature and thus use environmentally friendly techniques wherever possible. We experiment with new techniques and living arrangements (Dome, Yurt). We have a cultural association that organizes music, dances and other classes. We also like to celebrate the seasons. We have our own source and a turbine for electricity production.

→See also description of networks **"GEN Europe, Colibris"**
☺ 15 Euro/day during the time of getting to know each other (3 months), later membership of the association of the ecological cohousing project and pay rent for the individual living spaces

La Nef des Fous

D4

Jansiac | 04200 Chateauneuf-Miravail
+33-(0)492 620 261
www.jansiac.lautre.net

1971 | 300 ha (near Sisteron) | 20-80 % | 2-10

f, e, g | all

all who are concerned (expressed unanimity) | shared economy

non-profit organisation

"utopie topique" (local Utopia)

subsistence farming, crafts

ecological, anarchistic, self-sufficiency

☺ (✋✋)

To call yourself an environmentalist, anarchist or egalitarian is not enough to define oneself among other alternative people because that is just the minimum requirement. Situating oneself as utopian rather than militant means having questions rather than answers, sharing doubts rather than certainties, taking the necessary time for understanding one another. Situating oneself as "island people" rather than "movement people" means choosing to isolate a bit of the Earth in order to found a nanosociety on this islet that deviates more or less from the global society.
Founded in 1971, the community is isolated with 300 ha in the mountains at 1000 m altitude to experiment with a local utopia, an imaginary society organised around the place rather than people (which allows humans, among other things, to stop thinking of each other as owner-tenant, employer-employee, producer-consumer and buyer-seller). This place offers secure means of subsistence, communication, expression and culture for people who don't own anything, wherever they go. Children and elderly people are being looked after, where they are, by those who are there (teaching, care...).
Guests are welcome! (Please contact us before)!

☺ common political objectives

La Nouvelle Mayapour

C3

Domaine d'Oublaise | 36360 Lucay le Male
+33-(0)667 975 503, +33-(0)254 402 395
new-mayapura@orange.fr
newmayapur.com

1975 | 87 ha (near Chateauroux) | 25 % | ♀10 ♂15 | 25

g, e, f, i, sp, Portuguese, Lithuanian, Slovak, Aqui, Hindi, Telugu

administrative council (consensus) | shared economy

property of association

ecological, ahimsa, Gaudya Vaisnavism

garden, wood building, renovation, maintenance, cleanliness, kirtan

ecological, non-violent, spiritual, inner growth, meditation, yoga, organic diet, vegetarian diet, adult education, pedagogical, seminar-house, peace-work, animal protection, environmental protection, self-sufficiency, permaculture, organic agriculture, organic gardening, organic farming, animal friendly husbandry, forestry, alternative energy, ecological construction, construction, non-smoking, woodwork

☺

We welcome you to New Mayapur! It is located in the heart of France 3 hours south of Paris, surrounded by unique lush green forests. You can experience a pollution free natural life and a get away from the concrete jungle.
The Château Doublaise is a historic monumental building that shows the rich heritage of France. Now it is a home for spiritual and cultural seekers who value moral ethics as the way of life. Visitors can charge their prana through our forest walks. We encourage the cultural seekers to come and experience cultural life which has its roots in the east.
Community life revolves around simple living and high thinking. Ecological, chemical free farming is the essence of our traditional way of growing fruits vegetables and flowers for the community. Most of our community fruits and vegetables grown in our farms. Cows and pets are also part of our community and are treated like a family, as our community emphasis on ahimsa (non violent life style) animals are looked after with care and love.
As we are an international society we are multicultural. It adds the flavour of unity in diversity. Residents are expected to aid in the services include gardening, cooking, maintenance of the castle, cutting wood for heating the castle. In exchange we provide healthy 4 course cooked breakfast and 6 course cooked lunch, simple dinner, accommodation in single or shared room with all necessities, WiFi, possible yoga courses, mantra meditation classes, all our love and care as part of our family for a while, and a wonderful cultural discovery.

→See also description of network "**ISKCON**"
workakway
☺ vegetarian, no intoxication, interested in our local culture spiritual practice

☆ La rêverie du Loupier The dreamfactory of "Le Loupier" C2

Hameau du Loupier | 89340 Saint-Agnan ☎+33-(0)953 877 216
www.colibris-lafabrique.org/les-projets/la-reverie-du-loupier

2018 3.3 ha (near Sens) 5 % ♀2 ♂2 4

f, e, g 4 4
as we are still a small group, everybody (consensus, sociocracy, non violent communication)
individual financing property of one member
our ideology is to refuse ideologies, we are non dogmatic and we prefer common sense and to respect nature
ostheopath, smith, physiotherapist & psychomotorist, director of photography, science animator, druid, spirit realm walker
ecological, self-managed, anarchistic, feminist, inner growth, organic diet, pedagogical, seminar-house, environmental protection, self-sufficiency, permaculture, organic

gardening, animal friendly husbandry, alternative energy, ecological construction, humour, no cell phones/wireless, clown, non-dogmatic Stalinist, cat cult, board games player

☺

We are two couples who are creating a place where we will live in yurts in a field around an old farm. We aim at reinforcing social and ecological connection between people through various actions.
We use various tools as non-violent communication, mindfulness, psychology in order to function with simple rules. The idea is that if each individual takes care of his or her psychological and physical balance, then life together will be balanced. So each of us dedicates time to explore his or her own mind as we see that an excellent self-knowledge is key to a harmonic life.
We want this place to be open to the world, with concerts or debates. We enjoy singing, dancing, laughing or playing cards.
Our goal is to create a place where we are happy to feel the sun and the rain, a place to support each other, a place where inner growth is fed naturally by our interactions, without the need of an ideological system and without any dogmata.
By the way, we are so non-dogmatic that here vegan and hunters laugh together about the absurdity of life.

☺ understanding of own emotions and psychology, working on it, being truthful and able to question yourself and be able to share as smoothly as possible

☆ L'Arche en l'Île

C2

rue Vandana Shiva | 93450 L'Île-Saint-Denis
tinyurl.com/eurotopia8090

2014 | 1825 m² (near Paris) | ♀16 ♂15 21 64

- f, e
- all members whether they are owners or tenants
- individual financing for the flats and shared financing for the shared spaces
- 14 flats are property of each inhabitant and 12 flats are rented by a social landlord
- environment and solidarity
- ecological, self-managed, non-violent, permaculture, ecological construction, solidarity, inclusivity, diversity, implication, participation, empowerment, friendly, open-minded

Through our co-housing project, we collectively bring a different way of housing into the city, so it turns out more inclusive, more friendly and more respectful to the environment.
We wanted to preserve our privacy with a quality housing adapted to our needs and our financial means, but we chose to include many shared spaces, over 100 m² in the interior of the building and over 500 m² outside. This allow us to manage a permaculture project on the roof, thus producing locally and promoting awareness.
We reinvented the condominium by adopting a very inclusive and democratic governance system in which everyone takes part in the decisions, no matter they are owners or not: 1 person = 1 voice.

Coordinaction de l'habitat participatif

Le Broglio

C4

Le Bruel | 12700 Tournhac
sci.lebroglio@gmail.com

2012 2 ha (near Villefranche de Rouergues) 0 % ♀2 ♂1 3

f, e 1

everyone (collective intelligence and consensual management)

individual and shared economy — property of all

no

permaculture, eco-construction, common administration

ecological, self-managed, non-violent, meditation, organic diet, free school, permaculture, organic gardening, ecological construction, construction, humour, common administration, straw & clay-construction

We are a couple of stone builders living in a collective house rebuilt by us. We have 2 ha (2000 m² of permaculture garden), a green house, dry toilet, a clay house we built.
Our project is to share live and experience together. Soon we will continue constructing more buildings using alternative methods, which works very well here.
We use collective intelligence in our daily lives, this as well as music and an ecological way of living are very important for us. We are dynamic, open, mindful and we like to meet people. In the beginning we were three associates, one having left us recently and we are looking for new people who are prepared to invest and have similar goals.

→See also description of network **"Colibris"**

☺ no addictions, financial participation

Le Callisto

C5

1 rue Maurice Genevoix | 31130 Balma
jbe@dragon-creation.com callisto.link

2015 2000 m² (near Toulouse) ♀35 ♂25 16 60

f, sp 5

majority (vote) — shared economy

property of all

humanist

reunions, shows, workshop

ecological, environmental protection, organic gardening, alternative energy, accessible to the disabled

The project is quite open, not restrictive and based on voluntary work. We are moving towards the progressive acquisition of property (over 40 years) in partnership with a social housing agency. We keep in mind our aspiration for social diversity. "Rich" accessors are also present.
Our main wish is to be in good relationship with our neighbours and we share spaces such as a large common room (100 m²), and a laundry room (10 m²). A large garden of 1000 m² is managed in common. The common room is open to the residents of the neighbourhood who are members of our residents' association.

☆ Le collectif du Moulinage de Chirols D4

An experimental space for collective and solidarity living and activity.

Mairie de Chirols | 07380 Chirols ☎+33-(0)783 473 497
contact@lemoulinagedechirols.org lemoulinagedechirols.org

2018 (near Aubenas) ♀6 ♂7 2 15

f, e, sp 1

members of the cooperative (consensus, during monthly meetings and our AGM)

equal shares in the cooperative with no speculation + solidarity donations, investments and loans by members and supporters + government funding, donations + cooperative community loans = hopefully no bank loan needed !

community/cooperative

sustainability, solidarity, diversity and mutual respect

ecological and participatory renovation in practice and in training to create a space for arts and crafts workshop spaces, performance space, co-working, community canteen, well-being centre and growing space

ecological, left-wing political understanding, pedagogical, organic gardening, alternative energy, ecological construction, humour, performance, theatre, clown, music, dance, circus, well-being

☺

We are working to transform the moulinage - an old mill characteristic of Ardèche's industrial heritage - into a space for eco-citizen living and activity, with a focus on collaboration and solidarity with the dynamic local territory and village, Chirols.
The renovation (from 2019) will be carried out by the habitants and volunteers, with support from local professionals and craftsmen, specialists in ecological and participatory architecture.
The smaller building (1600 m²) will be dedicated to cooperative housing with some shared spaces dedicated to communal use. The larger building (2400 m²) on the riverside will be dedicated to social and economic activities with a focus on performance arts and local crafts. A training centre for ecological and participatory architecture will be launched during the renovation period and continue as a key educational activity.
We will experiment with new ways to share living and working spaces, skills, materials, tools and services between cooperative members and also with the wider public. All activity will be not-for-profit, with any income generated being invested back into the cooperative and its activities and in the creation of local employment.

→See also description of network "**Colibris**"

Habicoop

☺ financial and/or time investment for the building renovation and project coordination

Le Hameau des Buis Association du Hameau des Buis

D4

Route de Chaulet | 07230 Lablachère ☎+33-(0)665 059 799
chrislhop@yahoo.fr hameaudesbuis.org

2011 (near Joyeuse, Les Vans, Aubenas) 40 % ♀16 ♂11 14 41

f, e 4 12

part of the group (majority vote) individual financing

education, garden, goats, bakery

ecological, self-managed, non-violent, pedagogical, free school, environmental protection, organic agriculture, ecological construction, intergenerational

☺ ( )

The Hameau des Buis ("Boxwood Hamlet") is a multi-generational collective which was created around an alternative school, but is now independent. The village consists of 20 houses built according to passive house standards, constructed with 70% natural materials. Pensioners live in 12 of the apartments for which the construction phase ended in 2011. The rest is inhabited by families. The collective tries to live in solidarity and to develop the interactions between generations.

→See also description of network "**Colibris**"

☺ ideological and financial

Le Kolkhoze

C2

10, rue Léonardi | 91160 Saulx les Chartreux ☎+33-(0)637 082 617

1975 (near Paris) ♀5 ♂5 2 12

all shareholders (consensus)

broadening family environment, fighting against isolation, share daily life, solidarity

Under the strong impression of the events of May '68 in France, the project Le Kolkhoze arose from the deep desire to open and expand nuclear families, to share a warmer and supportive everyday life and to give our children the opportunity to grow up with many brothers and sisters. Le Kolkhoze was founded by six families (12 adults and 16 children) and is located in the heart of a village of 5,000 inhabitants, 17 km south of Paris. Since the beginning 40 years ago, new families have been established and new people have been added.
The living area is a co-housing project with several units and was built in 1978, it consists of private residential units of 145 m². About 30 % of the living space is community area. The architecture includes common spaces to promote daily gathering and communication: a gallery in the meeting hall (80 m²) serves as a corridor between the six apartments and is complemented by a common living room. Other areas are also shared: a DIY workshop, storage rooms (bicycles, refrigerators), cellars or a photo lab.

On the garden side (600 m²), the terraces are connected on the upper floor and have direct access to the garden via common stairs. In the garden our members grow their own fruits and vegetables or keep chicken there.
We are involved in social and local life (city) and our communitiy room offers space for various public activities and events.

Lebensgemeinschaft Bazian

B5

Bazian | 32230 Louslitges ☎+33-(0)562 645 183, +33-(0)56270-9513

die.gocls@gmx.net www.ecoledepresence.fr

1994 55 ha (near Auch) ♀2 ♂2 3 7

g, f, e

everyone in common (consensus) individual and limited common financing

common property

construction, vegetable garden and orchard, kindergarten, school

ecological, self-managed, non-violent, inner growth, yoga, tantra, organic diet, adult education, pedagogical, free school, seminar-house, animal protection, environmental protection, self-sufficiency, permaculture, organic gardening, alternative energy, ecological construction, non-smoking, training in Montessori pedagogy, Emmi Pikler, R. & M. Wild

☺ ()

Bazian is a place with 20 years of experience in child care, ecological building, group processes, conflict resolution and permaculture. The families that have lived here as homeschoolers have mostly gone away because the children have grown up and are going their own way. The two families still living on the square are looking for new people with personal responsibility and financial means to transform the square into an ecovillage with a training centre for Montessori pedagogy. The land has 55 hectares with its own springs, forest, meadows, fertile soil with vegetable gardens and fruit trees.

☺ consent of residents/members, financial independence, ecologically oriented, drug-free

Les Castors du Crêt de Roc

D4

Société Civile Coopérative de Construction

31 Impasse Desjoyaux | 42000 Saint-Etienne ☎+33-(0)753 527 589

sccc.castorscretderoc@orange.fr

2015 ♀14 ♂9 9 32

everyone (consensus and/or vote) each family owns their own flat or house

We are a co-housing project: 13 families who got together to plan and proceed the construction of flats and houses on a plot of land in an underprivileged neighborhood of our town Saint-Etienne (near Lyon and Grenoble). The first ideas date back to 2005 and we got into our dwellings at the end of 2015. It is a grouped settlement of passive housing built around a common garden with a vegetable garden. We own our flat or house and share a common room, a space for guests with bathroom,

a communal laundry and a small common workshop. We are very different in occupations and ages (varying from 83 to 8 months). We consciously pursue qualities like close neighborhood, mutual support or care. We share good times and bad times together: Two children were born here since we moved in but we also lost the oldest member of the adventure after a few difficult months for him and his wife. Everybody tried to help her as best they could. We do have collective decision making processes (monthly meetings) and we also get all together at least twice a year for a day of work in the garden and common areas.

Les Mûriers

D3

1367 route des Crêtes | 69480 ANSE
+33-(0)601 733 706
alisa.lienard@free.fr

1981 ♀5 ♂6 1 12

- f
- all adults (majority vote and consensus for big investments)
- common financing for common items (heating, water, common parts of the building)
- property of all
- outside jobs and/or retirement

The project was born in 1980. We were 5 families who did not want subdivisions and who wanted to share equipment and parts of life especially for our children. Thanks to the warm welcome of the local MHGA [movement for self-managed grouped housing] and the availability of large and old houses within a radius of about 30 km around the job place of at least one member of the families, our choice was quickly carried on a Beaujolais farm which was waiting for volunteers to be restored.
In order to better communicate internally, common areas (a kitchen, a playroom for children, a laundry room with drying area and 2 guest rooms) were distributed according to the respective choices of living on 2 levels including attics or ground floor. To move freely internally between the apartments and common areas, a long corridor has been built.
In 1988 it was decided to integrate a 6th family, because a part of the building was still unoccupied and the budget was low.
The northern courtyard embellished by the presence of 13 old muriers trees has hosted many parties and neighborhood meals.

Les Toit Moi Nous Habitat groupé solidaire

C1

9 allée de la Courtoisie | 59650 Villeneuve d'Ascq
+33-(0)609 121 116
contact@toitmoinous.fr
www.toitmoinous.fr

1400 m² (near Lille) ♀20 ♂10 15 45

- f
- monthly general assembly and an administrative council (majority vote, consensus is being sought)
- individual financing
- 1/3 rented (social housing), 2/3 private ownership
- none
- exchange between residents, solidarity and collaboration
- ecological, self-managed, animal protection, organic gardening, humour, accessible to the disabled

We are 22 families living in the same building and we share various facilities: a room for meetings and gatherings, an apartment for facilitating and welcoming friends and families, a laundry room and a workshop.
Our goal is to develop and support exchange and to facilitate social cohesion. One of our main characteristics is being intergenerational: 11 families consist of seniors and 11 others are younger, most of them with children, which contributes to a helping and life-improving exchange.

☺ agree with a charta of statements

MagnyÉthique

D3

Château de Magny | 69550 Cublize
☎+33-6-782 154 87
contact@magnyethique.org
www.magnyethique.org

2017 8 ha (near Amplepuis, Roanne, Lyon) ♀5 ♂4 8 17+

- f, g, e, sp, i
- the whole group, sub-decisions in commission groups, plenaries (agreement)
- property of the non-trading society
- renovation, gardening, livestock, reception, associative café/repair café, organising events
- ecological, self-managed, non-violent, organic diet, environmental protection, self-sufficiency, permaculture, organic agriculture, organic gardening, animal friendly husbandry, alternative energy, ecological construction, humour, co-working, carpentry, bakery, bike friendly, intercultural, intergenerational

☺

MagnyÉthique is a group of people who started their adventure in 2017/18. In May 2019 our community was actually founded and in June we bought a nice place in the Beaujolais Vert region. We want to live together in a respectful attitude and share 8 ha of land partly for permaculture and some livestock, run a café, welcome helping hands and other guests, organise trainings and events and live our dreams and passions. There will be three different areas: a private residential area, a shared area (meeting, studio, music, meditation rooms ...) and a public area (co-working, seminars, camping, guest rooms...).
Currently (Summer 2019) the group consists of four to five households, and we are planning to become ten members. Additionally, there will be guest rooms and camping/caravan places for temporary stays. Within the first months, there will be a lot of renovation work, set-up and gardening to do. We appreciate support of all kinds to manage this and all other daily work.
We will propose open days and present ourselves, our work and our project. We are open to international friend- and partnerships to exchange experience, support each other and to consolidate our openness towards the world.

- Fabrique des Colibris
- ☺ agree with our behavioural charter and share our ideas; some financial resources appreciated

Manoir de la Beaume

C3

Manoir de la Beaume | 03160 Couzon
+33-(0)680 881 711
labeaume@wanadoo.fr
labeaume.wixsite.com/monsite

2013 | 35 ha (near Moulins) | 60 % | ♀6 ♂1 1 | 8

f | 6

all together (consensus)
property of one member

ecological, self-managed, non-violent, animal protection, environmental protection, self-sufficiency, permaculture, organic gardening, organic farming, animal friendly husbandry, alternative energy, ecological construction, humour

La Beaume is a place to live in simplicity, to find meaning in one's actions, to respect the living, to live according to the rhythm of the seasons... it is the desire to be in the right place, in harmony with nature and with others, in a spirit of sharing and joyful sobriety.
La Beaume is an oasis of life, which tends towards autonomy and resilience. It is also a natural oasis, which protects biodiversity. La Beaume is also a welcoming place: Whether you come as a friend, visitor, host, trainee, WWOOFer, new resident... You are welcome!
Our motivation: To preserve this magnificent place which offers endless possibilities and to create a true shared living space, in respectful and ecological connection with nature, in search of autonomy and harmony, a place that preserves the privacy and sovereignty of its inhabitants while collectively creating another living together and a place full of life, hospitality and openness with permanent and passing residents.

MasCobado

D5

615 avenue de la réglisse | 34070 Montpellier
contact@mascobado.org
mascobado.org

2016 | 2440 m² | ♀19 ♂10 15 | 44

f

all inhabitants together (consent)
individual financing

rented & individual property

sharing, caring, participation, conviviality

very diverse, different occupations

ecological, self-managed, non-violent, environmental protection, organic gardening, ecological construction

MasCobado is a participative co-housing project in Montpellier (South of France) with 23 housing units. We have co-designed our residence and apartments with the architects and we manage our living space collectively. The buildings have been designed to have the least possible ecological impact, the lowest possible energy consumption and the best comfort in both summer and winter.
We take our decisions together in monthly plenary meetings. We have working groups that deal with different topics and propose resolutions for the plenaries. All decisions are made with the principles of consent.
We share many common spaces (multi-purpose room, guest rooms, laundry rooms, DIY workshop, gardens).
Photo on page 13

neighbourhood committee, local and national participatory housing network

Oasis de Belledonne

D4

Le Coudray | 38570 Hurtières
+33-(0)664 094 325
oasislecoudray@gmail.com
www.mariedenoix.fr/grainedoasis.html

2016 | 5000 m² | 50 % (hopefully soon) | ♀1 ♂2 | 3

f, e, sp, i | 1

group decision (consensus) | limited common financing

property of 2 people

personal development

ecological, self-managed, feminist, non-violent, meditation, yoga, organic diet, seminar-house, environmental protection, permaculture, organic agriculture, organic gardening, ecological construction, construction, accessible to the disabled, dance

The core of Oasis de Belledonne is a place of ecological, intergenerational and collaborative life, a place of welcome that is integrated into local life and promotes personal and collective development, artistic expression, experimentation and permaculture.
Le Coudray is a property consisting of several plots with a total area of 6000 m², including a 300 m² early 20th century building on 4 levels. It is located in Hurtières at 850 m altitude in the Belledonne massif, 40 km from Grenoble.
The aim is to create a living space for 3 to 6 households within the next 2 or 3 years, an activity space of 100 m² and a capacity for 15 trainees, vegetable gardens, permaculture plantations... It is planned to build the existing building (attic), to extend the existing building (greenhouses...) and to erect further buildings on the 1000 m² plot.

→See also description of network "**Colibris**"
Les Habiles, Twiza

Oasis de Lentiourel

C4

Lentiourel | 12400 Saint-Affrique
+33-(0)565 492 114
lentiourel@orange.fr
lentiourel.com

2011 | 35 ha (near Saint-Affrique) | 30 % | ♀2 ♂2 0 | 4

f, e, i | 2

3 associations (consensus or systemic agreement) | shared economy

open spirituality

permaculture, new architectures, music, pranism, meditation

ecological, self-managed, non-violent, meditation, organic diet, vegetarian diet, vegan diet, pedagogical, seminar-house, environmental protection, permaculture, organic agriculture, organic gardening, forestry, ecological construction, humour, pranism, therapy, zomes

This place of learning and experience, created in 2011 in South-Aveyron on the "Rougier lands" of an 11th century farm, encompasses 35 ha of diverse land, mostly cultivatable with 5 springs and a river. The farm has 560 m² of rooms and common areas, as well as 1000 m² of outbuildings. The acquisition of this magnificent estate was made possible by solidarity loans from both residents and friends of the group.
The permaculture and artistic project "Oasis of Lentiourel" consists in the construction of a

space for fulfilling relationships, encompassing care of the earth, the world and oneself:

- cultivate the joy of living in a happy sobriety and an open mind, in a quest for creative autonomy and sharing
- creating a collective place of alternative living respectful of everyone's needs.

The current group is now reduced to a few people while the capacity of reception of Oasis, in the medium term and as it develops, could be several dozen people. Today the community at the Oasis of Lentiourel continues its activities: home, internships, trainings, participative building sites ... while looking for a group, families or single people to join us on this very beautiful old big farm built on an exceptional place.
The buildings have space for 12 bedrooms, 2 dormitories, common rooms, bread oven, outbuildings, areas for light habitats, and are supplied with spring water. Several options are possible to revitalize the project: sale, structural changes, partitioning of buildings into different flats... In the meantime, the Oasis still welcomes new tenants or residents, as well as travellers.

→See also description of network "**Colibris**"
Intonation juste (music)
financial resources

Oasis La Source Argelouse

B4

2838 | 40430 Argelouse
contact@eco-hameau-argelouse.org www.eco-hameau-argelouse.org

2015 17 ha (near Bordeaux) 6

f, e all
all who are concerned (consensus) shared economy
property of association
agriculture
ecological, environmental protection, permaculture, organic agriculture, organic gardening, organic farming, forestry, alternative energy, new technologies, ecological construction, humour

The intergenerational eco-hamlet "Oasis La Source Argelouse" respects ecological and human values and wants to be a reproducible social and economic model. From the ethics of the "Mouvement Colibris" and the manifesto for "Oasis en tous Lieux" (published by Pierre Rabhi), it promises a GOOD LIFE TOGETHER in "happy sobriety" ("sobriété heureuse", after Pierre Rabhi). Currently, the site has a total area of 17 hectares in rural areas, covers 6500 m^2 of greenhouses, 7 hectares of acres and orchards, about 9 hectares of forest, a house of 520 m^2 and a well.
In order to leave behind the concept of individual ownership, the association is the sole owner of the place, ground and buildings, which he would lease to the residents. Apart from seniors and children, every inhabitant earns his living locally, at farm activities or in other contexts. Social relationships are based on attention for others, on mutual help between generations and places people and nature at the center of activities.

ideological & financial

Ramounat Peter Theile & Vincent Derenemesnil

C5

Haus Ramounat | 09320 Massat (Arige) ☎+33-(0)688 531 442
timputzke@web.de

1982 80 ha (near Saint-Girons) 95 % ♀2 ♂5 7

f, e, g

ecological, self-managed, non-violent, organic diet, self-sufficiency, organic gardening, organic farming, animal friendly husbandry, ecological construction

☺

The Ramounat house in the Pyrenees was renovated and build from a ruin by Peter Theile and his family 37 years ago. By now, there are two habitable houses, a large guest house, a workshop with a forge and leatherworking facilities, a small goat shed and a large tipi. We produce cheese daily from the milk of 80 goats that are herded at an elevation of 1750m every day and brought back every evening. All construction and agricultural work is done without machines and we work a lot with horse power. All work is done together, volunteers are welcome by prior arrangement, many find us via the WWOOF organisation. The stream that runs across the property offers drinking water, which is also used for washing. We cook on an open fire. An important part of our community work is social work. Groups of juvenile delinquents come to Ramounat with supervisors. Ramounat can offer young people in difficult situations a place to stay. Ramounat is not barrier-free, but nonetheless people with physical disabilities are welcome, as long as they are independent. The community is generally interested in people who can imagine staying long-term, as the necessary capacity and opportunities are in place. Whether the person fits into the community is decided on an individual basis. Ramounat is 95% self-sufficient, making a lot by hand and grows without state subsidy. To reach the nearest paved road you need to walk 20 minutes on a footpath. Many musicians come our way and long jam sessions in the evening, or even during breaks, are not uncommon.

→See also description of network "**WWOOF**"

☆ Solenvie

D3

Solenvie 1, Rue du Prieuré | 39100 Jouhe
www.solenvie-oasis.org

2017 (near Dijon) 6

f, (e)

owned by own SCIA (co-ooperative)

human and ecological values of the Colibris association, but no obligations

shared meals, construction and renovation work, gardening, organising (socializing) events, workshops and so on...

☺

Solenvie is a community housing project located in the heart of Jouhe, a small village in the east of France, half way between Besançon and Dijon. Close by you will find the eurovelo (=long distance cycle route) n°6 connecting the Atlantic Ocean and the Black Sea. Jouhe is surrounded by two forested mountainous regions, le Mont Roland and La Serre. This small village is also crossed by the pilgrim route to Santiago de Compostela.
The Solenvie project started in 2017 and is currently involving 6 people that came together

to share the fundamental values of the Colibris association created by Pierre Rhabi, a father of the agroecology in France (www.colibris-lemouvement.org). Solenvie is about creating a new way of community living and where we get our inspiration from ecology, permaculture, energetic sobriety and autonomy, resilience and personal development, social mixing and laicism.
We can host people temporarily as well as permanently if you are interested in joining our community. We offer occasional activities and workshops (permaculture, natural paint making...) and our doors are always open, so do not hesitate to come to say hello, we got a wide range of tools to make and create as well as a wide range of hammocks, deckchairs and loungers!

→See also description of network "**Colibris**"
local networks
☺ sign Solenvie-charter (our values and ethic policy), buy co-op shares

Sunflower Cohousing SCI du Tournesol B3

SCI du Tournesol, Lieu-Dit La Loge | 16350 Vieux-Ruffec ☎ +44-(0)7908-782 753
sunflowercohousing@gmail.com sunflowercohousing.org.uk

2013 3.5 ha (near Confolens) 30-40 % ♀2 ♂2 4
e, f, sp 0 4
all (consensus)
equal investment or rented (if available) + limited common financing
individual private dwellings
no
building new accommodation, tending animals, vegetable garden
ecological, self-managed, non-violent, organic diet, environmental protection, self-sufficiency, permaculture, organic agriculture, organic gardening, animal friendly husbandry, alternative energy, new technologies, ecological construction, construction, humour
☺

SCI du Tournesol (Sunflower Co-housing) comprises a group of 4 companions (to date) who have purchased a property (house+barns+land) in France near Confolens, and have started to convert one of the barns into two further self-contained terraced dwellings, with further similar dwellings planned as new members join. The intention is that new members will be given an equal shareholding, with equal rights, matching the investment of the founder members, as they join.
We welcome all nationalities.
By growing as much of our own food as possible, and through water and energy conservation methods, we hope to make full use of what nature provides, lowering our own costs, but without negatively impacting the environment.
The life balance we are looking for can only be achieved by the community finding its own balance between the privacy of individuals and households on the one hand, and the solidarity of community life on the other. We try to achieve a balance between hard work and play and to achieve a healthy and mutually supportive lifestyle - we believe this is best achieved through the fullest possible integration into French society, and our local community in particular.

☺ yes - please contact us for details *photo on page 6*

Terra Arte Association Participatif

B5

36 chemin de Loustaounaou | 64100 BAYONNE ☎+33-(0)682 935 961
www.facebook.com/terraarte.seque/

2017 (near Biarritz, Pamplona)

f, sp, e
all of us (vote) individual financing
ecological and sharing
ecological, self-managed, non-violent, permaculture, organic gardening, alternative energy, ecological construction, accessible to the disabled, non-smoking, shared garden

☺ ( )

Terra Arte comprises the residential area and the inhabitants of the participating dwellings of the eco-quarter "Le Séqué" in Bayonne. We have group meetings and once a month a daily group work. We also meet once a week for yoga class, often eat together and organise events for the children.

☺ to live there

Terra Flor

C5

Passage du Fauga | 31220 Saint-Michel ☎+33-(0)768 334 261
projet.terraflor@gmail.com terra-flor.org

2018 (near Cazeres) ♀2 ♂2 1 5

f, e, sp, Portuguese 4
all who are concerned (consensus) some common financing
property of all
ecological, self-managed, permaculture, organic gardening, ecological construction, construction, no cell phones/wireless

Terra Flor is the home of two families engaged in a nature-friendly and friendly-nature way of life. Our houses are made from natural and recycled resources with low ecological footprint. Total energy, food and water self-sufficiency are long-term objectives of all.
Since the summer of 2018, many volunteers have helped on various building projects: a Kerterre house ("Hobbit" style), a tiny house, a passive house, permaculture gardening, water recovery system and many others. We want to share our experiences on permaculture and eco-construction techniques in face of current environmental and social issues.

→See also description of network **"Colibris"**

Tifolk

A2

03, Croas Ar Coq | 22310 Tremel ☎+33-(0)611 176 712
contact@tifolk-habitat-groupe.com www.tifolk-habitat-groupe.com

2015 1 ha (near Lannion) 60 % ♀2 ♂3 5

f, e (a little), esperanto (planned) 3

permanent residents, at weekly meetings (majority decision)
Letting of land and/or apartments
property of a member made available to the community for 10 years.
dances and ecology
ecology
ecological, self-managed, non-violent, adult education, pedagogical, animal protection, environmental protection, self-sufficiency, permaculture, organic gardening, organic farming, animal friendly husbandry, alternative energy, ecological construction, humour, traditional dances

We are located in Trémel Côtes d'Armor, in the "middle of nowhere". Our nearest neighbors are a herd of cows! But we have a lot of visitors, especially in the summer: Volunteers, students, scouts, friends, etc.
Five years ago, I (Rémi) decided to live in the company as I conceive it, instead of constantly criticizing. So I left my previous job and bought the farm where I currently reside. I set up several community projects since, like the establishment of the Housing Group Tifolk, a food garden, a shared garden and a self-service garden, a new beverage based on ginger and beekeeping. I am professor of traditional couple dances: Waltz, Mazurka, Scottish, Polka. So dances (folk) are present all around us.
But now, being the only one permanently working on all these projects, I get a little out of breath. The idea is to share these ideas and space with motivated people... Why not you? The daily routine requires 3-4 hours/day to help. This can be gardening, bricklaying, joinery, renovation, kitchen, weeding, storage, dishes, etc. We are close to the sea (10 minutes by car) and I can introduce you to wild and savage places. At low tide and good weather, we stop all work and pick cockles or mussels at the beach. We provide WiFi, two Hang drums, a yurt, bonfires etc. Four retired caravans just wait for you to be repaired! I only speak French, so the exchanges are somewhat limited, but it promises many laughter!

Réseau Folk, Troc de plantes
☺ accept the rules of life of the place

Village Emmaüs Lescar-Pau Une alternative

B5

Chemin Salié | 64230 Lescar
☎ +33-(0)559 811 782
contact@emmaus-lescar-pau.org
www.emmaus-lescar-pau.com

(near Pau, Lourdes)

Our community is apolitical, secular and does not receive any direct or indirect subsidies. Our income only comes from our work, which is the recovery and recycling of objects that people no longer want: 250 to 600 vehicles per day come to deposit all kinds of objects, components, raw materials, green waste, rubbers... This represents between 300 and 400 m^3 of goods recovered every day!
All in all, 110 companions live in individual dwellings. In addition, dozens of volunteers stay there all year round. Village Emmaüs was created in 1982 and today, it covers 11 hectares. In addition to our individual housing, several workshops, auction rooms, recycling and waste disposal facilities offer a wide variety of communal spaces and activities.

→See also description of network "**Emmaus**"

Germany

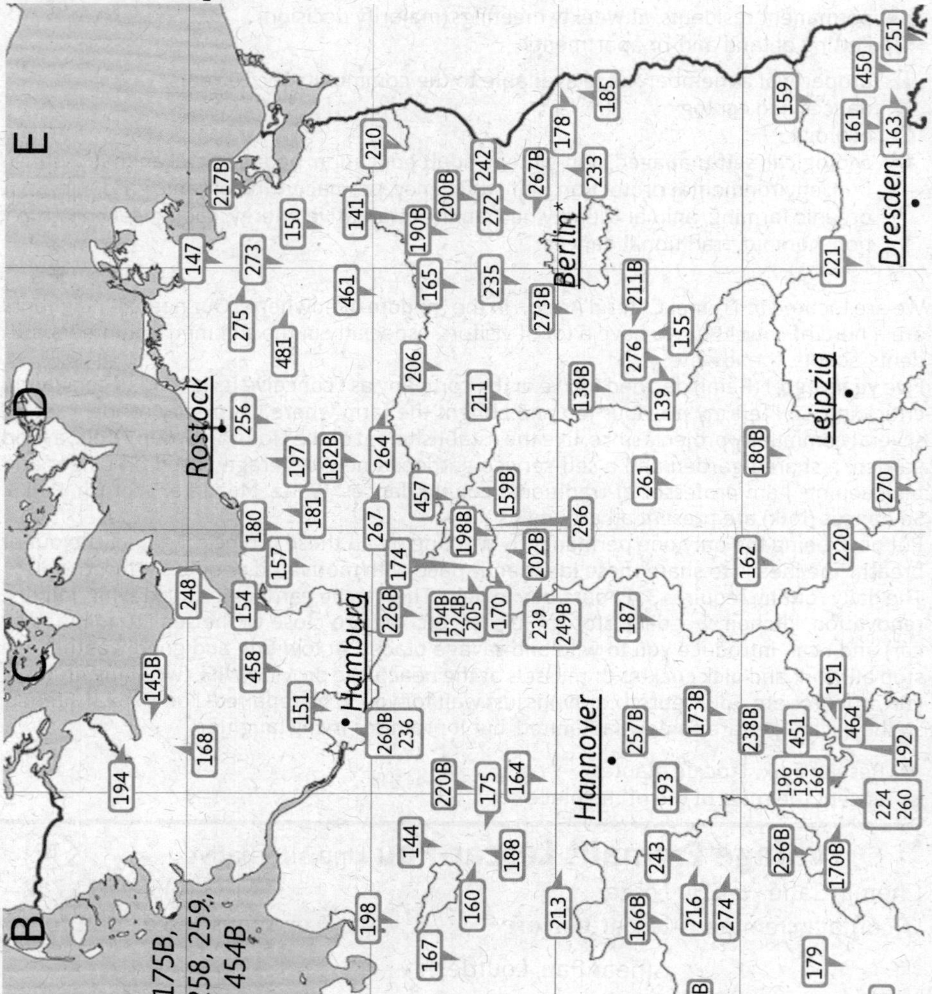

A
B
C
D
E
1
2
3
4
5
Rostock
Hamburg
Berlin*
Hannover
Leipzig
Dresden
Dortmund
*Berlin: 140, 145, 146, 175B, 227B, 243B, 254, 252, 258, 259, 262B, 265, 271B, 451B, 454B
217B
210
185
251
450
159
161
163
147
273
150
141
190B
200B
242
272
267B
178
233
461
165
235
275
481
206
273B
211B
155
276
221
138B
139
256
197
182B
264
211
457
159B
261
180B
270
180
181
267
198B
266
157
174
202B
248
154
226B
194B
224B
205
170
239
249B
187
162
220
145B
458
151
260B
236
173B
238B
451
191
464
192
168
194
220B
175
164
257B
193
196
195
166
224
260
144
188
243
236B
170B
198
160
213
166B
216
274
167
179
208B
232
186B
147B
137
263
237
459B
269
229
153

There's no denying that eurotopia is a German initiative: nowhere do we know more communities than in Germany. This surely has to do with the popularity of this directory in Germany (we publish a German edition, after all) - that feeds back to us with lots of community infos. It also seems that the conditions are favourable for communities in Germany, though: alternative lifestyles are widely acknowledged, there are few political or legislative obstacles and there's a variety of legal entities suitable for collaborative groups. This doesn't make intentional communities perceivable players in the German mainstream, but they do make a strong apperance in this directory: 195 existing communities and 27 planned ones, plus 11 additional Camphill communities and 70 (!) addresses from the last edition. Oh, plus one dummy. Exactly: one German description is too silly to be true. There are two strong community networks with GEN Deutschland (page 522f) and KommuJa (page 530f).

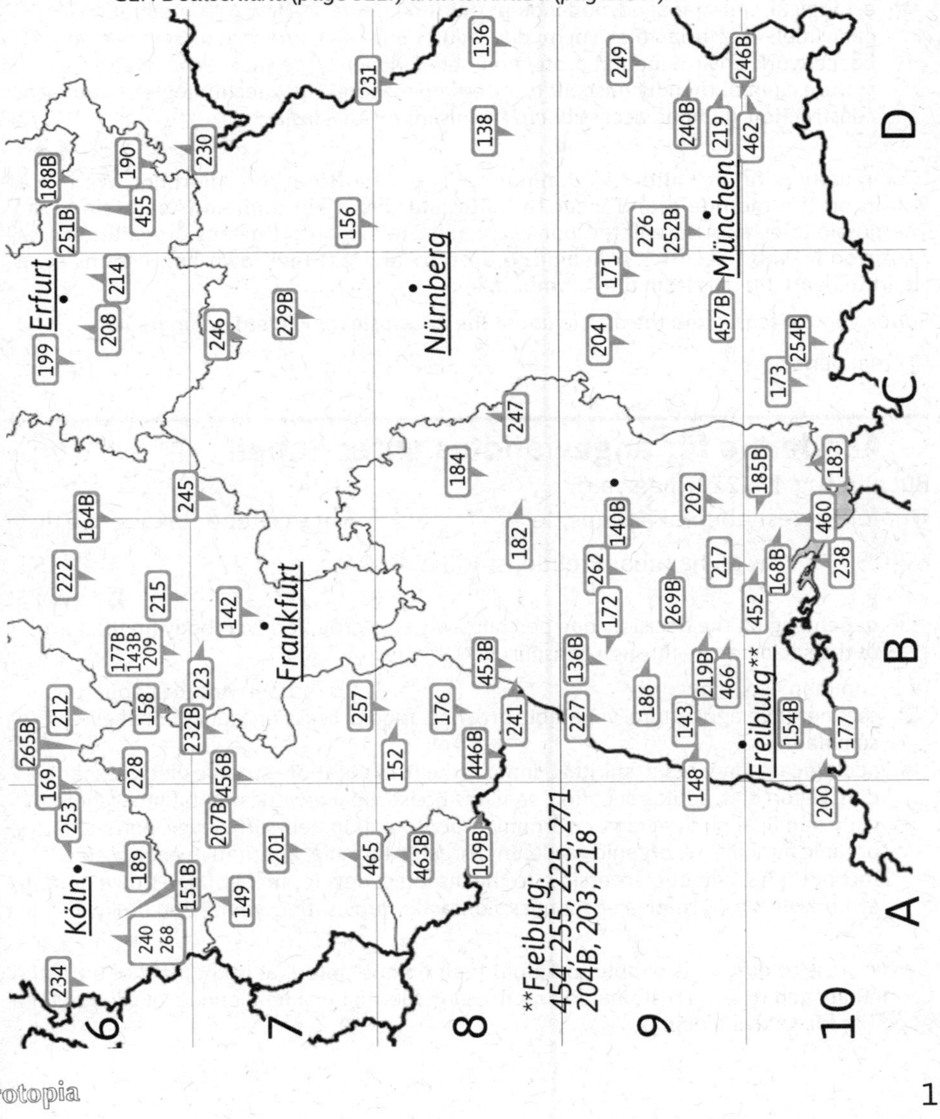

Ahornhof e. V.

D8

Ahornwies 3 | 94379 Sankt Englmar
+49-(0)9965-801 02 14
am-ahornhof@t-online.de
ahornhof.org

2011 5.13 ha (near Straubing, Regensburg) 25 % ♀3 ♂2 1 6

g, (e)

the community (talking stick)

our property

in harmony with nature, we are all divine

permaculture, gastronomy

ecological, self-managed, non-violent, spiritual, esoteric/New Age, meditation, organic diet, vegetarian diet, vegan diet, adult education, free school, seminar-house, peace-work, environmental protection, self-sufficiency, permaculture, organic agriculture, animal friendly husbandry, alternative energy, new technologies, ecological construction, humour, accessible to the disabled, Anastasia

☺

"Community is future, future is community" is our motto at the Ahornhof. We live at 900 m on the mountain and want to build and live a kin domain (Anastasia style). The possibilities exist to enlarge our estates, for this financial means are still necessary. Anybody who likes to is welcome to drop by and get their own impressions of us. Heart to heart, Marcus from the Ahornhof!

Editor's note: Please note the article about the Anastasia movement on pages 49f.

☺ trial period

Akademie für angewandtes gutes Leben

B9

Buchenweg 4 | 72297 Seewald
info@gutes-leben-akademie.de
gutes-leben-akademie.de

2015 65 ha (near Freudenstadt) 50 % ♀7 ♂8 0 15

g, e

5 7

depending on the decision: one person, a working group or everybody (in the terms of the sociocracy) - through consensus

common economy

property of the cooperative

regenerative agriculture, vegetable growing, food processing, educational events, seminars

ecological, non-violent, spiritual, inner growth, meditation, organic diet, vegetarian diet, vegan diet, adult education, seminar-house, peace-work, mentally/psychologically handicapped persons, environmental protection, self-sufficiency, permaculture, organic agriculture, organic gardening, organic farming, alternative energy, new technologies, humour, accessible to the disabled, development of one's own potential, bodywork, alternative economics, contact improvisation, young and funky!

☺

Our vision is to empower people to unfold their unique potential in the field of a trustful community and to learn sustainable practices in order to become pioneers of change for a good life for all living beings.

From 2019 on, together with the community of Sonnenwald, we will move into a former facility for people with disabilities in the idyllic north of the Black Forest. The area offers seminar houses, workshops, a canteen kitchen, a gym, a café-house, a farm and other special features. On the subjects of human beings, culture, ecology, technology and systems, we are building exploratory, productive and public welfare-oriented companies that strengthen sustainable economic cycles and make the region more sustainable. We start with the restructuring of the existing farm to a farm of regenerative agriculture, including bio-intensive vegetable production.
In September 2019, the "Wandelreise" (journey of change) starts for the first time and offers orientation for people of all ages. The programme provides experience and inspiration around socio-ecological change with a focus on personal development, community building and dealing with the challenges and opportunities of our time.

→See also description of network **"GEN Deutschland"**
- CSA network, Verband freier Bildungsalternativen
- shared values; interested in personal development, global challenges and political action

Akkoallee 40 ehem. Delta GbR

A5

Akkoallee 40 | 45659 Recklinghausen ☎+49-(0)2361-261 95

1982 (near Gelsenkirchen, Dortmund)

The 580 m² property in Recklinghausen (three-storey house with an extension, four stables, a yard and a garden) was acquired in 1982 by us, the couple Preukschat. In the same year we moved into the house with six underage children (three of our own, three foster children). In connection with the purchase of the house it was our concern to create open living forms from the existing rooms of the building. Together with the Verein für Gemeinschaftsbildendes Wohnen e.V. (= association for communal living) in Recklinghausen we founded the company Delta GbR. In the statutes we set our objectives as follows: "Guarantee the necessary living space to children and adults threatened by social uprooting, to give them a place of belonging". Between 1982 and 2003, we lived with people who needed support at different times, in the short or long term. We lived with foster children, people with addiction problems, people at risk of suicide, homeless foreigners as well as schoolchildren and students. In 2002 we initiated the donation of our property to the TRIAS Foundation. With the execution of these measures we wanted to set an example to take the acquired land out of speculative business once and for all. We then left the association and even dissolved the company Delta GbR.
In 2008 we rebuilt the three floors of the house for families. We currently live in the house with two other families. Living together takes place on a partnership level. All residents share common cellars, sheds, courtyards and the garden. In the recent years we have prepared the annex to the house so that one or two people can stay there for a short time - up to three months.

- trias Foundation

Allmeind - Alle miteinander Multigenerational Housing D8

Kirchfeldallee 59 | 93055 Regensburg ☎+49-(0)941-541 45
utahildt@web.de www.allmeind.de

2009 3000 m² ♀34 ♂12 13 59

g, Polish, i, sp, Portuguese, Russian
resident assembly, resident council (majority, 2/3 majority for delicate issues)
some common financing
property of the Catholic Housing and Settlement Association
garden-group, library-group, gathering of residents, current tasks
alternative energy, organic gardening, accessible to the disabled

☺

The Allmeind is a housing project with rental apartments. The residents have individual leases with the owner. Residents are members of resident association Allmeind e.V. and organize their social coexistence (for example, mutual support - children - illness - skills ...). All activities for neighbours or for the community are voluntary. Higher-level concerns are discussed in the monthly residents' meetings. Gardening, staircase cleaning, cleaning of the common room etc. are taken over by the community, which reduces costs and contributes to a sense of community. The Residents' Association represents the interests of the residents towards the homeowner and is authorized to designate subsequent tenants.

☺ available apartments, supporting the general principles

Aloha am See Zentrum für Lebensfreude D4

Mühlenstraße 1a | 14789 Wusterwitz
info@aloha-am-see.de www.aloha-am-see.de

2010 1.7 ha (near Brandenburg a.d. Havel) ♀5 ♂3 1 9

g, e 3 everybody
the founding circle, the community (consensus)
individual and common financing individual owners
spiritual, undogmatic
kitchen, gardening, building, artisan work, handicrafts
ecological, self-managed, hierarchical, leader/guru, spiritual, inner growth, meditation, yoga, organic & vegetarian diet, adult education, seminar-house, environmental protection, organic gardening, humour, no cell phones/wireless, non-smoking, seminar centre

☺

The Aloha community was founded in 2010 in Berlin/Potsdam. In 2016 we moved to an idyllic place in nature near Brandenburg an der Havel. Here we have established the spiritual community "Aloha am See - Zentrum für Lebensfreude" ("Centre for joie de vivre") and the non-profit association "Aloha Education and Health Centre e.V.".
The highest goal of the Aloha community is to increase peace and love between all beings. We promote mindfulness and appreciation, connecting communication, self-knowledge and awareness. Against this background, we want to live as a community in a way that allows for increased healing and heart opening through our own personal development.
Essential principles of the Aloha community are awareness and mindfulness in dealing with yourself and others. The basis of an adult and peaceful coexistence is the assumption

of responsibility for one's own thoughts, feelings, and actions, and the regular cleaning of negativities through shadow work and meditation practice. We meditate together for at least 1 hour every day. We practice Osho meditations, Sufi meditations, Vipassana, Chakra meditation and heart meditation, among others. The Aloha community is meant to be a place of joy, silence, singing and dancing.
We are looking for new members who want to evolve personally and spiritually. We assume an interest in common spiritual practice as well as a degree of self-awareness. We are looking for people who want to engage with love and energy in the construction of this spiritual project.

☺ accordance with our vision

☆ Alte Mühle Gömnigk

D4

Dorfstraße 1 | 14822 Brück ☎+49-(0)33844-149 970
altemuehlegoemnigk@riseup.net alte-muehle.org

2014 | 18.5 ha (near Bad Belzig) | 10 % | ♀6 ♂11 10 | 27

g, e, f, Dutch — differs from 3 to 10 persons
everyone (consensus) — individual financing + limited common financing
property of all/property of a company we founded and control together with more than 100 other similar groups ("Mietshäuser-Syndikat")
anarchistic
ecological building/restoration, gardening, natural forest management, permaculture, summer group housing
ecological, self-managed, anarchistic, left-wing political understanding, feminist, chaotic, vegetarian diet, vegan diet, adult education, seminar-house, environmental protection, self-sufficiency, permaculture, organic agriculture, organic gardening, forestry, alternative energy, ecological construction, construction

☺

We are about 20 bigger and 10 smaller persons as well as chickens, two dogs and three tomcats. Most of us already live here or will be living here after finishing our (first) expansion. We are united by our openness for diversity: From 0 to 47 years old, living vegan or eating omnivore, born in Europe, Africa or South America, practicing polyamory or living monogamous and married, spiritually focused, active in Christian community or radically political - all with one aim: a world in which all people can live freely according to their needs.
We prefer uncommercial working and acting and we decide consensually and on the basis of a respectful contact to each other. The size of our terrain (18 hectares of farmland, forest, forest garden and meadow) allows to have very different initiatives on one ground - also for those who do not live here but participate as well.
For electricity, we have a small water power station and a solar system. We have a small "guest mansion" with kitchen for guest group housing, another room with bar and a hall with gallery - usable on the basis of donations. We established two vegetable and herbage gardens, are transforming an old meadow orchard into a permaculture forest garden and share knowledge as free as possible, not in overpriced trainings or books you have to pay for. We opened a free shop and offer our wood and our metal workshop for everyone to make use of. Space for living is more than scarce, but sometimes a place is free as well.

☺ free space to live, willingness to contribute to the tasks and to decision-making by consensus and autonomy

Alwig - allein wohnen in Gemeinschaft

E4

Falkstraße 25 | 12053 Berlin
+49-(0)30-679 526 14
veronika@masdelchusco.com, risteberg@gmx.net
#6028 at eurotopia.directory/booklinks

2008 several dwellings in one large building ♀9 ♂3 12

g

plenum consensus (discussion and voting) all of us are pensioners

rented flats

a small community garden

ecological, self-managed, left-wing political understanding, feminist, non-violent, Christian, ecumenical, meditation, organic diet, vegetarian diet, adult education, humour, non-smoking

We are a group of people over seventy who live in individual rental apartments within a large block of flats in Berlin-Neukölln. As a group we meet weekly in plenary and do something together 1-2 times a month. In addition, we celebrate our birthdays together and travel once a year. If someone gets sick, we take care of him/her, but we do not provide long-term care. We consider ourselves a tolerant group, politically leaning to the eco-left, some of us are religious, others are not. We all are or were socially, educationally, politically active to some extent. Our intention is to give each other a sense of security and belonging.

between 50 and 70 years of age

Am Sternberg Multigenerational Housing and more...

B9

Stuttgarter Weg 1 | 72532 Gomadingen
+49-(0)7385-96900-0
info@amsternberg.de
amsternberg.de

2018 7 ha (near Reutlingen) ♀ca 12 ♂ca. 12 ca. 10 ca. 35

g, e ca. 15 ca. 20

all who are concerned together (striving for consensus, sociocratic approach)

individual financing and shared financing of some areas (food, construction and renovating, garden)

property of the cooperative

respect for humans, animals and nature

seminar venue, health house, garden and permaculture, ecological planning, building and renovating, organic food store, coffee shop, beekeeping, ecological products

ecological, self-managed, non-violent, spiritual, inner growth, meditation, yoga, organic diet, vegetarian diet, vegan diet, adult education, pedagogical, seminar-house, peace-work, animal protection, environmental protection, self-sufficiency, permaculture, organic gardening, alternative energy, new technologies, ecological construction, humour, creativity, music, singing, dancing, Dances of Universal Peace, seasonal celebrations, nature perception, geomancy, forest bathing, herbal hikes, herbal knowledge, women's circle, men's circle, youth projects, mindfulness, appre-ciation, time-out accompaniment, vision search, integral life practice, Spiral Dynamics, consciousness development, heart opening, emotion-integration, detox on all levels, vital food, raw food, fasting accompaniment, massages, body work, yoga, natural fitness training, apitherapy, 7-star beekeeping, bee air pavilion, nature-oriented construction, mostly WiFi-free terrain, edible village, tiny houses, sauna, indoor swimming pool, canoe, hiking, juniper heaths, winter sports, caves, climatic health resort, biosphere area, pure nature

☺ (✋✋)

In the 7th year of searching, our initiative from the greater Stuttgart area finally found what it was looking for at the beginning of 2018: in the former holiday village at Sternberg in Gomadingen and in the heart of the Swabian Alb biosphere region.
It was love at first sight. Living, working and growing at such a powerful place in the midst of nature and yet within reach of larger towns such as Reutlingen, Tübingen and Stuttgart. Soon more people joined us and in summer 2018 we founded our cooperative in order to tackle the purchase, the ecological renovation of the existing structures and planning of new construction with bundled forces. We are pleased about further active and above all also investing members, who would like to participate in the development of our holistic community project.
Under the motto CONSCIOUS COOPERATION WITH HUMAN AND NATURE we dedicate ourselves on the one hand to the development of our human potential - individually and collectively - and at the same time to the development of meaningful enterprises such as seminar venue and guest accommodation, health houses, construction, organic shop and culture-coffee-shop. We are also committed to permaculture and self-sufficiency.
The topic of holistic health is a focal point of our activities, research and offers. We would like to practice a healthier way of dealing with ourselves (self-care), with each other in the project (ability to live together), with our immediate surroundings (local networking), with nature and with the wider environment (peace work). We consider open and respectful communication, conscious conflict resolution, joint decision-making and mutual support to be of great importance. Our vision is to build a model for a sustainable way of life for the sustainable well-being of all.

- Artabana, good exchange circle, CSA
- ☺ approximation process, buy shares of cooperative, vegetarian/vegan diet

Amo e.V. Eigensinnige Gemeinschaft in der Uckermark

E3

Klaushagen 10 | 17268 Boitzenburger Land
claudeo@web.de   www.amo-klaushagen.de

(near Templin)   ♀1 ♂1-3 2 ca. 5

- g, e, Mandinka
- all (consensus)
- currently establishing a shared hosehold budget, the rest individual
- property of us
- tolerance, equality, humanity, solidatity
- living, gardening, construction
- ecological, self-managed, left-wing political understanding, feminist, non-violent, inner growth, yoga, organic diet, vegetarian diet, vegan diet, permaculture, organic gardening, ecological construction, construction, humour, dance, music

☺ (✋✋)

We are 5 people at the moment and we are open to growth. People with heart, openness, humour and the desire to be part of the group are welcome. We are 100 km from Berlin. There is not too much of living space left, but there is a house to rebuild (if money can be raised). One more trailer might be possible. WWOOFers are welcome. At the moment we mainly live here and don't have too many projects. Everybody has their own job besides. Usually we have one big party and a volunteering week during summer.

→See also description of network **"WWOOF"**
☺ trial period

antagon Theateraktion Kulturverein protagon e.V. B7

Orber Straße 57 | 60386 Frankfurt/Main ☎+49-(0)69-417 041, -941 477 19
office@protagon.net
www.antagon.de, www.protagon.net, www.sommerwerft.de

1990 ♀10 ♂10 5 25

g, e, Portuguese

owner (consensus) leased

office work, making costumes, fixing vehicles, illustration/graphics, welding, wood work

ecological, self-managed, left-wing political understanding, non-violent, peace-work, alternative energy, theater

"antagon theaterAKTion" is an independent Frankfurt performance theatre. Our theatre work is created in a process that takes place directly in a project group. Life and work are closely linked within the projects. We have weekly plenary sessions with consensus decisions and work together both in the different phases of production, tour, festivals (Sommerwerft.de, international female theater-festival IFTF.de and Winterwerft.de), laboratories and workshop periods, etc., as well as in the care, expansion and organisation of living spaces, work, storage and event areas.
The area has 6000 m² with the possibility for artist residences for groups up to 12 persons (more on request), outdoor areas with trucks, trailers and caravans for living, garden and performance areas, various buildings and relatively much greenery. There are living areas with common social rooms and a self-managed canteen: all cook in rotation for all, common food daily.

Recent comment: It looks just like it used to look in Berlin... Truly a cultural free space, and this in the middle of Frankfurt, the financial metropolis.
antagon works with performers from all over the world and with various theatre forms and methods.

"Power arises where people come together with their strengths and weaknesses. This is a long process that differs from the shortness of breath in this world and cannot be acquired with money. In organic theatre work, the human being must be at the centre if an honest and direct relationship to life is to be at stake."

Theaterallianz, Meridians

☺ interview, interest in full time theater work,
maybe participation in theater lab "hambre y vida"

Arche Haus Elztal Arche Haus Elztal e.V.

B9

Merklinstr. 20/Hindenburgstr. 2 | 79183 Waldkirch

+49-(0)7681-473 45 07/-494 73 76

www.diearche-freiburg.de/arche-haus-elztal/

2013 > 1000 m² (near Freiburg) ♀10 ♂4 1 15

g

community (consensus whenever possible, otherwise majority)

partially shared finances (housing association account) rented

administration, plants/garden, communal room

ecological, self-managed, non-violent, physically handicapped persons, organic gardening, accessible to the disabled, non-smoking, ethical, mindfulness, respect, tolerance, honesty

Mutual assistance, especially in emergency situations. Self-responsible, creative, ecological commitment to the roommates, homes and gardens.
12 own rental apartments, most of them wheelchair accessible (36-113 m²). A shared flat (with terrace) as communal space for meetings, celebrations etc. - and for overnight stays. It is also rented to initiatives (economic basis) or for free to voluntary initiatives. In addition, there are two more terraces, three dinky gardens and (basement-)storage-room for communal use (bicycles, laundry, drying).

ARNIA e.V.

B6

Weingartenstr. 1 | 35619 Braunfels-Neukirchen +49-(0)6445-5257

f.zinke@solawi-sonnenhof.de

gemeinschaftneukirchen.wordpress.com/about/

2015 30 ha (near Wetzlar) 10 % ♀3 ♂6 3 12

g, e 5

all (in plenary) individual financing, bidder's round for larger projects

still privately owned

none

organic agriculture and organic gardening

ecological, self-managed, non-violent, spiritual, organic diet, vegetarian diet, vegan diet, adult education

We are realizing a model that is based on the role models Heckenbeck, Hirschlanden and Flegessen, but also includes influences from other communities such as Tempelhof and Siebenlinden.
By now, we own three old half-timbered courtyards in the old center of Neukirchen (the fourth one is to be bought pretty soon). Heart of our work is our CSA (community supported agriculture), which has been active for five years and is currently supplying 90 shareholders. We have equipped the old bakehouse so that we can bake professionally in the future. The construction of a farm-/nature-kindergarten is in the final stage and to be opened in the summer of 2019. A worksite based on the principles of permaculture to construct tiny houses, construction trailers, yurts, and tippis is in an early stage. It is being developed on an approx. three hectare ground.

When the 80-year-old tenant will give up her business, we will also buy the last hostelry in town. Currently for sale is the village community center and we have loads of ideas to make good use of this house. One of our farms is to be transformed into a commons farm with three separate 2-room apartments, a regional cooking kitchen with juicer and various workshops. chickens and garden are already there.
There is a 'we-process' (Scott Peck style) and a plenary session at least once a month.

☺ written presentation, sponsorship, approach at least six months and 3 times participation in plenum resp. 'WE-process', entry fee.

Bahnhof e.V. Initiative für neues Wohnen

B3

Am Bahnhof 2 | 28870 Ottersberg

briefkasten@bahnhof-ev.org www.bahnhof-ev.org

1990 ca.1000 m² (near Bremen) ♀9 ♂8 0 17

- g, e, f, sp
- all together (consensus)
- some common financing/ individual financing
- liberated from the real estate market
- redemption of land, sisterly living together, enhancing art and culture
- cultural work, maintenance of the building
- self-managed, non-violent, chaotic, inner growth, humour, open space, creative chaos, direct democracy, celebrating parties, art, theatre, change, growing, exploring community life (living together)

☺ ()

In 1990 a group of students of the HKS Ottersberg (art therapy/theatre pedagogy) founded the Bahnhof e.V. - initiative for new living. The organisation bought the old Ottersberg railway station to create a non-profit-based space for residency and for cultural activities which is used by the residents on a self-administered basis. About 20 people live in the building which has 5 living areas. It is important for us to focus on community and that every individual can take part in creating forms of common life. Together we take the responsibility for the maintenance of the building, the interests of the organisation and our living together. The Bahnhof offers possibilities for different cultural events with its rooms/hall, such as theatre performances, concerts, exhibitions, seminars, workshops, etc. In this context we see ourselves as both promoters and artists. We invite all interested people to use our free spaces with us. We have a weekly meeting on Wednesday at 7pm and there is also room for guests and their requests. We welcome everybody and encourage you to get into contact with us!

- Vivenda e.V., SchönWedder
- ☺ introduction at community meeting and one week of probationary living

Basisgemeinde Prenzlauer Berg

E4

Dunckerstr. 14 | 10437 Berlin
☎+49-(0)30-444 07 00
berlin@basisgemeinde.de
www.basisgemeinde.de

1990

♀6 ♂1 4 11

g, e
members (consensus)
all all
full community of goods
rented
ecumenical
urban district work, daycare for children
ecological, non-violent, spiritual, Christian, ecumenical, inner growth, organic diet, pedagogical, peace-work, environmental protection, alternative energy, ecological construction, construction, humour

☺

Our call is to follow Jesus and to live out the gospel in community - as it is described in the sermon of the mount.
Together we experience, that following Jesus in discipleship is a radical and revolutionary power, which can change ourselves and the world.
We don't live isolated and we're definitely not an isle of different or better people. We are involved in different activities in our neighbourhood, we're running a clothing department and a local kindergarten.

→See also description of network "**Church and Peace**"

Diakonisches Werk, Ökumenischer AK
☺ openness for the way of God

Basisgemeinde Wulfshagenerhütten

C1

Zum Wohld 4 | 24214 Tüttendorf
☎+49-(0)4346-36800-0
info@basisgemeinde.de
www.basisgemeinde.de

1973 2.5 ha (near Kiel) ♀14 ♂18 15 47

g, e, f, sp, Russian
all
all members decide together (consensus)
shared economy (shared income and capital)
common property
we are Christians
wood workshop, household, common child-rearing
ecological, self-managed, left-wing political understanding, non-violent, Christian, ecumenical, adult education, peace-work, mentally/psychologically handicapped persons, environmental protection, alternative energy, construction, humour, joint property, pedagogy of motion according to Emmi Pikler and Elfriede Hengstenberg, handicraft, dance, political commitment

The community in Wulfshagenerhütten is a Christian community in the north of Germany. Adults and children, families and single people live together on one property and attempt to share all the spheres of their lives and all their property according to the example of the first Christians. The community members and guests all work inside the community in a workshop for wooden play equipment, in the household and in administration. The community is open for seeking people and invites them into its daily rhythm. All visitors take part in the common work and in the meals, and they are also invited to take part in the prayers.
The community is linked to a growing network of Christian communities, for example to a grassroots community in El Salvador. We experience the sharing of joy and sorrow, work and celebration, and listening to one another as the foundation of peace and justice. We invite visitors to live with us and to share our life. In 1990 a branch community was founded in Berlin (Prenzlauer Berg). Between 1996 and 2010, there was another community in the Ukraine (now ceased).

→See also description of network "**Church & Peace**"
☺ preparedness to be part of the common life and work

Bau- und Wohngenossenschaft Spreefeld Berlin e.G.

E6

Wilhelmine-Gemberg-Weg 10, 12, 14 | 10179 Berlin
buero@spreefeldgenossen.de
spreefeld-berlin.de

2014 | 0 % | 150

g, e | 0

all members of the housing cooperative or the board (vote)

individually financed | housing cooperative

no ideologic principles

construction, accessible to the disabled, housing cooperative

The Spreefeld is a project for communal living and working in downtown Berlin, on the banks of the Spree between the districts Mitte and Kreuzberg. The Spreefeld Berlin e.G. consists of three buildings with a total of approx. 6000 to 8000 m² of living/usable space, predominantly for the residential use of its members, supplemented by rooms for commercial, public and various internal uses. In addition, it is important to preserve the potential of the place in the course of the entire inner-city Spree bank development (interlocking uses, public access to the bank). Since 2014, the coop members have lived in three Spreefeld buildings. These have "passive house" standard and are supplied with heat and partly also with electricity by a heating house with block-type thermal power station and heat pump as well as PV systems on all three roofs. The members meet monthly to make joint decisions where necessary and possible. The cooperative cannot accept new members, no further buildings are planned.

Bauwagengemeinschaft Alt Ungnade - FreirAUm e.V. Alt Ungnade Trailer Community

E2

Kurze Straße 4+9 | 17498 Levenhagen
Freiraum_Kennenlernen@freudenkinder.de
www.honigholz.de/allerlei/wagenleben

2000 | 4 ha (near Greifswald) | ♀9 ♂9 3 21

- g, e, f, sometimes even more — 10
- all inhabitants (common decisions in plenary meetings (consensus or majority))
- some common financing — leased
- construction of trailers, gardening according to permaculture principles, ethical husbandry, enhancing the rural area by cultural events etc., cultural barn, clay stove for bread and pizza for participation, dreaming and experimenting
- ecological, self-managed, left-wing political understanding, chaotic, inner growth, organic diet, environmental protection, organic gardening, humour, cultural events: parties, concerts, exhibitions

☺

Life at the Alt Ungnade trailer community is enticing, exciting, straining, enriching and a lot more. Everyone and everything gets their space to develop and live a pristine life.
Round about twenty people live in self-made trailers on two properties left and right of the street leading through our tiny village. Personal paradises are created in between meadows, orchards, shrubs und wild herbs. Students, self-employed and regular employed people live here together with sheep, geese, chickens and ducks. It is everybody's responsibility to take care of the animals and maintain the property. For those interested in committing more to the community life, we have a huge barn which is currently used for festivities and has lots of maintenance to be done, there is a clay oven used for baking bread and at times pizza, more than enough space to do your own gardening and lots more.
We are working towards finding solutions for legalizing living in our trailers and have had some success already. For example all trailers have got electricity, sewage and LAN connection (internet). The sewage leads to our own reed-sewage-treatment-pool and we use composting toilets aspiring to lead a sustainable lifestyle. We are organized as a charitable association and organize a few educational and cultural events every year. Come and meet us!

- regional network of ecological initiatives is planned
- ☺ own trailer (partially possible to rent), agreement of all inhabitants, participation in the implementation of our development concept

☆ Beginenhof Dortmund Verein Beginen Dortmund e.V.

A5

Gut-Heilstraße 18 | 44145 Dortmund — ☎ +49-(0)231-395 42 74
info@beginenhof-dortmund.de — beginenhof-dortmund.de

2006 | ca. 2700 m² | 2 % | ♀31 ♂1 2 34

- g, e, f, Turkish, Greek — about 15, partly working at home, partly retired general assembly (partly in consensus, partly by majority)

individual financing — rented from private investors/landlady-lords

feminist

gardening

ecological, self-managed, left-wing political understanding, feminist, spiritual, Buddhist, chaotic, inner growth, meditation, yoga, organic diet, vegetarian diet, physically handicapped persons, animal protection, organic gardening, alternative energy, non-smoking, lesbian, women-oriented - like all keywords each applies only to a part of the inhabitants

We are or at least try to be a community oriented women's co-housing project, part of the "modern Beguine movement". We are to be found in the middle of big city Dortmund in the post-industrial Ruhr-area and gather very different women, quite some lesbians, a few women with a migration background or -foreground, very different life stories, very diverse education and training, also very different concerning social and political activism and willingness or options to be an active part of the community and its development. Truly a "social experiment", with a lot of failures but an amazing amount of successes too! And our garden is just beautiful!

→See also description of network "**Beginen**"

Bergfritzenhof Freiamt

B9

79348 Freiamt

info@fritzinnen.de — www.fritzinnen.de/

5 ha (near Freiburg im Breisgau) — <10 % — 14 — 34

g, e, sp — roughly half of us (10 adults)

plenary session (currently consent method)

individual with strong solidarity, e.g. rent through bidding process

owned by the community

running our seminar room (Waldraum e.V.), chores around house & garden.

ecological, self-managed, non-violent, seminar-house, animal protection, environmental protection, organic gardening, alternative energy

☺ (✋✋)

We are a group of adults and children who have been able to realize their strong desire for self-managed living and working space since 2017. We live a constructive, open and friendly life, which is organized in an ecologically meaningful, sustainable and social community. We want to counteract the rural exodus by making rural life feasible and more attractive. Sustainable use of nature and its resources is firmly anchored in our everyday lives and economies. We want to create a place that offers us and our children creative space for development and commitment, that has the potential for professional development - a place that we can live in together.

In the outdoor area of the Bergfritzenhof, a garden will be jointly managed, which is to grow even further in order to provide the residents with even more fresh food. In the long term, we aim to develop a holistic management and utilisation concept for parts of the outdoor area. Our professional backgrounds in various areas such as the environment, tourism, education and coaching are to be incorporated here.

Our 60m² multifunctional room "Waldraum" with its attractive outdoor area gives many

possibilities for use through seminars and can also be booked for vision weekends etc. :-) - See www.waldraum-freiamt.de.
The house project is realized after the model of the Mietshäuser-Syndikat and is part of the same. We share the objectives of the "tenement syndicate" to withdraw real estate from the speculative market in order to secure affordable and self-managed housing in the long term.

🌍 Mietshäuser Syndikat (www.syndikat.org/en)
☺ limited opportunities due to space constraints. Apart from that, time to get to know each other (a few weeks)

Beuerhof Kommunikationszentrum Beuerhof GmbH

A7

Beuerhof 1 | 54579 Üxheim ☎+49-(0)2696-283
info@beuerhof.de www.beuerhof.de

1990 | 12 ha (near Daun/Eifel) | 30 % | ♀3 ♂3 6

g, e | all of us
the whole team (majority principle) | salaries and communal kitty

private ownership, might become a foundation
we follow creation and the universal conditions
gardening, guest hosting, artisan crafts
ecological, self-managed, non-violent, spiritual, inner growth, meditation, yoga, organic diet, vegetarian diet, vegan diet, adult education, pedagogical, life learning space

☺

The Beuerhof is a historical place in the midst of the volcanic Eifel in a free location surrounded by woods and meadows without neighbours on an area of 12 hectares. It used to be a Celtic place of worship in the early days, then the Romans were here and after that it experienced an interplay of different cultures. For over 20 years now, Beuerhof has been a meeting place for people who want to experience nature and make contact with Mother Earth and the Spirits. We are not a living community, but an operating community that wants to lead this place further into the future. This also creates living together. People of all faiths come to us, and since we have also made teambuilding for companies a subject of our work, there are also companies like Telekom, Lufthansa, Yellostrom, Thyssen Krupp,

DM drugstores and many others. This challenges the team and turns everyday life into an adventure.
Since the actual owner has gotten older, the project will have a new management, possibly also new operators in the close future. Everything is possible, nothing has to.

☺ being fully present

BlackSwan Schwarzer Schwan

E2

L 723 | 17098 Friedland

alternativleben@riseup.net

2016 | 1.5 ha (near Neubrandenburg) | 40 % | ♀3 ♂2 0 | 5

g, e, f, sp, Portuguese | 4 actually all of us

plenum (consensus) | gem. Ökonomie

going to be property of the association

anarchistic, feministic

workshops - like wood, metal, serigraphy, house, garden (activities vary dependent on demand; everyone decides as per individual needs)

ecological, anarchistic, left-wing political understanding, feminist, non-violent, vegetarian diet, vegan diet, animal protection, environmental protection, self-sufficiency, organic gardening, ecological construction, humour

We are five people with some years of project experience and are living on an old farm in the Mecklenburg Lake District between Neubrandenburg and Anklam.
To us it is important to create a basis for solidarity with each other on equal terms. One of our principles for that is the sharing of our incomes and debts. We want to live in a stable group where everyone pays attention to everyone else; we want to grow "healthy" to a maximum size of 15 people.
The trailer space provides plots for another 4-5 vehicles and we would like to expand the living space in the house. It is also planned to create workshop-, meeting-, community- and safe spaces in the house. The shed in the yard provides space for our workshops. By spreading existing knowledge and skills at a flat hierarchy level, we are able to create and realize work projects together. Our serigraphy company "widerdruck" has already been established for 1.5 years now and we want to realize a tofu production and a trailer construction enterprise.
By building community companies as well as cultivating biological vegetables and fruits in the sense of permaculture, we are aiming to get more and more independent from the capitalistic system. Moreover, we are traveling with our cooking equipment for people's kitchen within the field of political actions, camps and events. The left-wing political activism has a high priority for us and is part of our principles – emancipatory, antisexist, non-hierarchic, antiracist, animal friendly.
Networks of political communities are very helpful to do something against the isolation especially in rural regions and moreover, we can share the focus on political issues and the conception of political networks.
We like to welcome guests, travelers, interested or helping hands to give an insight view of our utopia – but as we are all involved in our daily projects it would be great to receive inquiries of independent people.

→See also description of network "**KommuJa**"

☺ acceptance of and accordance with our principles paper

Brot & Rosen - Diakonische Basisgemeinschaft C2

"Haus der Gastfreundschaft"

Fabriciusstr. 56 | 22177 Hamburg ☎+49-(0)40-697 020-85
basisgemeinschaft@brot-und-rosen.de www.brot-und-rosen.de

1994 450 m² ♀4 ♂1 3 20

g, e, f, sp — all, part-time work outside
everyone in common (consensus) — shared income
rented/leased
Christian/ecumenical
solidarity with homeless refugees, simple living, solidarity and anti-nuclear work
anarchistic, left-wing political understanding, feminist, non-violent, Christian, ecumenical, peace-work, solidarity work: support of (illegalised) political refugees, deaconry work

☺

The basic ecclesial community Brot & Rosen (= Bread & Roses) in Hamburg is a small ecumenical-Christian partnership. Since 1996 we have been living in a rented house that used to belong to the church community. Our communal life is characterised by dedicated commitment to justice, peace, ecology, communal meals and devotion. We want to live and work with people who have been excluded from society and whose fundamental rights have been taken away. Up to 8 homeless refugees can be accommodated by us regardless of their legal status. The community's members pay for their own expenses as an income co-operative through part-time work. The expenses and rent of our "guests" is financed by contributions from our friends and supporter group. We want to live simply so that others may simply live. Through our consciousness-raising work, for example during our open evenings, we strive to be our "guest's" loudspeaker and to champion solidarity, non-violence and justice. At the same time we are trying to join forces with other initiatives to make a difference in public life and in the Christian parishes around us.

→See also description of networks "**Church & Peace, European Catholic Worker**"

Catholic Worker - Christian Anarchist Community Movement
☺ getting to know each other through a time of volunteering at our house

☆ Burg Disternich A6

Burgstraße 1 | 52391 Vettweiß ☎+49-(0)2252 830 91 03, +49-(0)2252 838 7142
kontakt@burg-disternich.de www.burg-disternich.de

2018 9 ha (near Zülpich) 10 % ♀9 ♂3 4 16

g, e — 6
everybody (sociocracy)
common financing and shared household — cooperative
ecological, spiritual, inner growth, meditation, yoga, organic diet, organic gardening, ecological construction, humour

☺ ()

Burg (= castle) Disternich eG is an association of people (eG meaning "cooperative") who explore an alternative model of coexistence in freedom and responsibility for themselves and each other. We believe that economic crises, globalisation, an aging society, irresponsible

consumption of natural resources, increasing loneliness, communication gaps between the generations and much more challenge us for new ideas and perspectives - and forward-looking, courageously lived experiments in implementation.
Burg Disternich eG offers and promotes life in a healthy environment. Whenever possible, food is grown organically, energy is generated in an environmentally friendly way and ecological materials are used. We aim for maximal independence in our basic needs. The castle's long history (16th century) is part of the project. It is respected and history and buildings are also being made accessible to the public. We cooperate closely with the department of monument preservation in order to make alterations and renovations as true to the original as possible as well as environmentally compatible.
Burg Disternich eG is cosmopolitan. We want to be free of dogmas and well connected to our environment. Our vision is of non-violent cooperation and lived diversity in accordance with one's own potential. Each member is also held accountable to the entire community. Examples of planned projects are a sewing studio for meditation accessories, a small shop for the residents of the area, a wild pasture and nature experiences for children and curious folks. Day care for toddlers, Zen seminars for guests, an architecture company with an ecological and sustainable focus and a large vegetable garden have already been implemented. Cultural and social projects are designed to connect with the outside world to inspire locals and visitors.

☺ acquaintance weekend, living with us for a while, one year probation

☆ Burghof Stauf Gemeinschaftliches Leben eG (B8)

Burgweg 1 | 67304 Eisenberg-Stauf
info@gemeinschaft-burghofstauf.de www.gemeinschaft-burghofstauf.de

2015 | 1.8 ha (near Eisenberg (Pfalz)) | ♀11 ♂8 10 29

g, e, f, sp, Turkish, Czech, Croatian
all concerned (consensus) — limited common financing
property of all (cooperative)
ecological, self-managed, non-violent, spiritual, meditation, yoga, organic diet, vegetarian diet, vegan diet, environmental protection, permaculture, accessible to the disabled, non-smoking, music, Zapchen, undogmatic

The special thing about our community is our diversity. People from all generations with different professions, personalities and world views come together in this powerful historical place. Our community house is directly located at the foot of the ruins of castle Stauf and has a beautiful view over the Palatinate Forest. Founded in 2015, we are now living together since 2017. Our community is therefore still under development and there is a lot of creative potential and opportunities to realise individual dreams. We have no certain vision, we just want to live together in mutual respect and friendliness, free of conventions, and use the power of community to create a wonderful place. Acceptance and tolerance are very important to us. Although everyone has their own personal flat, we love getting together to cook, play music, sit around the campfire and spin new ideas in spontaneous encounters. We support each other in our everyday lives, the children grow up like brothers and sisters, we go shopping for each other, and all of us tackle challenges with no doubt. We can rely on each other as good friends. We also experiment with ways of meeting on a non-verbal, meditative level. In the future, we would like to share more common property, to create work opportunities on our premises and to develop our garden cultivation towards permaculture.

De Vlierhof

Kleverstr. 115 | 47533 Kleve-Keeken
+49-(0)2821-970 242, +31-(0)6588-5077
info@vlierhof.org
www.vlierhof.org

2002 | 3 ha (near Kleve) | 30 % | ♀6 ♂8 2 | 16

e, Dutch, g | all
founder (consensus) | limited community financing

- a foundation
- working on spiritual growth and awareness
- hosting seminar house, working the land, workshops, maintenance, kitchen
- ecological, self-managed, non-violent, spiritual, inner growth, meditation, organic diet, vegetarian diet, vegan diet, adult education, self-sufficiency, organic agriculture, organic gardening, alternative energy, ecological construction, humour, craft: straw bale construction, earthworks, carpentry, art: print, communication techniques: talking stick discussions

☺

Vlierhof is an international community, where we try to work towards a more loving world, where we live together with nature in a more open and honest way, where we work on a new and better world. We focus on the following values: integrity, transparency and openness, interest and care for each other and the environment. What we encounter on this path is utilized for personal development. Vlierhof is located on the border of Germany and the Netherlands, close to the cities of Kleve and Nijmegen. We grow as much of our own food as possible in our organic garden. Many of the walls at the Vlierhof are plastered with clay from our own land (thus creating the swimming pond for the sauna). Much is recycled and upcycled. The kitchen waste, for instance, goes to the chickens, who would give us eggs in return.

The pride of the Vlierhof is the seminar house. Formerly an old barn, this beautiful group accommodation was built entirely with natural materials. The walls of the building are largely made of clay. This gives a fine warmth and energy to the rooms. The seminar house is a serene and stylish place for workshops, seminars or any kind of group work or recreation. There is a meditation and yoga room, a beautiful dining room and lounge. For maintaining everything volunteers are welcome for cleaning, construction, gardening and cooking.

→See also description of network
"GEN Nederland/Ecodorpennetwerk/Vereniging Ecodorp Nederland"

☺ we want to live together with respect and love and expect everyone to take his/her responsibility.

Deel 28 Mehrgenerationen Wohnprojekt Palingen GbR

C2

Hauptstraße 28 | 23923 Palingen

www.deel28.de

2003 | 1.5 ha (near Lübeck, Wismar) | 1 % | ♀9 ♂7 | 14 | 30

g, e, f, Turkish | changing, at least three people

all residents or all owners, depending on the financial consequences for those concerned (consensus, in case of conflict two-third majority is also possible)

community financing of community shares (café, garden, hall, garage), individually financed living space

organisation/private

none

ecological, self-managed, non-violent, ecological construction, regional supply, café, cultural events

We seem to be something between a co-housing project and a community. Since we are not united by a shared belief system or world view, we find that what unites us is mostly the work we share in the house, courtyard café and in our garden. Our project hardly provides any paid work, but there are different kinds of cooperation between individual members, which also improves our economic foundation. The living space is completely up to individual responsibility.

Delfin-Gemeinschaft - Wege zu Gott

B10

ÖKODORF-Institut für Wohnprojekte

Alpenblickstraße 12 | 79737 Herrischried-Niedergebisbach

+49-(0)7764-933 999

oekodorf@gemeinschaften.de | www.gemeinschaften.de

1995 | 1100 m² (near Schopfheim) | 10 % | ♀3 ♂2 | 0 | 5

g, e, f, Alemannic | 2 | 3

all concerned (consensus, inner voice, higher self)

individual finances (shared economy planned)

property of some community members

all religions are of equal value

retreat for people who are ill from cell phones etc, garden/permaculture, kitchen, prana-walking

ecological, self-managed, non-violent, spiritual, Christian, ecumenical, Buddhist, inner growth, meditation, yoga, tantra, sexuality, organic diet, vegetarian diet, adult education, seminar-house, animal protection, environmental protection, self-sufficiency, permaculture, organic gardening, alternative energy, ecological construction, no cell phones/wireless, non-smoking, Singing, research on community development, healing, radiation-free, www.pranawandern.de, snowshoeing, cross-country skiing, e-mail newsletter "Communities News".

☺ (✋✋)

Wiesental is a good starting point to build up our community here. The population is tolerant, ecological and spiritually open. We are surrounded by a very beautiful landscape and our children attend the Rudolf Steiner School.
Our central goal: we help each other along on the path to GOD, along with others who desire the same thing. One contribution to this is the creation of a larger interspiritual community consisting of smaller communities of various spiritual orientations. Its range is not arbitrary, but our framework is provided by the commandments which are the same in all world religions (see the world ethos of theology professor Küng).
Learning from history, society can only develop if we do not keep reinventing the wheel. That is why we are learning from other communities (failed and successful ones) and pass this knowledge on through seminars, consultation etc. for those seeking and founding communities of their own.
Commitment to the region: co-operation (e.g. hometown association, church, fire brigade). Discovering which of our region's needs are currently not being satisfied (e.g. ecological producer-consumer co-op). Then initiate this as a complement (not as competition) to existing enterprises.

Informal network of interspiritual ecovillages, Artabana
☺ spiritual openness and tolerance, interest in the simple life, work with everyday addictions

die Frieda Gemeinschafts- und Projekthaus

D4

Straße des Friedens | 14822 Brück ☎ +49-(0)3384-477 48 83
friedahaus@riseup.net die-frieda.org

2014 600 m² houses, 0.9 ha forest garden (near Berlin) 5 %
♀3 ♂3 1 7

g, e, sp, f all of us (in different extent) in working circles 4
all involved in the project
(decisions are made by simple consensus or systemic consensus)
shared financing of infrastructure and running costs;
earning money shall be organized together increasingly
owned by our association
building, renovation, forestgarden, administration, fundraising, seminar management
ecological, self-managed, non-violent, inner growth, organic diet, vegetarian diet, adult education, seminar-house, peace-work, self-sufficiency, permaculture, organic gardening, forestry, ecological construction, humour, Regional networking, sustainable development of life and work alternatives

☺ ✋✋

We are the community Frieda! We are currently 6 adults and 1 child, but about to get more. The "Alter Konsum" in Brück was bought in 2014 and founded as a project house. From the end of 2016 a community began to form here.
Since then we have been working diligently to get closer to our vision. That is the vision of resilient regions, which are subsidiary networked and culturally, socially and organisationally able to close material cycles and to enable a sustainable good life for all. In 2018 we were able to buy another house in the street and about 0.9 hectares of garden land, where we are now in the process of establishing a forest garden that will provide us with local food and timber in the future.

Our goal over the next few years is to build a coherent community model, which consists of well-functioning reference groups whose members, in turn, organize themselves into working groups and companies. And at the same time, to develop cooperation with communities, the neighborhood and other Fläming communities and to actively create a region worthy of life.
If you would like to visit us and work with us or if you would like to attend one of our events, then you are very welcome. We are very open for working guests (please contact us by phone or email!).

🌍 bildungs-bande.de
☺ slowly growing into group and project. Sympathy - Resonance - Common Vision - Good Collaboration

Dorf - Gemeinschaft - Münzinghof

D8

Münzinghof 9 | 91235 Velden ☎+49-(0)9152-9297-0
verwaltung@muenzinghof.de www.muenzinghof.de

1979 | 140 ha (near Velden) | 30 % | ♀70 | ♂70 | 15 | 155

- g
- property of the association
- anthroposophy
- social care & work, handicrafts
- ecological, self-managed, non-violent, spiritual, Christian, anthroposophic, organic diet, adult education, mentally/psychologically handicapped persons, environmental protection, organic agriculture, organic gardening, organic farming, animal friendly husbandry, forestry, alternative energy, humour

The village community Münzinghof is a community with commune characteristics and offers people in need a sense of belonging in a familial context and, if required, a lifelong home. We offer our residents living spaces, work in a workshop for handicapped people and a cultural life. Employments in the workshops are tailored to individual needs and ability. Workshops are: domestic economy, gardens, agriculture, cheese dairy, bakery, caretaker work, metal and wood workshops, as well as candle workshop and the communal group. The content and sequences of the work are meaningful, and the results from domestic work, bakery and gardens also benefit the community, the family and others. The communal life in the families and in the whole community, as well as our interactions with one another, are based on the values of Christianity and anthroposophy. We value rhythmic sequences throughout our days, weeks and years. Aside from that, there are certain rituals which guide us through the days, weeks, months and years, which are discussed by all members of the community and can be changed and kept alive (e.g. each Saturday there is a gospel evening, on Sundays there is morning celebration and we celebrate the annual Christian festivals).

Dorfgemeinschaft Klein Hundorf

C2

Klein Hundorf 6 | 19205 Gadebusch
moin@klein-hundorf.de www.klein-hundorf.de

1989 | 15 ha (near Gadebusch) | 20 % | ♀9 ♂9 | 7 | 25

g, e, f, Swedish, Latvian | 5-6

everyone in common (consensus, if necessary majority vote)

limited community financing and individual financing | common property

garden, fieldwork, building, renovating, making firewood, kitchen, forest kindergarten

ecological, self-managed, chaotic, organic diet, environmental protection, self-sufficiency, organic agriculture, organic gardening, alternative energy, ecological construction, humour, carpentry, blacksmith's shop, music, working with horses, chickens, sheep and bees, forest-kindergarten, CAM-practitioner, seminars, sewing, kitchen, voluntary serviceF

☺ (✋✋)

We are a cooperative that co-owns a farmhouse and land, as well several outlying buildings. The farmhouse is the hub. It contains our communal spaces and we meet here for weekly meetings and communal work. There are bedrooms upstairs in the farmhouse, and there are further living spaces such as caravans and straw bale houses scattered around the property. We lead a simple lifestyle with composting toilets, stove heating, solar showers and our own wetland wastewater treatment.

There's a blacksmith's shop, a carpentry and a arts and leisure center as meeting point for cinema, dance and celebrations. With our juice squeezer we produce fruit juice from our more than 100 fruit-bearing trees. A huge outdoor-bathtub and a sauna-trailer invite to chill. Our workhorses are helping to cultivate our gardens or to make hay. There is a community supported agriculture, a forest-kindergarten and a central kitchen, cooking for the kindergarten, seminars and workshops.
Our legal land-use plan (we've got a custom zoning regulation) sets the foundation for the future realisation of our further plans.
Visitors are welcome if they announce their visit prior to coming, and can use the communal facilities to cater for themselves, in return for a small financial contribution or in exchange for work. The winter can sometimes be a little lonely here, the summer is very busy. In order to get to know us, please come to our "building weeks" or "communal days", if possible.

→See also description of network **"WWOOF"**

☺ 1 year probationary period, entry fee (1.500 EUR) with co-op membership. After the request everybody of the community must agree to you moving in.

Germany

Ein Neues Wir A New We

B6

Wiesenstraße 37-39 | 65627 Elbtal-Hangenmeilingen ☎ +49-(0)6436-949 43 76
leben@ein-neues-wir.de www.ein-neues-wir.de

2011 1.3 ha (near Limburg an der Lahn) ca. 10 % ♀9 ♂8 12 29

g (e, sp, f, Russian) approx. 5-7

dynamic principle of "all leaders", consensus (in the circle (plenum), in working groups)

common household budget, otherwise individual

property of the cooperative

kitchen, garden, renovation

ecological, self-managed, non-violent, spiritual, inner growth, organic diet, vegetarian diet, pedagogical, peace-work, environmental protection, self-sufficiency, permaculture, animal friendly husbandry, ecological construction, construction, humour, cosmopolitan and full of joy of living, respectful communication, especially/also with children, creative and innovative, open heart, cross-generational, permaculture, self-responsibility, sports, cultural work

☺ ()

- We practice mindful and responsible interactions with each other and our environment.
- We network for and contribute consciously to social and ecological change.
- We develop a sustainable model for living and working together. We're open to people of all generations and cultures and respect everyone's personal approach to spirituality.
- We build a place where people can develop their potentials. The healthy and individual development of children is especially important to us.

Community building meetings since 2007, communal co-housing since early 2011, project headquarters since January 2013: 3 buildings on a nature-oriented property (12,800 m²), space for approx. 40–50 people. Social tools: supervision, respectful and authentic communication in daily life (nonviolent communication), family constellation, talking circles, council etc. Each adult pays the same rent, no rent for children.
Our children attend Montessori School in Westerburg.
We are happy to get to know new interested parties, collaborators and friends...

Artabana (some)

☺ 6-12 months probationary period, cooperative shares, earn your own living

Eine Spinnerei

E5

Spreewitzer Straße 5 | 02979 Neustadt/Spree ☎+49-(0)35727-579 341
fred@eine-spinnerei.de www.eine-spinnerei.de

2011 3000 m² (near Cottbus, Hoyerswerda, Spremberg, Weißwasser) 20 %
♀2 ♂1 2 5

g, e all adults
power of decision according to area of responsibility (direct democratic and consent)
mixture of individual financing (individual income and expenditures) and common financing (household budget and real estate)
property of founding members
garden, craft and building, self-sufficiency, eco-political citizens
ecological, self-managed, organic diet, adult education, pedagogical, free school, animal protection, environmental protection, self-sufficiency, organic gardening, new technologies, ecological construction, humour, non-smoking, recycling, sustainable living

☺

Since 2011, we (a small community of 3 adults and 2 children) have been living as closely to our principles of a steadfast sustainable life as possible. We try to be as self-sufficient as we can by eating, preserving, canning and drying the vegetables our own garden provides. We have built a cooling cellar/earth cellar in order to do without a refrigerator. We try to get everything we need second hand, for free or by bartering, and we even reuse and upcycle building materials. We take what we find and experiment. In doing so, we are inspired by traditional crafts and come up with unconventional solutions, e.g. insulating with old wine bottles and making tea lights from leftover wax and suet.
We want to share our vision and give others the opportunity to experiment with different life styles at the Spinnerei.
We organize a small number of workshops every year, such as Cooking with Native Plants. Acting in an ecologically and socially responsible manner doesn't stop at our property line: That is why a good chunk of our everyday life is political activism. For instance, we are engaged against planned new open lignite mines in the region and we publish a paper on this topic.

→See also description of network **"WWOOF"**
workaway, alliance "Strukturwandel jetzt - Kein Nochten II"

Elbehof und VITOS e.V.

D3

Am Elbdeich 68 | 39615 Wahrenberg ☎+49-(0)39397-4155-2
vitos@elbehof.de www.elbehof.de

1996 3.5 ha (near Wittenberge) ca. 20 % ♀2 ♂1 0 3

g, e, f 2
owners and inhabitants together (round tables) individual financing
property of the Elbehof GbR (3 owners), others pay rent
a peace-loving, cooperative and sustainable way of life
natural and environmental education courses, organic vegetarian café (open all year), ecological gardening, seminar house

ecological, meditation, yoga, adult education, pedagogical, seminar-house, animal protection, environmental protection, permaculture, organic gardening, animal friendly husbandry, alternative energy, ecological construction, humour, non-smoking, nature and environmental training

The former Fährgasthof in the triangle of the German states of Niedersachsen, Sachsen-Anhalt and Brandenburg was founded in 1996 as an ecological joint project and called by its location on the river Elbehof. A seminar barn, three guest rooms and a river café have been created. The hall was refurbished, solar outdoor showers and compost toilets as well as a camp kitchen were set up to hold river camps and festivities of all kinds on the Elbwiese. The regional cultural and environmental education association VITOS e.V. organizes sustainable and cultural education events with various partners and members.
The farm has two separate apartments, which are currently inhabited by 2 residents. The kids are out of the house. There is also a shared apartment with three rooms for new residents, who could found a new community project. These could get involved in the community garden with chicken keeping as well as in the whole farm and can find their place. People with new sustainable ideas for the farm and the region are welcome. Expansion possibilities are still available.

VITOS e.V.
probationary period, openness to join community creating processes and to contribute skills/knowledge/financially to the community growth

Finkenburg

B3

Eißeler Finkenburg 1 | 27321 Thedinghausen-Eißel +49-(0)4204-1497
kontakt@finkenburg.info www.finkenburg.info

1989 1 ha (near Bremen) ♀3 ♂5 1 9

g, e, Plattdeutsch 3
everyone in common (consensus) limited community financing
common property
ecological, self-managed, anarchistic, left-wing political understanding, feminist, non-violent, organic diet, environmental protection, organic gardening, alternative energy, ecological construction, cultural work, playing

The Finkenburg is an old, detached farm in the Wesermarsch region between Bremen and Verden. It has existed as a communal project since 1989, and at present it is occupied by nine permanent residents. We consider ourselves to be part of the political commune movement and see our manner of communal living as an alternative model for a more social, ecological and solidarity-based society. Self-determination and personal space for individuals, as well as the acceptance of differing needs are the foundations of this goal. We practise consensus decision-making with equal rights for all. The house and land are shared property and we manage them ourselves. Several residents are working in self-organised workspaces, others are employees or freelancers in the local area. We have a juice press (see www.mosterei-finkenburg.de), a pottery workshop (see http://keramikwerkstatt.blogspot.de) and the editorial office for the regional magazine "Överblick" (see www.oeverblick.de) which was co-founded by a resident. "Överblick" also provides placements for the government-run scheme "Voluntary Ecological Year" (FÖJ), offering people up to the age of 27 the opportunity to live with us for a year and get to know our project. In our gardens we are cultivating fruits, vegetables and herbs for our own consumption. Furthermore, we are a home for animals like sheep, several cats and bees.

→See also description of network "**Kommuja**"

☺ introductory visit, probationary week, probationary period (½ year)

Franziskusgemeinschaft e.V.

E5

Leutwitz Nr.1 | 02633 Göda OT Leutwitz ☎+49-(0)35953-8431
Franziskusgemeinschaft@web.de
www.franziskusgemeinschaft-leutwitz.de

1988 20 ha (near Bautzen, Dresden) ♀1 ♂2 0 3

g, e all

everyone in common (less important topics by majority vote, central issues by consensus)

shared economy partially rented/leased, partially property of all

ecumenical in the widest sense

farming, processing and selling agricultural products

ecological, non-violent, Christian, ecumenical, organic diet, peace-work, environmental protection, self-sufficiency, organic agriculture, organic gardening, organic farming, alternative energy, ecological construction

☺

Our community came into existence in 1988 as a result of our search for God as the centre of life. Since we carry the responsibility for God's creation, our world, our life must be a positive alternative to unfairness, fear, lack of peace, hate and destruction.
JOINT PROPERTY: We don't merely wish to share our material possessions, but our entire lives including everything that human lives are made up of: our joys and worries, our hopes and fears, our strengths and our weaknesses...
WORKING GROUP: We strive to live in harmony with the rest of creation and therefore we try to produce the most important foodstuffs ourselves. Any surplus is sold directly from the smallholding.
PRAYER COMMUNITY: The busy working life on our smallholding requires a good balance between action and contemplation. Twice daily, in the morning and evening, the community meets for prayer in our chapel.
HOSPITALITY: We are happy about people who wish to visit and share our lives. Anyone interested in getting to know our community is invited to announce their visit! We like to run an open house and this is how we understand our ecumenical alignment in the broadest sense, i.e. the ecumenical movement not only between churches and religions, but also between people who are not part of any church in particular.

☺ life-affirming, Christian, ecumenical attitude

Freie Feldlage - Gemeinschaft in der Heilstätte Harzgerode

C5

Freie Feldlage 9 | 06493 Harzgerode ☎+49-(0)39484-743 333
info@freiefeldlage.de www.freiefeldlage.de

2018 21 ha (near Göttingen, Magdeburg, Leipzig) 5 % 10

g, e 8

all who are concerned (consensus, systemic consensus principle)

individually, solidarity, common food purse, shared income possible

owned by individuals (for the time being), cooperative is being founded

project development, community building, renovation/maintenance, gardening

ecological, self-managed, non-violent, inner growth, organic diet, vegetarian diet, adult education, seminar-house, self-sufficiency, permaculture, organic gardening, forestry, alternative energy, new technologies, ecological construction, humour, non-smoking, sharing culture, solidarity-based economy, alternatives to commerciality, Foodsharing

☺

Since May 2018, a group of committed, energetic people has been coming together on the grounds of the former children's lung sanatorium, who have taken the project forward with giant steps in its first year. Our goal is to build a community on a cooperative basis. Approximately 50-80 people of different ages will live, learn, work and live together in shared flats but yet in a large community. We are now clearly in the pioneering phase. That brings imponderables, but also a lot of freedom.
The site comprises 21 hectares of land with various buildings (the main building as well as various residential and auxiliary buildings), meadows, gardens and woods. A multitude of initiatives and projects can be developed here, to serve the development of the individual and the community. A goal of these initiatives should also always be the interaction with the region and beyond, in order to enrich these, as well as the effect as role model for lasting, sustainable life.
Possible projects: Living, multi-generational housing, working in communal enterprises, free and shared spaces, self-determined learning in and from nature (wilderness knowledge, free school, forest kindergarten and outside of these categories, learning place, open spaces), open workshops, ecological horticulture and forestry, permaculture, energy supply, communal kitchen, event location for arts, culture and camps on the open air grounds and in the hall, monument protection (Bauhaus), repairing instead of throwing away, and much more.

→ See also description of network **"GEN Europe"**

☺ pioneering spirit, willingness for personal development

Friedensgarten

Haus Nr. 5 C | 02627 Pommritz ☎+49-(0)35939-806 02
uwewhaspel@yahoo.de, schenker_oeff@yahoo.de
www.global-love.eu, www.dieschenker.wordpress.com

E5

1998 1500 m² (near Bautzen, Dresden) 60 % ♀1 ♂2 0 3

g, e, i, Romanian 3

aim: all concerned (responsibility based consensus)

gift-economy, sharing the things we have depending on what people need

the premises are used in the sense of moral using rights, like organic cells in the organism of the world

"Givers-Movement", open for people of different ideological, spiritual or religious beliefs

public relations for support and development of global responsibility; spirituality; self-sufficiency

ecological, self-managed, non-violent, spiritual, inner growth, meditation, yoga, organic diet, vegetarian diet, vegan diet, adult education, peace-work, animal protection, environmental protection, self-sufficiency, permaculture, organic agriculture, organic gardening, ecological construction, humour, simple craft techniques, nutrition from the (wild) nature, cooperation with nature, conflict resolution, nonviolent communication, networking, health work

☺

We train to give and take with love and justice, as the people have and need - instead of exploitation and suppression. We train to communicate and handle conflicts with mind, heart, humour, playfulness and spirituality instead of violence. We welcome the public for visits, and we make courses and lectures here and outside. Regularly we engage in peace work (mediation, courses in nonviolent communication, global responsible life), spirituality (prayer, yoga, meditation, rituals, singing and making music), health work (wild herbs, therapy for free), office work, spiritual work - as a spiritual centre for inner life and meetings. Volunteering is possible - with telephone-call before - from April to October.

- We respect nature very much and try to make only minimal changes.
- We use permaculture and cooperate with nature.
- Many things for our use we make ourselves.
- The things we need shall be produced with respect and justice, directly and indirectly!
- The social conditions of other people and nations and the ecological systems shall stay in good function, or even be supported through our influence.
- We have no electricity support, only candles and LED.
- We get water from a well.
- We get firewood from our own garden and a forest nearby.
- We integrate modern technology for contacts and PR.

→See also description of network "**Givers-Movement**"

Holon

☺ sharing the ideals of the Givers-Movement and the fundamentals of the project

Friedenshof-Kommunität

C3

Hammerstein Str. 3 | 31535 Neustadt-Niederstöcken ☎+49-(0)5073-7022
info@friedenshof.org www.friedenshof.org

1991 3 ha (near Hannover) 25 % ♀3 ♂4 0 7

g, e, f, Dutch 5
everyone in common (consensus) shared income
organisation of community members
inter-religious, spiritual, non-violent
household, organic farming, animals, ecological building improvements, guests, educational work
ecological, self-managed, non-violent, spiritual, inner growth, meditation, yoga, organic diet, vegetarian diet, adult education, peace-work, environmental protection, self-sufficiency, organic agriculture, organic gardening, organic farming, alternative energy, new technologies, ecological construction, milk sheep, bees, chickens

☺ ( )

We live on a 280-year-old farm in Lower Saxony. The farmhouse with a thatched roof is located on 5000 m² of land with a garden, field, greenhouse and workshop in the middle of a small village, 35 km north of Hannover. The rhythm of our daily routine is composed of meditation and rituals, work and mealtimes. The work we do is divided between on-farm and off-farm work: help with organic vegetable sales, organic vegetable and fruit production, sheep-rearing for milk production, chickens, bees, renovation and upkeeping of our buildings, art therapy, yoga, education on developmental policy and other educational work. Our community finds its centre and the source of its strength within spiritual experiences. We take differing religious and spiritual paths, and we are interested in shared principles. In our lives, financial gain and consumption have lost importance compared to spiritual fulfilment, communal life and a meaningful occupation. We are affiliated with the French L'Arche-communities that go back to Lanza del Vasto, and their ethical foundation of non-violence. We offer various events, such as meditation evenings, mindfulness days, dances of universal peace, the arts and yoga. Alternatively, guests are welcome to share our communal life for limited periods of time. Those who are interested in sharing our communal life in the long term can be taken on as prospective members after an initial meeting period, and may become part of the community in good time.

→See also description of networks **"L'Arche de Lanza del Vasto, GEN Europe"**

☺ probationary period, desire for individual and spiritual growing

Galions e.V. Hofgemeinschaft Galions

B6

Brunnenstr. 3 | 36323 Grebenau-Bieben ☎+49-(0)6646-919 02-5
galions@web.de

1988 1.5 ha (near Bad Hersfeld, Fulda, Magdeburg) ♀3 ♂3 6

g, e, f, sp 4
all concerned together (majority vote, consensus in individual cases)
limited community financing association
we are varicoloured and appreciate that
taking care of house, animals and garden; office for house utilities planning; meditation and body work (rebalancing und TRE)

ecological, self-managed, left-wing political understanding, spiritual, Christian, Buddhist, inner growth, meditation, organic diet, pedagogical, environmental protection, self-sufficiency, organic gardening, alternative energy, ecological construction, humour, individual diet including vegetarian and vegan

☺

We live in a small village in the Vogelsberg near Alsfeld surrounded by meadows and forests at the foot of the castle Herzberg. We are a colourful mixed group and appreciate exactly this. There are now six people living here: three women, three men (one couple only). We live with two cats, a dog, three goats and three horses. We are cultivating a vegetable garden. Our association is committed to the environmental and natural protection (water ecology, planting of live buildings and hedges, regenerative energies) and to the improvement of the physical and mental condition of the individual through health and consciousness formation through meditations and courses. We also make public work for peaceful and social changes.

→See also description of network "**Kommuja**"

☺ ecological attitude, interested in natural nutrition, caring for plants and animals, spiritual openness

Ganesha-Hof Linde

D3

Meeting Place and Tantric Community for the Elderly

Linde-Dorfstraße 40 | 16775 Löwenberger Land ☎+49-(0)171-837 22 41

justus@tantra-wg.de tantra-agil.de

2008 7500 m² (near Oranienburg, Berlin) < 50 % ♀1 ♂1 0 3

2 2

g, e

all who are concerned (preferably by consensus) individual financing

property of some members, transfer to foundation is planned

tantra, left-wing alternative

leisure, garden, construction, Qigong, massage. We are not a business.

ecological, self-managed, left-wing political understanding, esoteric/New Age, chaotic, meditation, tantra, sexuality, free love, organic diet, environmental protection, self-sufficiency, organic gardening, alternative energy, construction, humour, non-smoking, terra treta, qigong, music, art, being, keeping company until death

☺ ( )

The Ganesha-Hof Linde ("Linde Ganesha-Estate") in the Löwenberger Land, 25 km north of Berlin is a tantric community project for older people. Respect, self-respect, serenity, sincerity, openness, tolerance and integrity unite the residents and their visitors. We regard closeness and touch to be fundamental human needs, even in later life. From our experience, a respectful and attentive tantric way of life is particularly suited for these needs.
Our large residential house is located on a 7,500m² property at the edge of a large forested area. The barn includes community and seminar spaces for up to approximately 15 people: group space, kitchen, communal space with a wood burning stove, sauna and bedrooms. The second house was just renovated and prepared for an extension with several apartments: From mid-2019 apartments will be ready for 5 people! The construction of two further buildings is permitted on the premises if required. Please do not visit without prior arrangement!

☺ minimum age 45 years, psychologically balanced, decent physical health, be comfortable with ecological life (e.g. compost toilets)

gASTWERKe e.V.

C5

Forstamtstr. 6 | 34355 Staufenberg/Escherode ☎+49-(0)5543-302 44 10
kontakt@gastwerke.de www.gastwerke.de

2007 10 ha (near Kassel) 25 % ♀10 ♂12 13 35

g, e, f — about two third of us
all of us (consensus) — shared economy
owned by our non-profit association
undogmatic, tending to the political left, we are open for diversity
organic garden incl. farm shop, gASTWERKe academy
ecological, self-managed, left-wing political understanding, non-violent, spiritual, inner growth, meditation, yoga, organic diet, vegetarian diet, vegan diet, adult education, pedagogical, seminar-house, environmental protection, self-sufficiency, permaculture, organic agriculture, organic gardening, alternative energy, new technologies, ecological construction, humour, culture

☺

We are a colorful community, some preferring manual, others mental work, quiet and loud, very young, quite old. This makes our life fresh and lively, buzzing and controversial.
We value undogmatic and authentic communication and practice it through various group methods and external support, especially in our monthly social plena and group weekends. Our daily meeting spot is the communal meal in the evening. We want everyone to be heard and to be understood empathically, precisely because our everyday life teaches us that.
We are just leaving the pioneers status behind: There is movement in our group. Now, the original question is asked again: Why do we live here? What are we committed to? Are we an alternative consumption-critical housing project - or what is our common denominator underneath the individual sensitivities?
We are looking forward to impulses from outside. Curious, alert people who want to get to know us and/or live here are invited to get an idea of our "practice", how we live and work and experiment with sustainability and solidarity through our info café or during our open guest-work-weeks. There is still plenty of room here for technical, artistic and creative work.
Over the years, different areas of work have emerged and grown up: Bioland-gardening, landscape building, childcare and freelance work (IT, education / social) ... We have also just started to create an outdoor environment right on the edge of the forest for seminars, breaks and cultural events, for our newly founded gASTWERKe academy.

→See also description of networks **"KommuJa, GEN Deutschland"**
☺ sympathy, intensive multi-level entry process

Gemeinschaft am Lambach

B4

Am Lambach 14 | 32051 Herford-Eickum ☎+49-(0)5221-334 38
verein.am.lambach@posteo.de www.lambachgemeinschaft.de

1995 0.66 ha (near Bielefeld) 15 % ♀3 ♂4 5-9

g, e, f
everyone in common (consensus) — shared income
property of the association "Am Lambach"
direct democratic, nonviolent orientation
one's own profession, garden, building

ecological, self-managed, left-wing political understanding, non-violent, organic diet, peace-work, environmental protection, organic gardening, alternative energy, ecological construction, supporting refugees, music, non-violent actions, "Lebenslaute" ("Sounds of Life") music group

☺

Up to eight people from the twenties into their fifties are living in a former farm house in the countryside of Herford. All of us are busy with jobs or studying part or full time. We like gardening, organic food, music and driving by bicycle. Once a week we cook together. We have a plenum every month. Some of us are active in political/social grass-root movements. We invite friends or guests several times a year for a day of home repair or gardening/harvesting. The house is surrounded by nature and fields. The city of Herford and the next train station are 15 minutes by bicycle or bus. Since 1995 we renovated the house with ecological materials. For a few years we've been having a heat pump with solar water heating and photovoltaics on the roofs.

- peace group, group for refugees, cooperative for renewable energy
- ☺ being able to look after oneself; supporting the societal involvement, the mainly ecological lifestyle and the co-housing of guests

Gemeinschaft Clan B

B3

Wöschenweg 28 | 26209 Hatten — +49-(0)4482-2566-001

gemeinschaft@clan-b.eu — clan-b.eu

2013 — 7.5 ha (near Oldenburg (Oldb)) — < 5 % — ♀9 ♂10 — 11 — 30

- g, e, Dutch — 10
- plenum (consensus)
- common financing for some areas (household budget, shared facilities)
- community property
- none (undogmatically spiritual)
- children, garden, kitchen, office, guests
- ecological, self-managed, non-violent, inner growth, vegetarian diet, non-smoking, down to earth, neighbourhood get-together, health centre

☺ ( )

We have cooperatively bought and converted a former youth hostel. There we live and work in the developed apartments, practices and offices. In addition to this, we try to organize our everyday life - also community life - in such a way that we create a space for new ideas and impulses and thus find ways to form a truly contemporary living community. We evolve its design by exploring and looking at promising ways in conversation. Stumbling blocks and setbacks are not excluded when learning to walk...

→See also description of network "**GEN Deutschland**"

- Netzwerk-Wohnprojekte
- ☺ a desire for communal living, a readiness to share the financial responsibility, and participation in community work

Gemeinschaft Feuer & Flamme

C1

Hauptstraße 34 | 24790 Haßmoor ☎+49-(0)4331-770 794
tonio58@gmx.de

1998 2000 m² (near Kiel (25 km), Rendsburg (15 km)) ♀3 ♂3 6

- g, i, e, Dutch
- concerning the property: the owner; otherwise everyone in common (consensus)
- individual financing — property of individual members
- interior construction of apartments, garden
- ecological, self-managed, left-wing political understanding, non-violent, inner growth, organic gardening, no cell phones/wireless, non-smoking, no television, humor, improvisational theater

Our community lives in the Westensee Natural Park, located in the middle of Schleswig-Holstein. It is particularly attractive for people who long for natural and simple things in life: Walking, cycling, working in the garden, enjoying the silence of the forest, collecting mushrooms... Our project idea is focussed on neighbourly living for young and old, as well as preferably ecological building and living. There are simply decorated, but generous communal spaces, including a kitchen. Our possibilities within the project are diverse. Intellectual and creative activities are maintained and can be extended. We meet about once per month in order to discuss questions regarding the house as well as personal issues. We have had good experiences with the supervision method. Within our circle of friends and with people who are open to the idea of community, we organise seminars about community themes three times per year. We are open to the possibility of new kinds of communication, such as forum work. Occasionally we offer choir, dance and regularly improvisation theatre and yoga (once a week).

- Ökodorf-Institut
- engaged, responsible, craftspeople

Gemeinschaft Haslachhof e.V.

B10

Haslachhof 1 | 88693 Deggenhausertal ☎+49-(0)179-737 50 67
poesieexpress@gmx.de gemeinschaft-haslachhof.de

2015 5 ha (near Ravensburg) 5-10 % ♀6 ♂5 6 17

- g, e — 1-2
- all of us in common and in a circle (consent (sociocracy))
- some common financing — property of our non-profit association
- we are diverse, all of us love the simple and close-to-nature life.
- working with the land, building preparations, organising events, inner work
- ecological, self-managed, non-violent, inner growth, organic diet, adult education, environmental protection, permaculture, organic gardening, organic farming, forestry, ecological construction, humour, construction sector

Our vision is to develop our love for the earth and for all its creatures and to express it in the world. We create a culture of mindfulness and a sincere and trusting togetherness. We are spicing up our everyday life with humour, live in close touch with nature and support ourselves

in the free development of our potentials. We cherish and cultivate spaces of silence, vitality and consciousness. We are part of the global community network and contribute through our researching, loving actions and our simple being to a world worth living in, peaceful and sustainable. The knowledge about the cycles of life helps us to do so.
We develop the Haslachhof into an inspiring living space - ecologically and collectively. We create a place of life and learning that invites to make music and dance, laugh and cry, dream and act. We work sustainably with the land that nourishes us. We dedicate ourselves responsibly to the creation of social spaces and the solemn celebration of life transitions. It is important to us to create a fertile ground where people of all generations can be now and in the future: in freedom, dignity and security.
We are on the way, gaining experiences, struggling for new possibilities, encountering our imprints and patterns, practicing loving acceptance, standing in the fire of our conflicts and experiencing more and more trust and solidarity. It is often exhausting and sometimes very, really very beautiful. A new trace is love, which we follow sniffing and groping.

→See also description of network "**GEN Deutschland**"
wirundjetzt e.V. ~ sustainability network in the Lake Constance region
☺ we can grow as soon as we have expanded our barn, enriched by people who fit to us...

Gemeinschaft Haus Bierenbach

A6

Living, Learning, Working, Growing Together
Freibadstraße 1 | 51588 Nümbrecht
info@gemeinschaft-bierenbachtal.de
www.gemeinschaft-haus-bierenbach.de

2018 3 ha land, 3000 m² dwelling etc. (near Köln, Bonn) 35-50
g, e, f, sp, Russian
share holders and members (consensus) individual financing
cooperative
none
gardening, kitchen, renovation, rebuilding
ecological, non-violent, inner growth, adult education, pedagogical, environmental protection, organic gardening, living sustainably and consciously
☺ ()

In the house Bierenbach in the municipality Nümbrecht, approx. 50 km east of Cologne, a community of people of different generations is being founded. This chosen extended family increases the quality of life for all, integrates into the surrounding village and has a positive effect on the region. About 50-60 people, children, singles, couples and families are finding their place in the former leisure and seminar house in common rooms that aim for the well-being of all, combined with individual apartments. Part of the community will be an education centre with an emphasis on family, consciously sustainable living, movement, and culture to accompany people from early childhood to old age. The offers will support sustainable learning, strengthen social skills and promote environmental awareness and creativity. A special focus of this community is dedicated to our children and young people. We create suitable internal and external developmental spaces for the self-determined development of their personality and the development of their own lifestyle. Children choose their educational activities and places according to their own rhythm. The outdoor area will be a landscaped garden with seating, open areas and plenty of room for playing and gardening.

Additionally, a café, a farm shop and cultural activities are planned. The result is a small settlement joyfully integrated into the neighbourhood of the village Bierenbachtal.

planned: foundation Schloss Tempelhof & CSA
☺ membership, cooperative shares

Gemeinschaft in Güstritz (Wendland)

C3

Landstraße 6 | 29462 Güstritz ☎+49-(0)5843-1574/-8868
uwedonarski@gmx.de

1993 7200 m² (near Salzwedel, Uelzen) 20 % ♀4 ♂4 5 13

g, e, f 6
everyone in common (consensus) limited shared financing
association property
authentic, mindful contact
gardening, house, wood, maintenance, honey farm
ecological, self-managed, anarchistic, spiritual, chaotic, inner growth, meditation, yoga, organic diet, vegetarian diet, vegan diet, environmental protection, organic gardening, alternative energy, ecological construction, humour, non-smoking, painting, music, Shiatsu, Shintai

☺ ()

In 1993 the community purchased a residual farm including outbuildings on a large property in Güstritz near Wustrow. It is possible to live in several separate units of different sizes. On the main building we have a solar energy system - combined with a wood gasification plant - for hot water and heating. Photovoltaics are in operation as well as a plant-growth waste water treatment system for four premises. We are involved in wind power stations nearby. Our community of people - at this time aged between five and 85 years - is connected with other communities in the neighbourhood and interested people.
We want to share spiritual and religious variety in their concrete forms accompanied by a free and creative dialogue about our experiences.

☺ probationary period

Gemeinschaft Lebensbogen

B5

Auf dem Dörnberg 13 | 34289 Zierenberg ☎+49-(0)5606-563 90 73
kontakt@gemeinschaft-lebensbogen.de
www.gemeinschaft-lebensbogen.de

2015 4 ha (near Kassel) ♀12 ♂10 1 23
16 16
different (consensus) common economy
ownership
conference House, café - restaurant, administration, facility management
please check our website

☺ ()

We have agreed on common values, strive for a sustainable lifestyle and a respectful and authentic togetherness. We decide by consensus and live a common economy, i. e. all income

and salaries go into one fund, from which everything is then paid.
The aim of the project is to create, design and maintain a lively meeting space in which community and guests can research individual and socially relevant topics, exchange ideas, create synergies and develop sustainable initiatives and activities.
Based on our values and attitudes, we operate a conference house and a café as a collective and organise a wide range of events.
With our project we would like to invite to participation, inspiration, joy, exploration and exchange of experiences. Through jointly designed visions of the future and bundled strength, we can actively work in society and the world.

→See also description of networks "**GEN Europe, KommuJa, GEN Deutschland**"
☺ see web page

Gemeinschaft Schloss Blumenthal

C9

Blumenthal 1 | 86551 Aichach-Klingen ☎+49-(0)8251-890 43 00
gemeinschaft@schloss-blumenthal.de www.schloss-blumenthal.de

2006 33 ha (near Aichach, Augsburg, München) 25 % ♀23 ♂19 18
60

g, e — few full-time, many part-time — depends
all concerned together (preferably by consensus) — individual financing
property of all residents
guest house, hotel, seminar centre, organic agriculture, art and culture, self-employment
ecological, self-managed, non-violent, spiritual, inner growth, meditation, yoga, seminar-house, organic gardening, ecological construction, humour, humor, undogmatic

☺ ()

We have been living in the castle grounds since 2007 and are still very occupied with further development. We run a 80 bed hotel with seminar business, corporate events and conventions in the spirit of the common good economy. In addition, we have a CSA, a vegetable garden, a few animals, a meadow orchard, a total of 27 hectares of land, forests and meadows for the development of our own agriculture. Our community is vivid - without dogmatism and with room for variety and diversity. We are ecologically minded in the sense of the highest possible level of sustainability, socially minded in terms of our responsibility for one another and our shared growth, we promote art and culture in terms of creative offerings and a lively cultural scene. We're economically minded in terms of providing opportunities to earn our living on-site and our aspiration for an economy for the common good. We create and maintain a space where we can live and work healthily and consciously and have an effect on the outside world. We see ourselves as a meeting place for different people, cultures, age groups and nationalities. In order to develop our project/business we have regular planning meetings. We own the property communally and live in our own apartments in order to safeguard necessary retreat options.
We hope for more people with a genuine interest in community development, WE-processes, non-violent communication, dialogues, consensus decision-making, open-mindedness and tolerance. Every two months, we host information days for interested guests.

→See also description of network "**GEN Deutschland**"

☺ admission process over half a year, after that half a year probationary period. Vision. Funds for personal living space

Gemeinschaft Sonnenwald Schernbach

B9

Haldenweg 17 | 72297 Seewald-Schernbach ☎+49-(0)151-212 143 71

info@gemeinschaft-sonnenwald.de www.gemeinschaft-sonnenwald.de

2019 8000 m² living/functions, almost 60 ha agriculture (near Freudenstadt, Straßburg, Stuttgart) ♀22 ♂14 1 37

g, e, f, i, Russian, Czech, Bosnian 4 7

the full community assembly (six-stage consensus decision making)

the whole community will finance specific areas and community supported agriculture (CSA)

to be owned by the community, only 80 acres of agricultural land will be leased by us

no

regenerative biological agriculture and seminar and training facility

ecological, self-managed, non-violent, spiritual, chaotic, inner growth, meditation, organic diet, adult education, seminar-house, peace-work, mentally/psychologically handicapped persons, physically handicapped persons, environmental protection, self-sufficiency, permaculture, organic agriculture, organic gardening, organic farming, animal friendly husbandry, forestry, alternative energy, new technologies, ecological construction, humour, accessible to the disabled, integration into the existing village, art, health care, meeting point café/farmyard shop

☺ ()

In the northern Black Forest, on a sunny plateau, lies Schernbach, a village of just fifty inhabitants, including five children - and soon, we hope, still more, when the Community Sonnenwald Schernbach moves into their new Ecovillage. At the moment we are a group of twenty-four full members and have acquired the buildings and agricultural land of a charity that operated a home for the handicapped. Together with twelve young members of the Akademie für angewandtes gutes Leben ("Academy for Applied Good Living") we are about to found a community for living and working. Here we will develop a seminar center, a place for silent retreats, health clinics, workshops - among many other things - and, above all, a community supported farm to supply not only our own needs but also the region with organically grown produce.

We have a total of 8000 m² of living and work space for ourselves and all who will come to join us. This space consists of four large buildings surrounding a 300 year old village lime tree and a light flooded center, where we will have our community kitchen, meeting point, cafeteria and gym hall.

This impressive complex stands in the so-called Upper Village. On the other side of the county road is the center of our farm, with all the usual farm buildings and a farmhouse, as well as most of the 60 hectares of tillable land, pasturage and forest.

Soon, we will be welcoming guests from all over the world here, young people on their journeys and older people interested in a contemplative retreat.

→See also description of network "**GEN Europe**"

☺ you have to support and agree with our concept/basic ideas on the website

Gemeinschaft Sulzbrunn

C10

Sulzbrunn 1-8 | 87477 Sulzberg ☎ +49-(0)8376 - 929 49 49
kontakt@gemeinschaft-sulzbrunn.de www.gemeinschaft-sulzbrunn.de

2015 15 ha (near Kempten) 50 % (vegetables) ♀23 ♂19 10
52

g, e 7 15
all together (consent or consensus) individual

belongs to our Foundation
none
gardening, hosting Seminars
ecological, self-managed, spiritual, inner growth, organic diet, vegetarian diet, seminar-house, peace-work, self-sufficiency, permaculture, organic agriculture, organic gardening, alternative energy, new technologies, ecological construction

☺ ()

We are a colorful mix of people between 0 and 79 years, who started to build a community in the Allgäu region.
We want to experiment with sustainable life models in which people can discover and develop their potentials.
Life models in which we are mindful with each other and with our environment.
We have no spiritual, political dogma or beliefs of any kind. But each one of us has recognized in his own way that everything is interconnected and there is a spiritual reality which is greater than our personal one.

→See also description of networks "**GEN Europe, WWOOF, GEN Deutschland**"
☺ a half year of approach, then one probationary year,
30.000 Euro as payment in the cooperative

Gemeinschaft/Netzwerk in Heckenbeck

C4

Loose, Partly Completely Open Closed Community
Schulstr. 5 | 37581 Bad Gandersheim ☎ +49-(0)5563-999 18 90/-999 492
netzwerk-heckenbeck@gmx.de netzwerk-heckenbeck.gmxhome.de

1984 (near Göttingen) 10 % ♀90 ♂65 120 275

g (e, sp) approx. 40
the groups and projects decide for themselves - the community has no formal structure
individual resp. project-related financing

family homes, rented apartments and house-groups
everything is possible
ecological, self-managed, non-violent, spiritual, inner growth, meditation, adult education, free school, seminar-house, organic gardening, organic farming, ecological construction, humour, therapy: different body and psychotherapy methods, midwifes, medical practice, salutogenesis, craft: joinery, carpentry, clay construction, strawbale building, felting, art: sculpting, painting, other: many communication groups

☺ ()

Since 1984, the small village of Heckenbeck (500 residents) and its surroundings have seen the arrival of newcomers (families and individuals), who have set up a number of alternative groups, initiatives and creative projects. Besides the traditional village, there are the following initiatives: an independent school (up to the 10th grade) and kindergarten, a toddler group, the "Klang der Stille" association with a house for meditation, the "Weltbühne" Cultural Centre, the "Heckenrose" association for ecological projects, the German section "Action for Happiness", a community supported organic shop and organic gardening, organic beekeeping, organic sheep and goat farming, fruit growing and tree care, a joint doctors practice, midwifes, building and craft firms, art-atelier, women's and men's groups, meditative dance, various forms of meditation, plenum. Every now and again, there is community singing, theatre improvisation, dance parties, contact improvisation, Tango, Qigong, Tai Chi, etc. In contact with the network is a "Zentrum für Salutogenese" (Salutogenesis) and seminar houses "Alte Mühle" and "Einbecker Sonnenberg".
The initiatives and projects are autonomous; some welcome new members, others are established groups. These groups organize themselves and are linked in various ways and to varying degrees with the network. For many of us, life in Heckenbeck is a continuous search for new ways of living together, ecologically and with spirituality.

- some projects (e.g., school) belong to networks, rsp. are the center of networks (Zentrum für Salutogenese)
- ☺ we hope that newcomers get on well both with the original villagers and with other newcomers

Gemeinschaftliches Leben e.V. Blütlingen

C3

Dorfstr. 30 | 29462 Wustrow ☎+49-(0)5843-7780
dorofrank@web.de

1995 | 0.8 ha (near Lüchow) | ♀4 ♂3 8 15

- g, e, sp, f, Russian
- all concerned together (consensus) — limited community financing
- association
- home economics
- ecological, self-managed, left-wing political understanding, non-violent, organic diet, organic gardening, alternative energy, ecological construction

Our community is joined together by the co-existence of small and large people (we have four families at the moment) on an old farm (in the Wendland) which we have expanded and remodelled according to ecological principles. We are striving for an ecological household and both design and care for the land adjacent to the house with its variety of gardens and several small animals. We seek equitable living conditions and are working to develop a resource- and environment-saving lifestyle which does not suppress the needs of the individual.

☺ sympathy, but we don't have vacancies currently

Gemeinschaftssiedlung Neumühlen

B3

Am Oxer 1a | 27283 Verden ☎+49-(0)4231-937 960, -667 724
info@ulrichsteinmeyer.de, a.wortmann@posteo.de neumuehlen.org

2004 6000 m² (near Bremen) ♀10 ♂7 8 17

g, e, f, sp none

joint decisions (aim for consensus)

individual financing for private living, common financing for common areas and energy supply

cooperative

ecological, self-managed, alternative energy, ecological construction, individual financing for private living area

☺ ()

We are an ecological community settlement on the edge of the town of Verden. The first of our houses were inhabited in early 2004. The settlement is a combination of private (houses/flats and gardens) and communal property. Some flats are also rented. Occasionally there are free rooms in shared households or flats become available.
The communal area consists of a main house with a large space for celebration, dancing and frolicking, yoga etc. It's a multi-purpose space with kitchen, guest rooms, sauna and the grounds include a football pitch, playground, terrace with fire pit and barbeque, as well as a technical area with a combined heat and power plant and grey water treatment facilities. The resident association manages the settlement. Apart from the financial participation and minimal working requirement for the communal areas, there are no obligations. Our social activities are entirely voluntary. (Almost) Nothing is compulsory, but everything can be!

Verbunt, Aller Wohnen e.G.

☺ funds for private ecologic building and for buying shares of the cooperative

Gemeinschaftswohnen im Wedding 4. OG West

E4

Lynarstraße 38 | 13353 Berlin
lynar@posteo.de

2014 342 m² ♀2 ♂5 7

g, e, f, sp

residents together with association
(consensus principle, otherwise qualified majority (75 %))

individual financing

cooperative ownership, permanent usage contracts

world ethos, ideologically neutral; mutual respect

external activities, jobs outside the community

ecological, non-violent, adult education, pedagogical, peace-work, one-world development, mentally/psychologically handicapped persons, physically handicapped persons, ecological construction, non-smoking, cross-generational

☺ (👐)

The communal living in this special residential building is characterised by the fact that we residents share central areas such as kitchens and living corridors. This corresponds most closely to the model of a classic, but generation-spanning residential community. Admission to our house project is only possible if the people who move in here have previously found each other as a group and want to live together to some extent.

☹ longer period of getting to know each other, joining the cooperative, recognition of the house rules

Generationenhof Landau

B8

Haardtstraße 7A | 76829 Landau ☎+49-(0)6341-54962-50
info@gehola.de www.gehola.de

2010 | 3000 m² (near Karlsruhe) | ♀36 ♂17 10 58

g, e, f | 15

all together (varies) | individual financing

property of the co-operative

maintenance of buildings and outdoor facilities, property management, body of the co-operative

self-managed, accessible to the disabled, generation-spanning

☺

"Living neighborhood is more than housing"
Under this motto, the LAWOGE eG as a building society realised its independent model of a co-housing project "GEHOLA" (generation court Landau). Structures, instruments and models were created to promote and to develop the communal aspect of co-housing. This is reflected in many details:

- in the architecture and equipment of the residential buildings which should promote meeting, communication and openness
- in the internal structures which are democratic, self-governing and clear
- in the juridical and economic constructions which are not aimed at short-term effects or personal profit but at long term social and human benefit
- in various concepts which should stand as models for residential forms and life forms of the future, incl. a generation-covering inhabitants mixture, active neighbourhood, car-free residential layout, barrier-poor, power-saving and climate-neutral house technology, integrated nursing apartment-sharing community.

The project has given itself the legal form of a registered co-operative to create a democratic basis for the arrangement of the project. The co-operative is owner of the buildings and flats. The inhabitants are by acquisition of co-operative shares owners as well as users of the available buildings, flats and common rooms.
The housing complex of the Generationenhof Landau includes two buildings in 3.5-storey construction method, underground parking, lifts, common room, workshop, child space and guestroom. There are approx. 40 residential units with approx. 3000 square metres of living space.

☹ material and social conditions

☆ Giegelhof Elvira Horn

B10

Giegelstraße 19 | 79736 Rickenbach-Hottingen ☎+49-(0)7765-748
elviragiegelhof@gmail.com

2012 | 12 ha | ♀2 ♂2

The sunrise makes the Alps appear in pink pastel shades, our cocks crow and want to get out, the ducks chatter and also want to finally go outside, on the vegetables lies the morning dew. It is cool in autumn, wood has to be chopped to heat up the oven. This cosy warmth delights the two- and four-legged inhabitants of the Giegelhof.
The Giegelhof is situated at an altitude of 750 metres in the southern Black Forest, close to the Swiss border, on the edge of the village of Rickenbach-Hottingen. For decades families, helping hands and various community members have farmed this farm ecologically. The farm has 12 hectares of land, including four hectares of forest and four springs. The house itself is 250 years old, the floorboards are creaking - it has charm and smells of childhood to many people who come and go here.
In the winter of 2012 I, Elvira Horn, took over the farm with my former partner. I was fed up with big city life, as well as with dependence on supermarkets and other producers. The focus for me and my community is still on self-sufficiency. Currently this community consists of two women and two men aged between 23 and 47. WWOOFers are always welcome, as are suitable roommates.
A film about us was shot in 2013: #8956 at eurotopia.directory/booklinks
Mindfulness, resilience, tolerance and independence are important for us so that our little paradise can continue to be cultivated. It is not only a dream, an idyll, but also hard work that makes us happy.

Gruppe Laufdorf des Laurentiuskonvents

B6

Ringstr. 21 | 35641 Schöffengrund ☎+49-(0)6445-5560
laufdorf@laurentiuskonvent.de www.laurentiuskonvent.de --> Laufdorf

1976 | 1000 m² (near Wetzlar) | 20 % | ♀5 ♂5 1 11

- f, e, g, i, sp
- all concerned together (consensus)
- 3 retired + 2 self-employed
- limited shared economy
- property of the Laurentiuskonvent
- ecumenical, non-violent
- peace work, refugee work, mediation
- self-managed, non-violent, Christian, ecumenical, organic diet, peace-work, organic gardening, non-smoking, meeting place, mediation, refugee work, integration work

☺

We live in an intentional neighbourhood community. Living in separate houses enables us to sensitively seek a balance between closeness and distance, and between our private lives and our life as a group. We come from different parts of Germany and elsewhere in Europe. And we come from a variety of religious backgrounds. Nevertheless, we experience again and again how our experiences, the influence we experienced, our gifts, and our dreams all come together in a new whole - thanks to our shared praxis and reflection.
We maintain a wide variety of contacts and relationships in the village, in the region and in ecumenical and other networks. Each of us plays an active part in the life of our own church

or congregation. We take joint responsibility for looking after houses and gardens. We enjoy food from our own gardens and purchase fairly traded, local/regional, and organic produce as much as possible.
We eat together several times each week. At weekends we spend time after an evening meal playing games, singing, making music, dancing, looking at photos or watching films.
We come together

- to pray - with Bible readings, songs, and silence
- to share - pictures, music, texts from various sources, through quiet times or through discussion
- to celebrate - with games, dancing, music, food and drink
- for Bible studies, allowing expressions of faith from long ago to speak to us
- to celebrate communion

Our everyday life as a group depends on frequent consultation and transparency. We make decisions about our life together through consensus.

→See also description of network "**Church & Peace**"

Laurentiuskonvent

be Christian/secure livelihood

Gut Gorgast

E4

Amthof 4 | 15328 Gorgast +49-(0)152-014 395 95, +49-(0)173-230 31 72
kontakt@gut-gorgast.de gut-gorgast.de

2015 ca. 6000 m² (near Frankfurt (Oder), Berlin) 0 ♀6 ♂5 3 14

g, e 2

the community (we decide in a consent procedure (according to sociocracy principle), either on-site or electronically)

a cooperative association

none

constructing in an ecological way, workshops for living and working together e.g. Yoga, alternative leadership, ecological gardening, personal development and living our multigenerational community

ecological, self-managed, inner growth, ecological construction

We are a multigenerational community. We are not explicitly focused on only living together, but being together. The focus is on joint action with mutual and comprehensive care in all areas of life. We are organised as a cooperative association. Our common life on-site and all decisions are made by consent in a sociocratic manner. Not all of our community members are members of the cooperative association. As a complement we have another non-profit association in formation which supports social-ecological living together.
We reconstruct our community house, which was built in 1840 as a mansion, in an ecological way.
The house is situated by the small river "Alte Oder" and the old parkway. Besides the 1500 m² for common rooms and individual living,

there is also an old upgradable boiler house and a gym for bigger events. At the moment we see our site as a possible center for our growing community in and around Gorgast and Berlin especially for coming together, conducting workshops and networking with other interested people. Our alignment is openly oriented towards social and ecological issues, but otherwise very flexible according to the interests of new coming people.

☺ no specific rules, we get to know each other and see whether it fits for both sides

Gut Möglich

B5

Arnold-Flues-Straße 2 | 59872 Meschede ☎+49-(0)2919-936 97 36

info@gut-moeglich-meschede.de www.gut-moeglich-meschede.de

2017 1.2 ha (near Dortmund) 10 % ♀4 ♂9 0 13

3 5

all concerned (consensus)

aiming for shared economy, currently in some areas (household budget, car pool, real estate)

private ownership currently

pragmatic idealists, a motley mix

arts, crafts and culture, gardening, beekeeping, extension and renovation

ecological, self-managed, left-wing political understanding, non-violent, organic diet, physically handicapped persons, permaculture, organic agriculture, organic gardening, ecological construction, humour, organic food, carpentry, beekeeping, car sharing, cosmopolitan, sharing, respectful togetherness, solidarity, art, arts and crafts, tiny houses, trailer conversion

☺

We live on an old manor in the Sauerland region of Germany's midwest. In about 3 km distance is lake Hennesee. Roughly 1200 m² residential and common area offer single apartments to live in as a couple, family or just to share a flat.
Approximately 2500 m² can be directly used in the form of stables and barns, or possibly converted. A small stream flows through our property.
A common income economy is the basis of solidarity. For this we are still experimenting with shared income models to find an appropriate one. The foundation of an association and / or a cooperative is planned.
Being aware of one's own environment is one of our core ideas. From beekeeping and vegetable growing to construction and conversion, we implement our ideas with our own energy and individual skills and abilities. We have no intention to cut ourselves off from society completely, rather, we want to participate in changing it for the better.
Together we can complement each other and learn from the people around us.
We keep privacy and community life more or less in balance.

Everyone shall feel welcome with us, age and origin do not matter. A balanced group needs the experience of older people, the curiosity of children and the strength of young adults

☺ aquaintance, probationary period, community, sharing our work

Gut Saunstorf - Ort der Stille Modern Monastery

D2

Gutshaus | 23996 Saunstorf

hausleitung@gut-saunstorf.de www.kloster-saunstorf.de

2010 (near Hamburg, Lübeck) 50

g, e 33

advaita, spiritual wisdom teaching

self-managed, hierarchical, leader/guru, spiritual, inner growth, meditation, yoga, organic diet, vegetarian diet, organic gardening, alternative energy, self-awareness, self-enquiry, enneagram

"Gut Saunstorf - place of silence" is a modern monastery, a place of retreat and recreation. Gut Saunstorf is accommodating a monastery community and gives space for wisdom teacher OM C. Parkin to teach. In this place the wisdom teaching and its quiet core are preserved and made practically experienceable for the people.
Gut Saunstorf is also a material place that needs care and attention in order to be a place of silence for people who are interested in their innermost being. People from the monastery community work together with people who support this place for a time.
Support (year-round or seasonal) is needed with: vegetables and orchard, park with outdoor facilities, manual work in house and garden, simple work in kitchen and housekeeping, research, technical support at events and much more.
For the group of Karma Yogis - an average of 4-6 live here - there is a weekly guided meeting which supports the research of emerging topics. If you are deeply interested in learning more about yourself, you can participate in the Sangha, a regular community meeting. We wish for motivated and self-reliant people with practical and flexible attitudes who would support us six hours a day.

GutAlaune

D5

Alaune 9 | 06193 Petersberg

gutalaune@riseup.net www.gutalaune.de

2016 1 ha (near Halle (Saale)) 10 % ♀5 ♂6 5 16

e, g, Turkish, Polish 5-8

the whole group (consensus) sharing food, car and household costs

association

ecological building, cultural events, workshops, gardening, childcare

ecological, self-managed, non-violent, inner growth, organic diet, adult education, environmental protection, permaculture, organic gardening, new technologies, ecological construction, humour, traditional crafts, art and aesthetic, open workshops, art and culture

We are a colorful bunch, living, working and realising projects north of Halle (Saale) in the northwest of Germany. Most of us work in arts, music, education or crafts.
We are establishing a place where we treat each other as well as mother nature with care and respect. Also it is a place for exchange and socio-ecological projects. We are busy creating space for living and working. There is a row of sheds waiting to be transformed into workshops and project spaces, and the main community building is in need of a second floor.

We are outdoors a lot, building and playing, and are always happy about helping hands - let us know before you show up ;-).

network of open workshops

Gutshaus Hermannshagen

D2

Verein zur Sanierung und Revitalisierung als Ort von Begegnung und Bildung

Untere Dorfstr. 1 | 18246 Bernitt

info@hermannshagen.de www.hermannshagen.de

1998 (near Bützow)

g, e

In-season 3-4 people for juice production

members meeting, board of the association, meeting of working groups (direct democratic, partially consensus-based)

some common financing (e.g. of food and household)

owned by association (Gutshaus Hermannshagen e.V.) + leased land

no ideologic principles

gardening, management of meadows, fruit tree pruning, fruit picking/processing (juice extractor), maintenance work (house and farm)

The association Gutshaus Hermannshagen e.V. was founded in 1998 by a group of people to renovate the old manor house and to revive the town and the region. The place was shaped by numerous volunteering events, whereby historical and ecological building techniques were used. In the surroundings of the manor house a vegetable garden was created, the old orchard is maintained and cultivated, new trees were planted, the historical manor house forecourt was paved and a plant purification plant was built. People from all over Germany and Europe participated in the projects. The association operates a mobile apple juice press to enable people from the surrounding area to process their own apples. In the last years preservation & care of the area bundled the existing capacities. Currently there are a few new heads in the association, with many ideas & visions regarding the manor house. In the center stands the completion of the house, in order to facilitate communal, solidary living & ecologically lasting management. The aim is to create a place for cultural, political life and children and youth projects in rural areas. Small construction sites will continue to take place, where guests are very welcome (you can find information about events on our homepage). We offer various long term volunteering possibilities. Should you feel like participating in the design & revival of the place, please contact us by e-mail!

Working Community Environmental Education/ANU MV e.V.
https://www.umweltbildung.de, www.anu-mv.de

Hasenhof

C8

Hasenhof 8 | 71540 Murrhardt ☎+49-(0)7192-8176, +49-(0)162-239 74 59
sempervivum@posteo.de

1998 | 0.3 ha (near Stuttgart) | 5-15 % | ♀2 ♂3 | 5

g, e, Swabian (dialect)
everyone, questions regarding the house itself, the owner (by consensus)
shared finances for some areas | owned by one person
multi-spiritual, agnostic
everyone works in her/his job in addition to the work in the garden and the household
ecological, feminist, non-violent, inner growth, yoga, organic diet, vegetarian diet, self-sufficiency, organic gardening, alternative energy, humour, undogmatic, friendly, community-oriented, loving

☺

We are a small house community on the outskirts of a town 50 kilometres north-east of Stuttgart (10 minutes walk to the closest bus stop, 30 minutes to the train station). We live in, clean, and take care of a small two storey house and try to grow some of our own food in our vegetable garden. We enjoy spending time together and often host events at our house or go out (to concerts, movies, lectures, etc.) together. We try to continually learn from our every-day life and improve ourselves as human beings. We live together, laugh together, work together, and discuss any arguments together. We have a cat, some chickens, nice neighbours, and Wi-Fi.

☺ undogmatic, hard-working, willingness to work on personality

Haus der Gastfreundschaft (HdG)

D2

Alte Dorfstraße 6(-13) | 19370 Dargelütz bei Parchim
☎+49-(0)38720 88558, +49-(0)157-850 656 76
schenker_oeff@yahoo.de www.global-love.eu, vfs-dargeluetz.npage.de

1994 | 1 ha (near Parchim) | ♀2 ♂6 | 8

g, e, f | 5-6
individually and consensus (with different weight of votes)
some make individual financing, others family-sharing, as the individuals agree
HdG as the center: the premises are used in the sense of moral using rights, like organic cells in the organism of the world
givers-(or gift-)movement
caring for the place (self-sufficiency, building work), creating a loving community-life, searching for edible left-overs
ecological, self-managed, anarchistic, left-wing political understanding, feminist, non-violent, spiritual, Christian, ecumenical, Buddhist, Taoist, anthroposophic, inner growth, meditation, yoga, free love, adult education, peace-work, animal protection,

environmental protection, self-sufficiency, permaculture, organic gardening, organic farming, ecological construction, humour, bestowed alternative social work in the sense of mankind's family charity

The "House of Hospitality" has been founded in 1994 as the first settlement of the "Schenkers- (Givers- or Gift-) Movement" (for idealistic background see chapter "Networks of Communities"). In the beginning we had only one house with a stable and a garden, without electricity, only rain-water; later our "giver-allies" managed to buy some of the neighboring real estate.
We try to practice "unconditional love", and "social work" in this sense: "Accepting every human being as humanity-family." According to our ideals, such "social-work" is a second priority besides "conscience work": How we as 'global consensus brothers and sisters' can learn to plan together what is necessary for the world - in free argumentation-culture...
The "House of Hospitality" has seen very "wild" times, with sometimes only one female "giver" and up to 10 very difficult and dangerous "guests".
As everywhere, it has been hard to create an atmosphere of open communication and free argumentation-culture, but nowadays a group of former guests has grown together with some of our giver-allies to a fascinating experimental-group. As a result of that and the "ally-projects" in the neighborhood, life has become much easier. Newcomers can - in the "ally-zone" of the project - find rather "normal" conditions of living (with internet etc...), as well as a slaughter-free and vegan zone, possibilities for self-sufficiency and for participation in the "social-work". From normal flat to earth-hole, there are lots of alternatives to choose from. The project is drug-free.

→See also description of networks **"Givers-Movement, European Catholic Worker, Emmaus, L'Arche de Lanza del Vasto"**

working with LIGHT and LOVE, internat. gift economy (moneyless.info), Gandhi-Movement, Terrania, Freiwind

fit into our (holistic) culture-ideal

Haus Mercurial - Multigenerational Housing (C10)

Schwarzwaldweg 20 | 88239 Wangen/Allgäu ☎ +49-(0)7522-912 310/-914 791
info@ig-lebensgestaltung.de, ingrid_feustel@web.de
www.ig-lebensgestaltung.de

2000 | 700 m² (near Ulm, Stuttgart) | ♀3 ♂1 2 6

g, Dutch | 6

all together (consensus) | individual financing

trust

ecological, self-managed, Christian, ecumenical, anthroposophic, adult education, free school, peace-work, environmental protection, alternative energy, non-smoking, editorial office of local magazine: "Die Welle", contact point for citizen engagement

The association "Integrated Living" ("Arbeitsgemeinschaft Integriertes Wohnen") was founded in 2005. Integrated living means living together in small houses that are connected to one another. Our motto is: "young and old under one multi-generational roof!" As such, it offers an alternative to the existing large housing complexes with residential carers or care homes.

When care and household help become necessary, one can make use of the outpatient care services and organised neighbourhood support available thanks to exchange systems. The "Mercurial House" has been in existence since 2000 and has been developing consistently. Our criteria for communal living are the following:

- to live at home and to grow older in a self-determined manner
- young and old, i.e. multi-generational, own household, individual living, house community, elective affinities, mutual support
- new legal agreements and alternative financial arrangements, as well as a new way of handling property in order to remove it from speculation
- sharing of other people's experiences according to necessity

Interessengemeinschaft für Lebensgestaltung e.V., LETS Lindau/Wangen, GM-free region of Lake Constance

Heller Wohnen Mehrgenerationenhaus

C8

Auf dem Katzenkopf 15 | 74523 Schwäbisch Hall +49-(0)7919-782 17 85
heidrun.schaumann@t-online.de www.heller-wohnen.de

2011 2500 m² ♀19 ♂14 5 38

g
community plenary (vote or consensus)
private property and rented
none
social and cultural activities in the house, the neighbourhood and the town district
ecological, self-managed, non-violent, pedagogical, peace-work, alternative energy, humour, accessible to the disabled, social commitment, supporting migrants

everyday tasks in teams
shared household budget

☺

Personal report of a resident:
"For seven years now, I have lived in our intergenerational housing project HELLER WOHNEN, in which I rely on a wheelchair to get along wonderfully in the house and garden. I am 76 years old and getting old brings so many inconveniences. So suddenly I needed to be in the hospital for a while. I talked about it in the check-in session of our community meeting, which takes place every 14 days. Immediately someone was ready to help me with packing; another accompanied me to the hospital. There I received daily calls and visits from residents of our house; surprisingly, also from people with whom I have little to do otherwise, because they themselves are pretty busy with their everyday lives. So M., a young single mother from Jordan, visited me with her five year old daughter. For the first time, we talked about our cultural differences. At the final day, there were also people who helped me and took me home. A pot of fresh soup awaited me in front of my apartment door. I'm back home, I feel well, safe and secure in our large, colourful house community.
Despite all diversity, the desire for a solidary and self-determined life unites us in responsibility for each other."

☺ support our concept

Hof Hackenow

E4

Hackenower Weg 6 | 15328 Alt Tucheband ☎+49-(0)157-735 181 82
hof@hackenow.de hackenow.de

2016 2 ha (near Frankfurt/Oder, Berlin) 10-15 % ♀1 ♂4 1 6

g, e, f 1-2

all concerned together (plenary/consensus)

shared financing for some areas property of all

pragmatism, anarchism

working in the garden, building things, taking care of animals

ecological, self-managed, anarchistic, left-wing political understanding, feminist, non-violent, chaotic, vegetarian diet, pedagogical, animal protection, permaculture, organic gardening, alternative energy, construction

☺

We are a housing and farm project in the southern Oderbruch (close to the Polish border east of Berlin). Currently, there are six people living there. When more living space has been created, around 10 people will live and partly work together on 2 ha of land. The houses are in need of much renovation.
We try to grow our garden according to permaculture principles, we have two compost toilets, we produce our own biochar. The fruit stock provides a lot of freshness in the summer.
We are undogmatic and punk. We receive occasional visits of helping hands, organise construction weeks and cultural or musical events several times a year.
It is important to us to increase the share of self-sufficiency in food and energy. We also want to be involved in ecological networks locally and translocally.
In 2019 we will build a wetland and a yurt. In addition to us humans, some cats, chickens and ducks live on the farm (as well as many wild animals). There is also room for more of them.

☺ we will have to find or build more living space first

Hofgemeinschaft Guggenhausen e.V.

C10

Hauptstr.15 | 88379 Guggenhausen ☎+49-(0)7503-916 877
hofgemeinschaft-guggenhausen@web.de

1993 4 ha (near Ravensburg) ♀8 ♂5 3 16

g, e, f, sp 2

everyone in common (consensus) limited community financing

owned and leased

Christian/ecumenical

individual professions, garden, kitchen, orchard, forest

ecological, self-managed, left-wing political understanding, non-violent, Christian, ecumenical, organic diet, adult education, one-world development, environmental protection, organic agriculture, organic gardening, forestry, alternative energy, humour

Since the summer of 1993 we are living as a Christian life partnership in three reconstructed farm houses of former family farms: we are currently 13 adults and 3 children- some of us as core of the partnership, some of us connected in a looser way. We earn our living by engaging in professional work in a variety of advisory and social areas.

Our common life is expressed in regular shared meals and daily prayer times, weekly organizational meetings and running our small farm as a Bioland (organic agriculture) operation.
Our two guest rooms provide those interested in community living and also people in life crises an opportunity to live with us for a time. As individuals we are involved in a variety of social, ecological, development and ecumenical initiatives.

→See also description of network "**Church & Peace**"

Hofgemeinschaft HOFbauernHOF

B9

Reinerzauer Str. 13 | 72290 Loßburg
+49-(0)7446-916 047
info@hof-bauern-hof.de
hof-bauern-hof.de

2005 30 ha (near Freudenstadt) 5 % ♀5 ♂3 9 17

g, e, f, sp 5 5

5 (consensus) shared income

leased

farming, vegetable growing, bakery, adventure education, construction

ecological, self-managed, non-violent, anthroposophic, organic diet

Above all, or at the basis of everything, we as a community of locally active people would like to give IMPULSES to the world – in the most varied of subject matters, imbedded in our vision to live and work in a self-determined manner. This is the origin of our MOTIVATION to be an active part of the non-materialistic ecological agriculture and all that comes with it. Goals:

- ecological lifestyle, simple life
- "different" communication/sense of community
- impetus for a more nature-aware, holistic perception
- vitalisation of the farm and its environment ("vibrancy", "farm aura")
- bio-dynamic farming
- cultural impulses on the land

demeter

Hofgemeinschaft Kotthausen

A5

Kotthausen 1-5 | 42399 Wuppertal
+49-(0)202-250 43 50
info@hof-kotthausen.de
www.hof-kotthausen.de

2001 ca. 30 ha (near Remscheid, Wuppertal) 10 %

♀15 ♂13 12 40

g, e

everyone concerned in common (consensus) individual financing

common property (association)

organic farming, assisted living, biological food delivery service

ecological, self-managed, non-violent, organic diet, pedagogical, seminar-house, mentally/psychologically handicapped persons, environmental protection, self-sufficiency, organic agriculture, organic gardening, organic farming, animal friendly husbandry, forestry, alternative energy, humour

Our community farm was born from a registered non-profit association.
The farm is primarily collectively and voluntarily run by the people living on it. Some residents take care of the vegetable gardens (for self-supply), others of animals or forestry.
Besides residential buildings, barns and stables, 30 hectares of land (field, grassland and forest) belong to the farm. We also have sheep, a goat, horses, chickens, dogs and cats.
Apart from that there is a "Biokiste" (delivery service for organic products) including a small nursery (under development) as well as assisted living for people with mental stress on the farm.
We also offer hippotherapy, pedagogical programs for groups, an annual international workcamp in summer, wwoof stays as well as numerous farm festivities and activities in the course of the year.

→See also description of network **"WWOOF"**

Hofgemeinschaft Lindenhof Allmende e.V.

C4

Presseweg 6 | 38170 Eilum ☎+49-(0)5332-3547
lindenhof@eilum.de www.lindenhof-eilum.de

1979 100 ha (near Braunschweig) 50 % ♀5 ♂5 3 13
e, f, g, sp 10 10
jointly (consensus) some community financing
rented/leased
organic farming and vegetable gardening (Bioland certified), bakery, direct marketing (weekly and special markets, own food shop)
ecological, self-managed, left-wing political understanding, organic diet, environmental protection, organic agriculture, organic gardening, organic farming, alternative energy, ecological construction, music

Together we run a large organic agricultural business (100 ha) with farming, animal husbandry, vegetable garden, bakery and direct marketing. We have two residential houses: one family runs its own household and the others live in a flat-sharing community. Important decisions are made by the entire group on a consensus basis in our weekly plenum. Our life is based on producing everything we need with our own operation. Because of economic constraints, we have less personal freedom than we'd like and little time for cultural or political activities. Still, together with the non-profit Allmende Association, which owns the farm, we keep managing to organise a varied programme which makes its mark on the region. Anyone who would like to combine living with us with an internship or training, or who has already had training, professional experience or other competences to help develop one of our work areas and our community life, is welcome to inquire about the possibilities.

aufpAssen e.V., GM-free region of Braunschweig
☺ available living space; willingness to live and work with us

Holunderhof e.V.

B3

Holunderstr. 8 | 27321 Morsum

info@holunderhof-ev.de | www.holunderhof-ev.de

1992 | 3.5 ha (near Bremen) | 10 % | ♀5 ♂5 | 2 | 12

g, e, Turkish | 3

plenary assembly (consensus, democratically) | individual

non-profit association

sustainability

maintenance/operation of house and farm, work in association, spiritual work, group work

ecological, self-managed, non-violent, spiritual, esoteric/New Age, inner growth, meditation, yoga, organic diet, seminar-house, animal protection, environmental protection, permaculture, organic agriculture, organic gardening, animal friendly husbandry, ecological construction

The aim and purpose of the association is to contribute to the creation of a self-determined, non-hierarchical and ecologically responsible society, against the exploitation of humans, animals and nature, aiming to create cultural, artistic, political and social activities.
In the last two years, the residents of the Holunderhof have changed fundamentally due to many newcomers. We are currently a young group of 12 people, from the age of two to the early fifties. The farm offers many opportunities and spaces for artistic, artisan or gardening activities. Studio rooms and community gardens are being created, and we are working to preserve and maintain the existing buildings.
As part of our activity as a nonprofit association and in order to implement the goals set out in our articles of association for the construction of alternative living and living spaces in harmony with nature, we organise public and private activities, celebrations and opportunities to meet throughout the year.
The Holunderhof was established in 1992. Since then, the former residents established a farm shop, gardening and agriculture, beekeeping, planted trees and many more things. As a new group we strive to preserve and develop these traditions. For 2019, the completion of our straw clay construction is planned, which will be used as a seminar house.

→See also description of network "**Kommuja**"

☺ probational week/period

Holzlandgemeinschaft

D6

Talweg 18 | 07639 Bad Klosterlausnitz | +49-(0)36601-922 988

holzland@bruderhof.com | www.bruderhof.de

2004 | 1.5 ha (near Jena, Leipzig) | ♀7 ♂8 | 2 | 17

e, g | 8

all members (consensus) | shared economy, common ownership of property

common property

Christianity, non-violent

garden, woodwork, workshop, house

non-violent, Christian, humour

☺ ✋✋

We are a Christian intentional community, which belongs to the Bruderhof Movement. This movement consists of more than 2700 people living in over twenty settlements on four continents. We are a covenant of families and singles who want to follow Jesus without compromise and in the spirit of the first church in Jerusalem. We gladly renounce private property and share everything with each other. We in Holzland often work as trainees in companies nearby or study at one of the universities in the area. Others work in the workshop, in offices or in the garden.
Regardless of what we are doing concretely, we want to follow Jesus with all our heart and with everything we do.

→See also description of network "**Bruderhof**"
☺ live with us

ISKCON Köln e.V.

A6

Taunusstr. 40 | 51105 Köln
☎+49-(0)221-830 37 78
✉kontakt@gauradesh.com
💻www.gauradesh.com

1969 ♀2 ♂7 9

- g, e, Russian, Hindi
- shop, monthly supporters, donations, selling books and sweets
- yes
- shop, monthly supporters, donations, selling books and sweets
- ecological, self-managed, non-violent, hierarchical, leader/guru, spiritual, inner growth, meditation, yoga, vegetarian diet, humour, non-smoking, Krishna temple, Krishna consciousness, Vaishnavas, Bhagavad-gita

☺ ✋✋

Gauradesh is a spiritual centre with the goal of reminding and encouraging people to see their spiritual identity as a soul with an eternal relationship with God. For that purpose we offer the millennia-old practice of Bhakti Yoga on the basis of the Bhagavad Gita as it is.

→See also description of network "**ISKCON**"
☺ following the 4 regulative principles, chanting daily 16 rounds on the Japamala, participating in the daily morning programme, able to live in a community

Jakobgut

D6

Jakobgut | 07952 Ranspach
kontakt@jakobgut.de jakobgut.de, oadien.de

1997 8 ha (near Plauen) ♀2 ♂1 3

g, e 3

everyone in common or the person who introduced a subject (consensus)
contribution to common economic activities by conduct or means of barter
property of individual members
Christian-undogmatic; very open to non-Christian people
individually motivated projects, text art, maintenance of buildings, leisure, possibility of self-sufficiency
ecological, self-managed, left-wing political understanding, non-violent, spiritual, Christian, ecumenical, Buddhist, inner growth, self-sufficiency, permaculture, organic agriculture, organic gardening, forestry, alternative energy, ecological construction, construction, humour, non-smoking, pluralistic, drug critical, able to be ascetic, inner peace

We live in a rural estate with adjoining ground with clay soil suitable for building, but since a while ago we optionally dwell in rooms in the town of Plauen. Anyone who understands lively, spontaneous community as a stimulating framework for a self-motivated life could be in the right place here. We live where others go for their holidays. And by bike it is 2 km to the nearest supermarket.
We create living spaces for people who yearn to spend a great parts of their days on mental retreats, tranquility, their own interests or community with their roommates, rather than being haunted by the repressive cult of consumption. As a community, we are fundamentally oriented towards a materially simple existence, since it facilitates us with this desired freedom. Therefore, the idea of self-sufficiency plays an important role in our community. Our specialty: Communicating on community issues must happen in written form. Moreover, it is possible to live with us without money. Particularly important are our values, which should be shared as a personal goal, for example:
Altruism, joy, peace, patience, inwardly friendly attitude towards others, kindness, loyalty, reliability, gentleness, self-reflection and self-determination over/against the worldly issues. We cultivate a frank, objective direct contact with each other. And we are a Christian community. However, we pay less attention to lip service than to the innermost ideal orientation of an individual. Since the beginning, you could also live here well if you'd say "I am not a Christian", but share those mentioned values.

☺ sharing our values as a goal, written communication on community issues

Karla*hof

E3

Karlshof 1 | 17268 Templin ☎ +49-(0)3987-200 96 50
karlahof@gegenseitig.de www.karlahof.de

2013 25 ha (near Templin) 60 % ♀7 ♂4 11

g, e, f, sp ca. 8

all persons concerned (consensus)
common financing for some areas, partially shared income
de-privatized and now owned by a foundation
farming, gardening, food processing, building and construction
ecological, self-managed, left-wing political understanding, feminist, non-violent, inner growth, organic diet, vegetarian diet, seminar-house, environmental protection, self-sufficiency, permaculture, organic agriculture, organic gardening, forestry, ecological construction, social skills: weekly "emo plenary", regular supervision and facilitation for conflicts and other topics

Karla*hof is a community project in the Uckermark region of Germany's northeast. Our community experiments with a collective form of self-sufficiency, non-commercial agriculture and social non-violent communication.
In addition to the permanent core group that consists of ten people, others live temporarily in Karla*hof as well. All together we are taking care of Karla*hof and are developing our project. Our buildings are surrounded by farm land where we mainly cultivate grain and potatoes. Furthermore, a big garden including a greenhouse provides us with vegetables throughout the year. Our community is involved in food processing, small-scale beekeeping and other activities at our locksmithery and carpentry. We seek to cover our daily needs by producing as much as we can by ourselves. Beyond that a few of us have either a regular employment or do casual jobs. In the near future we want to organise our finances more collectively.

Our living together is not only a common organisation of the project but also an intentional attempt to develop a social culture that helps us in conflict situations and allows personal growth. All together we feel connected by the spirit of commons and non-commerciality.

Projektewerkstatt auf Gegenseitigkeit (PaG)

Kastanienhof Marth Lebens- und Agrarkulturelle Initiative e.V. C5

Am Rusteberg 21 | 37318 Marth ☎+49-(0)36081-600 53
info@kastanienhof-marth.de www.kastanienhof-marth.de

1994 10 ha (near Göttingen) 50 % ♀0 ♂3 3 6

g, e, f 2
everyone in common (consensus) limited community financing
association, partially leased
ecological, Christian, Anthroposophic
self-sufficiency from organic farming and animal husbandry, orchard and garden, renewable energy, organic building, conservation, wood collection, political activity, fun
ecological, self-managed, anthroposophic, organic diet, adult education, pedagogical, environmental protection, self-sufficiency, organic agriculture, organic gardening, organic farming, animal friendly husbandry, forestry, alternative energy, new technologies, ecological construction, humour, organic ice-cream, ecological and historical building, renewable energy

☺ (✋✋)

"Sowing the future together." The Living and Agri-Cultural Initiative Kastanienhof e.V. seeks to find and realise sustainable forms of living. We are running an ecologically oriented farming and garden culture.
Joining the notion of progress with respect for the past, we want to link traditional crafts and simple technology with modern-day ecological awareness. The 1.5 ha meadow behind the farmyard is used as grazing land for goats, a donkey, a horse and pigs. One of our concerns is to preserve the diversity of old plant varieties and endangered breeds of farm animals. We have set up a half-timbered exchange in which used building materials can be purchased and returned and where new organic materials can be sold (www.fachwerkboerse-eichsfeld.de). We have begun offering educational work in the form of seminars, school class visits and workshops. We want to network with other living and commercial communities. The farm community provides a framework for socially dedicated and supervisory tasks. An important part of our work is working with clay, wood and metal. We use these craft techniques in the renovation of the farm. We are aware of our responsibility towards energy forms and raw materials as life creating and life limiting products of nature. This is why we strive for closed energy and material cycles. The farms gate opens onto the centre of the village - a symbol for our concern to integrate our farm project into the life of Marth.

→See also description of network "**WWOOF**"

- www.noeb-eic.de
- ☺ orientation period, immediate participation in common till, self-initiative e.g. in developing an area of activity

☆ Klangraum Witzenhausen

C5

In der Aue 6 | 37213 Witzenhausen ☎+49-(0)5542-507 05 22
info@klangraum.online www.klangraum.online

2007 200 m² living, 2600 m² garden (near Kassel, Göttingen) 20 %
♀1 ♂4 1 15

- g, e 3 5
- as stipulated in the statutes of the association; very important decisions are negotiated with all (consensus)(consensus minus one)
- joint financing of the household budget, individual financing, possibly membership fees of the association
- rent and property
- spiritual, liberal
- events, seminars, sound massage, wood workshop (musical instrument construction), festivals, association work, garden
- ecological, self-managed, non-violent, spiritual, esoteric/New Age, inner growth, adult education, peace-work, organic gardening, new technologies, ecological construction, humour, art and culture, sound massage, sound research, creative crafts, musical instrument making

☺ ✋✋

We, that is the people who live in the Klangraum ('sound space') and all the siblings who support our actions. We wish to live in a circle of people and grow together in charity.
It's about creating a level of love and understanding. But what does love actually mean? The word love can also be translated with the word connection / oneness. When we love

someone, we try to learn a lot about them. Thus, the personal circle widens around the circle of the beloved and both together form a larger circle. So there is an expansion of the self, an extension until we love everything, as the goal of our self-development.
The community here offers us an optimal practice field. In this process, we can support each other in our love circle.
We invite you to find a place of healing in the sound space, to be accepted as an individual, to find peace and school yourself in love.
We wish for you to find your creativity and joie de vivre here. We want to help you realise your dream here and hope for your help with our activities.
Come visit us!

☺ commitment and willingness to learn to love

Kommune Buchhagen

C4

Buchhagen Nr.4 | 37619 Bodenwerder ☎+49-(0)5533-1400
buero@kommune-buchhagen.de www.kommune-buchhagen.de

1997 3.5 ha (near Hameln/Holzminden) 10 % ♀2 ♂4 0 6

g, e, German sign language 2

everyone in common (consensus) shared economy (shared income and capital)

common property

leftist and humanistic

hostel, cultural work, coffee shop, hand weaving, vegetable gardening and permaculture, building our own house

ecological, self-managed, left-wing political understanding, feminist, non-violent, organic diet, environmental protection, permaculture, organic gardening, ecological construction, humour, cultural work, homo-, hetero- and bisexual life designs

☺ ()

Since 1997 we have been living within the beautiful Weser Mountain Region. Here we inhabit the former council houses of an old estate. There is also an old watermill, a barn and stables as well as about 7 acres of land with a big vegetable garden and a forest garden, which we farm according to permaculture principles. Inside the house and in the garden is also where our animals dwell: cats, Indian runner ducks, rabbits, goats and a few bee colonies.
The mill building hosts both a culture and event centre and a small café. The cultural project brings together people living in our community and various locals. Our barn is home to a carpentry and a joinery to which we own the rights of use. Furthermore, we run a little guesthouse with 12 beds, an apiary and a hand-weaving mill. People with various social backgrounds, from East- and West-Germany with homo-, hetero and bisexual visions of life, involved in all kinds of occupations and social commitments, are living here.
What unites us is our joint economy and responsibility for our living together, the willingness to get involved with and within the group and to grow with it and reach decisions by consensus, as well as a socially and ecologically responsible way of life. Our home could accommodate up to 20 people and we wish to share our house and live with more people. If you would like to get to know us, you may visit for one of our activity weeks.

→See also description of network **"Kommuja"**

☺ mutual sympathy as well as interest in ecologically and socially responsible lifestyles and community based economy

Kommune Hof Rossee

C1

Rossee 1 | 24360 Barkelsby ☎+49-(0)4351-880 812
kommune@hof-rossee.de www.hof-rossee.de

2001 1.2 ha (near Eckernförde, Kiel, Flensburg) ♀3 ♂2 1 6
g, e, Danish, Norwegian, (f) 1 1
everyone in common (consensus) shared economy
common property
grassroots activism, (queer) feminism
outside of the community; ecological holiday homes
ecological, self-managed, anarchistic, left-wing political understanding, feminist, non-violent, organic diet, vegetarian diet, seminar-house, environmental protection, alternative energy, ecological construction, humour, heating with wood

☺

Our farm is located on the outskirts of Eckernförde, 60 km south of the Danish border, and it consists of a residential house, a cow barn with two thirds of remodelling done, an old stable, a hall and 1.2 hectares of land. It has been communally owned since 2001 and there are currently four adults and one child permanently living here. However, we would like to become a larger group and to build the remaining living areas together. We are looking for people who understand communal living as a possibility to grow together and to break new ground together. We are looking for people who like living with children, who find atomic energy and heteronormativity stupid, and who want to participate with their ideas and wishes. We, as "old hands", want to be open to new people and new ideas!
It is important for us to get the balance right between personal needs and commitments to the group. We take joint responsibility for the organisation of our daily life and for project matters. We have weekly meetings and 1-2 conventions per year. We live in the farmstead (7.5 rooms), 2 trailers are available and we have an additional 80 m² communal room with a kitchen. We have also constructed two ecological holiday apartments (for max. 12 people). If all of the space were fully remodelled and used, approximately 15 people could live here (large and small) in about 3 housing units.

→See also description of network "**Kommuja**"
member of the CSA Schinkeler Höfe
☺ we like each other, willing to take part in shared economy, probationary period

Kommune Karmitz

C3

Karmitz Nr. 8 | 29482 Küsten ☎+49-(0)5864-288
kommune.karmitz@web.de

1990 4.5 ha (near Lüchow) 80 % ♀5 ♂7 2-3 14
g, e, sp 9-10
everyone in common (consensus) shared financing of all areas of life
partially rented/leased, partially common property
self-sufficiency, gardening, fruit press, work with food in general
ecological, self-managed, anarchistic, left-wing political understanding, feminist, non-violent, organic diet, environmental protection, self-sufficiency, organic agriculture, organic gardening, organic farming, animal friendly husbandry, alternative energy, we don't like keywords

☺

We share money and responsibility and the aim for self-sufficiency. Our high political ideals include respect for each other and acceptance, avoidance of hierarchies, the principle of consensus, active criticism of mainstream society and we turn against everyday and structural violence. We believe ourselves to be part of the undogmatic political left wing movement. Sometimes we reach our ideals, often enough we play catch-up with them. There are people who eat chips, fancy ketchup, who are smokers or eat meat and others, who are not quite happy about that. Together we take responsibility for our way of living, for our money, our place and our wellbeing. Some of us dig themselves in for weeks at our farm, others go out into the wide world.
But we don't want to live like on an island, so we are connected with other communes, friends and neighbours. Even though all of this is pretty exhausting, we are still happy to be able to live a small part of our utopias. In this we want to contribute to alternatives to capitalism and to a liberated society.

→See also description of network "**Kommuja**"
☺ active participation in the communal project

Kommune Lossehof

C5

An der Losse 13 | 34260 Oberkaufungen
☎+49-(0)5605-927 17 13
kontakt@lossehof.de
www.lossehof.de

2012 | 1 ha (near Kassel) | 1 % | 24

g, e, sp, f | 1 | 6

all (consensus) | completely shared economy

property of all

undogmatic, anarchistic, radical left, ecological, feminist, anti-fascist

house construction, garden, management, members shop

self-managed, non-violent, inner growth, organic diet, vegetarian diet, vegan diet, organic gardening, alternative energy, ecological construction, humour, with inner growth we mean personal development, social plenary (methodologies from: non-violent communication, radical therapy...), fun and games, emancipatory

☺

We live in houses and wagons divided up between two properties that are close to one another. For us, our commune is an emancipatory project, a peninsula with which we make a stand against the various forms of domination in our society, with which we search for the good life together and from which we can act in solidarity in both small and large matters. The commune should be a space where each of us has the opportunity to unfold him-/herself. Part of this is that we reflect on dominant behaviour and group dynamics.
By tram we can get to the centre of Kassel within 30 minutes. It doesn't feel like either a city nor a countryside commune. There are several communes in our area, to which we are connected socially, politically or economically. We celebrate together, help each other out, organise anti-racism activities together and are joined together via collective or solidarity-based agriculture. We envisage living together on the farm with up to 40 people. Little by little we are setting up our own collective. The renovation of our holding, run by our building collective, offers an opportunity to integrate craftspeople as well as untrained people interested in wood and clay building. We run regular information afternoons for interested parties. Itinerant tradesmen are most welcome anytime!

→See also description of network "**Kommuja**"
☺ currently: women, lesbians, trans

Kommune Niederkaufungen

C5

Kirchweg 3 | 34260 Kaufungen ☎ +49-(0)5605-8007-0
info@kommune-niederkaufungen.de www.kommune-niederkaufungen.de

1986 8 ha (near Kassel) 35 % ♀29 ♂27 20 76
48 48

g, e

everyone in common (consensus) fully communal (income sharing) economy

communal property of registered association

ecological and social sustainability, non-dogmatic left, emancipatory

seminar centre, non-violent communication trainers, educational work with electric vehicles, integrated kindergarten, day care centre for elders with dementia, community kitchen with organic catering service, organic vegetable and seed cultivation

ecological, self-managed, left-wing political understanding, non-violent, organic diet, adult education, pedagogical, free school, seminar-house, one-world development, environmental protection, organic agriculture, organic gardening, organic farming, alternative energy, ecological construction, emancipatory, collective enterprises, seeds, CSA, crafts

☺ ( )

We live in the centre of Niederkaufungen (near Kassel) in a complex of buildings, courtyards, gardens and half-timbered houses. Our intentional, undogmatic left-wing worker-, consumer- and housing-cooperative exists since 1986. The founding group formulated a statement of aims and principles in 1983 which is still today the basis of our communal life: communal economy, consensus decision making, reduction of gender specific and other power structures and the opening up of nuclear family structures, emancipation, common responsibility for the whole project, ecology, a resource and energy saving lifestyle and economic mode, global justice.
In our supplementary text, "The dream is over, but...", we describe the present daily practice in the commune. We live in 15 small living groups of various sizes and meet for our shared meals in our commune common room. There is a daily vegan meal, mostly also a vegetarian variation and sometimes meat from one of our pigs or cows. Once a week we hold our plenary meeting and twice a year we hold a commune retreat in order to discuss certain themes in more detail. Constructive methods of dealing with conflicts and various methods such as non-violent communication (Marshall Rosenberg) which help us to develop our personal and collective communication skills are an important part of our communal life. Political and cultural events are organized by those who are interested or who feel like it, and these are often open to the public. Through our various activities and contacts, we are connected in regional and international networks. Every summer we hold a commune festival which is open to all. For those interested in joining us, we offer information afternoons (open days), commune seminars and "getting to know us" weeks. We had the idea for the first "Los-Gehts" community start-up meeting in 1999 (which is still a regular event in Germany) and we participate in the Commune-Info-Tour which regularly travels through the country.

→See also description of networks "**Kommuja**"
☺ commitment to our principles

Kommune Olgashof

D2

Rabenzweig 1 | 23972 Dorf Mecklenburg OT Olgashof ☎ +49-(0)3841-793 337
info@olgashof.de www.olgashof.de

1997 | 4 ha (near Wismar) | 5 % | ♀7 ♂7 9 23

g, e, sp, Bulgarian, Hungarian | approx. 11

all together (consensus) | joint financing of certain areas

common property

garden, carpentry, architecture, renovation work, woodchopping, childcare

ecological, self-managed, anarchistic, left-wing political understanding, feminist, non-violent, chaotic, anthroposophic, inner growth, organic diet, adult education, seminar-house, peace-work, organic gardening, ecological construction, humour, handicraft, mediation, architecture, arts, music

The Olgashof is a colorful, bio-diverse oasis in the midst of a monoculture-farmed agricultural desert. In the distance we see the towers of the Hanseatic city of Wismar and feel the Baltic Sea. We live in a manor house on a large, park-like area with an orchard and an adjacent old LPG area. We have several pets (cats, 1 potbelly pig, chickens etc. and no dogs).

We are organized as a GbR (civil law association) and have a food- and a house-account, in which all adults pay in equal parts. Once a week we meet in plenary, decisions are made by consensus or kitchen table doodle. The shared kitchen offers the opportunity for further discussions and encounters. Our world views are different (from anarchist to anthroposophical) and enrich our coexistence.

As work areas we have the architecture workshop, the wood collective, the seminar or holiday house and the garden and landscape care. Since 1997, people have been living together at the Olgashof to find out if another life in this society is possible and what conditions are required for it. We strive to ensure that the synergy effects of our coexistence are not eaten by friction losses. At the moment our focus is on our current roof construction site. There we want to create urgently needed living space for our small and large residents. We do not want to grow at the moment, but hardworking visitors, who contribute for food and lodging, are welcome. Of course, guests can also book in our guest house.

→ See also description of network **"KommuJa"**

CSA, Artabana

Germany

Kommune Uthlede

B3

Moorstraße 16 | 27628 Hagen im Bremischen OT Uthlede

☎+49-(0)4296-748 225

info@kommune-uthlede.org

2002 | 2.4 ha (near Bremen) | ♀1 ♂2 0 3

g, e, f, sp — 3 1-3

all together (consensus) — shared economy

property of all

undogmatic left/radical left

farming, dealing with goods of communes and collectives, others under construction

ecological, self-managed, anarchistic, feminist, organic diet, organic agriculture, organic gardening, animal friendly husbandry

☺

Our commune is situated in Uthlede, a small village to the North of Bremen; we live in a former country inn with its outhouses and a hall. The project is intended to be free from hierarchies and structured in a grassroots-democratic way. Issues concerning the project as a whole are discussed together and decisions taken by consensus. We aim to live with each other in solidarity and to so create a field for everyone to take part according to her*his possibilities and needs. Everyone should be equally responsible for the project. House and land are property of all. We reject private property of production facilities or money. That's why we have a shared economy: we put all our incomes in a common purse, from which all project costs as well as (consumption) needs are being met. We value all different forms of work equally – be it child education, housekeeping, project work, political work or wage labour.
We see the project as a possibility for emancipatory change, not only within the project, but also for acting together against the prevailing "order". Therefore we fight together with other left or radical left oriented people for political and societal changes and want to integrate the project itself in the conflict with dominating structures

→See also description of network "**Kommuja**"

☺ sharing of political ideas, trial period

Kommune Volzendorf - AufBruchLandung

C3

Volzendorf 13 | 29485 Lemgow ☎+49-(0)5883-989 95 10

kv13@systemausfall.org

2011 | 1 ha (near Salzwedel) | ♀5 ♂2 5 12

g, e, f — 50 %

all (consensus) — shared economy

property of an association

fruit-growing, repairs

self-managed, left-wing political understanding, feminist, non-violent, adult education, peace-work, self-sufficiency, permaculture, organic gardening, humour, unconventional

☺

We live in the Wendland in Volzendorf/Lemgow between Salzwedel (Saxony-Anhalt) and Lüchow (Lower Saxony) on the edge of the green belt. We bought the beautiful farm of a former root crop farmer. The main house has been renovated so that 13 people can live in it. In the large, well maintained outbuildings there is a 180 m² seminar room and an additional 600 m² of not renovated space. The property encompasses approximately 1 hectare, we have created a large garden and a small parking space for trailers here. There are five chickens on our meadow. We still have lots of space: renovation and development possibilities for the site, rooms and buildings. We are therefore looking for further communards. All attempts at giving ourselves a shared caption failed. In the process we found that our quality lies in our diversity. We are well integrated into the village and part of the commune network in the Wendland, which practices free exchanges of goods and services.

→See also description of network "**Kommuja**"

Kommune Waltershausen KoWa e.V.

C6

August-Bebel-Str. 4 | 99880 Waltershausen
+49-(0)3622-200 504
info@kommune-kowa.de
www.kommune-kowa.de

2003 | 1.7 ha (near Gotha) | 20 % | ♀8 ♂8 | 2 | 18

g, e, f, sp, Portuguese | 6 | 70 %

everyone in common (consensus)

shared economy (including property/fortune/debts) | common property

leftist, undogmatic, ecological, pacifist

seminar house, vegetable garden, seedling production, bar, political and social work, workshops - and construction sites

ecological, self-managed, anarchistic, left-wing political understanding, feminist, non-violent, inner growth, organic diet, adult education, seminar-house, environmental protection, self-sufficiency, organic gardening, alternative energy, new technologies, ecological construction, humour, big events/camps

☺

We live in a former doll factory on the edge of the Thuringian Woods. By now, we have renovated one of two buildings and created space for about 30 people, including seminar rooms, community space, offices and workshops. The commune is in the middle of a small town, well connected to the public transport system. We live together because community and social interchange is important to us, and in order to be able to contribute to political and social change of our society and economy. We want to take active control of our lives, treating both human beings and nature with responsibility and consideration. This means challenging our consumption habits and creating space and time for human and social needs. Once a week, aside from the organisational plenary, we hold a meeting designed to cultivate our communication and relationships and also to tell each other how we feel.
Alongside our construction work, we have gradually developed our own collective work areas (seminars, seedling production, café, vegetable garden). Our basic principles include a shared economy, social interaction, the rejection of hierarchies, and decisions taken by consensus.

→See also description of network "**Kommuja**"

☺ willingness to share in common economy and to learn

Kommunität Beuggen

B10

Schloss Beuggen | 79618 Rheinfelden ☎+49-(0)7623-748 746

info@kommunitaet-beuggen.de www.kommunitaet-beuggen.de

2005 ca. 2000 m² (near Basel, Lörrach) ♀4 ♂4 12 20

g, e, sp it depends

all together (consensus)

cash box for the household (food), cooperative for housing expenses

collective property

Christian, ecumenical, ecological

ecological, self-managed, non-violent, Christian, ecumenical, organic diet, vegetarian diet, adult education, peace-work, environmental protection, non-smoking, families, daytime prayers, common lunch

☺ ()

The community Beuggen is a Christian community in which people of different ages, marital status and denominations live together in a responsible way. Many of us have been or are involved in the peace, environment, world shop movement or working with refugees. The community currently includes 8 adults and 10 children aged 5-64 years. Since our foundation in 2005, we have lived in Schloss Beuggen near Rheinfelden in Südbaden.

☺ to be spoken about - to live with us for a certain period, the membership in a Christian church is not a condition, for membership candidates, however, it is

Kommunität Grimnitz e.V.

E3

Grimnitzer Str. 11 | 16247 Joachimsthal ☎+49-(0)33361-645 133

info@kommunitaet-grimnitz.de www.kommunitaet-grimnitz.de

1997 1.5 ha (near Eberswalde, Angermünde) 15 % ♀4 ♂3 1 7

g, e, sp 2 3-4

everyone in common (in consensus when possible)

limited community financing, guests, refuge in times

association property

ecumenical/Christian

international encounters, regional community economy, youth and adult education, working with and for refugees, lodging groups and individuals, ecological farming, involvement in climate protection, spirituality and art

ecological, self-managed, left-wing political understanding, non-violent, spiritual, Christian, ecumenical, meditation, organic diet, adult education, seminar-house, peace-work, one-world development, environmental protection, self-sufficiency, organic gardening, alternative energy, ecological construction, art exhibition, open monument

☺ ()

We live near the beautiful lake Grimnitz in the heart of a village on the outskirts of a small

Brandenburgian town in the former forestry headquarters of Schorfheide. Our key activities are: we organise international movements in peace via justice to maintain God's creation and human rights (including overseas projects, particularly in Colombia), as well as meditation and arts seminars. Our regional education work concerns further education in theology, including questions about the climate and organic gardening. Furthermore we house people who need time for contemplation ("cloisters in time") and who are curious about communal co-housing. Our central project is tolerant co-housing in the spirit of Jesus Christ with corresponding spirituality, in order to help us on our quest to find our own meaning in Christianity and to be dedicated Christians, which may result in reforms for the church and society. We are searching for supporting members of differing generations and civil status. In our small town we are accepted as a "cultural factor" and given political co-responsibility.

Ecumenical net in Germany
☺ recognising our articles of association and community rules, active participation

☆ Königsfarm

A7

Königsfarm | 54470 Bernkastel-Kues ☎ +49-(0)6531-7757

1991 4 ha (near Bernkastel-Kues, Trier) 25 % ♀2 ♂3 0 5

g, e 3

everyone in common (consensus) property of individual community members

none

gardening, repairs, seed production, all the activities of a life close to nature

ecological, non-violent, spiritual, inner growth, organic diet, vegetarian diet, self-sufficiency, permaculture, organic gardening, ecological construction, no cell phones/ wireless, non-smoking, seed production

☺

The Königsfarm ("king's farm") is located in the Hunsrück (a low mountain range in Germany's mid- to southwest) above the river Mosel in a secluded location in the forest without electricity, without water pipe connection, accessible via a gravel road. On the property are two self-built clay houses, several construction trailers and a composting toilet.

In our gardens thrive a wealth of old and unusual crops and edible wild plants. These are mainly used for seed production as our main means of income and the preservation of genetic diversity. Our self-sufficiency activities is also looking forward to some active hands. Numerous fruit trees and berry bushes are available. Three hectares of pasture await grazing fellow creatures.

The absence of electricity leads to a natural, decelerated life rhythm. We draw our water from a spring, which requires careful handling of the gifts of Mother Earth. Conventional detergents, shower gels, perfumes, etc., which almost always contain nanoparticles, microplastics and synthetic fragrances, are taboo for us. Thus, sensitive people and our wastewater treatment plants can exist with us.

Our everyday life, often shared with a colourful variety of different guests, is characterised by mutual help and loving understanding, and usually a common meal.

An important role for us plays our contact with invisible fellow-creatures: elementals, fairies, fire devas ... Here is room for sacred chants during planting, shamanic healing, and the Agnihotra fire ritual. We look forward to seeing you

Dreschflegel
☺ open heart, open mind, enjoying inner growth, understanding and compassion, a willingness to live in loving togetherness

Kooperative Dürnau

C9

Im Winkel 11 (mail)/Braunenweiler Straße 1 (visitors.) | 88422 Dürnau

☎+49-(0)7582-9300-93

erstkontakt@kooperative.de

www.kooperative.de, www.ferchervonsteinwand.org, www.dorfuniversitaet.org

1980 (near Ulm) ♀7 ♂12 4 23

d, e, f, sp, Polish, Danish, Italian, Russian, Dutch — all — all

all who are concerned

print shop, joinery, publisher, distributor, agriculture, retail, nursery, all-round service, international workcamps, assisting youth with social problems, village university

self-managed, chaotic, anthroposophic, adult education, pedagogical, environmental protection, organic agriculture, organic gardening, animal friendly husbandry, construction, humour, education, seminars, workshops, summer university

☺

Dürnau Cooperative is a living and working community which evolved from the information service of the alternative cooperatives in the Wuppertal area. From the community's beginnings within the disorganised, undogmatic and politically disinterested alternative movement of the early 1970s – especially since the move to Dürnau – has evolved a stable community of currently 23 people, who try to achieve their goals by working together. The expression of these ambitions include several small businesses, the "Fercher von Steinwand e.V." and a branch of the anthroposophic society, both of which are media for cultural and social interests, as well as the most recently founded village university, which offers an alternative to conventional studies. Achieving a satisfied communal life with the most different kinds of people and therefore varying endeavours requires a high degree of social imagination and technology. The list of daily things to do not only provides us with variety, but also with a constant stream of new challenges, which our means cannot always completely satisfy. A large number of friends from completely different schools of thought provide us with the necessary support to keep us going and to improve life in the province immeasurably. Visitors and interested parties are mostly welcome, but we are not fond of surprise guests, since we do not always have the necessary time to give such guests the attention they deserve.

☺ none, please inquire for details

Kranich e.V.

C3

Pretzierer Str. 25 | 29410 Salzwedel OT Groß Chüden ☎+49-(0)3901-829 49

kranichueden@web.de

www.kranichueden.de

1993 4.5 ha (near Salzwedel) ♀4 ♂3 5 12

g, e, f, sp — 2

everyone in common (consensus) — individual financing

property of all

seminar house, bees, food co-op, mail order bookshop, garden, (therapeutic) horse riding, children, environmental education

ecological, self-managed, left-wing political understanding, inner growth, organic diet, seminar-house, self-sufficiency, permaculture, alternative energy, seminars, books, horses, organic shop

☺

Our communal life is based on sympathy and individuality, not on fixed guidelines. The holding comprises 450 m² of living space and various parking spaces for mobile homes etc. All inhabitants are members and tenants of the association that owns the place. "We" run an educational centre ("the colourful four"), a bookshop ("Bücherwinkel"), keep bees ("Schwarmzeit"), a food cooperative ("EVG Immersatt e.V."), offer therapeutic riding on our farm, and some of us also work elsewhere. In addition we are busy with circus, children, gardens, animals, music, politics and the search for "the good life" (sometimes we even find it). We are aware that we are not perfect and do not strive to be, we're always open to compromise. The aim is a patchwork of alternative diversity in and around us.

networking with local communities
☺ sympathy

Kratzbürste e.V. (B10)

Haus für Bildung, Kultur, Handwerk und ökologischen Landbau
Untere Gasse 22 | 79244 Münstertal ☎+49-(0)7636-1614
info@kratzbuerste.de kratzbuerste.de

1980 | 4000 m² (near Freiburg) | 25 % | ♀7 ♂6 5 18

g, e, sp | 4 10
house congregation (democratically) | common financing in some areas
association
ecological
seminar venue, garden, animals
ecological, self-managed, left-wing political understanding, non-violent, organic diet, adult education, seminar-house, one-world development, environmental protection, self-sufficiency, organic agriculture, organic gardening, organic farming, animal friendly husbandry, alternative energy, ecological construction, humour

Kratzbürste e.V., the Association for Education, Culture, Organic Farming and Handicrafts, was founded more than 30 years ago on the idea of living together with several families self-governed and in an ecologically oriented way. Today, the association has over 50 members, made up of residents, alumni and interested friends.
At regular house meetings, upcoming tasks and the further development of the community are discussed. Each resident takes on individual tasks in our project according to their abilities and the time available to them.
Thus, in the course of extensive renovation work, a former brush factory has evolved into eight apartments and several companies as well as a venue for seminars. All the remodelling and conversion is guided by ecological and holistic building aspects.
House and land belong to the association. Endeavouring not to be dependent on banks, we finance ourselves only through personal loans from friends and club members.

Krautkreise e.V. KonZentrum für Permakultur & Gemeinschaft C9

Schmiedgasse 6 | 86494 Emersacker ☎+49-(0)162-796 14 18
krautkreise@posteo.de krautkreise.de

2016 4800 m² + 3.6 ha (near Augsburg) 50 % ♀2 ♂3 0 5

g, e, sp, i minimum of 2, varies with the seasons everyone

all permanent residents (consensus)

collective funding of collective concerns, barter trade, second-hand

rented with purchase intention

permaculture design, farming, beekeeping, Gestalt therapy, seminars, café, building with clay, terra preta

ecological, self-managed, non-violent, spiritual, inner growth, organic diet, adult education, pedagogical, environmental protection, self-sufficiency, permaculture, organic agriculture, organic gardening, ecological construction, humour, non-smoking, beekeeping, Gestalt therapy, food saving

We are a project that combines permaculture with living in a community. As a community we are 5 people from age 25 to 62. We practice a lot with communication and group processes to stay conscious about conflicts and our relations. We are very open to people who wish to stay with us and work with us as WWOOFers. We love to travel ourselves, and welcoming travellers here is always an inspiration.
We have a big permaculture area with raised and crater beds. We plant a lot of exotic and local vegetables and do the whole process by ourselves. We are also beekeepers and we do seminars and workshops about permaculture, conserving and fermenting, personal growth and community building. There is a practice for Gestalt therapy included in the project.

Permaculture Academy Germany, SoLaWi Augsburg, Food Sharing, Slow Food

Krone gGmbH B10

Hauptstr. 11 | 79295 Sulzburg
krone_sulzburg@yahoo.de www.syndikat.org/de/projekte/krone/

1978 780 m² (near Freiburg) 10 % ♀5 ♂3 4 13

g 1-2

all residents and tenants from 18 years (majority and consensus)

shared finances for some areas common property

carpentry, juice factory, sculpturing, jam and syrup making, office

ecological, self-managed, left-wing political understanding, organic diet, environmental protection, organic agriculture, organic gardening, alternative energy, ecological construction

The Krone ("crown") in Sulzburg, a 200 year old inn, was bought by our association in 1978. At the beginning, it was still mostly about working together, but now it has become a self-managed housing project.

In the yard and in the outbuildings, four businesses have now found their place, a joinery, a juice factory, a cider factory and a biologist's office. In addition, there is a sculpting studio and in the vaulted cellar, there is wine from our own small vineyard. We have just completed the overhaul of the roof of our main house - that had been a long time in the making and now it's finally done, a great pleasure for us! The former dining room with large kitchen and visitor toilets are common areas that are used by all. These premises can be rented and cultural associations, initiatives and private individuals make use of it.
We ourselves regularly host cinema evenings, concerts, readings and other events. It is important to us to further develop the cultural monument "Gasthaus zur Krone" as our place of life and to preserve it for future generations of residents.

Krumme Eiche e.V.

C6

Krummasel 31 | 29482 Küsten
+49-(0)5864-986 795
krumme.eiche@zeromail.org

2006 3.5 ha (near Uelzen, Lüneburg, Salzwedel) 5 %

♀3-5 ♂3-5 3 9-13

g, e, f
all members (consensus)
3 3-10, varies daily
we share our income
common property
some, but not all of us are: grassroots, anarchists, radical ecologic, feminists
housework, children, farm and gardening, joinery, carpentry, pottery
ecological, self-managed, anarchistic, left-wing political understanding, feminist, non-violent, chaotic, organic diet, vegetarian diet, vegan diet, environmental protection, forestry, alternative energy, ecological construction, construction, humour, some are pedagogues, psychologists, engineers, artists of life and engaged in various self-organized initiatives

We want:

- to preserve the former guesthouse (from 1911) and its outbuildings in accordance with the monumental requirements and to create the necessary living space.
- to ecologically preserve and maintain the Erlenbruch biotope, the wet meadows and orchards
- reduce our ecological footprint through tiny mobiles, near-natural living in construction trailers, etc.
- be socially & politically effective
- push renewable energies, turn off nuclear power, leave natural gas, oil and coal in the ground
- cooperative management, deceleration and outgrow capitalism
- spread healthy food and seeds for all: organic & free from genetic engineering
- counter right-wing populism
- regional trade

We juggle between harvesting, cooking, baking, celebrating, making money and farm & family management. We process home-grown and storm wood into furniture and construction timber and accommodate the organic food co-op "Rabattz". We realize most of our professional construction work ourselves. We try to promote understanding and communication in the village. We use the old dance hall and the former guesthouse as a private venue for concerts, meetings, celebrations and culture in the village - and as a play- and living room.

→See also description of network "**KommuJa**"
- partly in Artabana, sharing networks, movement against nuclear power
- it has to fit

☆ Kuckucksmühle

D3

Ausbau 7 | 16909 Heiligengrabe
kuckucksmuehle@ecohackerfarm.org
wiki.ecohackerfarm.org/kuckucksmuehle:start

2016 | 1 ha (near Wittstock, Berlin) | ♀1 ♂3 0 4

e, g, Maltese, i | 1 2

permanent residents (100 % consensus) | rented

non spiritual, non religious, eco friendly, technologically minded

permaculture living and hackerspace, automation

ecological, self-managed, left-wing political understanding, feminist, non-violent, adult education, environmental protection, self-sufficiency, permaculture, organic agriculture, organic gardening, organic farming, animal friendly husbandry, alternative energy, new technologies, ecological construction, hackerspace

Kuckucksmühle (Cuckoo Mill) is a historical water mill in the North East of Germany (Brandenburg Federal State). We aim to restore the mill and its outbuildings, set up a Hackerspace/ Hackbase and permaculture farm.
We would like people to come, share their knowledge and participate in community activities linked to restoration of the old mill and its outbuildings, gardening, digital and hardware infrastructure, organizing and planning the future.
Kuckucksmühle is part of the Eco Hacker Farm project.

- Eco Hacker Farm (headquarter)
- ideological and financial requirements to be discussed

Kulturfabrik Meda

E6

Hainewalderstr. 35 | 02763 Mittelherwigsdorf ☎ +49-(0)3583-510 735
info@kulturfabrik-meda.de kulturfabrik-meda.de

1997 7500 m² (near Zittau, Görlitz, Dresden, Praha, Wroclaw)
♀4 ♂4 2 10

e, g 1
everyone in common (consensus minus one) shared and individual financing
property of Kulturfabrig GbR (civil law association)
none
cultural work
ecological, self-managed, non-violent, inner growth, organic gardening, alternative energy, new technologies, ecological construction, humour, art actions, cultural events, cinema, festivals, dance and music events, men's and women's groups, forum, dialogue

☺

This is what we're planning in the Kulturfabrik:
We want to counteract the growing isolation in our society (almost half of all households today are one-person households) and try out new forms of communal living. We want to treat our sewage ourselves, minimise and compost waste, use our own well, and in the future we want to generate part of our own electricity. We want to use ecological, recyclable building materials. We want to obtain our food from organic farms as far as possible or grow it ourselves. We see ourselves as a cultural project organising film, music and art events.

☺ prospective members should take time to get to know us and accept guest status, a probationary period, and open communication

KulturTransistor e.V.

A7

Unterstraße 16 | 56829 Kail ☎ +49-(0)2672-913 895
kulturtransistor@yahoo.de www.kulturtransistor.de

2007 3700 m² (near Koblenz, Trier) ♀2 ♂1 4 7

g, e 3
owners (financial issues), all together (organisational matters) (straightforward by talking to each other)
individual financing property of some members
seminars, cultural work
ecological, self-managed, feminist, non-violent, adult education, seminar-house, peace-work, environmental protection, ecological construction, humour, art, literature and culture

☺

The CultureTransistor ("KulturTransistor") is a remote place where we implement our own artistic and creative ideas. This involves the organisation of exhibitions, seminars, cinema evenings... and work on personal projects. For more information please visit our website.

☺ people who want to work with art/culture are welcome

Kunesto Cobstädt eG

C6

Schulplatz 11 | 99869 Drei Gleichen/OT Cobstädt

+49-(0)36202-787 65/-903 69/-772 82

kunesto.cobstaedt@gmx.de kunestocobstaedt.jimdo.com

2013 2400 m² (near Erfurt) 5 % ♀4 ♂2 6 36

g, sp, e 3

all members together (consensus minus one)

mainly individual financing, joint financing of certain projects

property of all

garden, maintenance and repair of buildings, firewood, kitchen

self-managed, spiritual, inner growth, organic diet, adult education, pedagogical, peace-work, physically handicapped persons, organic gardening, alternative energy, ecological construction, construction, humour, sound massage, yard kitchen, crafts: wood work, art of life

☺ ( )

We are a small but growing community and relate to each other as a family with all that that entails: individual responsibility for yourself, for the others and for the wider world, to support each other and yet to learn to stand alone, to learn from each other...
Our home is a four-sided farmyard enclosing a central courtyard. The site and some of the buildings have over a thousand years of history and there is also an old garden.
This is a place for dreaming and settling down and it's never dull. We are undogmatic but not haphazard.

→See also description of network "**GEN Deutschland**"

WohnStrategen e.V. (Weimar)

☺ to be on good terms with each other

Laakenhof

B4

Pöling 5 | 59320 Ennigerloh- Enniger +49-(0)2528-405 00 14

laakenhof@arcor.de laakenhof.de

1995 30 ha (near Münster, Hamm, Bielefeld) 90 % ♀4 ♂3 1 8

g, e all all

everyone in common (consensus) shared economy

rented/leased

politically left

vegetable growing, fruit, milk cows, cheese making, chickens, marketing

ecological, self-managed, left-wing political understanding, feminist, non-violent, organic diet, organic agriculture, organic gardening, organic farming, animal friendly husbandry, ecological construction

☺

We leased the "Laakenhof" in Neubeckum in 1995.
From 2017 to 2020 we move to our new farm, which is near to the old Laakenhof. Our commune is an open-minded group of people who wish to live a non-hierarchical and self-determined life. Consensus decision-making and a shared economy are important components

of our communal living. We try to be consistently ecological and resource-sparing in our behaviour (e.g. organic agriculture, solar energy use, building biology).
We stand up for a fair, non-violent society and wish to participate in the development of real alternatives. Networking with other groups is important to us. On the long run we would like to live in a commune with 10 adults plus children.
Communal agricultural work is important to us. At present all of us work in agriculture, each independently within his or her field of activity. We have 11 cows and all the milk is processed in our cheese dairy. Aside from that, there is vegetable cultivation, as well as direct marketing by means of veg boxes, a farm shop and market stalls. We are currently seven adults and one child. Life and work on the farm are complemented by further colleagues, two agricultural apprentices and regular interns.

→See also description of network "**Kommuja**"
☺ sympathy, probation period

Landhüter Wohnungs- und Landbaugenossenschaft eG

B6

Wetzlarer Straße 31 | 35647 Waldsolms - Griedelbach ☎+49-(0)6085-986 87 73
info@landhueter.org www.landhueter.org

2013 0.5 ha (near Wetzlar, Frankfurt a.M.) 30-50 % ♀2 ♂5 0 7

g 4

board of directors, supervisory board, general assembly, plenum of the residents (majority decision according to statutes and a plenum of the residents for house affairs, in which we strive for consensus)

bank loans and individual financing, limited common purse

property of the association

renovation and extension of the farmstead, timber, loam and stone construction, biological garden and agriculture with permaculture approach, greenhouse, planning and conversion of the large meadow property into a living garden, kitchen and cooking

ecological, self-managed, organic diet, adult education, seminar-house, self-sufficiency, permaculture, organic gardening, alternative energy, new technologies, ecological construction, construction, humour, classical homeopathy, bio cybernetics, raw food, primary therapy, Reiki, handicraft

☺ ()

Landhüter ("Land keeper") offer and live community. We organise ourselves to realise goals. We offer help and support. An interesting conversation often develops when we eat together. The feeling of community can be experienced in addition to living together. A large ex-gastro kitchen serves to prepare food for the community and for events. A common guest room is a meeting point at different times and also serves for eating, playing and being together. Like another large guest room, it is

used for lectures, events and celebrations. There is always something to do in the garden. We offer two concepts: 1) living together in one house, in single rooms with common rooms and 2) living in single apartments with approx. 40 m² living space and use of the common rooms. This is made possible by the expansion. We strive for a good mixture of woman and man, would like to unite many generations and live together.
Joie de vivre, with wit and mutual respect want to be lived: We are looking for people who are down-to-earth, have tidied up their inner cellars and are willing to invest. We do not only want to grow here as a community, but offer like-minded people who want to live the community to use our experience and the framework of our cooperative to build their own project.

- Deutsch-Europäischer Genossenschafts- und Prüfungsverband e.V. (German-European Cooperative and Auditing Association)
- ☺ getting to know each other, membership in our cooperative, conversations with local residents if you intend to move in

Landpartie Ziemkendorf

E3

Ziemkendorf 10 | 17291 Randowtal ☎+49-(0)152-556 057 46

2014 | 1.5 ha (near Prenzlau, Stettin) | ♀4 ♂4 4 12

g, f, e | 1-5 people

- consensus or delegates (talking circle, consensus)
- private households individually financed; houses, land and garden are common property
- property of all
- Mietshäusersyndikat
- ecological, self-managed, left-wing political understanding, feminist, non-violent, spiritual, chaotic, inner growth, free school, organic gardening, ecological construction, humour, non-smoking, no alcohol

☺

A tiny, unbending village, not far from Szczecin (Poland), a good hour away from the Baltic Sea or to Berlin. Project of the housing syndicate (www.syndikat.org/en). Left, green, tolerant, anti-discrimination, anti-racism, anti-all-that-is-stupid. We are the "good people", we want sustainability, environmental protection, own food from the garden.
We want it all! We want to question our habits. Alcohol and cigarettes, media junk etc., we do not want to live as "normal" in everyday life. No one is evangelized, but everyone is asked for mutual consideration. We want to have fun!
Uckermark! Wide and wild, sky spectacularly beautiful or exciting. Horizon in almost every direction. The fields are very large and in order to reach the next forest you need a bike. In the vicinity there are some very beautiful bathing lakes. There's no way around windmills. When I want to describe the beauty of this landscape, "rough", "powerful", "wide and clear" comes to mind.
Facebook group "Ziemkendorf!" with the exclamation mark!

- Mietshäuser Syndikat (www.syndikat.org/en)
- ☺ driving license

LandRausch Gesellschaft für gemeinsames Wohnen

D3

Roddahner Dorfstr. 18/20 | 16845 Neustadt, GT Roddahn

landrausch@googlemail.com gemeinschaftlandrausch.wordpress.com

1992 1 ha (near Rathenow, Neuruppin) < 10 % ♀3 ♂3 2 8

g, e, f, sp 4 5

everyone in common (consensus) shared income

common property

ecological construction, building with clay, ecological sewage technology, psychotherapy, social work, humanities, garden

ecological, self-managed, left-wing political understanding, organic diet, adult education, free school, organic gardening, alternative energy, ecological construction, cultural and political engagement

☺

LandRausch is a small community with a large yard, approx. 100 km northwest of Berlin. We share the desire to live in solidarity, collectively and as ecologically as possible outside the pattern of the nuclear family. Our community exists since 1992 and since 2017 there is a generational change in progress that especially involves the search for new people. We are gladly welcoming anybody who wants to participate, work with us, live with us! Currently our group consists of 4 "old" and 2 "new" adults and 2 children (of kindergarten-age). There already is some housing space for new people, more could be created in consideration of individual needs.
We have specific plans concerning construction work, to create more housing space for our growing community. We will start realizing them in 2019. Volunteers wanting to work with us on that project are very welcome. There will be special "construction-weeks", but there is also space for volunteers who want to visit us individually. We are working with ecological building material and are very experienced in clay-building-techniques. Besides we own a large garden, that always needs tending to, six chickens, two ducks and a cat.

☺ probationary period

Lauter Leben eG

D4

Dorfplatz 10 | 14547 Beelitz OT Kanin +49-(0)151-202 630 73

info@lauter-leben.de www.lauter-leben.de

2009 2500 m² (near Beelitz, Potsdam, Berlin) 5 % ♀8 ♂7 19 34

g, e, sp, f, Polish approximately half of the adults

all of us together (consensus) limited shared financing in some areas

registered cooperative

ecological, self-managed, left-wing political understanding, feminist, non-violent, pedagogical, mentally/psychologically handicapped persons, physically handicapped persons, animal protection, environmental protection, organic gardening, alternative energy, ecological construction, humour, music, graphics, illustration, documentary, puppet show, art

We are currently 15 adults and 19 children between the ages of 10 months and 14 years. We live together in an old four-sided farm. Common to all of us is the desire to live in community and in the countryside. We enjoy the many benefits that are offered to us here in everyday life:

- meeting in the evening without appointment stress or babysitting with nice people, around the campfire, at the court cinema, at parties or game evenings...
- watching the children play together and developing friendships
- sharing many things (e.g. cars, tools, toys) and accomplishing tasks together (e.g. purchasing via food co-op, repairing of living and common areas) and thereby saving costs, resources and time.

We work in a wide variety of professions, but have a lot in common, which cannot be defined by the job title.
What we work for, professionally and through our way of life and what is important to us personally: environmental and climate protection, renewable energies, an inclusive society, good music, reliable foundations of buildings, good, healthy food and a resource-efficient agriculture without toxic pesticides, work-life balance, gender justice, responsible parenting, good home-made jam, visual design and illustration, documentaries, social education youth counselling, global justice, lots of time for our kids, fun in life and much more!

Photo (above) by Carolin Kott

Lebendiges Haus

B6

Melanchthonstr. 61 | 57074 Siegen ☎+49-(0)271-231 66 79
lebendiges-haus@gmx.de lebendiges-haus.de

2003 0.25 ha (near Gummersbach, Marburg, Köln) ♀16 ♂14 5 35

g, e, Chinese 1 (part time)
initiative group formed from the association (consensus)
individual financing & limited common financing cooperative
approx. 2/3 of the anthroposophical and the Christian Community philosophy, undogmatic, in New Christian Community consciousness
bazaar and festival organization circle, housekeeping, public relations work and event circle of lectures, courses, artistic and therapeutic activities
ecological, self-managed, non-violent, Christian, ecumenical, anthroposophic, inner growth, adult education, mentally/psychologically handicapped persons, environmental protection, alternative energy, new technologies, ecological construction, mutual help, chirophonetics, eurythmy, woodwork, hiking, painting, music

The long-stretched multicoloured "Lebendige Haus" (="Living House") is situated on the slope of the "Häusling" directly below the forest, right next to the church of the Christian Community. The Waldorf Kindergarten (Rudolf Steiner's anthroposophy) is our neighbour. The city centre with the shops, the train station etc. can be reached in 15-20 minutes by foot. A modern wood pellet heating system supplies the house with heat. Rainwater is

collected and prepared for washing machines and toilets. Green electricity flows through our houses' cables.
The Living House was created from retirement home of the Christian community. The beautiful property with pleasant living atmosphere accommodates 21 residential units on four floors, between 25 and 145 square meters. Several generations live here so that a coexistence of young and old is promoted. The self-governing building and housing cooperative "Lebendiges Wohnen" e.G. is the owner and manager of the facility.
Furthermore, there is the association "Lebendiges Haus e.V.", which has set itself the goal to creatively shape the life in the house. The social life consists of everyday encounters, regular residents' meetings, spontaneous activities and celebrating common festivities in the course of the seasons. The common rooms are available for artistic and cultural activities of the house community as well as for the interested neighbourhood and the public. For us as residents, mutual awareness is an integral part of living and being together.

- trias
- ☺ openness to and/or interest in anthroposophical ideas in social community life, in the arts and in the understanding of Christ (Christian community), waiting list for applicants

Lebensgarten Steyerberg Social, Cultural & Ecological Visions B4

Ginsterweg 3 | 31595 Steyerberg ☎+49-(0)5764-2370
info@lebensgarten.de lebensgarten.de

1985 | 4 ha (near Nienburg/ Weser) | 20 % | ♀79 | ♂40 | 29 | 148

- g, e, sp, f, Dutch, Danish, Swedish, Gaelic — 30
- all together (limited consensus) — individual financing
- ⅔ rented/leased, ⅓ property of individual owners
- seminars, arts and crafts, various therapies and natural health care, permaculture, non-violent communication, mediation
- ecological, self-managed, non-violent, spiritual, Christian, ecumenical, Buddhist, esoteric/New Age, inner growth, meditation, yoga, organic diet, vegetarian diet, adult education, pedagogical, free school, seminar-house, peace-work, environmental protection, self-sufficiency, permaculture, organic agriculture, organic gardening, alternative energy, new technologies, ecological construction, humour, nonviolent communication, mediation, coaching, therapies: homeopathy, acidosis therapy according to Collier, physical therapy, eurhythmy therapy, psychotherapy, occupational therapy, family constellation, holding therapy, speech therapy, yoga, shiatsu, shin tai, orgodynamics and other, Zen

☺

The non-profit association Lebensgarten Steyerberg was founded 33 years ago. Our guiding ideas are: tolerance and respect in dealing with ourselves, our fellow human beings and nature, ecology in everyday life, mutual support and joint decision-making processes.
We live our spirituality in common on a daily basis (meditation, singing in the chapel, circle dances), we maintain neighbourly contacts and can shape our everyday life, leisure time and special life phases together. There are many things which we share and just as many which distinguish us. Community-related and individual growth crises bring us closer together and show us our own responsibility. We don't have a guru.
We try to make our decisions according to the consensus principle. Members can decide for themselves how they want to work on the association's tasks and goals. Our seminar business

passes on our own experiences and those of seminar leaders affiliated with us, and also grants us some jobs.
Beyond that, there is the school for understanding and mediation, the center for non-violent communication, practices for naturopathy, occupational and physical therapy and psychotherapy, a shop for ecological construction materials, an ecological planning office, a permaculture park (PaLS), car sharing with electric cars, and also freelance artists and musicians. The Lebensgarten seeks to be a place that encourages its residents and guests to face themselves and inspires them to unite with others.

→See also description of networks "**GEN Europe, GEN Deutschland**"

🌍 Money Network Alliance (MonNetA), ARTABANA, Klimaschutzagentur Nienburg e.V., Bund Solare Mobilität e.V.

☺ spend some time here several times beforehand! You have to arrange your accommodation and work yourself (we can give you some tips but our flats are largely rented out)

Lebensgemeinschaft auf Schloss Tonndorf eG

C6

Das Schloss 156 | 99438 Tonndorf — ☎+49-(0)36450-440 54

✉gaeste@schloss-tonndorf.de — 💻www.schloss-tonndorf.de

2005 | 17.5 ha (near Weimar, Erfurt) | 20 % | ♀17 | ♂19 | 28 | 64

g, e, sp — ♻3-4 | 17

all adult members (consensus)

shared investments (from our own rents) in various areas (construction, outdoor area, events, etc.) + individual

collective ownership

environmentally conscious, spiritual, with different individual characteristics

ecological construction, horticulture, agriculture, utilities, catering, forestry, beekeeping, forest-nursery, seminars, guest accommodation

ecological, self-managed, left-wing political understanding, non-violent, spiritual, inner growth, meditation, yoga, free love, organic diet, vegetarian diet, adult education, pedagogical, free school, seminar-house, environmental protection, self-sufficiency, permaculture, organic agriculture, organic gardening, organic farming, animal friendly husbandry, forestry, alternative energy, ecological construction, humour, festivals and markets, forest kindergarten, community-run school (in development)

☺ ()

Our community exists since 2006. The age of our members ranges from 0 to 60. We all share the desire to live all aspects of life in a manner which is consistent and sustainable for us and for this planet. This includes aspects such as the birth and growth of our children, their learning in respect and freedom, the creation of work places in our midst, the reflection on one's own vocation and the creation of a self-determined life, the development of a

constructive, benevolent togetherness, an ecological way of life or the involvement of and the care for elderly people. We gradually learn to listen better and to pay attention to each other. Our decisions are made by consensus.
Together we enliven this beautiful estate with its various buildings, fields, orchards, meadows and woods with new life, create homes and workplaces (for example: beekeeping, a catering kitchen, a dairy, horticulture, agriculture, a forest kindergarten etc.), we plant and harvest, organize seminars, festivals and markets. The castle and our community have become an attraction for the people of the neighbourhood and our region, with whom we have developed a good relationship. We strive to live with an intelligent humility and to offer hospitality to our guests. We are happy to support people who want to develop further communities in our surrounding area. First examples already exist.

→See also description of networks **"GEN Europe, GEN Deutschland"**
☺ 12 months probationary period/10.000 EUR cooperative share

Lebensgemeinschaft Berstadt

B6

Bismarckstr. 9 | 61200 Berstadt ☎+49-(0)6036-904 023/-983 342
info@pravahi.de www.gemeinschaft-wetterau.de

2005 1500 m² (near Friedberg, Giessen) ♀4 ♂4 0 8

g, e, d, f, sp, Portuguese 5
all who are concerned (consensus if possible, otherwise majority vote)
common financing of certain areas (common household budget)
property of single community members
environmentally conscious and spiritual
healing, kids, love, man-woman relationship
ecological, self-managed, non-violent, spiritual, inner growth, meditation, yoga, tantra, sexuality, organic diet, vegetarian diet, environmental protection, alternative energy, ecological construction, humour, massage, gestalt therapy, coaching, organic nutrition

☺

We are a small community in the Wetterau, at the foot of the Vogelsberg. We live a balance of community and individuality. We are all interested in self-awareness without having a common spiritual orientation.
At the moment we are looking for more people who want to bring their creativity to us.
The community has owned our farm since 2005 and has changed its essence throughout with creative ecological renovations, artistic interior design, reconstruction of the former chicken house, a sunny yellow lick of paint on the façade, a loft conversion in 2012 and a new roof – and we still have more ideas. We see ourselves as a future-focused housing and living community, as well as a regional healing biotope, which provides and celebrates space for meeting, community formation and inner growth. Our attitude is spiritual, humorous and affectionate, and it values individuality and community. We are always pleased about visitors and people who would like to link into our network in a loving and active way.

→See also description of network **"Soshona Freundeskreis"**
friends with ZEGG, Tamera
☺ especially families with children are welcome

Lebensgemeinschaft Dalborn eG

B4

Hauptstr. 28 | 32825 Blomberg-Dalborn
willkommen@lebensgemeinschaft-dalborn.de ☎+49-(0)5236-382
lebensgemeinschaft-dalborn.de

2010 3 ha (near Detmold, Bielefeld) 35 % ♀5 ♂7 4 16

g, e, Dutch 4

all members of the housing cooperative (depends: 2/3, 3/4, consensus)

housing cooperative; common household budget; determination of individual rents according to the solidarity principle

registered housing co-operative

no specific political party or religion; intention: fundamental change of economy, lifestyle and the cooperation of all beings

extension of housing space, agriculture, cultural, social and political projects

ecological, self-managed, left-wing political understanding, non-violent, spiritual, Buddhist, inner growth, meditation, yoga, organic diet, vegetarian diet, adult education, environmental protection, self-sufficiency, organic gardening, alternative energy, ecological construction, think globally - act locally, intercultural, deep ecology

☺ ( )

We have committed ourselves to living differently and within a community: together we are inhabiting two neighbouring farms within a small village and within a greater neighbourhood, associated with the surrounding region. We are embedded in a network whose centre (cultural pub) is in the immediate vicinity, see www.lebendiges-miteinander.de.

We want to live simply and share a lot among ourselves. Further goals are: Ecological housing and healthy food that should increasingly originate from our vegetable gardens. Self-sufficiency is not part of our concept, because we want to live within a (regional) network. Most important is to live in friendship with nature and to increase our sensitivity and connectivity towards the Earth - our home planet. We do participate in initiatives to give the world a more friendly, peaceful and fair-minded shape.
There are about 2 meetings a month, we separate organisational and group process meetings. There is also an intensive weekend three times a year.
We are looking for individuals, couples, families or even an already existing (small) group who would like to join us.

Community Supported Agriculture (CSA)

☺ consistent life perspective, social/political/ecological/spiritual/financial commitment

Lebensgemeinschaft Freiheithof

B10

Freiheithof | 88605 Sauldorf-Roth ☎+49-(0)321-212 530 13
verein@freiheithof.de www.freiheithof.de

1981 (near Stockach)

g, e
anthroposophy
care
self-managed, spiritual, anthroposophic, meditation, mentally/psychologically handicapped persons, organic agriculture

The "Freiheithof" ("Freedom Estate") is a site of the St. Luc * Freiheithof * Vaudesy community. The members of the community are active in several places in Germany and Switzerland. The tasks and responsibilities are borne by a cross-locational council. The community was founded on the desire to develop methods of human co-housing and shared work, without the stimulus of addiction to personal gain. Where people who would otherwise be excluded can find their own space. Where communal and individual development is possible simultaneously. The Freiheithof is located near Lake Constance in Germany. The name goes back to its first owner: The name "Freyheit" or "Freiheit" ("Freedom") can be traced back as far as 1883. The following fields of activity are represented:

- The social sector incorporates the communal life of the people at the Freiheithof.
- The agricultural sector is run according to bio-dynamic principles and sub-rented to a Demeter-farmer.
- In terms of the economic sector we are working on the development of a company.

☺ interested in Anthroposophy

Lebensgemeinschaft Klein Jasedow

E2

Am See 1 | 17440 Klein Jasedow ☎+49-(0)38374-752 27
cs@eaha.org www.zukunftswerk-kleinjasedow.de

1977 20 ha (near Anklam, Wolgast, Greifswald) 60 % ♀13 ♂7 8
28

g, e 18 all
everyone in common (consensus) shared economy (shared income and capital)
common property and property of the European Academy of Healing Arts
none
music production, ecological agriculture, ecological magazine Oya, free democratic school, musical instruments, internet service, gardening
ecological, self-managed, non-violent, inner growth, organic diet, adult education, pedagogical, free school, seminar-house, peace-work, environmental protection, self-sufficiency, permaculture, organic agriculture, organic gardening, animal friendly husbandry, alternative energy, new technologies, ecological construction, accessible to the disabled, gong manufacturing, sound house, musical facilitation, free school, ecological farming, Oya magazine

( )

Our community began in the mid-1970s", at the dawn of the New Social Movements" in Germany. 28 people make up the community core at present. We are students and tea-

chers, musicians, authors, businesspeople, media specialists, craftspeople, therapists and practitioners in the fields of clay work, gardening and ecological agriculture. Until 1993, we lived south of Munich near Wolfratshausen and were the core of a socio-cultural initiative, the "Temenos Project" which in its heyday was an encouraging model of self-determined thinking and action for many people.
Since 1997 we've been living in Klein Jasedow (Pulow), in the beautiful hilly landscape on the mainland across from the Baltic island of Usedom. We are active in the region's "Renaissance": music, socio-culture, communication, ecology and education are our core areas of expertise. Within the framework of the "European Academy of Healing Arts" (www.eaha.org) which we founded, a meeting and educational space has been created in Klein Jasedow, the "Sound House on the Lake", as well as a multi-generational house, a free school and other beautiful projects. We produce gongs and musical instruments for music therapy in our instrument workshop called SONA.
Our economic foundation are several small businesses, the academy and our foundation "Zukunftswerk". Our magazine "Oya" (distributed nationwide) exists since 2010. Communication and music play central roles in our lives, and find expression in the "Sound&Body Music Project" ("Klang&Körper Musikprojekt").
Please see the following useful websites for further information: www.stiftung-zukunftswerk.de, www.eaha.org, www.siedler-film.de, www.drachenverlag.de, www.kleine-dorfschule-lassaner-winkel.de, www.kraeutergarten-pommerland.de

local networks, EUDEC and DPWV

Lebenshaus Heitersheim e.V.

B10

ökomenisch-solidarische Gemeinschaft

Staufener Straße 1a | 79423 Heitersheim ☎+49-(0)7634-551719

mail@lebenshaus-heitersheim.de www.lebenshaus-heitersheim.de

1997 (near Freiburg) ♀3 ♂3 7 13

g, i

core community with the association

leased

living and organizing the day by day life in a self-responsible manner

ecological, non-violent, Christian, ecumenical, organic diet, mentally/psychologically handicapped persons

( )

The Lebenshaus Heitersheim is a small community of three families with the common goal of supporting people in difficult life situations. The offer to live in the Lebenshaus gives the necessary attention and security. Our communal and solidary way of life makes it possible to give concrete support and thus new courage to live. In this way we can accompany people for a period of up to two years in the Lebenshaus. The support of the roommates by the core community is honorary. The association accompanies the project and ensures the financial and organizational framework.

→See also description of network "**Lebenshäuser**"

Lebenshaus im Chiemgau

D9

Ökumenisches Netzwerk für soziale Integration e.V.
Siboling 2 | 83361 Kienberg ☎+49-(0)8628 - 986 985
mail@lebenshaus-im-chiemgau.de www.lebenshaus-im-chiemgau.de

1999 12 ha (near Rosenheim, München) 80 % ♀2 ♂7 2 +-11

g, e all
team of the farm
Christian
people in crisis, farming
ecological, Christian, ecumenical, inner growth, organic diet, mentally/psychologically handicapped persons, self-sufficiency, organic agriculture, organic gardening, organic farming, animal friendly husbandry, alternative energy, humour, arts

☺ ()

The "Chiemgau Living House Association" ("Lebenshaus im Chiemgau e.V.") offers a temporary family home and human support for people in difficult life situations. It is based on ecumenical Christianity and is open to people from all nationalities, religions and social backgrounds.
Our life in the rhythm of nature has a healthy effect on the soul. The large manor offers many possibilities to develop artistically, musically and ecologically.

→See also description of network "**Lebenshäuser**"
☺ trial week

☆ Lebenshaus Trossingen

B9

Ökumenische Gemeinschaft für soziale Integration e.V.
Kirchstraße 23 | 78647 Trossingen ☎+49-(0)7425-1446
lebenshaus@t-online.de www.lebenshaus-trossingen.de

1985 (near Tuttlingen) 12

board
financed by donations and rent
Christian
independence, security

☺

We have set ourselves the task of giving people in crisis situations the necessary human comfort. Therefore the association "Lebenshaus - Ökumenische Gemeinschaft für soziale Integration e.V." was founded in 1985. The founding members were almost all engaged in peace circles and in various churches and were members of the "International Fellowship of Reconciliation". The aim is to provide a framework for people seeking a protective community for social or psychological reasons. During the temporary stay of each "guest" we support him/her in his/her efforts to overcome his "life crisis" and search with him for another possibility of life. In 1987, a listed factory building from the turn of the century was converted into a residential building with special charm. On average, the house is inhabited by two families and six guests of different ages. We live together under the motto "freedom and security", because the chance for development for every resident is very important to us.

→See also description of network "**Lebenshäuser**"

Lebenslernort Am Windberg

C5

Am Windberg 1 | 99625 Beichlingen/Thüringen ☎ +49-(0)3635-602 71 07
info@amwindberg.de www.amwindberg.de

2014 | 10 ha (near Kölleda) | 10 % | ♀11 ♂9 9 | 29

German, e | 15 18

orga-plenum (consensus (for now)) | at the moment: individual financing

owned by Windberg trust

seminars, school, kitchen, office, kindergarten (planned),

ecological, self-managed, non-violent, spiritual, inner growth, meditation, yoga, organic diet, vegetarian diet, adult education, pedagogical, free school, seminar-house, peace-work, environmental protection, self-sufficiency, permaculture, organic agriculture, organic gardening, alternative energy, new technologies, ecological construction, humour, non-smoking, silence, meditation, family constellations, feelings constellations, community building according to Scott Peck, transparent communication, process engineering, yoga, appreciative communication, mediation, therapy, couple therapy, sharing the presence, handicraft: woodwork, journalism, cook, kindergarten teacher, piano tuner, culture management, editing, tailor, paper-work, social pedagogics, technical draughtswoman, photography

☺ ( )

Core of our "life learning location" is a community of 29 people aged between 2 and 65 years who strive for individual and collective efforts to raise the level of consciousness and who support each other's spiritual and emotional growth.
After we have founded the community 2014 and a seminar venue in 2015 we started a school at 2018 and we will start a kindergarten within the next time. Part of our vision is connection to a higher self and to the highest creative potential of each person as well as to our collective intelligence.

→See also description of network "**GEN Deutschland**"

Paritätischer Wohlfahrtsverband, Bundesverband der Freien Alternativschulen (BFAS), Artabana

☺ approach the community with the following steps: Get-to-know-weekends, workcamps, Windberg intensive weekends, one year trial period

Lebensräume e.V.

C3

To Experience Community - To Live Hospitality - Engagement For Others

Lönsstr. 19 | 27374 Visselhövede ☎ +49-(0)4262-956 91 60
info@lebensraeume-online.de www.lebensraeume-online.de

2009 | (near Landkreis Rotenburg (Wümme)) | 1 % | ♀5 ♂1 0 | 6

g | 1 3

board of society (consensus) — individual financing

property of society

principles of the "Deutsche Evangelische Allianz"

social and missionary local involvement; care of houses and premises

Christian

☺ (✋✋)

In 2009 a city villa in Visselhövede was made available to Lebensräume e.V. In this house, four apartments, two guest rooms and common areas were established for people, to whom community is important and who want to invest in it. In 2016, a second house with four more apartments was added in the immediate neighbourhood.

Each inhabitant leads his independent life and is responsible for himself. Nevertheless, no one is living alone, and we want to encourage each other and support the motto "What can each one contribute to make the other doing well?". We want to shape our life together according to Christian principles. The "household community" forms the basis for the work on site. We want to share our community with others. That's why guests are welcome. They should get the opportunity to experience our community and to recover. We want everyone to know that community can do good and that every single person is precious.

Therefore, various offers have been developed, to which everyone is welcome. Common areas (kitchen, living room and dining room) are available for joint meetings and celebrations. There is also a large garden. What matters to us in everything is that Christian fellowship can be experienced. We want to get involved where help is needed and work with our gifts.

Solo & Co

☺ common Christian basis, getting to know each other quite well

Lebenstraum Gemeinschaft Jahnishausen

E5

Jahnatalstraße 4a | 01594 Riesa - Jahnishausen ☎+49-(0)3525-517 851

www.ltgj.de

2001 — 3.5 ha (near Riesa) — 30 % — ♀30 ♂13 8 51

g, e, f

everyone in common (consensus) — individual, limited community financing

common property

seminar house for communication, health centre, art, guesthouse, ecological garden

ecological, self-managed, non-violent, spiritual, inner growth, meditation, organic diet, vegetarian diet, adult education, seminar-house, peace-work, self-sufficiency, organic gardening, alternative energy, ecological construction, art projects, Project "Adventure Ageing", healing space

☺ (✋✋)

Our community project is th realization of a long-cherished dream, in which the interests of individuals and the needs of a thriving social community support one another. We are committed to new forms of communal life, of dealing with nature, technology and health. In our project, all generations live together and learn from one another.

We want to live in such a way that respect for all living things and respect for diversity and variety become genuinely lived values. Areas of trust develop in which we can encounter each other in a more authentic way. When conflicts arise, we consciously seek to clarify and resolve them. Self-responsibility, truthfulness, mutual support, empathy and an eye to the well-being of the whole community become a reality in our daily lives.

We also feel strongly about:

- ecologically responsible technology
- organic foods – an 0.6 ha organic garden supplies us with fruits and vegetables
- healing and health – of people, the environment and nature
- undogmatic spirituality
- art, culture and creativity

The framework of our community is formed by the Jahnishausen manor (3.5 ha) with its historic buildings. It provides room for up to 75 persons. In order to make it the property of all community members, we have founded a co-operative.

→See also description of networks "**GEN Europe, GEN Deutschland**"

☺ time to get to know each other, becoming a member of the cooperative Gut Jahnishausen e.G. (capital contribution), sustainable lifestyle, participation in the community process, active participation in different work areas

Lilitopia Ganzheitliche Bildungs- und Zukunfts-Werkstatt B6

Eulenweg 16 | 35260 Stadtallendorf

☎+49-(0)6428-441 135, +49-(0)170-435 03 94, +49-(0)157-850 656 76

✉rochelta@yahoo.de, schenker_oeff@yahoo.de 💻www.lilitopia.de

2010 5000 m² (near Marburg) 10 % ♀1 ♂2 1 4

g, e, f 2 4

owner, in steps consensus-level (individually, in part consensus)

some make individual financing, others family-sharing (love-communism)

property of individuals, in steps opened for sharing

givers-(or gift-)movement, working with light and love

holistic community-work, educational activity, natural living, nonviolent dog-school for humans and animals

non-violent, spiritual, inner growth, adult education, seminar-house, peace-work, animal protection, non-smoking, holistic sustainability

☺

Our core orientation very briefly: As global consensus-siblings, we are planning everything necessary together - and in a free arguments culture... We would like to find out everything else together with you in this argument search. There are no taboo topics (state, money, property, private life, etc.). More at www.global-love.eu! We experiment with "holistic education work" in value orientation: in radical self-responsibility we form a way to a global consensus democracy (better: demogracy) through non-violent sharing or donating global LOVE (consensus communities of global sharing). Free earth for free sharing people!

The core family (Anke Rochelt, Öff Öff aka Jürgen Wagner and our son Aljoscha) seeks to be the germ cell for the formation of such consensus communities. We aim for holistic healing processes in the world by giving examples of interior and togetherness work through our lifestyle, seminars and actions. More on www.lilitopia.de, also with a free download of our scientific book "Education for Holistic Sustainability and Global Responsibility: Global Identification Instead of Fragmentary Thinking or Reconciliation" by Anke Rochelt.
To grasp the different aspects of our ideals, it is also beneficial to read about other projects of the Schenker (= giving) movement: Haus der Gastfreundschaft ("House of Hospitality"), Biotopia or Friedensgarten ("Peace Garden")... Our commitment extends to the design of an alternative system to the concept of "State": Our "Terrania" is associated in Räte-Republik Friedens-Garten RRFG ("Council Republic Peace Garden RRFG"), for which we chose the "Earth Charter" as a constitution.

→See also description of networks **"Givers-Movement, L'Arche de Lanza del Vasto, European Catholic Worker"**

Gandhi-Movement, internat. gift economy (moneyless.info)

☺ fit into our holistic culture-ideal

Maibacher Schweiz Lebensgemeinschaft und Seminarhaus

B7

Alt Maibach 12 | 35510 Butzbach OT Maibach ☎+49-(0)6081-443 517

info@maibacher-schweiz.de www.maibacher-schweiz.de

2003 2200 m² (near Usingen, Butzbach, Bad Nauheim, Frankfurt)

♀7 ♂7 5 19

2 6

g, e, f, sp, Dutch

all persons concerned together (consensus and small groups)

limited community financing

property of individual community members - own registered association to take over soon

seminars, working with teenagers, ayurveda-portal.de, yoga, outdoor-trainings, maintenance, cooking

self-managed, non-violent, spiritual, esoteric/New Age, inner growth, meditation, yoga, organic diet, vegetarian diet, pedagogical, seminar-house, humour, non-smoking, Ayurveda, non-smoking house, seminar venue, EFT, Hooponopono, BaLaance Board, Kundalini yoga, dance, theatre, veterinarian, hatha yoga, prana cereal, swinging massage, sweat lodge, longest private zipline

☺

Our community and seminar house, a former country hotel, is located in a beautiful small village surrounded by forest in the Taunus, about 45 km north of Frankfurt a.M. Some community members live in the village. 1,500 m² space offers us many possibilities for development. Our goal is the promotion of our creative and spiritual development. Our community is characterised by the freedom which we have as a group and as individuals. We are united by the desire to live in a spiritual community in which we grow by our challenges with one another and continue to develop. In our daily life we share communal lunch and childcare. There are weekly organisational meetings, quarterly communal days, regular movie nights and singing evenings. Some members meet regularly for meditation, yoga or dance sessions. The women meet in the red tent on the new moon.
Professional projects of the community members include: website ayurveda-portal.de,

"Theater Herzstück", a brewery, canoe and climbing tours, visual arts, forest exploration, "swinging massage", social work, Kundalini Yoga, Dancing For Birth™. Other activities are: external leases for seminars, workshops, biannual community festival, various network meetings and parties. We hope for more extension into the village.

→See also description of network **"Soshona Freundeskreis"**

Artabana (some)

☺ personal and group responsibility, getting to know each other, then moving in for a trial period (6 months), then applying for membership

MartiniQ genossenschaftliches Wohnprojekt eG

B5

Parkstr. 38 | 34119 Kassel ☎+49-(0)561-602 908 90

info@martiniq-kassel.de www.martiniq-kassel.de

2017 | 2200 m² | ♀18 ♂11 4 33

g, e, sp, f, i, German Sign Language

the general assembly (democratically (majority principle))

individual finances | own cooperative

ecological, self-managed, animal protection, environmental protection, alternative energy, ecological construction, accessible to the disabled

☺

We are a cooperative housing project in the construction phase, planned move-in is summer/autumn 2020. We have been united by the desire for a form of living that enables us to make our home in a community-friendly residential area cost-conscious, ecologically sustainable and intergenerational. Once all flats are occupied, we will be 40-45 people. Our house is designed to offer as much communal and meeting space as possible: we have a common room, a guest flat, a communal garden, arcades with common balconies, a patio with seating and a laundry room. As we are building in the midst of a newly developing district on the site of a demolished brewery, we also hope to extend our network beyond our house with the other residents of the neighborhood.

☺ mutual sympathy, becoming part of the cooperative, living in one of the flats

Meuchefitzer Gasthof und Werkstätten GmbH Gasthof und Tagungshaus

C3

Meuchefitz Nr. 19 | 29482 Küsten ☎+49-(0)5841-5977

gasthof@meuchefitz.de www.meuchefitz.de

1980 | 2 ha (near Lüchow; Salzwedel) | ♀5 ♂5 0 10

g, e, f | 4

all (consensus) | various

common property

anarchist

conference centre, bar, cooking, garden, building and craftsmanship on a DIY basis

ecological, self-managed, anarchistic, left-wing political understanding, chaotic, organic diet, seminar-house, peace-work, organic gardening, ecological construction, bar/restaurant, conference centre

We are a group of ten persons whose composition changes from time to time. We run and administer the restaurant and guesthouse Meuchefitz by ourselves - The Place is often called "Gasthof des Widerstands" (Guesthouse of Resistance). Looking back at the history of the house, we still see Meuchefitz as a place of confrontation: first in our resistance against this inhuman system, and second in our attempts to live more humanly and more creatively as self-determined human beings. The guesthouse is used by a broad spectrum of groups, ranging from constructive criticism to open resistance. We strive to stay in continuous contact with a variety of groups from the cities and to remain in dialogue on a wide range of issues. We don't want to content ourselves with just hiring out the rooms to others. For children and youth groups, a countryside holiday with us should arouse curiosity and a desire to learn more about animals and plants, about landscapes and agriculture.

Interkomm Region Wendland, Freier-Fluss-Netzwerk Wendland/Altmark
common political goals

Milan-Vaubanaise Miteinander leben - Füreinander sorgen B9

Lise-Meitner-Str. 14 | 79100 Freiburg

+49-(0)761-476 8051, +49-(0)761-709 888

wohnen@mitundfuer.de www.mitundfuer.de

2011 ♀11 ♂5 7 23

g 3

all in common (consensus, if possible) individual financing

housing cooperative in which all residents are members

no

self-managed, non-violent, accessible to the disabled, inclusion, cross-generational

Since June 2013, the Milan-Vaubanaises have been living in ten residential units (between 45 and 120 m^2) in the inclusive VAUBANaise project in the Vauban district of Freiburg. We are 16 adults between the ages of 24 and 81, and seven children and adolescents between the ages of three and 14.

Our guiding principles for communal living are the willingness...
- to live in a community and to participate constructively in shaping it
- to accept the other in his individuality
- for open dialogue and honest negotiation of concerns
- to confront conflicts and strive for fair solutions
- to accept new experiences as a positive personal challenge
- to neighbourly care for people and their homes - not only in case of vacation or illness.

We rarely have vacant apartments. Nevertheless, if you are interested in our housing project, please contact us. Who knows..

network for communal living (Gewo-Netz)

Miravillage-Zentrum

D9

Danzigerstr. 1 | 82131 Gauting ☎+49-(0)89-850 87 51
miravillage@mirapuri-enterprises.com www.mirapuri-enterprises.com

1970 1000 m² (near München) ♀13 ♂11 7 31

g, e, it, f 25

shared income association

free spirit, understanding of the world as a unity of spirit and material

various professions, IT, digital and offset printing, graphic design, multimedia, digital video, music

ecological, self-managed, anarchistic, feminist, non-violent, spiritual, chaotic, inner growth, meditation, yoga, tantra, sexuality, free love, organic diet, vegetarian diet, adult education, pedagogical, free school, seminar-house, peace-work, animal protection, environmental protection, self-sufficiency, new technologies, humour, Health and New Age seminars, various craft, health training, various media

☺ ( )

Mirapuri in Italy as well as its first satellite, Miravillage in Germany, have been created for people who like Sri Aurobindo and Mira Alfassa, their ideals of human unity and spiritual consciousness progress. Mirapuri and Miravillage are environments where people can live together and work together to realise these ideals. Everyone of goodwill who likes Sri Aurobindo and Mira Alfassa and their ideals can join Mirapuri and Miravillage.
There are no financial or other material obligations. If you want to live in Mirapuri or Miravillage you can do so on a private or contributing basis. That means you may buy your own house or apartment which is your property, or else you may contribute financially to the growth of the general Mirapuri property. In this case, Mirapuri and Miravillage provide accomodation and services for you during your lifetime at Mirapuri or Miravillage and the property belongs to Mirapuri or Miravillage and its future development.
Everyone works and learns on a free and independent basis. The growth and respect of individuality are understood as the basis and flowering of real communal life. There is no institutional spiritual authority. Each individual follows his/her own spiritual evolution according to his/her aspiration and inclination.

☺ knowledge of Sri Aurobindo and Mira Alfassa and their integral yoga, as well as good will and genuine co-operation

Mittendrin leben eG

C3

Im Dorf 2 | 21368 Dahlem-Harmstorf ☎+49-(0)5851-944 39 62
mittendrin-leben@posteo.de www.gemeinschaft-mittendrin-leben.de

2016 1.2 ha (near Lüneburg) ♀9 ♂7 5 21

g 6-8

all in common (or in task groups) (consensus minus one)

individual financing building: association; ground: foundation 'trias'

Some of us follow a spiritual path - others do not. All are actually ecologically oriented (what is that, though?). Everyone is also interested in intensive experience processes. We experiment to find new and open ways of living together

ecological, self-managed, spiritual, inner growth, alternative energy, ecological construction, humour, Forum, periodical group works

☺

We are experimental, trying to transform our inner images from "as it should" towards "as it is", but without losing sight of the vision of a new world in whose creation we actively participate. We are still at the beginning, working on our basics - as well as on other buildings. Gardens grow, people emerge and leave, projects in the project take shape ... or slip away again. We learn a new we and accept and welcome what happens on the way there.

→See also description of network "**GEN Deutschland**"

Mehr Leben e.V, CSA

☺ it needs to fit - and - 1/3 of the construction costs need to be covered due association shares

Mühle - Wohn- und Baugenossenschaft e.G. B9

Weidenstr. 30 | 77871 Renchen ☎+49-(0)7843-841 46

info@muehle-renchen.de www.muehle-renchen.de

1985 6000 m² (near Strasbourg) 5 % ♀3 ♂7 10

g 3

plenary session (discussions, consensus) cooperative

socialist

print shop

self-managed, left-wing political understanding, non-violent, chaotic, adult education, peace-work

☺

A few politically active people bought the mill in Renchen in 1985, in order to develop an alternative to the wide-spread individualisation, to create a space in which life and work, organisation, resistance, learning and fighting can be possible. We are very serious about the collective implementation of this concept. We don't see ourselves as an island – our trade union commitments, as well as our participation in the "Association of Nazi Regime Persecutees" ("Vereinigung der Verfolgten des Naziregimes")/"Association of Antifascists" ("Bund der Antifaschist*innen") and other social movements, are part of our basic concept of self. We have been a cooperative since the year 2000.

→See also description of network "**KommuJa**"

☺ time for getting to know each other/probationary period

Mühlenhof E3

Woltersdorfer Str. 26 | 16567 Mühlenbeck ☎+49-(0)151-153 381 87

2019 940 m² + allotment 5000 m² (near Berlin) soon

♀3 ♂3 4 10

e, g, sp 2

all (majority vote)

individual financing with common financing of several sectors

owner- and tenant community

refurbishment and extension of adjoining buildings

ecological, self-managed, left-wing political understanding, inner growth, permaculture, ecological construction, construction, humour, workshop, yard with fireplace

We are a newly founded community with an intentional character on the outskirts of Berlin that started implementing community life in February 2019. As three couples with children, we live on three floors of a lovingly renovated house (built in 1908) with a large courtyard and will soon extend the two adjoining buildings. We plan to implement different ideas: a therapy/meditation/community-room plastered with clay will be created. A sauna is planned as well as a room that can be used as a workshop and rehearsal room. At the center of our beautiful courtyard grows an old walnut tree, around which we can live, work and celebrate together in the future.
On an adjoining property we would like to start with permaculture - if the use allows it. We start with what is there and then grow bit by bit as a house and living community.
We are open for helping hands and offer board and lodging from summer 2019.

☆ Mutter Erde Gemeinschaft

B6

Holper Str.1 | 57537 Forst-Seifen ☎+49-(0)2742-8251
info@mutter-erde-gemeinschaft.de mutter-erde-gemeinschaft.de

1983 | 6 houses on 2 ha (near Köln) | 30 % | ♀12 ♂4 3 19

g 9

Individually after consultation with the others (consensus)

private property, membership fees, donations | owned by individuals

humanistic

seminars, publications, gardening, vegetables, beekeeping, baking

ecological, self-managed, spiritual, inner growth, meditation, yoga, vegetarian diet, vegan diet, adult education, pedagogical, free school, peace-work, animal protection, environmental protection, organic gardening, humour, non-smoking,
We want a healthful world!

☺

The Mother Earth Community - also known as the "Seifener Modell" - is based in the Oberbergische Land, about 70 kilometers east of Cologne. Here, people can come into contact with their innermost, their soul, their light. A model of life in which the best we have is allowed to grow and people can develop their potential. Where one learns to recognize barriers to thinking, to perceive emotional blockades and to reduce prejudices, to be careful with oneself and everything else. We are all the light of the world and we want to carry this awareness outwards.
We are a community in which everyone lives their lives responsibly, but supporting one another, when help is needed. In which one knows that inner structure and harmony also show themselves outwardly as order and cleanliness and in which simple life and high thinking are appreciated. We are people who put the good, the beauty and the joy of life first, appreciate the silence, reflection and meditation, who can actively get down to work, do not feel committed to a dogmatic faith, think solution-oriented and grow spiritually,live vegetarian/vegan and enjoy good food. There is plenty of space for humanistic, spiritual, educational and ecological projects here. Land for horticulture and space for young families is available. We even got fast internet lately.

☺ mindfulness, lust for life, love for creation, power, zest and pioneering spirit

Nachbarschaftlich Wohnen in Dormagen eG A5

Latoursgarten 1 | 41542 Dormagen ☎+49-(0)152-387 137 63
mail@nawodo.de www.nawodo.de

2018 3200 m² (near Köln, Düsseldorf) ♀21 ♂18 13 52
g, e, f, Dutch 8
plenum, sometimes working groups (consensus) joint property
none
ecological, self-managed, non-violent, physically handicapped persons, organic gardening, alternative energy, new technologies, accessible to the disabled
☺

Neighborly, intergenerational living means more than a mere collection of residents of different ages. We promote communal living and mutual support without hindering the individual need for freedom and privacy. People of all ages, with and without handicap, families or single parents, couples and singles, gay and lesbian people - regardless of world view, nationality or religion - are welcome.
If you want to live with us, you should like people, like community living, be prepared to help shaping the development of the group and be considerate.
This is what we offer: the recognition of the individual personality, an attentive interaction with each other, decisions by consensus, encounter at eye-level regardless of origin, income, age. Support and protection by the group and a stable social environment.
We have planned our apartments ourselves and manage them ourselves. We are happy to share our knowledge and experience with other groups and people involved in similar projects. Visitors, whether as a group or individuals, are always welcome.

neighbourhood network

Nachbarschaftliches Wohnen in der Ökosiedlung am Cherbonhof C7

Tarvisstraße12 | 96049 Bamberg ☎+49-(0)951-641 63
g.henzler@web.de oekosoziale-siedlung-bamberg.de

1987 1.4 ha ♀54 ♂37 11 102
g
our legal entity
indiv. financing & communal property (garbage collecting place, table tennis hall, arcades) and rented facilities (community room and garden)
individual property and rented (apartments for elders)
ecological, self-managed, ecological construction, model for eco-social living: in good neighbourhood, ecological and well designed

On the initiative of BUND (Friends of the Earth Germany), the eco-settlement was planned by architect Theodor Henzler as an example of a holistic urban development that is human and environmentally compatible. The project had to assert itself politically in a laborious planning process from 1981-1985. The settlement concept is characterised by a varied sequence of streets, alleys and squares. The private interior gardens, which should be protected from view as far as possible ("living rooms in the green"), are contrasted with residential courtyards and street spaces that can be used in a variety of ways and extend right up to the buildings. As in old towns and villages, the squares are formed by closed buildings. The "little village" with 25 single-family houses and 21 rented apartments suitable for the elderly is shielded by a belt of garages with outbuildings built above them and by commercial rooms facing north to the busy main access road. In the south, a communal fruit tree belt and vegetable garden adjoin on an urban leased area. Both serve not only the self-sufficiency, but also a meaningful leisure activity with lively communication.
All houses were designed by the owners and their architects on their own responsibility within the framework of the development plan and the special private design and neighbourhood conditions. A middle course was deliberately chosen between the monotony of housing estates and the often creative chaos of detached single-family houses.
In the national competition "Familienwohnung und Familienheim 1983-84" the project was awarded for an outstanding solution. The settlement celebrated its 30th anniversary in 2017.

See photo on page 4

Nameless: Shared House in Plauen

D6

(Please inquire address) | 0852x Plauen
exit@buerotiger.de

2015 ♀1 ♂1 2

g, e

individual financing — community property

inner growth, self-sufficiency, construction, humour

Specialty: Communicating on community issues must be in writing (for more information: http://www.jakobgut.de/erdnuss.htm).
The "WG-Haus" is an independent project whose rooms are partly used by residents of Jakobgut (see page 190). The orientation is similar to that of Jakobgut. It shares the values, though the project is not explicitly Christian. At the moment we are still in a finding process about the future direction, in which interested people are welcome to participate in.
The building is a 150 year old multi-family house with a wide view over Plauen. The city center can be reached by foot in about 15 minutes, a tram stop in about 5 minutes, also various park-like areas are close. Rooms in different sizes are adjacent to central hallways. Much of the house is currently not renovated and vacant.
As a community, we are fundamentally oriented towards a materially simple existence. We cultivate an open-hearted, direct, attentive, mutually-oriented relationship with each other, leaving each other a lot of space when asked.
In the long run, it is possible here to live an urban life without money. One means towards this is the idea of self-sufficiency. It is possible to live temporarily in Jakobgut (no automatism!) about 15 km away and to take care of "own" garden areas there.

Nature Community eG

St. Hubertus 1 | 92539 Schönsee
+49-(0)9674-319 09 97, -922 90
info@nature-community.de
www.nature-community.de

D8

2014 4 ha, near Nabburg (30 km) and Regensburg (90 km)

♀14 ♂20 15 49

g, e, sp, i, Portuguese, Czech — about 30

permanent guests and members of the cooperative (consensus)

shared finances — possession of the cooperative

hotel, kitchen, seminar centre, childcare, maintenance/building, garden

ecological, self-managed, non-violent, spiritual, inner growth, meditation, yoga, tantra, sexuality, free love, organic diet, vegan diet, adult education, pedagogical, free school, seminar-house, peace-work, animal protection, environmental protection, self-sufficiency, permaculture, organic agriculture, organic gardening, alternative energy, new technologies, ecological construction, humour, body therapy/massage, sound healing, music, instrument lessons, coaching, carpentry, bike workshop, milling machine, graphics

☺ ()

A world in which we all live in peace, freedom, trust, joy and connection... no less than that is our vision! A former holiday village near the Czech border in the middle of pristine nature offers living space to our ideas: We live vegan, ecologically and sustainably, connected to humans and animals, honestly and consciously.

We strive for self-sufficiency through permaculture and our own energy and water supply, we are building sustainably and ecologically. Free education for humans of every age, deep, authentic partner- and relationships, mutual support and solidarity as well as transparency and honesty are important to us – building trust means building peace.

A new culture. A culture of trust and cooperation. A culture of peace in a global sense. We, the people of Nature Community, are working on a concrete and functioning place for this culture to arise. And we invite community-interested people wanting to support us, wanting to carry our vision powerfully into the world. People wanting to join a growing community in creation. We are looking for sponsors, investors, supporters. Our most important current project is upgrading our biggest seminar room into a meeting place for the new culture. Nature Community wants to be a nourishing environment for social change.

→See also description of network "**GEN Deutschland**"

Artabana

☺ life according to our vision, co-creating our community according to our vision, entrance fee by entry into the cooperative

See photo on page 5

Germany

Neubeginn e.V. Hilfe in psychischer Krise

B5

Mintenbecker Str. 16 | 58515 Lüdenscheid ☎+49-(0)2351-7005
info@neubeginn-ev.de www.neubeginn-ev.de

1984 1 ha (near Lüdenscheid) ♀4 ♂5 9

g, e 2

everyone in common (consensus) limited community financing

property of Neubeginn e.V. (association)

spiritually open

advice/support of people in mental health crises, small seminars (psychotherapeutic, spiritual and ecological themes)

non-violent, spiritual, inner growth, meditation, adult education, seminar-house, consultation and assistance in psychological crises

Neubeginn ("New Beginning") is an association which offers help to people who are going through a psychological crisis to find their way back to an independent life, i.e. helps people to help themselves. We offer consultation, self-help groups and seminars. In addition, persons seeking help can live in our community for a shorter (a few days) or longer (up to half a year) period of time and pursue their reorientation, their personal "new beginning". Through their own experience, the project members are genuinely interested in dealing with psychological crises and are in a position to understand and assist persons in crisis.
Since 1984 our association has owned a residential and seminar centre in Mintenbecktal near Lüdenscheid. It is located in a peaceful nature area and has a chapel (non-denominational!) on its own property which is open to all people as a place of silence.
The project's members largely work outside the project during the day, some of them in the project. Thanks to years of living and working together with psychological and spiritual growth issues, the community has become very real and close to life. At the same time, it is open for the individual paths of individual "New Beginners".

☺ financial independence, emotional stability, interest in working with people in crisis situations (voluntary), one probationary year

Neuwagenmühle

B7

Neuwagenmühle | 56370 Kördorf ☎+49-(0)6486-6686
info@neuwagenmuehle.de www.neuwagenmuehle.de

1995 1 ha (near Katzenelnbogen) ♀2 ♂2 1 5

g, e, sp 2

depending on whom it may concern by consensus

individually owned by individuals

arts, garden, wood

ecological, self-managed, anarchistic, left-wing political understanding, feminist, non-violent, inner growth, adult education, organic gardening, alternative energy, humour, no cell phones/wireless

☺

The owners of Neuwagenmühle are a couple over 60. There are three apartments, two of which are rented out by the owners. One is suitable for shared use because of its size,

the other one has its own garden. The apartments are nice and very simple (including stove heating).
The Neuwagenmühle is an island location with its own electricity, water (source) and wastewater supply (reed wastewater treatment plant). Electricity is generated by the in-house waterwheel. Since it can also be scarce with electricity, depending on the weather.

We do not want to keep large animals (dog is ok, cats only marginal). Our only neighbours are trees, plants and animals from the surrounding forests and waters. There is enough space to create a garden. Every year we make our winter wood supply by ourselves.
Media consumption: We prefer to have WiFi switched off during the nights and switched on as little as possible.
Apart from the couple, the inhabitants of the project change every few years. Many people from the area here have once lived on the mill which makes a loose familiar character.
The mill is called Neu(=new)wagen(=daring)-Mühle(=mill), which stands for venture and new beginnings and new paths. In fact, living here is in some ways a risk for many. Self-chosen simplicity and ecological awareness are necessary for this. There is no other way, because here nature is dominant and not man. More information at: www.neuwagenmuehle.de

ÖkoLeA

E4

Hohensteiner Weg 3 | 15377 Oberbarnim, OT Klosterdorf

☎+49-(0)3341-359 39 30

info@oekolea.de oekolea.de

1990 | 1 ha (near Strausberg) | 10 % | ♀6 ♂5 4 15

g, e, f, sp, Persian, | 5 7

everyone concerned (consensus) | 55% of personal income to common till

common property

non-dogmatically leftist

garden, seminar centre, guest floor, household, woodstove bakery and café

ecological, self-managed, left-wing political understanding, feminist, non-violent, spiritual, meditation, yoga, organic diet, adult education, seminar-house, mentally/psychologically handicapped persons, environmental protection, permaculture, organic gardening, alternative energy, new technologies, ecological construction

☺ ( )

The Ecological Living and Working Community (ÖkoLeA for short) has been in existence since summer 1990. We have developed a project on the outskirts of Berlin in which people live and work on a self-determined basis within a framework of ecological and social responsibility. In 1993 we bought a farm in Brandenburg which we have been restoring and expanding in an ecologically responsible manner. We live in two of the houses, the third gives space for an educational centre and our bakery. We have a big, wild garden, where also some seminars take place. We eat together, meet once a week to discuss and have some parties together. Once a month we meet to take care about our social relations and processes.
What else to say? We decide in consensus, have communal property, a shared economy and a communal life with children.

Our group is fairly diverse, we have very different professions and interests. Some of us work on the farm, others outside. As a group we have no particular ideological or spiritual orientation, but you find a wide range of leftist positions and alternative ways of live at our place. Currently, we face generational change and look for four to five additional community members. We bought the neighbours property to provide new living spaces.

→See also description of network **"Kommuja"**
ANU, FÖL, VERN and other regional networks
☺ we require quite a long period of getting to know each other and try out to live together

Ökologische Gemeinschaftssiedlung Alte Windkunst

A6

Haus-Heyden-Straße 201 | 52134 Herzogenrath windkunst@gmx.de

1988 4000 m² (near Aachen) ♀11 ♂7 7 25

g, e 4

majority vote (consensus aspired) in general assembly
individual financing/joint fund for community institutions

home ownership
ecological
self-managed, non-violent, meditation, yoga, organic diet, peace-work, environmental protection, organic gardening, alternative energy, new technologies, ecological construction, humour, common house, central common heating system, common workshop, washing machine room, choir, dance & therapy, art courses, community building seminars

'Alte Windkunst' was created to be communal, ecological, child-friendly and inexpensive. Nine residential buildings and a community house are grouped around an inner courtyard. During the construction of the timber-frame-buildings, biological materials were used as far as possible according to the standards of the construction period. In order to make the project financially viable for the founders - mainly young families - the project was planned in a way that many things could be realized with the help of the founders themselves.
The property, the outer shell, the central heating, the community house and the solar plant are common property. The interiors are privately owned according to the German Home Ownership Act. In an additional contract the inhabitants secure themselves among other things mutual co-determination in case of owner changes.
In the first years the needs of the children and their care were formative for organization and common activities. A day-care center for children was built in the community house, first for pre-school children, then for schoolchildren. In the afternoons the parents took turns looking after the small kids. In the meantime, almost all of the first generation children have left the settlement. After changes of ownership, the residents today consist of three young families with a total of seven children, one single young adult and twelve people around the age of sixty. This has also changed their needs. A cosy square has recently been integrated into the inner courtyard as a meeting place. Social brackets continue to be regular meetings, work weekends to care for the outside facilities and the community center, parties, occasional excursions with the whole group and neighbourly support in everyday life.

Ökologische Gemeinschaftssiedlung Moldenhauer Hof

D3

Döringsbrücker Weg 2 H | 16766 Kremmen, OT Hohenbruch
+49-(0)33051-262 51/-262 65
dagmar.graef@posteo.de
www.moldenhauer-hof.de

1999 57 ha (near Kremmen, Oranienburg) ♀9 ♂6 0 15

g, e, f 7

everyone in common (consensus when possible) individual financing

common property

horses, rabbits, sheep, garden, preservation of the countryside, main employment outside

ecological, self-managed, left-wing political understanding, organic diet, free school, environmental protection, self-sufficiency, organic agriculture, organic gardening, alternative energy, ecological construction, producing own solar electricity

☺

We are:

- a group of people who came together via various life paths, with different backgrounds and perspectives
- 8 parties so far: 15 adults between their late-40s and mid-60s
- partially employed in Berlin and in the vicinity (e.g. one works as a non-traditional practitioner in Eden/Oranienburg), some have already retired.

We have decided on:

- a socially and ecologically compatible lifestyle
- a happy middle course between a suburban row house estate and a rural community
- the principle: as much community as each individual desires.

We have:

- purchased 57ha of land in Hohenbruch, 12 km North-West of Oranienburg, including 56 ha available for agriculture
- built 2 residential wings in the middle of our property with 4 privately financed houses and have been living in them since 1999
- a community wing
- erected the houses according to organic building principles
- a plant-growth purification system, a well, orchards, gardens, animals and various sports opportunities
- up to 4 free building lots and community shares in farm and meadow land
- lots of room for new ideas.

We are looking for people eager to build:

- who like our project: couples, singles, single parents, families, flat-sharers of various ages, origins etc.
- who can spend around 60,000 Euros for the land and related costs and who can finance their houses through the usual channels.

We still have 4 free lots (houses 9-12)!!! We aren't allowed to build any more houses.

☺ money for construction, interest in ecology and community; willingness to support concept

Ökologische Siedlung Braamwisch e.V. C2

Braamwisch 10 | 22175 Hamburg ☎+49-(0)152 018 251 83
info@ecoact.de www.wohnhof-braamwisch.de

1998 1.3 ha 100-150

g, e
all who are concerned (we strive at consensus) individual financing
private proberty with community areas
ecologic
appropriate technologies

Saving water and energy, healthy living and mutual neighbourly support is the motto of the project. In addition to ecological building materials, low-energy house standard and solar energy generation, compost toilets without water consumption and our own wastewater treatment in plant-based purification plants, as well as rainwater use are further ecological building blocks. The special element of the ecological settlement Braamwisch is the possibility of the individual living space extension through green areas in places where in conventional settlements parking lots have to be built. Through the cooperation with "Greenwheels" (car-sharing), even those who do not have their own car are comprehensively mobile. With our Hamburg-based association ecoact we carry out educational work in the fields of ecology, environmental and climate protection and implement projects. Our first project at that time was the ecological settlement Braamwisch. We are a member of the International Network for Sustainable Energy and since 2008 we participate as NGO observers in the annual UN climate conferences.

International Network for Sustainable Energy (INFORSE): www.inforse.org

Ökumenische Gemeinschaft/Laurentiuskonvent B5

Hintere Steinstr. 4 | 34474 Diemelstadt-Wethen ☎+49-(0)1511-866 13 86
gaestegruppe@oeg-wethen.de oeg-wethen.de

1975 (near Warburg/Kassel/Paderborn) ♀20 ♂17 2 37

g, e, f, Portuguese 3 persons in associated offices and 4 honorary
plenum, working groups (consensus)
revenue community, household budget, individually
owner is Laurentiuskonvent, the inhabitants pay a rent
ecumenical- interreligious
justice, freedom, integrity of creation
ecological, self-managed, left-wing political understanding, non-violent, spiritual, Christian, ecumenical, inner growth, meditation, adult education, seminar-house, peace-work, environmental protection, supervision, intervision, integration of refugees

☺

The keywords "Justice", "Peace" and "Preservation of Creation" describe the goals and values we strive for both within our own community and in the one "World Family." They make up our Christianity. With the Christian tradition as our background, we seek a contemporary spirituality that promotes life and the One World.

We are open to people of other faith, including those who see themselves as agnostics. In our community, we want to test ways of a peaceful coexistence that does not follow the social trend towards competitive and growth ideology and individualization. The relationship between community life and our environment - from the village of Wethen to the worldwide (ecumenical) level - is very important to us. We want to contribute to changing the church and the world through peace work both internally and externally. We are committed to sustainable development (e.g.: car sharing, change of clothes and books) and regional purchases.
We host three national or pan-European offices for peace ("Non-violent action", "Church and Peace") and lifestyle issues ("ÖIEW/Earth Charter"), in which also members of the local community work. In the neighboring village Warburg-Germete we operate a conference center, the "Zukunftswerkstatt Ökumene" (e: "future workshop for/of the ecumenical movement"). Our average age is high. That's why we're especially looking forward to younger people who want to live with us.

Laurentius-facilities in Laufdorf & Hamburg; council & head office of Laurentiuskonvent
☺ see website

Open Sky House Internationale Satsang und Kunstgemeinschaft A5

Rheinstraße 54 | 51371 Hitdorf am Rhein ☎ +49-(0)2173-409 92 04
office@openskyhouse.international www.openskyhouse.international

2004 ca. 2500 m² (near Köln, Düsseldorf) ♀11 ♂7 2 18

e, g, f, Danish, Russian, Dutch all residents all residents and volunteers
all residents in communal community-meetings (without formal structure)

common household budget, common economy, individual financing rented
spiritual teacher John David (based on Advaita)
cooking, guesthouse, garden, construction, handyman work, seminar house, administration
self-managed, leader/guru, spiritual, inner growth, meditation, yoga, tantra, sexuality, organic diet, vegetarian diet, seminar-house, humour, satsang, creativity (music, visual arts, dance, theatre, photography, design), bodywork, Osho meditations, karma yoga
☺

Open Sky House - a community for conscious living. It is open to all who want to know themselves at the deepest level. It's not about our stories and dramas, but rather our true nature. The question "Who am I?" is the focus of the community.
Daily life in the community - seminar business, guesthouse, publishing company, garden, art, creativity, play and intimacy - is used to see the habitual nature of our actions. When we are free of our habitual responses and patterns, the mind can become still. Simply by not believing the noise in our own head anymore, and just being present in the moment, our lives can be transformed. The community offers a weekly program; Friday and Monday evening Satsang with John David, and Wednesday Energy Darshan with mantra singing. There is also a weekly open theater group, daily yoga and meditation.
Creativity arises from silence. The Community supports all forms of creative expression. The Open Sky band and other musicians play on various occasions, there are exhibitions of painting, sculpture, pottery and jewellery, and there is an art and video studio.
Open Sky House welcomes everybody. We invite you to visit us. We invite you to become still. We invite you to be as you are. You can come at any time as a guest or volunteer.

☺ looking for insight, that you want to work on yourself with the help of our spiritual teacher

Paradies Wohn-GmbH Frauenwohnprojekt

B10

Rosenlächerweg 2A | 78462 Konstanz ☎+49-(0)7531-365 96 83
info@paradieswohn-gmbh.de paradieswohn-gmbh.de

2010 1400 m² ♀10 4 14

g, e, sp 10 (non-paid)

whole group (consensus) property of the corporation (GmbH)

ecological, grass roots democracy, mutual help

self-administration of house and site, mutual help

ecological, self-managed, feminist, non-violent, organic diet, peace-work, one-world development, mentally/psychologically handicapped persons, environmental protection, alternative energy, new technologies, ecological construction, accessible to the disabled, non-smoking

☺

Passive house for active women!
The idea was clear: "We would like to create the process of getting older within a self-determined group of women. We wish to do this by means of encounters and joint actions, mutual inspiration and supportive solidarity, fun and challenges in communal life – all this is known to maintain the fitness of mind and body."
Three women had this vision in 2002. This developed into an association in 2004. The women of this association then looked for a house to rent, but could not find a suitable place. So they decided to build their own house. This gave them the opportunity to create the living space according to their own imagination: suitable for the elderly and handicapped, a high ecological standard, individual and separate apartments and a communal space.
In order to make the implementation of these plans possible, such as purchase of the property, architect contracts etc., the limited liability company "Paradies-Wohn-GmbH" was founded. The group of 8 ingenious women between the ages of 52 and 69 began the building project in 2009 in a part of Konstanz named Paradies ("paradise"). 10 women moved into this colourful house in the early summer of 2010.
Since then we had many changes - one woman died, other women moved to other places. 3 younger women came to us with 4 children. We make many good experiences together und we have a lot to work in house and garden and in our community.

Mietshäuser Syndikat (www.syndikat.org/en)

☺ acquiring a share of our corporation, taking on tasks within our self-administration

PermakulturHaus Göttingen e.V.

C5

Initiative für Kooperation zwischen Mensch und Natur
Kerllsgasse 4 | 37083 Göttingen ☎+49-(0)551-797 56 95, +49-(0)174-396 68 63
kaja@erdklang.eu www.permakulturhaus-goettingen.de

2012 450 m² 1 % ♀7 ♂5 12

g, e 3, but until now nobody lives off this

all together (consensus) limited common financing;

association, land and houses should belong "themselves" (later possibly to a trust)

transition movement

building the community-house and working in the garden

ecological, self-managed, left-wing political understanding, non-violent, spiritual, inner growth, organic diet, vegetarian diet, vegan diet, adult education, seminar-house, physically handicapped persons, animal protection, environmental protection, permaculture, organic gardening, alternative energy, ecological construction, cooperation with nature, sustainability, Transition Movement

We want:

- to live communally in old half-timbered houses in Göttingen, which belong "to themselves", in cooperation with nature, which we involve in all our decision-making
- to attain the highest possible level of sustainability in our daily lives with the use of permaculture principles and the re-use of "rubbish"
- expression of solidarity and spirituality via our lives
- with the implementation of urban gardening ideas, in a highly populated urban setting, to test partially biological self-sufficiency and concurrently create habitats for animals, and to explore how all this can be implemented
- to find solutions as a model project – reproducible by anyone
- to test the feasibility of house restoration, with work swaps rather than money, with material exchange and recycled materials
- to relearn how to do necessary things ourselves
- the connection of town and countryside with a communal house
- to document and share our experiences in the process.

The community consists of people who live in our houses, but also of people who live in the surrounding area, who support us; it consists of the resident animals, plants and natural beings. Four of us form a living group in a half-timber building. Others in our group live in rented apartments around and about. A second communally purchased house is to become the community house and is currently being renovated. In the future we would like to inhabit the whole district (at the edge of town) with community, in living groups, individual apartments and cooperatives. In the beginning of each year a small Dragon Dreaming will take place in order to include new people and grow together. We rejoice in people who would like to create this project with us.

Transition Town Göttingen, Urban Gardening Network, LETS, CSA
buying/building the houses in cooperation with nature

Poppau 21

C3

Poppau 21 | 38489 Beetzendorf ☎ +49-(0)39000-905 954

1999 (near Klötze, Salzwedel) 60 % ♀9 ♂7 6 22

g, e

all association members (consensus minus 1) individual financing

the community via an association

everyday life with everything that comes with it, living together

ecological, non-violent, organic diet, humour, cross-generational

We are a cross-generational housing project in the old centre of Poppau; 1 km away from Sieben Linden Ecovillage. We live a common everyday life with all our similarities and differences. This is often very harmonious, sometimes chaotic and always lively.

PrymPark Düren Cohousing- und Klimaschutzsiedlung A6

Philippstr. 4 | 52349 Düren +49-(0)2421-698 33 42
info@prympark.de www.prympark.de

2006 3 ha

g, (e)

Residents own their buildings. The plot ist owned by the city of Düren and made available on a hereditary lease

no

ecological, self-managed, non-violent, environmental protection, alternative energy, ecological construction, accessible to the disabled, car-free center, intergenerational, coworking spaces

☺

Planning of the housing estate has been ongoing since 2006 in a public participation procedure with interested people. Plots are provided on hereditary lease basis and preferably allocated to groups of private persons who plan and build their projects together. Since December 2018, two building projects are being executed. Additional plots are available for more project groups. See also entry of "Wir Mittendrin!" on page 268. The aim is to build a lively neighbourhood based on solidarity by providing home ownership in combination with shared facilities, communal areas, gardens and an attractive environment. Decisions on communal facilities will be taken by the community and costs for maintaining them and the workload for their upkeep will be shared by all residents of the estate. PrymPark is a multi-generation housing project and one out of 100 climate action estates in North-Rhine Westphalia. Social and ecological aspects as well as economic efficiency are high on the project's agenda.
Düren is a medium-sized market town that provides everything necessary in day-to-day life: good shopping opportunities, child care facilities and schools, health care facilities and an active cultural life. Please view our PrymPark clip (#7631 on eurotopia.directory/booklinks), which will give you an idea of the surroundings and the project's underlying concept.

☺ willingness, to become involved in a neighbourhood based on solidarity, to invest in and benefit from community life; financial involvement and participation in the maintenance and upkeep of shared property

Puls der Erde D9

Gemeinschaft zur Erforschung & Verbreitung nachhaltiger Lebensweisen

Lenzwald 2 | 84570 Polling +49-(0)8630-986 86 48
willkommen@pulsdererde.org www.pulsdererde.org

2015 1.8 ha (near Mühldorf am Inn) 90 % (vegetables) ♀6 ♂3 4
13

g, (e) 3 3-5

the community (consensus or ¾ majority)

common household fund, solidary agriculture

owned by Puls der Erde e.V.

love to mother earth

CSA agriculture, personal development, seminar- & guesthouse ecological, self-managed, inner growth, organic diet, adult education, seminar-house, animal protection, environmental protection, self-sufficiency, permaculture, organic agriculture, organic gardening, ecological construction, humour, soil humus development and human humus development

☺

We live on a planet where all beings are connected through a global relationship-network. Our concern is to consciously reconnect with the Earth and its living beings, to let ourselves be affected by the state of the world created by ourselves, to feel the pain and use it as fuel to act, explore and to experiment, how a sustainable coexistence among us humans, but also with the earth and its living beings can look like and work.
It is us who lay the seeds in the ground so that our next generations can reap. Let us provide our children with this kind of food on a daily basis and as a matter of course, so that they grow into responsible adults who help turn our earth back into a thriving oasis.

We do this by working on the soil humus (regenerative & solidary agriculture) and human humus (personality development) and by networking with other people and organizations who care about that as well.
Link to the film about regenerative agriculture at Lenzwald: #7586 at eurotopia.directory/booklinks

→See also description of network "**GEN Europe**"
CSA agriculture, Artabana, SauRiassl Syndikat

☺self-responsible people who " love themselves" into our community

Quartier Am Albgrün

B8

Auf den Eiswiesen 1 | 76185 Karlsruhe
wohnen@am-albgruen.de am-albgruen.de

2016 (near Stuttgart)

house communities, Albgrün-Verwaltung, Albgrün-Verein (majority vote)

property of the project groups

community rooms, design, garden, community building, neighbourhood assistance, public relations, internal communication

ecological, self-managed, ecological construction, accessible to the disabled, inter-generational living

The quarter "Am Albgrün" is a multi-generation housing project in Karlsruhe-Grünwinkel. Since 2016, more than 150 people have been living in seven residential buildings and are shaping community life in the residential project - communally, self-determined and ecologically. Living and housing in a multi-generational housing project means starting something new with like-minded people, i.e. rediscovering the lost cohesion between young and old.

Our project structure is diverse: owners in building groups, cooperative housing (GeniAl e.G.) and housing for rent (in cooperation with the communal Volkswohnung GmbH). What all groups have in common, however, is that they want to live together and treat each other mindfully. We want to get away from the previous image of anonymous apartment blocks and revive the positive aspects of the former village structures. Different working groups take care of the organisation of the three common rooms, the design of the outdoor areas, the maintenance of the green areas and the community garden, topics of community building, neighbourhood assistance or public relations and internal communication. Administrative aspects are regulated by Albgrün-Verwaltung GbR, social and cultural aspects by Albgrün-Verein e.V.
We connect individual freedom and living community. Community life in the project mainly takes place in the common outside and green areas and the common areas. Each project partner (in building groups, cooperative, social housing) contributes a common room to the project. Our mission statement regulates, as a self-commitment, a good living together and a good communication in a diverse and heterogeneous housing project.

☆ **Quecke** linkes Landprojekt mit Schwerpunkt Antidiskriminierung E3

Cöthen 8 | 16259 Falkenberg
info@quecke.net quecke.net

2013 | 2 ha (near Bad Freienwalde) | 9-10

g, e | 5

the plenary meeting (consensus decision)

various mixed models based on solidarity | non-private property

leftist, emancipation, non-heteronormativity, anti-patriarchal

running seminar house, care for buildings and site, political participation

self-managed, left-wing political understanding, feminist, adult education, emancipatory, anticlerical

☺ (✋✋)

The Quecke is organised in a collective structure. Not all members of this structure live on site at the project. All activities of the project are organised in working groups, some of which it is mandatory to participate in.
We live in two residential groups, including one in wagons outside. Most of the rooms inside belong to the seminar house which is created and maintained mostly through unpaid labour, or to common spaces like the workshops. One of the two residential groups defines itself as a LFT*I (lesbian/female/trans*/inter) space. The project still has a lot of construction work and involves organisational and practical work.

From the self-definition:
"The Quecke grew out of queer and feminist contexts as a semi-open place (...). The project is critical of heteronormative and patriarchal structures and is a space where we reflect on gender and this is an important part of the project which should be preserved and promoted. We aim to try to understand different life realities and not reproduce repressive structures, which we recognise as integral parts of capitalist society."

☺ membership in working groups and self-organization, contribution to project building

Rawaule Anders Leben und Lernen e.V. B4

Rawaule 5 | 32694 Dörentrup ☎+49-(0)5265-955 8277
kontakt@rawaule.net rawaule.net

2015 5000 m² (near Lemgo/Detmold/Bielefeld) 20 % ♀3 ♂5 8

g, e ca. 50%
all people together (consent, sociocratic processes)
limited common financing; goal: common economy
in ransoming from private property > transmission to a foundation
no
ecovillage development, ecological activities (building, permaculture...)
ecological, self-managed, non-violent, Christian, Buddhist, anthroposophic, inner growth, meditation, yoga, organic diet, vegetarian diet, adult education, free school, seminar-house, peace-work, environmental protection, self-sufficiency, permaculture, organic gardening, alternative energy, new technologies, ecological construction, construction, humour, awareness, nonviolent communication

☺ ()

We see the Rawaule as a meeting point for freedom of thought, creativity and knowledge transfer. Thus we are also a place for people who are concerned with spirituality and philosophy for the healing of social conditions. This leads to practical projects in areas such as permaculture, buildings for the senses, dance, play and culture.
In the future, we would like to act as an initiator of the Transition Town movement in the village and strengthen the neighbourhood with an exchange & gift room, a food cooperative or the reconstruction of the baking house. Nature is our teacher. We want to create an aesthetic place, in which the needs of human, animal and plant are coordinated harmonically.
Our collective acting is carried by humor, vitality and the desire for a peaceful, respectful interaction. In doing so, tools like nonviolent communication and circle way are helping us. We are aware of the importance of working with our inner shadows.
Everyone contributes their abilities and gifts. Healthy, fair and ecologically grown food from our large communal kitchen is important to us as well as closeness to nature and self-sufficiency. We are inspired by travelling groups like Wilde Karawane and WanderUni (travelling university), to which we offer our space gladly. The interconnection with other communities and initiatives is giving us fertile impulses for a different living, loving und learning.

→See also description of network "**GEN Deutschland**"
CSA, Transition Town Movement
☺ active engagement for transition in all areas of living

Regenbogenfabrik E4

Kinder-, Kultur- und Nachbarschafts-Zentrum

Lausitzer Str. 22 | 10999 Berlin-Kreuzberg ☎+49-(0)30-695 795-0
info@regenbogenfabrik.de www.regenbogenfabrik.de

1981 ca. 100

e, f, sp, i, Arabic all
all (plenum, by consensus) solidaric, i.e. common financing
hostel, cinema, bike rental outlet, cafeteria, self-help workshop, day-care centre

self-managed, left-wing political understanding, non-violent, seminar-house, humour, collective, hostel

Regenbogenfabrik ("Rainbow Factory") started during the squatting movement in West Berlin. More than 30 years ago about 50 young people protested against housing speculation and demolition, which consequently led to squatting the empty factory building. The ramshackle houses were lovingly refurbished; the yards were converted into a green children's playground. More and more our idea of living and working in solidarity, sharing responsibility in the local area and developing a social project with the neighbours took shape. An assembly of old sheds frames the yard of Regenbogenfabrik. Along with the old brick chimney of the former steam powered sawing factory they are protected monuments.
Today 34 of us live together, another 60 or so participate in the project to some individual extent. In the great cinema hall we host feasts, concerts and other cultural events. Parties and bazaars are scheduled throughout the year. Self-help is the magic word in the carpenter's and bicycle workshop (also for kids). We are running a canteen, a kindergarten, a café and a cinema. We offer rooms for rehearsals and inexpensive accommodation in our hostel with space for 36 people. All projects as well as the entirety of Regenbogenfabrik are organised as a grass-roots democracy. We would love to see you here!

Samana Seminar-Haus & Osho-Platz

E6

Niederpöbel 27 | 01744 Dippoldiswalde/ OTSchmiedeberg
+49-(0)35052-224 56
info@samana-erzgebirge.de www.samana-erzgebirge.de

1999 2.5 ha (near Dresden) ♀6 ♂3 9

g, e 4 2

all concerned (consensus) common till

ownership

Osho movement, Buddhist teachings, awareness

maintenance of the place, caring for seminars (cooking, room service etc.)

self-managed, non-violent, spiritual, esoteric/New Age, inner growth, meditation, seminar-house, construction, humour, seminar house, seminar rooms, guest rooms, holiday, creativity, living together, meditation, nature, silence, retreat, individual retreat, culture in Dresden

Samana was founded in 2013 on the foundation stones of former Osho Manjusha Meditation Center. The wide space with several houses includes 25.000 square meters within the natural reserve of Osterzgebirge with forests, meadows and streams. Samana Seminar House offers a diversity of event rooms and guest rooms. Our trademark and special treasure is the large oval meditation hall (200 m^2) which we really love to rent and share for seminars. Sometimes there are events with guests from all over the world who come here to Schmiedeberg. We are looking after events like Yoga, Tantra, Healing- Fasting, Buddhism, Schamanism, Bodywork, Spritual Awareness Enhancements und many more. Everybody living in the community is invited to support (e.g. as a cook). In this way the all-together-caretaking of seminars is a connecting project which contributes to the preservation of the place.
Between seminars in even calmer times community life is an adventure with a lot of possibilities: meditations, living together and celebrations... One who wants to be creative is able

to realize many ideas here. A pool of materials, music instruments and rooms are available.
Individuality and personal growth and a fulfilling and conscious life as well is an important fact for all of us. Sometimes the ways seem to be different, but on those concerning our huge beautiful place with its calm, lovely and lively energy, we all agree.

☺ financial independence, interest in community life

Sannerzgemeinschaft

C7

Sannerzhaus - Lindenstraße 13 | 36391 Sinntal-Sannerz ☎ +49-(0)6664-402 498
sannerz@bruderhof.com www.bruderhof.de

1920 — 3 ha (near Fulda, Frankfurt a.M.) — ♀7 ♂7 3 17

g, e — 10

everyone in common (consensus) — shared economy, community of property

common property

Christianity, non-violent

forest, garden, field, house, workshop

non-violent, Christian, humour

☺

We are a Christian intentional community, which belongs to the Bruderhof Movement. This movement consists of more than 2700 people living in over twenty settlements on four continents. We are a covenant of families and singles who want to follow Jesus without compromise and in the spirit of the first church in Jerusalem. We gladly renounce private property and share everything with each other. While most of us live in larger, mostly rural communities, we are a small community right in the village of Sannerz.
In 1920 the Bruderhof movement began in our house. Here the Pentecost meetings of the Youth Movement took place, the struggle for peaceful living together and the intense expectations and enthusiasm for a new way to put faith in God into practice.
We live this way because following Jesus should not only be a part, but our whole life.

→See also description of network "**Bruderhof**"

☺ just live with us

Saor Thoil Clan Clankolonie Altenstein

C7

Wilhelm-von-Stein-Straße 28 | 96126 Altenstein

2007 2000 m² (near Coburg) 20 % ♀3 ♂3 0 6+

g, e, f variable as needed

all who are involved in a matter - if nothing works out, by the Thane or his/her 1st assistant (discussion)

most similar perhaps to an early form of tribal feudalism - individually customized

allodial property

all too complicated

agriculture, industry, trade, gastronomy, nomad stuff (hunting, gathering, mining)

ecological, self-managed, anarchistic, organic diet, animal protection, environmental protection, self-sufficiency, permaculture, organic agriculture, organic gardening, alternative energy, new technologies, ecological construction, construction, humour, no cell phones/wireless, non-smoking, authority-free, archaic, sovereign

☺

We are the Saor Thoil - a tribe in reconstitution. We had enough of isolation and of the enslaved life within the system, which is caused by weakness and ineptness of an individual to fight and defend against. We decided to ally with each other first as a group of friends, to be able to repeal the endless alien attempts of sucking out of our life's time and energy. We did this in the way like in former times tribes tried to master the dangers and challenges of nature.
After some hard years we finally managed to make landfall and founded our colony, where we now try to surpass the diaspora and develop our own culture.
Foremost important is it to get rid of the foul alien tech and its slave things, and replace it with our unique own, so we may be able to free ourselves from the slave service to these short lived fashions. We are planting a consistent techtree to our developing culture. We have to master our lives by our own thoughts, under our own power and at our own ends. There is no such thing as progress, but only everlasting change.
It would be futile to describe everything here in detail, because till this book will be printed, many things will already have changed.

☺ logical thinking, minimalistic lifestyle, no narcissism/ego-bullshit

Schloss Oberbrunn

D9

Lebensgemeinschaft und Zentrum für intensive Gruppenprozesse

Seeoner Str. 4-6 | 83132 Oberbrunn/Pittenhart ☎ +49-(0)8624-879 59 62

gemeinschaftsbildung@t-online.de www.schloss-oberbrunn.eu

2011 1.6 ha (near Obing, Bad Endorf (Bahnhof)) ♀7 ♂3 6 16

g 3 3

all concerned (consensus) shared food costs

private company

interest in intensive group processes, verbal/non verbal; commitment

running seminars, seminar organisation in the German speaking area

ecological, self-managed, non-violent, inner growth, vegetarian diet, seminar-house, environmental protection, organic gardening, alternative energy, ecological construction, humour, non-smoking

☺ (hands)

Really deep community has a price (also in terms of time invested). Those who have once experienced how fulfilling intensive community can be know what we are talking about. It requires a supportive vessel, just like an engaging framework is necessary in a workshop in order to ensure the emergence of something. We would like to explore the balance between freedom and commitment verbally and non-verbally. Verbal processes are important to us in this context, we like to sit in a circle according to Scott Peck's principle of community building as a "group of all leaders". Where there is a good communal energy, material issues are more easily resolved.

☺ interest in community building based on Scott Peck

Schloss Tempelhof eG

C8

Tempelhof 3 | 74594 Kreßberg ☎+49-(0)7957-9239-030

info@schloss-tempelhof.de www.schloss-tempelhof.de

2010 32 ha (near Crailsheim, Dinkelsbühl, Würzburg) 70 %

♀55 ♂45 45 145

g, e, sp, Polish, Romanian, f 50 60

consensus, coordination circle of project representatives (consensus of 6 steps)

diverse: need based income, income sharing groups, giving circle

property of our charitable foundation

diversity, honest communication and encounter, ecological, everyday life as practice

agriculture, garden centre, cheese dairy, bakery, beekeeping, kitchen, seminar and guest house, free school, kindergarten, youth work, carpenter

ecological, self-managed, non-violent, inner growth, meditation, organic diet, vegetarian diet, vegan diet, adult education, pedagogical, free school, seminar-house, mentally/psychologically handicapped persons, self-sufficiency, permaculture, organic agriculture, organic gardening, alternative energy, ecological construction, WE-Process, regular community building work, Tempelhof Intensive processes

☺ (hands)

Since September 2010 the "Gemeinschaft und Zukunftswerkstatt (=community and future-workshop) Schloss Tempelhof" has been growing and thriving in Hohenlohe, Swabia. The area consists of 8 hectares of building land and 24 hectares, which are cultivated according to regenerative principles and which are inspired by permaculture. We run a free school with around 80 pupils. In a relaxed and prepared environment, we accompany the children and young people on their own learning paths. In addition, there is a forest kindergarten, a seminar facility with a guest house, a gardening and agriculture project that works closely together with our community kitchen, a village shop, a castle cafe, the first Earthship in Germany, a trailer workshop, as well as many private businesses and self-employed people. The vision of our village is to be a transferable, sustainable model to support social change. At the centre of the community are our basic values of solidarity, diversity, all-leader, relationship and communication culture, responsibility for me and the whole - and a potential-promoting economy and sustainability in social, ecological and economic action. We live, work and celebrate together - everyday life becomes an exercise. All decisions are made by consensus. Transparency is close to our hearts, which is why structural meetings are open to our guests. We come from different spiritual traditions and focus on authentic encounters between people. Our primary community-building process is Scott Peck's community building process. We have many guests at the place: seminar guests, supporters, guest helpers. We are happy that more and more young people come to us and spend some time with us in different volunteering contexts. With our land trust we advise and accompany initiatives throughout the German-speaking world in their group formation, structural questions and also concretely in their purchase. The present calls for exactly what we do: To understand our own longing as an opportunity and to develop heart courage - for ourselves, for a lived WE and for our earth.

→See also description of network "**GEN Europe**"

🌍 Artabana - solidaric network for health, Mehr Demokratie (="more democracy"), Transition Town, CSA Germany

☺ weeks for helping guests, 1 year of getting to know each other, cooperative share

Schöpferisches Zentrum OASE

C2

Schaarweg 70 | 23730 Neustadt-Rettin ☎+49-(0)4561-52536-10

✉info@schoepferisches-zentrum-oase.de 💻schoepferisches-zentrum-oase.de

1987 | 2.900 m² (near Lübeck) | 5 % | ♀2 ♂3 0 | 5

g, e, f | ♻5

board and active members of the associations (in the team)

property of two own associations

creative consciousness work, in order to be able to live in harmony with humans, animals and nature

house, kitchen, garden, construction workshop, public relations, guest facilities

ecological, self-managed, spiritual, inner growth, meditation, organic diet, vegetarian diet, adult education, pedagogical, seminar-house, organic gardening, alternative energy, ecological construction, non-smoking, exceptional architecture, arts, culture

We are a team of people from different professions. With the construction of the OASE Healing House we have realized a dream of a place where a life in harmony with creation becomes possible and where people can regenerate and work on their healing.
What connects us is our common desire and effort to live in peace with ourselves, with all people and with nature. Each of us has worked many years in his profession and has gone through many stations in his life to understand one thing above all: the essence of life is very simple. Only the way there is sometimes complicated and difficult, and it requires work on ourselves.
With all our heart we thank our spiritual teacher Johannes Heinz Löffler (1942-2004), who accompanied us for many years with his clarity, patience, love and joy of life. This place was created in 1987 through his tireless work. He established it and led it until his death. His love for all living things can still be felt here.

→See also description of network "**GEN Europe**"

🌍 Artabana

Seminarhaus Engl Verein für buddh. Leben und Handeln e.V. D9

Engl 1 | 84339 Unterdietfurt ☎+49-(0)8728-616
info@seminarhaus-engl.de www.seminarhaus-engl.de

1989 1.5 ha (near Eggenfelden) ♀2 ♂4 0 6

g, e 6
board of directors (in agreement) individual financing
property of one person
Buddhism
cooking, office work, caring for guests, gardening, housekeeping
self-managed, non-violent, spiritual, Buddhist, inner growth, meditation, yoga, vegetarian diet, seminar-house

The "Seminarhaus Engl" association was founded in 1990 to organise events based on a variety of Buddhist traditions. Engl is located amidst fields and forests in Lower Bavaria. We view ourselves as a Buddhist forum where various schools can become acquainted with one another and where exchanges and mutual encounters can take place. The "Engl" house project developed from the Tibetan Aryatara Institute in Jägerndorf. We have been supported and inspired by Fred von Allmen and other teachers from the Vipassana tradition. For us the encounter with the various strands of Buddhism has proved to be both interesting and challenging. In our house we offer events such as lectures, seminars, and meditation courses with the following content:

- Various forms of meditation, especially Buddhist insight meditation (Vipassana) and Zen Sesshins, as well as the teachings of the Tibetan tradition.
- Physical work and related areas (such as yoga, breath therapy, Hakomi).
- Events from the field of involved Buddhism (such as care for the dying, dealing with incurable diseases, courses in holistic ecology).

The community's main areas of work are the kitchen, the office, the household, and gardening. We try to link meditation, self-communion, and contemplation with responsible activities. It is our wish that we may actively help reduce suffering in the world: with an alert and open spirit, and with the goal of preserving a spirit of new beginnings which sees everything with new eyes.

DBU (German Buddhist Union)
☺ interest in Buddhism, readiness to take on all work as it arises

Sieben Linden C3

Poppau, Sieben Linden 1 | 38489 Beetzendorf ☎+49-(0)39000-512 35
info@siebenlinden.org siebenlinden.org

1993 100 ha (near Salzwedel, Wolfsburg) ♀55 ♂45 40 140

g, e, f, i, sp, Portuguese 30 75
5 councils, general assembly (GA) (delegation to councils with various ways to object or (in GA): consensus minus one)
limited common financing, mostly individual financing
co-operative = common property
holistic, social-ecological

- self-administration and self-sufficiency, communal culture, ecological building, seminars, care of guests, public relations (quite a lot of office work)
- ecological, self-managed, left-wing political understanding, feminist, non-violent, spiritual, inner growth, meditation, yoga, free love, organic diet, vegetarian diet, vegan diet, adult education, free school, seminar-house, animal protection, environmental protection, self-sufficiency, permaculture, organic agriculture, organic gardening, forestry, alternative energy, new technologies, ecological construction, humour, no cell phones/wireless, non-smoking (smoker's corner), partly raw food, wild herbs, forest kindergarten, diverse relationship culture, deep ecology, forum, systemic constellations, therapies (ShinTai, Craniosacral), work horses, straw bale building, various crafts, music, choir, community consultation, art & community, violin making, media, and: for each topic there is always someone who is interested...

☺ (✋✋)

Sieben Linden is a community and a village. All decisions regarding property, provisions, structures etc. are made by the community. Thereby we have created an alternative mini-society: mobile phones are not in use here, smoking is only permitted in the smoking area, there are no cars in the village and much more. We try to practice and live a healthy, sustainable, ecological lifestyle with honest, appreciative communication, but at times we fail. For example, we are much to busy, which makes us nervous.

As a village, Sieben Linden also has spaces where individuals can create and maintain their personal lives – everyone has their own living space, some smaller groups and families have their own kitchen. As of 2019, there are ten rather large, modern low-energy houses (mainly strawbale construction) plus common/seminar house, guesthouse (in construction) and workshop. However, most of our houses still need to be built; therefore there are currently almost 50 inhabited trailers.

If you visit Sieben Linden - there is plenty of opportunity to do so via our seminar programme - in view of the relaxed, barefoot atmosphere, you may not immediately realise that our project is also a high standard construction project and that we are in need of substantial funding for our building projects. There is some disagreement on whether our life is still very affordable or if we are already too expensive. Self-sufficiency is a grand concept, but we have come to believe that we can contribute more to the ecological change if we use modern methods to advertise our events and ideas (internet, flyers, office infrastructure) and to accommodate guests, rather than if we spend all day on our fields or in the workshops. Even though that would be great as well.

Sieben Linden is an example for how comfortable a sustainable lifestyle can be, and it is a test case for communal culture, with immense potential for personal development, research and insights. We ask you to please not to come as surprise visitors – if you are interested, then check our website or contact our offices for more information on visiting opportunities.

→See also description of networks **"GEN Europe, GEN Deutschland"**

☺ sharing of our "vision of a holistic village", probationary time, admittance to the co-operative and financial participation. Qualification: We need more craftspeople, planners, technical skills.

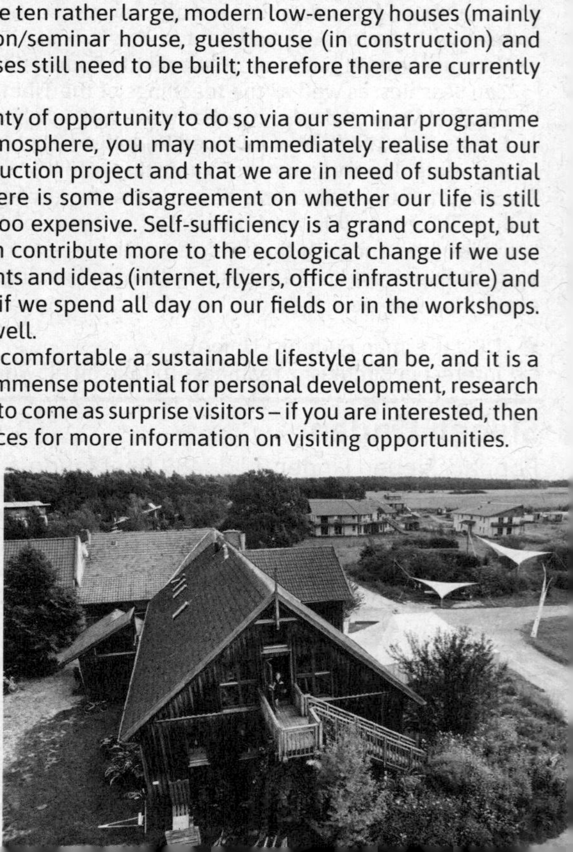

SiebenZwerge Landprojekt Seven Dwarfs Housing

C3

Im Waldhain 1 | 09548 Achternberg

kommerzfeind@yahoo.de www.fairies.de/snowwhite.html

1816 35 m² (near Dresden) ♀0 ♂7 0 7

g, e 7

everyone in common (we used to do what Snow White said)

shared economy (shared income and capital) all

romantic, polyamory

mining, jewellery, glassworks

self-managed, feminist, non-violent, leader/guru, inner growth, sexuality, organic agriculture, intergration of midgets and women, mining, fairy tales, singing, hats

☺

The project behind the seven mountains, which has existed since 1816 and has been described many times, is in a phase of reorientation. After years of growing relatively solid community structures, a traumatic experience has taught us a lot about our hidden emotions: After a new community member has run off with a prince from elsewhere, we have - after roughly a century of mourning - decided to open ourselves again in the future. Basically there is the position of a Snow White to be reoccupied. It is clear that a "new one" must create her very own place for herself. Combining individuality with a sense of community is one of our most important goals.
It is important to us to sing and work together, but we have certain principles - respect for private property is important to us. After long working days we crave for harmony. A gemstone mine that we have developed helps us to enjoy life in financial self-sufficiency. With our very own traditions - in terms of both fashion and content - we see ourselves as the crystallization core for a culture of tradition in the middle of modern Europe.

fellow fairy tales

☺ female, leadership ambitions

Sisiphusa e. V. Pfarrkesslar

D6

Am Angerberg 35 | 99444 Drößnitz ☎+49-(0)36422-20935, -20936

Pfarrkesslar@web.de www.pfarrkesslar.wordpress.com

2002 4.5 ha (near Weimar, Jena) 25-80 % depending on season

♀4 ♂4 6 16

g, e, sp 2

everyone, partly in delegated groups (mostly consensus)

the househould and house and land are financed together in equal parts

the community owns land and buildings

building, animals, garden

ecological, self-managed, left-wing political understanding, non-violent, organic diet, adult education, environmental protection, self-sufficiency, organic gardening, organic farming, ecological construction, construction

( )

Pfarrkeßlar is a community of 8 adults, 6 children, 1 volunteer (one year ecological volunteer service), sometimes WWOOFers, 1 donkey, 8 milk-goats, 1 cat, 2-12 geese, 12 chickens and their cock, many insects, wild bees, birds and fairies and neighbouring mouflons, orchids and

Germany

fireflies, who bring this ancient enchanted settling place back to life since 2002. The lacking infrastructure in the region brings freedom and problems which the people here learn to handle in cooperation with their neighbours and small and bigger networks.
After the long vacancy big parts of the main building are reconstructed meanwhile and in the so-called "Ostblock" from 1972 there is a carpentry and seminar- and guestrooms. The individuals and the group are permanently inspired by their wish to live holistically and find creative solutions in all fields of life. They also try to give impulses for the public and inspire interested people by organizing diverse activities.
To keep the open non-dogmatic and friendly character of the house the inhabitants expressly ask not to be surprised by unannounced visits.

Sonnenhaus e.V.

E4

Flutstraße 1 | 12439 Berlin - Schöneweide
info@sonnenhaus-berlin.de www.sonnenhaus-berlin.de/

2003 ♀9 ♂10 0 19

g — no one (anymore)
three-headed board (majority vote in the community member assembly)
individual financing — ground leased, house owned by association
no
living
self-managed, left-wing political understanding, accessible to the disabled, intergenerational, support of local social carriers

The purpose of the association is to practice equal and solidary living, housing and working forms and, if possible, to combine them in the same place. The intergenerational coexistence of children, women and men in families and family-like communities is seen as essential. The solar house offers space for social and cultural activities of the residents, social bearers from the quarter (family café at the ground floor) and the neighborhood. The design of everyday life is characterized by the economical, conscious and sustainable use of resources.

Wohnprojekte Berlin, eXperimentcity and Netzwerkagentur, trias

SÖWP-Allach e.V.

D9

Selbstverwaltetes sozialökologisches Wohnprojekt
Storchenweg 33/35 | 80997 München +49-(0)89-892 249 82, -(0)178-672 00 10
kontakt@soewp-allach.de www.soewp-allach.de

2007 20 % ♀5 ♂5 1 11

g, e, f — 1-6
association (partly by majority vote, partly by consensus)
partly common housekeeping till,
partial financing from the rent paid by the residents
property of the association
politically left
renovation of the houses, gardening, group processes
ecological, self-managed, feminist, non-violent, environmental protection, organic gardening, alternative energy, ecological construction, humour

☺

We are a small community on the northwestern outskirts of Munich, living together in two small houses with a large garden with chickens. We build and renovate our houses according to the principles of building biology and strive for a peaceful and respectful coexistence. We want to live in Munich in a different way, with each other instead of in competition, and create alternative forms of living.

☺ willing to participate in the project and to work with the people living there

SSM - Sozialistische Selbsthilfe Köln-Mülheim

A6

Düsseldorfer Str. 74 | 51063 Köln
+49-(0)221-640 31 52
info@ssm-koeln.org
www.ssm-koeln.org

1979 | 1600 m² | ♀9 ♂16 2 27

g, e | 20

everyone in common (majority vote, consensus when possible)

economy in solidarity | rented/leased

"new work" - common day jobs + our own work + politics

clearing flats, relocations, second hand shop, hall rental, building, political activism, renovations, project development

self-managed, left-wing political understanding, non-violent, seminar-house, mentally/psychologically handicapped persons, environmental protection, self-sufficiency, construction, "new work", second hand, commune, CONTRASTE editors Cologne, economy in solidarity, concerts, holistic social work

☺

The systemic crisis of capitalism brings social exclusion and the threats of climate change. Alternatives are needed. In the "new work" people are only active in the market with part of their working time. They earn the other part of their living costs together through joint self-employment. They build their common living space, they recycle and re-use, run gardens and accumulate knowledge of many kinds. People support each other and fight for peace and human rights. All activities that are important for the community are considered to be work: crafts, cooking, writing a leaflet, managing home clearances, childcare, group meetings, etc. All this is work at the "Sozialistische Selbsthilfe Mülheim" ("socialist self-help Mülheim", SSM). In the midst of this diversity, mentally handicapped, unemployed and other marginalised people have or can find their place. "New work" is inclusion, it combines the strengths of the individual and balances out their weaknesses: "New work" à la SSM may not be as productive as usual companies, but on the whole, it is very effective, because it renders nursing homes and rush hour, social work and welfare redundant in the long term.

See article about SSM with photos on pages 42f

→See also description of network **"Kommuja"**

CONTRASTE (newspaper)

☺ people of good will

StadtGut Blankenfelde e.V. Multigenerational Housing E4

Hauptstraße 24-30 | 13159 Berlin-Pankow ☎+49-(0)30-400 581 82
kontor@stadtgut-blankenfelde.de www.stadtgut-blankenfelde.de

2004 5.2 ha 3 % ♀36 ♂21 36 93

g, e, f, sp, Danish, Hebrew, Polish 5 15
managing committee and leading members of all groups (so far: majority vote - we work on new structures)
individual economy
owned by the foundation "Trias" - leased by the association/the cooperative
environmental conservation, preservation of historic buildings, community building, sustainable living, beekeeping, different trades
ecological, self-managed, left-wing political understanding, non-violent, organic diet, free school, mentally/psychologically handicapped persons, environmental protection, organic agriculture, organic gardening, animal friendly husbandry, alternative energy, construction, humour, accessible to the disabled, conservation station with exhibition and café, seminar rooms, 2 social carriers: work with (former) mentally ill, Free Nature School Pankow, FoodCoop, therapy rooms/office/practice, studio, workshop for play equipment made of natural wood, carpentry, beekeeping, home design workshop

☺ ()

In 2004 the society "StadtGut Blankenfelde e.V." was founded with the main idea of conservation of the environment and of the historical buildings here. We have now over 100 members, but not all of them live here. Since then, a lot of work has been done: Maintaining the area, disposal of old neglected deposits of toxic waste, building and demolition, working together with other projects, a school and a kindergarten were founded here, we started sheep-keeping...
Only in 2009, we started to collaborate with the cooperative "Selbstbau e.G." that made the renovation of the houses possible. In 2012 we moved into our 26 flats. In 2014 we renovated one more building with 6 more flats, offices, surgery, food coop, studios and a touristic station with an exhibition and a Café.

trias
☺ only if free flats are available, we look for people that fit in with us, that like the work here and are able to pay what is needed

Stamm der Likatier Likatien Tribe C10

Magnusplatz 6 | 87629 Füssen ☎+49-(0)8362-38993
stamm-der-likatier@likatien.com www.likatien.de

1974 65 ha, several houses (near München, Kaufbeuren, Kempten)
♀83 ♂77 131 291

g, e, f, i, sp, Portuguese — everyone in an appropriate age
everyone in common (consensus) — common and individual financing
tribal property
neo-tribal movement, community movement, bio-regionalism, depth ecology
a number of service businesses, retail/crafts businesses, farming/forestry, gardening
ecological, self-managed, hierarchical, leader/guru, spiritual, ecumenical, esoteric/

New Age, inner growth, organic diet, vegetarian diet, free school, mentally/psychologically handicapped persons, physically handicapped persons, animal protection, environmental protection, self-sufficiency, permaculture, organic agriculture, organic gardening, animal friendly husbandry, forestry, alternative energy, ecological construction, humour, celebrations, matriarchal culture, autarky, authenticity, holistic healing, therapy, consciousness research

☺ (✋✋)

In our view, tribes and communities are the basis of understanding and genuine human dynamism. One of our central goals is the development of an autonomous tribal culture harmonising its relationship to itself and to the world on the basis of a spiritual world view and a numinous-intrinsic and external realisation of the world (healing); which creates culture-specific communication opportunities (e.g. its own language); and which examines adequate opportunities for dynamic development and makes it possible to live them.
We are linked to neither religious nor political institutions, ideologies or traditions. We are striving for increasing autarky in all areas. At the same time we seek intensive co-operation and friendship with other communities and tribes. We are active in a number of very different economic areas, in which on the one hand we try to make sure that we cover our own needs and, on the other, that we remain as crisis-resistant as possible. The tribes' members own 47 houses and operate various businesses. We are currently beginning work in farming, forestry and gardening. Everyone who would like to get to know us is sincerely welcome. We are looking for new members.

→See also description of network "**GEN Europe**"

Neue-Stämme-Bewegung

☺ 3 months trial period, positive attitude towards the tribe

SUSI-Projekt

B9

Vaubanallee 2a | 79100 Freiburg ☎+49-(0)761-457 00 90
post@susi-projekt.de www.susi-projekt.de

1990 | 1.5 ha | ♀ca. 105 | ♂ca. 105 | ca. 70 | ca. 280

g, e | approx. 15

everyone in common (consensus-oriented, direct grassroots democracy)

individual financing, direct private loans, public funds, loans of the GLS bank, rent of the residents (cost recovery principle), same wage for all workers

common property

office, administration, planning, building (wood, metal, clay, mortarless construction, electrical, plumbing, heating), outdoor, maintenance

ecological, self-managed, left-wing political understanding, adult education, alternative energy, ecological construction, wood construction (cabinet-making, carpentry, handicrafts), metal processing (welding, screwing, recycling, ...), wall construction, tiles, plumbing, pottery, art: all sorts, cultural events, consensus-oriented, grassroots democracy, self-organised

The project name SUSI stands for Self-Organising Independent Settlement Initiative ("Selbstorganisierte Unabhängige Siedlungs-Initiative"). The SUSI-Project in Vauban, a typical urban quarter of Freiburg, originated in the summer of 1990, shortly after the "fall of the Berlin wall" and the big change in the East. Bobby Glatz, who was studying architecture at the time, had the idea to transform the "Vauban barracks" (which had been built by the Nazis in 1937 and were subsequently taken over by French forces) into an autonomous and colourful housing project according to social-ecological criteria. A politically and culturally active initiative developed, which was soon joined by people from the most diverse backgrounds, in order to jointly implement the concept. They included single parents, individual people and families, students, craftspeople, artists, hedonists, jobseekers and people looking for the meaning of life. The main common denominator was probably an interest in self-determination, collective and affordable housing, as well as the creation of socially responsible employment. Nowadays, SUSI is inhabited by approximately 280 people in 45 housing units (with a total of about 7,500 m² of living space) and around 25 trailers, circus wagons and caravans on the 1.5 ha site. Alongside association membership and the payment of cheap rent by Freiburg standards, all adults contribute a so-called "muscle mortgage" of 105 unpaid working hours for communal projects and chores (to be delivered within the first three years of residency). SUSI has a diverse socio-cultural and technically solid infrastructure at its disposal: Resident meeting space, various workshops, food co-op, youth room, pottery, office, drinks storage, boulder room, gym etc.

Mietshäuser Syndikat (www.syndikat.org/en), Der Paritätische Gesamtverband
☺ WBS (=certificate of housing need) or proof of student registration

SZG Sozial Ökologisches Zentrum Goldewin

D2

Goldewin 16 | 18276 Goldewin
szg.info@riseup.net
szg.blogsport.de/

2017 300 m² living, 2000 m² land (near Rostock) 60 %

♀1 ♂2 2 5

g, e
the child-caring person (changing)
all members of the association (consensus)
some common financing
property of our association
socially/ecologically
house renovation, childcare, ecological, social and political work in the region and throughout Europe
ecological, self-managed, left-wing political understanding, organic diet, adult education, seminar-house, animal protection, environmental protection, self-sufficiency, permaculture, organic agriculture, organic gardening, construction, accessible to the disabled, non-smoking

☺

The SZG is a project house and, to us as a small group, also a place of finding ourselves, in order to letting our community grow. We are located between Rostock and Güstrow, 30 min. from the main train station of Rostock. We have a 2000 m² plot and about 300 m² living space. SZG stands for Social-Ecological Center which means that we want to work for socio-ecological change. We are currently only three adults and two children in the place. We are looking for people who feel like sharing and creating community - in a life with children and at the same time with political work - and shaping our project house with us.

Transition Güstrow (headquarter), local CSA, foodcoop Rostock
☺ similar attitudes, probationary stay, more on our website and by personal contact

Technologie und Ökologie Ebertsheim
B8

Eduard-Mann-Str. 1-7 | 67280 Ebertsheim ☎ +49-(0)6359-405 27 23
info@alte-papierfabrik.de www.alte-papierfabrik.de

1985 5 ha (near Kaiserslautern, Mannheim, Worms, Ludwigshafen) 50 %
♀23 ♂20 15 58

g, e, f, sp, Arab — in the community only on voluntary basis
all residents/shareholders (4/5 majority vote) — individual financing
common property of our limited partnership, shareholders are the project members
various businesses: education initiative with eco-station, wood manufacture, landscape planning consultants, singing lessons, studio (painting and sculpture), creative workshops, Ecofonds-Shop, pottery art
ecological, self-managed, organic diet, adult education, pedagogical, environmental protection, permaculture, alternative energy, eco-station, art, joinery, pluralistic, educational initiative, culture

☺

Since 1983, the Alte Papierfabrik ("old paper mill") exists as a self-organized housing project. At the moment, more than 40 adults and 15 children and adolescents are living here together.
The factory halls were and are gradually being rebuilt into living space, offices and workshops. Regular concerts and other cultural events take place here, and our large outdoor area is used for various ecological and creative courses.

☺ young people with manual skills and ecological thinking

Trillke-Gut
C4

Steinbergstraße 42 | 31139 Hildesheim
genoss@trillke.net, verein@trillke.net www.trillke.net

1996 1.5 ha 1 % ♀26 ♂26 17 70

g, e, sp, Polish, Arabic, Portuguese — part of the group
plenary session (consensus)
various forms of common purse agreements, joint decision about cooperative funds
property of all (cooperative shares) and of the municipality
undogmatic
cultural association, IT, garden, construction, administration, energy, pub
self-managed, left-wing political understanding, non-violent, chaotic, adult education, organic gardening, alternative energy, construction, humour, arts, culture, world music, martial arts, IT, comics

The Trillke property ("Trillke-Gut") is a self-governing project which combines living, working and culture in the former agricultural college in Hildesheim, built in 1928/1929. It consists of a five storey main building, an outbuilding and approximately 1.5ha of parks and gardens with a small river: the Trillke. The first of us moved into the house in 1996. It is in communal

ownership of the inhabitants since 2000. 70 adults and 18 children live in the living groups, families and single accommodation. Countless artists, musicians, cultural scientists and IT-specialists live here, many are socio-culturally or politically active. Our concept sees three main pillars within the basic framework of cooperative self-governance: living, culture and work (e.g. music school, martial arts school, recording studio, graphics). There are regular cultural and informative events in our event spaces, which form an important part of the Hildesheim social life. We are particularly proud of our phenomenal house bands and the Folk 'n' Fusion Festival, which became an insiders' tip within the folk scene in a matter of just a few years.
Decisions that affect the community are taken according to consensus decision-making at fortnightly meetings. Any jobs that need doing in the house are largely divided up amongst us. Management working groups, a cultural association and communal building days are essential elements of our self-governance. We value a balance between community and individualism, as well as our role within the community of Hildesheim. Our annual highlight is the summer party, in which we celebrate our communal life with many friends and lots of music and dancing!

☺ payment of cooperative shares, involvement in work areas in the house

ufaFabrik e.V. Internationales Kultur Centrum

E4

Viktoriastr. 10-18 | 12105 Berlin
☎ +49-(0)30-75503-0
info@ufafabrik.de
ufafabrik.de

1976 | 1.9 ha | 5 % | ♀14 | ♂14 | 3 | 30

g, e | 200 | around 20

everyone in common (consensus)

shared financing of the 12 organisations under the umbrella of the ufaFabrik

rented/leased

we are open to everyone with a peaceful approach

culture, guest house, workshops, courses, seminars for all age levels

ecological, self-managed, non-violent, inner growth, adult education, free school, seminar-house, environmental protection, alternative energy, new technologies, ecological construction, cultural events, music, children's circus

☺

This comprehensive cultural, working and living project on the 18,000 m² grounds of the former UFA film processing laboratories has been in existence since 1979. The theatre operation, the "International Kulturcentrum ufaFabrik", maintains two stages year-round, an open-air summer stage, organises guest performances and its own in-house productions with artists from Berlin and from around the world.
There is an artist guesthouse, an extensive workshop programme for young and old, a children's farm, an organic bakery, an organic foods shop, the Café Olé with a summer garden, the centre's own UFA Circus, the circus school and the Terra Brasilis samba band. Extensive ecological projects ranging from roof greening to solar energy systems have been realised from beginning and can be viewed in an exhibit and in seminars.
The heart and driving force of the UfaFabrik and its 200 members is the commune, which is made up of ca. 30 people who live and work on the grounds. In this way the various working areas are linked together and can complement one another. Important decisions are made in consensus at a plenary meeting of all residents.
The ufaFabrik is an active member of various cultural, social and ecological networks. It

enjoys international recognition as a successful example of how life in a large city can be developed in an interesting and humane way.

- Trans Europe Halles, res artis...
- ☺ long-term acquaintance; friendship, sympathy, shared understanding of work and life

Vereinigte Varben WaWaVox

E4

Kastanienallee 77 | 10435 Berlin ☎+49-(0)30-4494330
ndeah@riseup.net, koch_andreas@gmx.de www.k77.org

1992 0.1 ha ♀9 ♂11 11 30

- g, e, f, i, sp, Japanese, Greek
- all together (consensus)
- individual financing, common financing of living and food
- rented/leased, partially common property
- an existential passion for freedom & responsibility, autonomy & solidarity
- self-managed, non-violent, organic diet, hierarchical

We are a living community in the central district of "Prenzlauer Berg" in Berlin. The house was built after an artistic squatting as a social sculpture in the early 90s and was renovated with our contribution in accordance with both conservation regulations and our needs. We only have one kitchen and one dining room for everyone. The residents each have a room and share common rooms and small bathrooms on the floors. There is a large bathroom for everyone, a laundry room, a library and many terraces and walkable roofs. Our supply is jointly financed and organised, daily changing cooking groups prepare dinner once a week. Otherwise, the daily requirements are organised by the nearby organic shop or by large orders. An important module is the rotation every two to three years, which means that we exchange all the rooms. Weekly we meet at the plenum and discuss. We decide by consensus. Further projects located in the house are the video platform AK Kraak, the cinema Lichtblick, the dance and movement floor K77-Studios, a ceramic workshop, a comic drawing floor and a carpenter's workshop, which are organized externally.

→See also description of networks "**KommuJa, Come Together**"

☺ one month probationary period

Villa Locomuna Gemeinsam Leben eG

B5

Kölnische Str. 183 | 34119 Kassel ☎ +49-(0)561-879 479-0

info@villa-locomuna.de www.villa-locomuna.de

2000 | 3015 m² | ♀12 ♂7 4 23

g, e, f, sp 12

everyone in common (consensus)

shared economy (with individual exceptions) owned by all of us

variety of offices, health care practice, translation, architecture, yoga

ecological, self-managed, left-wing political understanding, feminist, non-violent, inner growth, yoga, organic diet, vegetarian diet, vegan diet, adult education, seminar-house, alternative energy

☺

Villa Locomuna is an urban community with currently 15 adults aged between 23 and 61, one teenager as well as 3 younger children. Our project was founded in the autumn of 2000, and emphasizes our efforts of enhancing interpersonal communication. In these times of fragile biographies and precarious social and living conditions we pursue a vision of social sustainability. We support each other in our personal growth, but also on a practical, hands-on level, and we try to accept criticism as an enrichment. For our children, we try to create an environment in which they can develop in accordance with their talents and at their own speed.

We question traditional gender roles and create room for self-determined identities. All of these processes are important and intriguing contributions to communal life. We meet regularly, in order to decide on organizational or other matters, or to share on a deeper personal level. Most group members also share financial resources, and we are currently experimenting with treating "time" in a similar manner – as a resource to be shared with others. Our joint economy is based on the principle of "each according to his or her need", of balancing inequality of payment for external jobs and of appreciating unpaid labor.

On an everyday-level, we meet for dinner on five evenings per week, and for all kinds of activities in and around our two buildings.

As an urban commune, we do not aim towards self-sufficiency, but rather focus on creating and using networks, such as CSA (Community-Supported Agriculture).

→See also description of networks
"Kommuja, GEN Europe, GEN Deutschland"

CSA, Essbare Stadt, Transition Town, Interkomm Region Kassel

☺ participation in regular meetings (social and organizational), sharing in responsibility, participation in joint economy

Villa Magdalena K. e.V.

C2

Queer-feministisches Wohn- & Werkstattprojekt

Bernstorffstr. 160a | 22767 Hamburg

villamagdalenak@gmx.de www.villamagdalenak.de

1989 | 370 m² | ♀5 ♂0 0

g, e, f, sp

everyone in common (consensus) — limited community financing

rented/leased

queer feminist, left-wing political, undercutist (viral)

silk-screen printing, book printing, photo, exhibits, events

self-managed, anarchistic, sexuality, vegetarian diet, alternative energy, printing workshop, screen printing, event room, exhibitions, arts

A queer-feminist housing and working cooperative.
Once upon a time, "the villa" was an auto garage and was slated for demolition. A handful of women* had a better idea and finally made it happen: in 1989 it was given to the Verein Magdalena K. e.V. to use (and renovate!). The house soon became the "Villa Magdalena K." with the declared purpose of "artistic and craft autonomy for women*". Since then a great many women* have come and gone, have lived and worked here, argued and laughed, created art or just had fun, developed, discussed, or simply sat in the garden and enjoyed the sun. Much has changed, much hasn't, but one thing has stayed the same: Villa Magdalena is a SpaceDream (or the other way around) open to all women* who can or want to make something out of it. With 107 m² the workshop is the villa's heart and we always welcome women* who can bring even more life into it. We provide a partially furnished room with an optional flat-sharing connection for up to 3 months for artists who would like to develop an idea and would like to take advantage of the project's space, time and atmosphere.

☺ artistic interest, interest in workshop use

Vitopia

D4

Herrenkrug 2 | 39114 Magdeburg — ☎ +49-(0)391-408 293 29

mail@vitopia.de — www.vitopia.de

2012 — 3000 m² — 10 % — ♀3 ♂3 4 10

g, e — 1 5

all who are concerned (consensus) — limited common financing

owned by all (cooperative)

sustainable, ecological, intercultural

café with garden, seminars, permaculture, ecological building

self-managed, non-violent, organic diet, vegetarian diet, adult education, pedagogical, seminar-house, peace-work, environmental protection, permaculture, ecological construction, non-smoking, renewable energies, grassroot democracy, cooking classes, flea markets, annual cycle rituals

☺

Vitopia is a living-community with 6 adults, 4 children and 2 cats and a connected co-operative of 90 members. We are committed to the creation of a sustainable living and project location in the middle of the Middle Elbe habitat. We see ourselves as an open platform to implement socially sustainable projects. In our co-operative Café Verde we bake and cook with regional and organic food. Our association 'Kultopia - Lebensraum am Fluss e.V.' is hosting and giving a wide variety of educational and cultural events.
We are currently building a further house, which should be ready in 2020. We are happy that we will then accommodate more roommates and guests!

→See also description of network "**KommuJa**"

☺ probationary period, 1000 EUR share (cooperative), self-responsibility, participation in the community process, voluntary work

For the following two descriptions of trailer communities, see article & photos on page 47f.

Wagendorf Bambule e. V.

B9

Description for trailer communities Bambule e.V. und Kuntabunt e.V.
Eisenhutstr. 66 | 72072 Tübingen
wagenburgtuebingen@posteo.net www.kububabu.eu

1992 0.5 ha ♀15 ♂15 10 40

g, i, Turkish, e, sp, f
all concerned together (majority or consensus or just by taking action)
limited community financing rented/leased
more or less or not left-wing or ecological
ecological, self-managed, left-wing political understanding, alternative energy, ecological construction, cultural events

☺ ()

The Wagendorf KUNTABUNT and the neighbouring Wagendorf BAMBULE were founded in 1991 and 1992, respectively. Both groups were nomadic at the beginning, as they were forced from one site of squatted land to the next. Now there are about 40 adults and children (as well as guests, cows, dogs, cats, peafowls...) living in done-up wooden wagons or trucks on the site just between town and forest. We have no running water and generate our electricity mainly by solar panels. As a collective we run meetings from time to time, the famous "Culture Days" once a year as well as a small collective fund. Anything beyond this is done by smaller groups within the broader collective, or just by interested individuals. The connection between members is formed less by formal structure than by the building of affinity and friendship. We are involved in the neighbouring town, other collectives in town and stay in contact with other wagon places across Germany. You are welcome to drop by, and space for your tent, vehicle or housing in one of our guest-wagons will be made available. For more information and pictures see our homepage.

Wagendorf Karow Pankgräfin e.V.

E4

Pankgrafenstr. 12d | 13125 Berlin ☎+49-(0)177-540 36 16
info@pankgraefin.de www.pankgraefin.de

1993 5 ha 150

e, sp, g, Polish 5
all who are concerned together or part of the group (majority vote)
individual financing, membership fees rented/leased
development, environmental education in the summer
ecological, self-managed, non-violent, adult education, pedagogical, environmental protection, permaculture, organic gardening, library, internet café, soccer, volleyball, gym, band rehearsal room, nature kindergarten, sweat lodge, forest garden, healing garden, bicycle workshop, wood workshop

☺ ()

The "Karow Trailer Village" housing project ("Wagendorf Karow") provides a home for about 150 big and small people living in trailers and huts on almost five hectares. A large part of the leased premises, which are in close proximity to a nature reserve, serve as project areas

for the promotion of environmental education - especially the nature kindergarten (founded by the residents), the permaculture forest garden, the healing garden and the plant-based sewage treatment system serve to implement the statutes of the non-profit association "Pankgräfin e.V.", which was established by inhabitants in 1995 and which has been the leaseholder of the property at the edge of Berlin ever since.
Our emphasis is on alternative living and families. Decisions concern all members and are made at monthly meetings, and projects and ideas are mostly implemented by small groups or individuals. Community members and visitors use the internet café, the library, the band rehearsal space and the clothing room; the large party hall is regularly used for parties and there are weekly yoga sessions free of charge, as well as football and volleyball.
We have space for visitors with or without vehicles. Information about available visitor spaces as well as further information, pictures, current projects and offers can be found on our website.

☺ 6-month trial period, ideally people who want to get involved

Walnusshof Leverkusen

A5

Zentrum für Gelassenheit & Kreativität
Odenthaler Str. 65a | 51375 Leverkusen ☎ +49-(0)214-500 710 00
info@walnusshof.org www.walnusshof.org

2017 1 ha (near Köln, Wuppertal, Düsseldorf) 10 % ♀4 ♂4 8

g, e, sp 5

all concerned together (consensus) individual financing

rented

permaculture, arts & crafts, holistic bodywork, dance & movement

ecological, self-managed, non-violent, spiritual, esoteric/New Age, inner growth, meditation, yoga, vegetarian diet, vegan diet, adult education, pedagogical, environmental protection, self-sufficiency, permaculture, organic gardening, humour, non-smoking, art, community building (Scott Peck), social work

We are a group of peaceful, nature loving and creative humans. Right now we are living on a walnut and fruit farm, right outside of Leverkusen (midwest Germany). Next to more than 200 walnut trees there are also plenty other fruit bearing trees, herbs and vegetables. We are located close to a forest and a river but are still not far from the city and main traffic roads.
We want more than simply living besides each other. Our dream is to create a beautiful living space for people who want to be creative in community and who want to discover and grow their potentials.
If that sounds interesting to you and you like to be active in areas such as for example music, art, dancing, gardening, cooking, meditating, body work, energy awareness, and want to learn or teach about them you are welcome to live with us.
We are regularly hosting activities like mantra singing, meditation, Om chanting, yoga, Biodanza, drum circles, women and men circles, concerts and dancing events, always adding to the list, and are happy if you feel you can add something to it, too.
We are looking forward to meeting open minded people who want to support and become a part of our project.

Wangelin - die Werkstatt des guten Lebens

D3

FAL e.V., Am Bahnhof 2 | 19395 Ganzlin +49-(0)38737-202 07
info@fal-ev.de fal-ev.de, wangeliner-garten.de, earthbuilding.eu

1990 ca. 3 ha (near Plau am See) 30 % 15

g, e 10 5

working groups (majority vote) individual financing

partly owned, partly leased from the municipality

garden work, café, food processing, environmentally sound construction, cultural events, seminars and guest accommodation

ecological, self-managed, non-violent, inner growth, organic diet, adult education, seminar-house, animal protection, environmental protection, self-sufficiency, permaculture, organic agriculture, organic gardening, forestry, alternative energy, new technologies, ecological construction, loam construction

Wangelin, a small village in the South-West of the Mecklenburg Lake District, is home to an open-minded and creative village community. The past 30 years saw the creation of innovative and forward-thinking projects such as the Wangelin Garden, the European Training Centre for Clay Building, the Clay Museum, the Wunderfeld Cooperative and an exchange- and give-away-house, where everyone can bring and take whatever they like.
The work and ideas of people in and around Wangelin are beneficial for the development of the viability of living in this small region. All businesses are committed to sustainability and plan for the future. This is how Wangelin has become a place where people of all generations live and work together. The "Wunderfeld Cooperative", the "Wangeliner Garden", as well as our inhabitants (craftsmen, beekeepers, farmers or artists) work towards the goal of a regional economic cycle within our community.
Long-term accommodation is quite scarce at the moment, that's why we are currently developing a building site where we want to find new possibilities in sharing, living, working and building together. We rejoice in people who would like to participate in dreaming up, living and creating a future together with us.

Neulandgewinner Netzwerk

interest in active participation in the community projects

WEG am Rundling

E4

formerly Lebens(t)raum Johannistal c/o Hartmut Glänzel
Am Rundling 5 | 12487 Berlin ☎ +49-(0)30-681 79 85
glaenzel@t-online.de www.rundlinge.de

2004 | 8000 m² | ♀23 ♂20 ca. 20 | 63

- g, e, f, Russian — 2
- all who are concerned in a monthly assembly, delegated subgroups can decide on their own (high majorities vote, aiming for consensus)
- individual financing + solidarity payments in case of financial problems of single cohabitants
- individual property and a proportional share of communal property
- interested in community-life
- we are co-housing project and work outside, but there is a part-time position for administration, technical issues and housekeeping
- ecological, self-managed, non-violent, alternative energy, new technologies, ecological construction, free of cars, car sharing, food co-op, yoga group, table tennis group, choir

The "Lebens(t)raum Johannisthal" (now called "WEG am Rundling") has been in existence since 2004. With the concept of "young and old", there are now approximately 65 people aged between 8 and 76 in twenty houses immediately adjacent to the landscape park "Berlin-Johannisthal". We want to create a model that opposes the typical city loneliness, which provides equally for the needs of individuality as well as belonging. We are now in the process of developing a communal life in which we do not have to give up private space. This means that on the one hand there are clearly defined individual spaces (apartment/house alongside small individual garden space) and on the other hand a larger communally created and inhabited common space (including a communal house). We want a tolerant co-living of old and young. Our co-living should be defined by acceptance of different needs and the willingness to benefit from one another. We are not ideologically determined, but some running themes have now emerged, e.g. there are no cars on the premises, we prioritise the use of alternative energies, ecological materials and healthy food, we practice solidarity and support in difficult situations, love the nearby nature reserve and the readiness to get involved in communal processes on site. Other aspects, such as food-coop, yoga group, choir and car-sharing are left up to subgroup initiatives.

☺ usually living space is to be financed individually

Weiler Hundscheidt

B6

Hundscheidt 2 | 51598 Friesenhagen ☎ +49-(0)2294-909 492
weiler-hundscheidt@t-online.de

2008 | 5 ha | 60-80 % | ♀2 ♂3 0 | 5

- g, e
- primarily individual financing — individual property
- holistic and sustainable lifestyle, subsistence, local politics
- ecological, inner growth, meditation, adult education, self-sufficiency, permaculture, humour

☺ (✋✋)

Our two-house property in the silence of the wooded Wildenburger Land is the ideal place for us to unfold and flourish "the good life" in the service for all beings.
There is no shortage of external conditions: the wonderful quiet location in the low mountain range, the large swimming pond fed by the brook, the land with the orchards, the gardens with the beginnings of forest gardening, the fountain with the best drinking water, the large barn that provides enough space for cultural and festival activities, and much more - a wonderful place whose possibilities are far from exhausted.
We regard two aspects as elementary: on the one hand the inner work to clarify and heal one's own personality and on the other hand the importance of the degree of self-sufficiency. We are convinced that without the healing of one's own emotional wounds, the necessary change of tendency in civilization can hardly succeed.
In addition to the development of our personal life on the farm, we also see ourselves as a nucleus and impulse generator for our environment. We have initiated citizens' initiatives, run for the municipal council with our own list and invite people to our regular events. With a little luck - that's our goal - it will all lead to a lively "New Village Network" inspired by the spirit of new beginnings.

☆ Wendland-Hof e.G. Lomizil e.V.

D3

Alte Dorfstraße 24 | 29491 Prezelle/Lomitz ☎+49-(0)5848-1288
tagungshaus@lomizil.de www.lomizil.de

1987 3/40

Property of all
seminar-house

It all began when some young people from Berlin organised themselves in the Gorleben resistance against the nuclear waste projects in Wendland. They wanted to create a permanent structure with 100 people, and after all, 30 of them finally did. In 1987 they founded a cooperative and bought the old farm together. The property - a four-sided courtyard with a former restaurant and big hall - was gradually renovated and extended by a new building including a holiday apartment and sauna.
Today, the cooperative has 40 members, three of them living all year round on the "Wendland-Hof". On occasions such as New Year, Easter or the summer festival, a large part of the group comes together with family and friends.
In addition to the housing project, the association "Lomizil" was founded to dedicate itself to various tasks: It focuses in particular on the areas of exchange and encounter, including the support of disadvantaged people. In order to finance the Lomizil, the conference house (with its spacious kitchen, dining room and five multi-bed rooms with shared showers) and the hall are rented to groups, including children's groups, dance groups, theatre groups, sports groups and choirs.

→See also description of network "**KommuJa**"

Widugard

D3

Waldweg 1 | 19372 Stolpe OT Granzin ☎+49-(0)174-752 31 87
widugard@googlemail.com

1998 ca 5000 m² (near Parchim) ♀2 ♂2 0 4

g, e 3

everyone in common (consensus)

limited community financing, working towards shared economy

still property of individual community members

non-violent, natural religion

garden work, craft work in the house, clearing the property, wood cutting

ecological, self-managed, non-violent, spiritual, inner growth, meditation, organic diet, animal protection, environmental protection, self-sufficiency, permaculture, organic gardening, ecological construction

We are a small community of people who love the simple life in the countryside and enjoy handicrafts and gardening. Our ideological orientation is diverse, what unites us is our love for nature, animals and plants. We see ourselves as guardians of creation, as gardeners in God's wonderful garden. Out of respect for life, we prefer to eat vegetarian food. Raw food is also important to us every day. In order to live in harmony with nature, one must reduce one's personal consumption and limit it to the essentials. Less is more and the key to healing the world. Our community has now existed for 20 years with changing members, there have been ups and downs, but Widugard has not perished. We are still looking for people who love a life in the silence of nature and are willing to give up superfluous things. Against the isolation of the consumer society we set the community. As the North American Indians so aptly said with regard to our future: "Only tribes will survive". We have no left-wing political convictions! Who feels addressed, can contact us and visit us. Free board and lodging is provided.

☺ desire for a simple life, accepting the principles of the Giver Movement

Wilde Gärtnerei

E3

Landbau & Selbstversorgung - Lernen & Selbsttransformation
Bernauer Straße 6 | 16321 Rüdnitz ☎+49-(0)176-341 106 28
wildegartnerei@posteo.de wildegartnerei.blogspot.de

2010 10 ha (near Berlin) 70 % ♀1 ♂2 3 6

g, i, e, Finnish 3 3

everybody (consensus) individual economy/partly shared economy

private property of individuals

self-sufficiency, vegetable farming, handcrafts, construction work, CSA

ecological, self-managed, non-violent, spiritual, inner growth, meditation, yoga, free love, organic diet, vegan diet, adult education, pedagogical, free school, peace-work, environmental protection, self-sufficiency, organic agriculture, organic gardening, alternative energy, ecological construction, construction, autodidactic, alternative living, regenerative agriculture

☺

We are a small, open working and living community, which is taking care of our vegetable farm and life-learning-place, and runs a solidary agriculture, largely self-sufficient and operating in a radical soil regenerative approach of cultivation. Our Children are visiting a neighbouring free nature school, in whose development we are participating. We open our door for all kinds of guests who either are looking for first experiences in agriculture or want to extend their limits in life: Agriculture, vegetable gardening or the self-caring everyday life offers us the framework to operate, challenge and grow: The actual harvest is in the development of consciousness, which ultimately is in focus of all that we do; we are taking care of it through thoughtful communication as well as through practicing Isha Hatha Yoga.

We are continuing to expand the yards infrastructure on our small area and are looking for expansion opportunities in the surrounding area in order to round off the farm and community concept.

See photo on page 13

CSA, Bündnis Junge Landwirtschaft, VERN, Saatgut-Erhalter-Netzwerk-Nord-Ost

☺ a conscious decision of all

Wir mittendrin! im PrymPark

A6

Philippstraße 4 | 52749 Düren
+49-(0)170-445 22 66
wir-mittendrin@prympark.de
wir-mittendrin.prympark.de

2007

g

everybody (we are organized in sociocratic structures) — individual

owned by all

none

self-managed, environmental protection, inter-generational house, cohousing

☺

Wir mittendrin! (meaning "right in the middle!") is one of seven subgroups of the intergenerational cohousing project "PrymPark" (see entry for "PrymPark Düren) - a climate protection estate on a 30,000 m² natural area with several joint projects - unique in Germany.

Our house contains 40 apartments for families, couples, singles and a shared flat. Everyone has his/her privacy in their own home and the opportunity to participate in community life. For more quality of life through mutual support in everyday life, there are spacious common areas such as a communal dining room with kitchen, workshop, laundry café and a garden with seating and swimming pond.

Children have plenty of space on the car-free terrain with a grove and playgrounds and inside there is a room to play and romp around. We are currently looking for young families with children. Completion: Summer 2020.

www.wir-mittendrin.prympark.de

www.facebook.com/WirMittendrin

☺ support co-housing idea, keep the mix of ages in the multi-generation community

Wir vom Gut e.G., Gut Mydlinghoven

A5

Mehrgenerationenprojekt
Mydlinghoven 4 | 40629 Düsseldorf
info@wirvomgut.de www.wirvomgut.de

2015 7.7 ha 102

g, e
working groups, plenum (systemic consensus principle or consensus)
property of association members
co-housing, intergenerational, environmental protection

We are an alternative housing project on the outskirts of Dusseldorf in a natural reservation area. Within an area of 77,000 square meters, 102 people aged 0-88 years live in a former listed stud farm that has been converted into 45 apartments and 1500 m² of common area. The resulting work on the estate and in the garden are planned and done together. The care of eleven sheep and four bee colonies is one of the chores. For children there are many possibilities for playing indoors and outdoors. A day care allows small children under the age of three a loving care in the familiar environment. The common area includes seminar rooms - which can be rented internally and externally - a communal kitchen with dining room, media room, sewing studio, workshops and creative room, as well as a guest rooms. Additionally, there is also a fitness and exercise room, the room of silence and a coworking space.

WIR-Projekt

B9

Recksteinstr. 14 | 72393 Gauselfingen +49-(0)7475-951 144
bernd.kappel@gmx.de www.wir-projekt.de

1986 0.5 ha (near Tübingen) 5 % ♀5 ♂9 2 16

2
all of us (voting in a plenary meeting) individual financing
association "Wir-Projekt e.V."
open spaces, garden, rehearsal rooms and recording studio, conference centre and carpentry
ecological, self-managed, chaotic, organic diet, seminar-house, environmental protection, self-sufficiency, organic gardening, alternative energy, construction, humour, welcoming criticism, atmosphere of constructive debate, intergenerational, partly vegetarian, rural

We are a intergenerational (3 to 65 years) cohousing project consisting of three intentional communities with a lot of space, our own garden, rehearsal rooms with a recording studio and our conference center (conference rooms, sleeping spaces for 35 people and a former restaurant with kitchen). Our strengths are our self-administration and the autonomy of our decisions that follows within (apart from economic constraints). The WIR-Projekt ("WE project") is part of a future shaped by people who:

- are self-responsible
- give the available space and freedom life, shape, and a use
- want to learn and pass on knowledge
- are allowed to fail in their efforts.

LAKS

Germany

Wishmar - Wünschhausen

D5

Dorfstraße 04 | 07774 Wichmar
wichmarzwerge@riseup.net wichmar.eu

2011 (near Jena, Naumburg, Leipzig) 2 % 1 12

g, e, f, Portuguese 1-2
as a group in a plenum assembly (consensus)
solidarity-based financing of living, food and equipment
mostly rented, property of association
progressive, ecological
garden, music, festivals
ecological, left-wing political understanding, feminist, non-violent, spiritual, inner growth, meditation, yoga, organic diet, vegetarian diet, vegan diet, adult education, pedagogical, environmental protection, organic gardening, conventional gardening, animal friendly husbandry

☺

We are a diverse group of open-minded people aged between 0 and 55. We live in 6 different flats/huts/wagons in our village, all within 5-10min walk of each other. This space gives room for individuality.
What joins us is our love towards our quiet, little village; the lively river and the surrounding nature, as well as our desire to live in peaceful solidarity and sustainability with each other. We buy vegetables from local farmers and harvest from our gardens and surrounding trees. Other foods are obtained from wholesale or other reliable sources. Bread is baked weekly. We come together in regular plenary meetings. Besides that we often meet for meals, play music, construct, meditate & practice yoga.
In SPRINGTIME we celebrate a small festival with workshops and in AUTUMN we host an annual festival to press apple juice collected from surrounding orchards. During SUMMER we are often visited by guests. This is also the time of year when we like swim in the river, have picnics, campfires & do gardening. Wichmar is a wonderful place to dream, to experience community-living and alternatives of social & political understanding as well as to promote personal growth.

☺ 14 days probationary living

Wohnprojekt Ponyhof

B10

Kontakt bitte über E-Mail | 79189 Bad Krozingen

wgponyhofbk@aol.com www.diearche-freiburg.de/ponyhof

2004 1000 m² (near Freiburg) ♀6 ♂5 0 11

g

all together (consensus) rented

self-managed, non-violent, organic diet, vegetarian diet, helping each other, ecological values

We are 11 adults in the second half of our lives, and we live in a house with a garden where everyone has a self-contained apartment. Much of our communal life is based on voluntariness. We regularly meet for movie evening. Communal living presents us with beautiful enrichment in terms of spontaneous situations such as a skat game in the garden, somebody reading something out loud, a chat in the hallway or going on a spur-of-the-moment walk together. On a three-weekly basis we meet at our round table, where we use consensus decision-making.

Wönnich 103

E4

Wönnichstr. 103 | 10317 Berlin

kontakt@woennich.de www.woennich.de

2001 1050 m² ♀9 ♂5 4 18

g, e, f, Portuguese 4

all the people living here (consensus) shared financing for some areas

owned by an association of the residents

ecological, self-managed, non-violent, organic diet, vegetarian diet, vegan diet, environmental protection, organic gardening, alternative energy, ecological construction

☺

We are a socially and ecologically oriented community house in Berlin-Lichtenberg. With our project we realise a resource-saving, sustainable and communal way of life in the middle of the city.

The house is an old building with a lot of eco-technology: We reduce our resource consumption with water saving fittings, compost and 1-litre toilets, solar power and hot water production on the roof and in-house biological water treatment. Many tomatoes and berries grow in our large organic garden with pond.

We are a heterogeneous group of people with different lifestyles who are united by the pursuit of a life in community according to ecological and solidary principles. In addition to our living space, we have a large communal kitchen, a room for events, a library living room and a conservatory / greenhouse. Every week we meet in plenary. We make decisions by consensus. Necessary work gets done through hierarchy-free division of labour. Values such as mindfulness, tolerance, solidarity and open discussion are important to us in communication.

We open our house for events, guided tours and parties for the neighbourhood and all interested people and maintain contact with neighbouring projects and neighbourhood initiatives. Every month we organise people's kitchens with vegan-vegetarian organic food and cabaret or information programme.

☺ interested in an environmentally friendly and community orientated way of living

Wukania Projektehof

E3

Ruhlsdofer Str. 45 | 16359 Biesenthal
info@wukania.net www.wukania.net

2008 3 ha (near Bernau, Berlin) 2 % ♀15 ♂15 15 45

g, e, sp, f, (Swedish, Norwegian, Somali, Arabic) some 20
all together (consensus)
some shared financing (household budget, common terrain, building costs...)
collective property
solidarity, equal rights for all, against exploitation of humans, animals, environment, defusing private property
workshops (wood, metal, bicycle), nature nursery, bio catering, non-commercial seminar area with library & gift room, gardens, firewood, construction sites, video editing workplace, solidarity, social, ecological projects, local & supraregional work
ecological, self-managed, anarchistic, left-wing political understanding, feminist, organic diet, vegetarian diet, vegan diet, adult education, seminar-house, environmental protection, permaculture, organic gardening, alternative energy, ecological construction, construction, vegan/vegetarian communal meal, partly barrier-free, smoking only in the smoking area outside; game evenings, campfire and sauna are popular

☺

In summer of 2004, a group of committed people came together to found a community project. Since they had been continuously active in environment- and socio-political as well as in social issues, they wanted to share this solidary cooperation also in everyday life and habitation. In 2008, they found an old sports school in the small town Biesenthal (35 km northeast of Berlin), right next to a lake, with an area of 3 ha and several buildings.
In the last 10 years a lot has been growing here. Until now, three living groups in two houses became established, besides there are e.g. workshops, a nursery, a non-commercial seminar space and an organic food canteen kitchen for schools, gardens, hens, butterflies, compost toilets and solar showers. More than 60 people are here actively working, acting and effecting, living, cooking, eating, playing, meeting, discussing, laughing and fighting together, in collectives, small and large groups or project teams, as one likes and might do, consensually and undogmatic.
We don't want to be a bubble in the periphery but small seeds for a solidary world, regional and supra-regional, we want to act and effect together, open to participate and to contribute. So by the variety of actively participating people we developed very diverse projects and events: children's circus and free market, welcoming work and community café, environmental projects in schools and summer party, screenings on climate change and migration and a lot more.

PAG, regional networks
☺ to get to know each other, sympathy, to fit together, to actively contribute, to enjoy our project centre

Wunder Wald Wietzow

E2

Wietzow 17 | 17129 Wietzow
wunderwaldwietzow@gmx.de
wunderwaldwietzow.com

2006 30 ha (near Greifswald, Neubrandenburg) 60-70 %
♀4 ♂2 2 8
3 4

g, e, Czech, Polish
part of team depending on the tasks (consensus)
collective financing
ownership of some community members
organic farming, education, self-sufficiency, construction
ecological, self-managed, inner growth, yoga, organic diet, vegetarian diet, vegan diet, adult education, pedagogical, free school, self-sufficiency, permaculture, organic agriculture, organic gardening, ecological construction, humour, non-smoking, forest garden, sustainability, education

☺ ()

We live on the edge of an exceptional village in the beautiful countryside of Mecklenburg-Vorpommern, close to the river Tollense. Our place includes expansive natural areas, organic farming, a nature education project with bees and a forest garden.
Step by step, we are creating a life of living sustainability there. A place for people who want to be involved with fun and warmness to the emergence of a truly great adventure - a paradise.

☺ vegetarian

Wurzeln & Wirken

D4

Friedrich-Rumpf-Straße 16-18 | 14641 Wustermark
wurzelnundwirken.de

2015 1 ha (near Berlin) 5 %
♀8 ♂8 8 24
4

g, e, Greek, sp, f
everybody in plenary (consensus)
partially shared economy (food, rent...)
shared property
ecological, self-managed, left-wing political understanding, non-violent, organic diet, vegetarian diet, vegan diet, environmental protection, organic gardening, alternative energy, ecological construction, construction, accessible to the disabled, non-smoking

We are Wurzeln & Wirken, a project between city and countryside. Our property with two houses, a large barn, and a big garden is located in the village of Wustermark. But with our direct rail connection to Berlin, where most of us work, we are quite close to the big city life.

We see ourselves as a social-ecological housing project which aims to live environmentally conscious and in community. Our new building is a straw bale house which was built with as many ecological and regional materials as possible. We are also working on the ecological renovation of the existing old building. Our nutrition is bio-vegan-vegetarian, we maintain our own FoodCoop with vegetables from a CSA and small contributions from our garden and hope to open it for the neighbourhood in the future. We share finances for many things (food, rent, large purchases, etc.) in order to counterbalance different income situations. Decisions are made together and by consensus at weekly meetings. Childcare is not only the direct parents' business.
In short, we try to live together in a way that is solidary, friendly and ecological. We also envision a stronger impact on our neighbourhood and broader environment as soon as we are no longer busy with building and renovation works!

Mietshäuser Syndikat (www.syndikat.org/en)

Yoga Vidya e.V.

B4

Yogaweg 7 | 32805 Bad Meinberg ☎ +49-(0)5234-87-0
info@yoga-vidya.de www.yoga-vidya.de/gemeinschaft/

1992 (near Detmold, Paderborn, Bielefeld) ♀68 ♂63 14 145

g all

all concerned (majority vote)

limited common financing, work for pocket money and social security benefits

owned by the association

spiritual, holistic yoga in the tradition of Swami Sivananda and Vishnu Devananda

kitchen, household, child care, IT, online marketing, reception/telephone switchboard, carpentry, electric, seminar coordination, address administration, PR

ecological, self-managed, non-violent, hierarchical, leader/guru, spiritual, inner growth, meditation, yoga, organic diet, vegetarian diet, vegan diet, adult education, seminar-house, peace-work, physically handicapped persons, permaculture, alternative energy, non-smoking, yoga therapy, psychological yoga therapy, ayurveda, satsang

Yoga Vidya is a spiritual community which is dedicated to the life and teaching of holistic yoga in the tradition of Swami Sivananda and Swami Vishnu-Devananda. Humanist-democratic principles apply in the implementation of these teachings within our spiritual community. In doing this, Yoga Vidya aims for 3 main goals:

- Widespread distribution of yoga in Germany and Europe for health, personal development and spiritual growth as a service towards other people. The Yoga Vidya houses and city-based centres are to be spaces of light and strength, where people can develop health-wise, energetically, personally and spiritually.

- The creation of opportunities for rapid spiritual growth for serious aspirants through life in spiritual community in the tradition of Swami Sivananda/Swami Vishnu-Devananda. Colleagues find optimal prerequisites for personal and spiritual development on all levels.
- Growth of the powers of peace and mutual understanding of Earth via a build-up of light spots and light webs on Earth in connection with other spiritual and ecological traditions.

Yoga Vidya Network with 4 seminar houses, 6 urban centres and 50 cooperating centres
☺ vacancies; qualification; spiritual interest and practice; willingness to abide by the basic rules of the Yoga Vidya tradition

Zar Nekla die Zweite Im Grünen einfach leben

D2

Zarnekla 19 | 17121 Loitz ☎+49-(0)39998-104 87
Zar.Nekla.II@t-online.de www.coforum.de/?2370

1998 17 ha (near Greifswald) 28.9 % ♀2 ♂1 3 6

g, e half

all who are concerned (shot from the hip)
limited community financing and individual
the building is private, otherwise association property
aimed in the right direction
self-sufficiency, garden, forest, renaturation, culture & art, (cob)building, jugger, youth camps, antisocial networking, hosting, die PARTEI
ecological, self-managed, left-wing political understanding, yoga, organic diet, vegetarian diet, vegan diet, adult education, seminar-house, environmental protection, self-sufficiency, permaculture, organic gardening, forestry, alternative energy, ecological construction, humour, non-smoking, Maquis, free cooperation, non-commercial economy, long live mycelia and rhizomes!

☺

Zarnekla – endless skies above our flowering landscape: forest, fields, park, moorland, hedgerows, gardens, fruit trees, fallow, solar showers, trailers, barn, farmyard, house. Virtually isolated, cranes nest with us. Deep soil for rooting, nice and functional neighbourliness in the village, children welcome!
Remain agile: Together with our friends we search for creative answers to capitalism and we act upon them in order to lead a nature-friendly, healthy and joyful life, free from domination. We emphasise simple reproduction and art. We are a cooperative and solidarity-embracing part of the maquis in our region (Western Pomerania), and we have particularly close connections to the "[LAND]Kombinat" in Gatschow. We like to share our knowledge and experiences, have a good library and run a small organic shop. Our property can be used for various purposes, e.g. camps, seminars, trial dropouts. It is also possible to come here for a holiday, to find yourself or for healing purposes. We live a vegetarian, drug- and dog free lifestyle, with composting toilets, without fridge or TV... and we take in young volunteers on the FÖJ scheme.

→See also description of network "**WWOOF**"
Sudbury-Ini, regional initiatives
☺ ecos, veggies, non-druggies, non-doggies, endowed with reason

ZEGG - Zentrum für experimentelle Gesellschaftsgestaltung

D4

Rosa-Luxemburg-Straße 89 | 14806 Belzig
+49-(0)33841-595-100
info@zegg.de
www.zegg.de

1991 | 16 ha (near Berlin) | <60 % | ♀42 ♂38 13 93

g, e, f, sp, Greek — 40 — 40

plenary in sociocratic model (sociocracy)

limited community financing and individual — owned by a charity

ecology, cooperation

seminars and guest accommodation, maintenance and construction, gardening, several independent businesses

ecological, self-managed, left-wing political understanding, non-violent, spiritual, inner growth, meditation, tantra, sexuality, free love, organic diet, vegetarian diet, seminar-house, environmental protection, permaculture, organic gardening, forestry, alternative energy, ecological construction

ZEGG community comprises 100 people experimenting with cooperative ways of living and working together. We have a focus on social, spiritual and ecological issues. We see ourselves as a living experiment, as there are no ready recipes for the building of new ways of living. Besides the changes needed on the outer level like sustainable technologies, new circumstances must be created to support inner growth. We want to bring these two aspects of change together. Many ZEGG-members seek new ways of dealing with love and sexuality issues and live open relationships. Experiences and insights gained are then passed on to our guests during the events organised here. Four large festivals and numerous seminars take place in ZEGG each year. With our programme we want to offer a space for experiences in the area of love, community, communication and creativity. Additionally, the community offers the chance for guests to come and work alongside its members, as an opportunity to learn about gardening, permaculture and soil improvement.

Working together is an important part of living in ZEGG. The communal business has adopted a sociocratic model of organisation, allowing for flexible and adaptable operating structures. Big financial and other decisions are made by the whole community.

One of the main concerns of the ZEGG community is to ensure an ecological way of living. Practical examples include the CO_2-neutral energy supply, a closed water cycle with its own constructed wetland sewage treatment system, and a garden growing fruits and vegetables according to the standards of organic agriculture.

Our existence has sparked an alternative development in our region. Many like-minded individuals came to live next to us and several intentional communities were started.

→See also description of networks "**GEN Europe, GEN Deutschland**"

Attac, Artabana

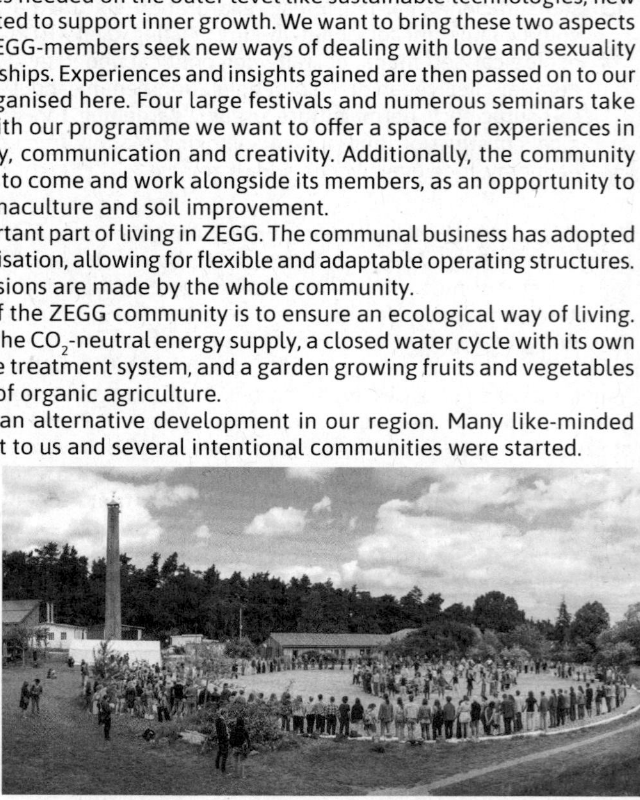

Greece, Hungary see pages 418ff

Ireland

Except for the following, we list one more Irish intentional community in planning stage on page 466f and 17 Camphill communities on page 512f.

Cloughjordan Ecovillage Sustainable Projects Ireland

WeCreate, Cloughjordan, County Tipperary | E53 VP86 Cloughjordan

+353-(0)505-428 33

office@thevillage.ie www.thevillage.ie

1999 | 22 ha (near Limerick) | 35 % | ♀54 ♂46 | 30 | 130

e, Irish | 4 | 20

community members (consensus) | individual financing

privately and cooperatively owned

education, ecological

education, training, research, livelihood creation, rural regeneration

ecological, self-managed, non-violent, ecumenical, adult education, pedagogical, seminar-house, one-world development, animal protection, environmental protection, permaculture, organic agriculture, organic gardening, organic farming, forestry, alternative energy, new technologies, ecological construction, humour, accessible to the disabled, community resilience, rural regeneration

☺ ( )

Cloughjordan Ecovillage (CEV) is a member-led, mixed-use development, extending an existing rural village in the Irish Midlands.

Members founded Sustainable Projects Ireland in 1999 as an educational charity to develop this sustainable community, designed to serve as an education, enterprise and research resource for all. Planning permission for two community buildings and 130 homes was granted in 2005. Building began in 2009 and fifty-five households were established in CEV by 2019. All are built to the high standards set by our ecological charter and supplied with heat from a member-run, wood-fueled, district heating system.

Businesses thriving in CEV include a community farm providing healthy food to 60 families, all locally grown on organic and CSA principles. WeCreate gives the community access to a "Fabrication Lab" in a co-working space for the eco-entrepreneur. Riot Rye Bakery uses locally-grown wood fuel to produce award-winning sourdough bread, while Django's Hostel provides year-round visitor accommodation.

The project won a first prize in the global, UN-sponsored, Livable Communities Awards in 2013 and was listed among the 23 leading European low-carbon projects in a 2016 international research study. The ecological footprint of a typical eco-villager has been assessed at less than half the Irish average of 4.3 global hectares: just 2.0 gha.

→See also description of network "**GEN Europe**"

Transition Network, Tipperary Green Business Network, Irish Environmental Network

☺ be able to buy a site and build a home; very few homes for rent at present

photo on page 16

Israel see page 423ff

Italy

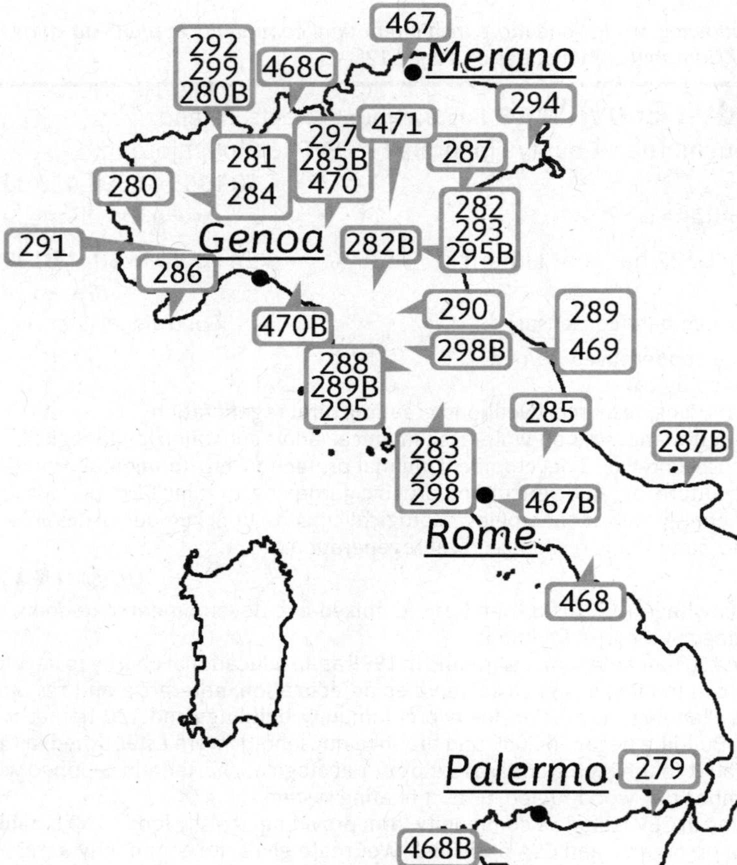

For this edition, we could collect 28 existing intentional communities and 9 planned ones (pages 467fff) from Italy. There is a lively network of ecovillagi (ecovillages), whose representatives have, unfortunately, not managed to formally get back to us. We have seen them around and they are very friendly, though. Their website is ecovillagi.it
They told us that the ecovillages in their network are almost always intentional communities living in a single house - so remember that the term ecovillages is used in all kinds of ways. There is also an Italian communities directory published by Vivere altrimenti:
COMUNITÀ INTENZIONALI, ECOVILLAGGI E COHOUSING, *see bibliography, page 568.*

Adventure Hill Eco-Adventure.org

Contrada Lupa | 98066 Scala di Patti ☎ +39-347-252 47 40
info@eco-adventure.org
www.eco-adventure.org, www.erlebe-abenteuer.de, www.eco-avventura.it

2008 | 20 ha (near Patti) | 25 % | 5-25

g, e, i | mostly 10-15, sometimes more, sometimes less

the Executive Board and all together (majority decision, consensus, as native tribes used to find decisions)

individually as well as collectively | eco-adventure.org

free like the birds

woodwork (Treehouses), Permaculture, therapy with animals

ecological, self-managed, non-violent, inner growth, meditation, yoga, free love, organic diet, adult education, pedagogical, free school, peace-work, one-world development, animal protection, environmental protection, self-sufficiency, permaculture, organic agriculture, organic gardening, organic farming, forestry, alternative energy, ecological construction, construction, humour, social work, experiential pedagogics, animal-based therapy, handcraft, carpentry, eco-tourism

☺ ( )

Eco-Adventure.org offers adventure travel, animal therapy, team trainings and a self-sufficient community in Sicily. Anyone who gets involved in the self-sufficiency community can stay indefinitely. On the Adventure Hill overlooking the sea to another seven islands you can also build tree and mound houses. We are glad about dozens of guests from all over the world each year, who join us for several weeks or months.
We will create an adventure playground and a free school. Children, adolescents and adults are free in their development. Montessori schools are already looking forward to co-operation. And we look forward to YOU! Just have a look... Pioneering like Robinson Crusoe, way better, without cannibals, in good company :-).
We still sleep in tents, cook over the campfire, wash our laundry by hand and ourselves with simple solar showers, or "fresh" bath in the pool as well in the lovely sea. Compost toilet in the "Sicilian jungle". We rebuilt the 110 year old stable as the first house/living room. Next we will build tree houses. First, a compost toilet is waiting to be inaugurated by you in the summer, followed by the construction of a tree house eco village and Hobbit houses... as well as a journey through Europe with our great donkey Hercules and the typical Sicilian cart.

☺ respect for nature, animals and people

☆ Agape Centro Ecumenico

Borgata Agape 1 | 10060 Prali (TO) ☎+39-(0)121-807 514
ufficio@agapecentroecumenico.org, campolavoro@agapecentroecumenico.org, direzione@agapecentroecumenico.org, agapecentroecumenico.org

1950 (near Torino) ♀7 ♂4 11

it varies depending of the period and necessities
all who are concerned, group, various committees
Valdesian Church
education, hospitality
ecological, self-managed, feminist, non-violent, spiritual, ecumenical, sexuality, adult education, pedagogical, seminar-house, discussion/dialogue, religion, philosophy

☺

Agape is an ambitious project that was born halfway through the last century from the ideas of a Waldensian pastor, Tullio Vinay. Its main goal is to be a meeting place. The founding values of Agape, i.e. reconciliation and reconstruction, are at the basis of all its activities. From the community life and volunteer work to the daily life in the centre, priority is given to being and living together in an inclusive and sustainable way.
The camps held at Agape provide a space for discussion about different subjects: current political and social topics, gender-related, spiritual and theological issues, as well as training and education.
Agape belongs to the centers and organizations of the Waldensian Church in Italy.

Valdesian Church
☺ to be motivated to live and volunteer in a community

Bordo Karma Dechen Yangtse

Fraz. Bordo | 28846 Borgomezzavalle (VB)
bordo@bordo.org www.bordo.org

1982 5 ha (near Domodossola) 5 %

g, e, i, f 0-20
board of cooperative, president and vice president of Dharma association
individual financing, no remuneration for decision-makers, sponsoring
Swiss cooperative with common ownership, activities made by Italian Dharma association
Tibetan Buddhism of the Karma Kagyu lineage
working in the guest area, Dharma courses, other courses, reconstruction of village, country care
ecological, self-managed, non-violent, spiritual, buddhist, inner growth, meditation, vegetarian diet, mind training, buddhist retreat place, historic village

☺ ( )

Bordo is a historic mountain village which was abandoned in the middle of the 20th century. From 1982 on a relatively stable community developed, with roughly 15 grown ups and 15

children (only from spring to autumn, number varying greatly). They strived to create a living space for people wishing to work with themselves and who were oriented on the values of Buddhist teaching. With the departure of the families an intermediate phase began, more centred on the course activities and maintenance of the village.
From 2015 on we had new concept. From spring to autumn intense collaboration with long time residents, responsible people of the Cooperative and Dharma association and volunteers, who with their work permit the functioning and maintenance of the place, and in collaboration with the others present work on themselves. Excellent possibility to learn and deepen knowledge of Buddhism and meditation.
During the whole time an orientation to the Karma Kagyu lineage was important. The foundation of the actual village was supported by the late Shamar Rinpoche. Many important teachers, among them Karmapa, Shamar Rinpoche, Kalu Rinpoche, Gendun Rinpoche and many others have visited the place and given their blessing.
For legal reasons only members of the Dharma association can be in Bordo and participate in activities. If in accordance with the fundamental principles of Bordo it is possible to make the membership on the spot. In winter Bordo is closed for activities.

🌍 Bodhi Path Europe, Italian Buddhist union, Swiss Buddhist union
☺ long probation period, strong interest for Buddhism and meditation, readiness to participate in all work required, participation in living costs

Co-housing Numero Zero

Via cottolengo Torino | 10152 Torino
info@cohousingnumerozero.org www.cohousingnumerozero.org/

2013 | 1000 m² | ♀8 ♂5 4 17

none none
all the members, we decide all together (consensus) — individual financing
property of all
no
we share services, common spaces and activities with the neighborhood
self-managed

☺

The cohousing Numero Zero was born within the CoAbitare association, the neighborhood and the city, which allows habitants to rediscover sociality and cooperation between neighbors. Co-housing is one of the answers to the need to live in a less individualistic and more social way, less consumerist and more creative, less expensive and more peaceful.
The objectives of the project Numero Zero can be summarized as follow:

- to realize a cohousing project in which collective spaces and private spaces interpenetrate and place themselves in a continuous relationship, while keeping their own connotation clear
- to form a heterogeneous group of people who plan the residence and common activities together, implement actions of mutual help and solidarity
- experimenting with an innovative living model of collective residence

- to experiment with new constructive solutions, inspired by bio-architecture concepts, allowing energy savings and the use of innovative technologies to reduce consumption and costs for individuals, the environment and society.

The building consists of about 750 square meters of residence, divided into 8 apartments from 50 to 130 square meters, about 200 square meters of outdoor and indoor common spaces and about 180 square meters of commercial premises.
The spaces in common use include: a living room with kitchen and wood oven on the ground floor, a workshop for DIY, a basement room for multi-purpose use, a cellar, a garden and a large terrace.

☺ no conditions

Cohousing SanGiorgio Associazione Cohousing Solidaria

via Ravenna 228 | 44124 Ferrara ☎+39-320 862 22 89
www.cohousingsolidaria.org

2009 (near Ferrara) 10-20

solidarity

SanGiorgio Cohousing was started in 2009 in Ferrara as a project of the Cohousing Solidaria association, which has set itself the task of spreading the culture and way of life of communal living together. We started a cooperative when we decided to buy land. Subsequently, in the context of transparent and participatory processes, tenders for architectural drafts and construction companies were carried out and then implemented.
Our community is located on the southern outskirts of the famous city of Ferrara, right on the banks of the river Po (di Primaro). The buildings are A4 standard certified, built entirely of wood and awarded the prestigious Green Building Awards in 2016. Today we try to realize beautiful and liveable projects for us as well as for the city of Ferrara.

Comunità della Guedrara

Baconi, 1 | 41029 Sestola (MO) ☎+39-0536-610 62
guedrara@interfree.it

1980 14 ha (near Modena) 20 % ♀3 ♂3 6

i, e, f 3
everyone in common (consensus) shared income
property of one member
Christian, ecumenical, nonviolent
dairy cows, social work
ecological, left-wing political understanding, non-violent, Christian, ecumenical, peace-work, one-world development

☺

We are a community of three families with five adult children who live with us. The community is inspired by the first Christian communities that lived in poverty, service, hospitality, prayer, work, nonviolence and with common goods.
We have a farm with dairy cows for our Parmigiano Reggiano. We have a common till, share our incomes and even cars. We have only one kitchen and we always eat together. The

house has common areas and spaces reserved for families. It has been entirely self-built.
Evenmore, we manage a little Bed and Breakfast. We welcome people in need and are active in political and social issues and movements.

☺ willingness to accept our concept

Comunità di Etica Vivente di Città della Pieve

vocabolo Pian di Mattone, 14 | 06062 Cittá della Pieve ☎+39-328-902 90 06
fiorenza.bortolotti@gmail.com comunitadieticavivente.org

1981 · 40 ha (near Città della Pieve, Perugia) · 25 % · ♀30 ♂15 · 5 · 50

i, e, g · 10 · alle

board of directors, divisional managers, general assembly (vote, if possible consensus)
some common financing
psychosynthesis, creative Meditation and Agni Yoga (Living Ethics)
Education for adults in the main topics of the psycho spiritual teachings of the past and the new century
ecological, hierarchical, spiritual, esoteric/New Age, inner growth, meditation, adult education, pedagogical, seminar-house, peace-work, organic agriculture, organic gardening, ecological construction, non-smoking, psychosynthesis, art and beauty, Agni Yoga/Living Ethics, education

☺

Comunità di Etica Vivente is a non-religious cultural and scientific association. Study, meditation and service are the three roots of the community.
The community was founded in 1981 by Sergio Bartoli and a group of co-workers as a centre for creative meditation, spread on more than 20 km on the hills between Umbria and Tuscany. Over the years it has developed and grown through the creative and practical skills contributed by an ever increasing number of members and friends. All activities are vivified by the study and application of the principles of psychosynthesis and living ethics. Programs in psycho-spiritual education have fostered the emergence of an integrated and harmonious group willing to experiment in daily life with new models intuited through meditation and inspiration.
As a result of over 20 years of intense and qualified group activity, the community has now become a cell of new life. This vitality is reflected in various fields: economics, education, culture and art, healing, permaculture, youth programs. At present a large proportion of members and co-workers live locally. Others come from all over Italy and the rest of the world. All take part as participants in the schools and as collaborators in the community life.

→See also description of networks "**GEN Europe, RIVE**"
World Service Intergroup
☺ to share a spiritual research and openness to work on your personality and in group

Damanhur

Via Pramarzo 3 | 10080 Baldissero Canavese (TO) ☎+39-0124-512 226
welcome@damanhur.org www.damanhur.org, www.damanhurblog.com

1975 520 ha (near Turin) 50 % ♀57 % ♂43 % 23 % 600 resident

i, e, g, f, sp, Japanese, Croatian 20 % 30 %

we have a complex system of elected bodies/organism (inside the communities we call our decision making "by majority", but indeed it is more similar to a consensus minus a tolerant minority to achieve fast decisions.)

individual and limited community financing shared property

Damanhurian philosophy is based on positive thinking, promotion of diversity, interior refinement and change. Playfulness and humor are used as a strategy against habituality.

building temples and sacred spaces, creating experiences and tools for the new alliance between humans and the plant world, developing tools and methods for healing & well-being, education

ecological, self-managed, spiritual, inner growth, meditation, organic diet, adult education, free school, seminar-house, peace-work, one-world development, environmental protection, self-sufficiency, organic agriculture, organic gardening, organic farming, forestry, alternative energy, new technologies, ecological construction, humour, accessible to the disabled, non-smoking, temples, healing, art, consciousness

☺ ()

Damanhur is a Federation of over 25 spiritual communities (nuclei) located North of Piemont, between Turin and Aosta, with a radius of 15 km that includes in the center an area named Valchiusella, a valley that is still green and clean. The 600 resident citizens of Damanhur and about 500 non resident citizens have given rise to a multilingual society, open to exchange with the world and to different cultures and peoples.
Damanhur was founded in 1975 under the inspiration of Falco Tarassaco, Oberto Airaudi (1950-2013). His enlightened and pragmatic vision created a fertile reality based on solidarity, sharing, love, and respect for the environment, so as to obtain recognition in 2005 as a model of sustainable society by the Global Human Settlements Forum of the UNO.
Damanhur is perhaps best known for its extraordinary subterranean work of art and architecture, a cathedral known as the Temples of Humankind, the spiritual heart of the community. Respect for the environment is one of the foundations. Damanhurians cultivate organic food and livestock, restructure and build according to green building principles, have design and installation companies in the field of renewable energy and prefer natural healing methods and a holistic view of medicine. This respect goes beyond an ecological vision - it recognizes the existence of a conscious and sensitive life in the plant and animal worlds, and researches contact with intelligences that inhabit this universe.
In 2017 we founded Damanhur Foundation as not for profit organization. Damanhur Foundation is active in fundraising initiatives to support Damanhur's main projects in the areas of:

- Respect for the earth, the environment and exploration of inter-species communication
- Fostering of spiritual awareness through the arts, education and the creation of Sacred Spaces
- Inter-cultural dialogue and support of Indigenous Peoples' wisdom

→See also description of networks "**GEN Europe, RIVE**"

Gaia Education, Green Phoenix, Conacreis

☺ temporary citizenship for three months available *Colour photo on page 9*

E.V.A. - Ecovillaggio Autocostruito

Straw bales houses of Pescomaggiore
Strada della Conserva - Fraz. Pescomaggiore | 67100 L'Aquila
+39-328-943 61 36
eva@pescomaggiore.org www.pescomaggiore.org

2009 (near L'Aquila, Rom (100km)) 2

i, f, sp, e 1

all who are concerned (consensus, majority vote if consensus not reachable)

individual financing

property of the Committee for the Rebirth of Pescomaggiore

ecological construction, revitalization of the mountain village, landscape maintenance and protection, traditional agriculture, civic participation

ecological, self-managed, environmental protection, alternative energy

☺ ( )

The "Self-Built Eco-Village" ("Ecovillaggio autocostruito", E.V.A.) initially stems from the experience of the Committee for the Rebirth of Pescomaggiore, which was founded in 2007 in order to counter the depopulation of the mountain village. Construction began in the summer of 2009, to quickly provide a local and self-managed housing option for displaced people after the earthquake on April 6 in 2009. Inhabitants, activists and volunteers from all over the world gathered in the MISA Association, built four ecological wood, steel and straw bale homes served by a phytodepuration plant. All without public money, but with donations from individuals and the contributions of the inhabitants.
After a phase of community life, today two people are living here permanently, the other associates and guests use it temporarily. The MISA Association and the Committee for the Rebirth of Pescomaggiore work for the preservation and enhancement of the historical settlement of Pescomaggiore, its territory and its environmental and cultural heritage considered as common good and for the dissemination of an ecological culture.

local cultural and civic activists network

☺ ideological and financial

Ecosol Fidenza

Via Simone de Beauvoir 47 | 43036 Fidenza
+39-349-392 13 88
www.ecosol-fidenza.it

2013 1500 m² (near Parma) 1 % ♀13 ♂14 9 36

i, sp, e, f

all who are concerned (consensus) individual financing

property of all and single property

social, school

ecological, self-managed, non-violent, peace-work, mentally/psychologically handicapped persons, environmental protection, alternative energy, ecological construction, accessible to the disabled, non-smoking, fair trade, social volunteering

The co-housing project Ecosol is characterized by its focus on relationships and participation. The aim is to strengthen the well-being of residents and the earth. Well-being is when the

place is sustainable. Sustainability includes environmental, social and economic aspects, such as zero carbon emissions, mutual assistance, sharing of common spaces, infrastructures and resources, consensus-making, outward-opening and participation in networks.
One housing unit accommodates projects for social inclusion. The diversity of residents is an important resource for neighborhood assistance, while radical ideologies and the rule of stubbornness are seen as potential conflict points for relationships. Ecosol allows all residents to adjust the level of sharing, through independent housing units and the provision of spaces, activities and services. It is a project that originated from below and does not receive any financial support from the authorities.

Ecovillaggio Torri Superiore

Via Torri Superiore 5 | 18039 Ventimiglia (IM) ☎+39-0184-215 504
info@torri-superiore.org www.torri-superiore.org

1989 2.3 ha (near Ventimiglia, Imperia) 25 % ♀7 ♂6 5 18

i, e, f, g 5 6

everyone in common (consensus for residents, majority in association)

individual funding plus a communal food share

common property and/or individual property

vision of an ecologically friendly and socially just way of life

ecotourism, education, ecovillage training, restoring medieval village, farming

ecological, self-managed, organic diet, adult education, seminar-house, environmental protection, permaculture, organic agriculture, organic farming, ecological construction, creative: yoga, pottery, natural soap making, networking, volunteer programmes

☺ ()

Torri Superiore is a medieval village (XIII. Century) that has been progressively restored and inhabitated again over the last 30 years by a community of Italians and Germans. The village is made of 160 rooms, all built in local stone and lime, most of which still feature the traditional arches and vaults.
The Cultural Association has developed the project since 1989, it includes non-residents and owns about 50% of the village, while the other half is individually owned by the members. Only members of the association are entitled to buy property in Torri Superiore.
The restoration of the buildings respects the original character of the beautiful hamlet while using ecological materials and appropriate technology wherever possible.
The co-operative company runs an educational centre teaching sustainability, hands-on training and ecotourism. Courses are organised mostly in the summer and range from permaculture design to facilitation, consensus decision-making, yoga classes, ecovillage creation, pottery and dry walls maintenance. Organic agriculture activities focus mainly on olive oil production and vegetable gardens to supply the community centre.
The proximity to the beautiful river and the sea, the easily accessible walks on the surrounding mountains and the mild climate make Torri Superiore an attractive guest-house for year-round visits by ecologically sensitive guests who appreciate the informal, friendly style of the group. Advance booking requested.

→See also description of networks **"RIVE, GEN Europe"**

Accademia di Permacultura, Legacooperative

☺ personal contact, 12 months trial period, independent financing

Fattoria solidale Brugine

Via Arzerini 44 | 35030 Brugine - Padova ☎+39-(0)049 997 609
fattoriasolidale@gmail.com www.facebook.com/fattoriasolidalebrugine/

1 ha (near Padova) ♀7 ♂4 14 25

i, g, e, sp, f 4
all who are concerned (consensus) shared income
the owner is a non-profit association
none
agriculture, wood products
ecological, non-violent, ecumenical, organic agriculture, alternative energy, accessible to the disabled, family, sharing

☺

The project "Fattoria Solidale" has the aim of creating an understanding and welcoming space where singles and families can live the values of sharing, open-mindness and sobriety.

→See also description of networks "**RIVE, WWOOF**"
MFC

Giardino della Gioia

Contrada Torre Mileto Bar Pertosa | 71015 San Nicandro Garganico (FG)
☎+39-329-984 26 08
giardinodellagioia@gmail.com www.giardinodellagioia.wix.com/eden

2011 3 ha (near San Nicandro Garganico) 40 % ♀4 ♂5 0 9

i, e
community circle (consensus) donations and courses
property of 2, cooperative ownership planned
harmony with nature, psycho-physical and spiritual wellness, open to compassion, tolerance and truth by living through a Circle Way, using consensus and deep listening to everyone
sustainable agriculture, eco-buildings, community buildings, alternative schools (steiner, libertarian etc.), meditation, yoga and awareness, theatre, art works, knowledge sharing, ancient tradition revival, new technologies, open to the local people
ecological, self-managed, non-violent, spiritual, meditation, yoga, tantra, free love, organic diet, vegetarian diet, vegan diet, pedagogical, free school, seminar-house, peace-work, one-world development, environmental protection, self-sufficiency, permaculture, organic agriculture, organic gardening, organic farming, forestry, alternative energy, new technologies, ecological construction, humour, non-smoking, social works, social promotion association, therapies, art, artisan works

☺

Giardino della Gioia is an ecovillage with the purpose of reconnecting with nature, the awakening of consciousness and the creation of more self-awareness. And we are an ecovillage that would like to become a community! There is already enough structure to host ten people: We have three yurts, a straw house, a caravan and a draft of a tree house, compost toilets, a kitchen, a circus for meditation & yoga and a bath-house with showers and sauna.

In 2019, just two people live here, but we would like to grow. We built everything with natural materials and with the principles of permaculture. We live in an olive garden in a national park in front of the Adriatic Sea (800 m walk). We consider this place to be a healing area that invites us to work on the awakening of consciousness and to work on ourselves. In these last seven years, many people have given us their time, effort and energy. Together, we found new concepts for relationships, based on trust and transparency. We believe in gift economy, whenever it is possible.
We believe in the way of the circle, to share emotions and to take decisions. We live our lives with healthy values, without drugs or alcohol (there is a smoking area) and all our food is always vegetarian and often vegan.
This is what we have achieved until now and all the rest will be developed together. The ecovillage lives with the awareness that everything is already here, waiting for you and your energy.

→See also description of networks **"RIVE, GEN Europe"**
☺ interested in ecovillage lifestyle, willing to work on ego and integration with the group, willing to work on attachments and dependencies, willing to not take any drugs, smoke or drink alcohol inside the community - there is a place for tobacco smokers

Istituto Osho Miasto Institute for Meditation & Spiritual Growth

Pod. S. Giorgio, 16 | 53012 Chiusdino (SI) ☎+39-0577-960 124
oshomiasto@oshomiasto.it www.oshomiasto.it

1980 | 130 ha (near Siena) | ♀18 ♂18 0 36

g, e, i | some

one part of the group (consensus) | common property

we are disciples of Osho

meditation, therapy groups and training, meditation camps, everything aimed at personal and spiritual growth

ecological, self-managed, non-violent, spiritual, inner growth, meditation, organic diet, vegetarian diet, organic gardening, therapy

Istituto Osho Miasto is one of the few existing Osho Communes in Europe. It is a commune for personal and spiritual growth and meditation. Each year we organize several hundred therapy groups, workshops and meditation seminars.
Furthermore, we are a community of Osho followers/disciples who live here as an experimental laboratory. The structure and form of the commune is constantly changing. Three meditations are held daily – Osho Dynamic Meditation, Osho Kundalini Meditation & White Robe (evening meditation meeting). The meditations are largely free for members. An annual association membership fee of 18 EUR is requested.

links to other Osho centres
☺ independent accommodation, spiritual search, interested in Osho

La Città della Luce

Via Porcozzone, 17 | 60012 Trecastelli, frz. Passo Ripe (AN)
+39-(0)071 795 90 90
info@lacittadellaluce.org
www.lacittadellaluce.org

1996 | 7 ha (near Ancona, Senigallia) | ♀18 ♂7 3 28

i, i, sp, f, Portuguese | half of us | all of us, some of us often travel to teach course
all of us (consensus) | shared economy

property of all
our common field for working on ourselves is Reiki; we are a community made of Reiki masters
personal and spiritual growth, holistic wellness and products, environmental protection
ecological, self-managed, non-violent, spiritual, inner growth, yoga, adult education, pedagogical, free school, seminar-house, peace-work, one-world development, animal protection, environmental protection, permaculture, organic agriculture, organic gardening, organic farming, animal friendly husbandry, alternative energy, new technologies, humour, non-smoking, Reiki, family and systemic constellations, the hero's journey, astrology, runology, holistic psychology, enneagram, sacred feminine, sacred masculine, childhood education, healing art, ayurveda, yoga, chinese medicine, reflexology, permaculture, ceramics, ethic jewels, natural clothing

☺ ()

We are a group of men and women who have decided to live and work together, sharing our passion for life and for our project – together with our children and the beloved ones who joined us.
La Città della Luce is a Permanent Holistic Laboratory, a place for growth and sharing, open for everybody. Our mission is to promote the practice and diffusion of bio-natural disciplines. We support each person in finding her or his personal connection to the Divine, however it may manifest itself. Through our teachings we promote wellness and consciousness for a balanced life, with ourselves and the whole ecosystem we live in. To support community life, all members share their growth path through Reiki, the Systemic Theory, the knowledge of Chakras and Psychosomatic theories, Yoga and the Ayurveda philosophy or Conscious Nutrition.
Relationships are at the top of our value system. We share time, property, talents and emotions, together with the common spaces of our home. We grow together, day after day, by taking care of our tasks, all different but complementary. Each decision is taken through the method of the consensus, in honesty and by freely expressing our inner feeling – aiming at respecting everyone and the common good.

→See also description of networks **"RIVE, GEN Europe"**
Associazione Conacreis Marche
☺ be willing to adopt our mission and lifestyle
Photo on page 3

La Comune di Bagnaia

Via Strada di Bagnaia, 10 | 53018 Ancaiano-Sovicille (SI) +39-(0)577-311 014
lacomune.bagnaia@libero.it

1979 | 80 ha (near Siena) | 80 % | ♀7 ♂8 15

e, f, i, sp,d | 6, others have jobs or are retired

every adult member (consensus) — shared economy

common property

social-political awareness, solidarity

organic farming, environmental, pacifist, political and cultural involvement

ecological, self-managed, left-wing political understanding, non-violent, yoga, organic diet, peace-work, environmental protection, self-sufficiency, organic agriculture, organic gardening, organic farming, forestry, alternative energy, ecological construction, equality, good cooking, self-building

☺

The commune of Bagnaia was founded in 1979. Its aim is to share human and economic resources and to experience group living based on the principles of reciprocal respect and comprehension, collaboration, equality. All property is communally owned. All decisions are taken by consensus. At present the group consists of about 15 people. The farm and living structures are located near a small hill town about 10 km from Siena. The building, part of which dates from the 14th Century, has been partially renovated for the use and needs of the commune. Other structures include the barn and hay loft, wine cellar, machine workshop, cheese and preserve making room, carpentry workshop, art studio, multi-purpose library and cultural room. The farm has been completely organic since 1990. It consists of 50 ha of woods, which is a source for our fuel and 30 ha of cultivated land which includes vineyards, olive groves, cereal crops and vegetable garden. We have cows, pigs, chickens, bees. The commune promotes and supports political, social and cultural activities as individuals and collectively, according to our ideals, interests and possibilities. Bagnaia is a non-profit organisation.

→See also description of networks **"RIVE, GEN Europe"**

CONACREIS

☺ interest in communal life, shared economy and autonomy, trial period of 1 year

La torre di mezzo

località Torre di mezzana 1, Cantagallo | 59100 Prato ☎ +39-(0)333 644 42 55

latorredimezzo@gmail.com — facebook.com/latorredimezzo/

2013 house: 300 m², land: 8 ha (near Firenze, Bologna, Prato) 30 %

♀4 ♂2 0 6

i, g, f, sp, e — work in progress 3

the core group makes all the decisions, for some specific decisions the approaching/on probation members are involved also (consensus, sociocratic consensus)

individual financing for house expenses, food, energy, services/common financing through events and workshops, bread and bakery products

rented

rural lifestyle, feminism, shamanism, holism, local traditions, etc... not radically, but like seeds for our human diversity.

bakery products, cheese, chestnut flour, vegetable garden, production of cosmetics and organic soaps, yoga and treatments, facilitation, laboratories, self-sufficiency

ecological, self-managed, anarchistic, non-violent, inner growth, organic diet, seminar-house, peace-work, environmental protection, self-sufficiency, organic gardening, humour, Zegg forum, councelling, nvc, facilitation, sociocracy

☺ ( )

La Torre di mezzo is an association and an ecovillage in progress, which aims to expand the number of the core group to become a proper ecovillage. We are based in Tuscany at 600 meters a.s.l., in the wild mountains of the Apennines but easily connected to big cities of culture like Florence and Bologna. The basic group is small but surrounded by friends and supporters who live with us for medium-long periods during the year.
La Torre di mezzo is well rooted in the territory as well as linked to projects and experiences of national and international community networks (RIVE, GEN) and the national networks of ecological activists.
We aim at personal growth through sharing circles and various methods to establish an authentic relationship and clear communication, trying to improve the world we live in. The community supports the growth of personal talents and abilities, it aspires to self-determination and self-sufficiency, to share with others, also spread a message of possibilities and courage for change. We care about the origin of the products we use especially through a short production chain and local exchange.
La Torre di mezzo spreads the principles of ecology by encouraging a direct contact with nature and a simple lifestyle. Improving our shared economy and working synergies, together with reducing our environmental impact are very important matters to us.
We offer a spontaneous and warm welcome. We love singing songs all together around the fire with the mountains embracing us.

→See also description of network "**RIVE**"
- local exchange trading system (gruppi di acquisto solidale), local neighbourhood network
- a daily small contribution to finance common expenses, volunteering work. The new member must deeply share our vision, mission, goals, agreements.

LaCasaRotta

Fraz. San Bartolomeo 94/96 | 12062 Cherasco (CN) +39-(0)339 381 29 50
casa.rotta@yahoo.it www.lacasarotta.com

8 ha (near Turin) 6 14

- sustainable agriculture: vegetables, old varieties of wheat/corn, orchards with peaches and more than 300 new fruit trees, 4 dogs, 5 cats, 20 hens, baking bread (clay oven), workshops, seminars, renovating, construction, collaborating, networking

LaCasaRotta is an association and sustainable agricultural project founded by a group of young people to restore an abandoned farmhouse and turn it into a cohousing project. We are the first ecovillage being developed in the Langhe hills, Piemont, famous for excellent wines and gastronomy. LaCasaRotta is a place where people of different ages and backgrounds can meet and talk, but also learn, plan, dream, express themselves. We want to move in new directions as well as retracing those forgotten; create new meeting places; a real exchange between generations and cultures; opportunities for solidarity to combat exclusion and return to simply being together: the community. The farmhouse is already a co-housing project for sharing space and time with different people, and after the restoration, the ground floor will become a bistro and a cultural circle.
LaCasaRotta organises small laboratories open to people of all ages and backgrounds to spread the knowledge of local traditions and other cultures, marked by high quality products from small local producers or from fair trade. Over the last six years, we established various annual workshops, eg. "Jam Day", "Tajarin Night", "Agnolotti Day... Plin!", "Make the Bread, Make the Oven" or "What Gnocchi". We also organise educational workshops for children to teach them about nature, respect for the environment and manual skills.

Visit our homepage or find us on Facebook ("LaCasaRotta")!

→See also description of network "**WWOOF**"
🌍 Workaway, Slow Food

Mirapuri - Peace Village in Europe

c/o Via Monte Falo 8 | 28011 Mirapuri/Coiromonte (Prov. di Nov.)
☎+39-0322-999 009
david.b@mirapuri-enterprises.com www.mirapuri-enterprises.com

1970 1 ha (near Stresa) 5 %
♀11-16 ♂12-16 5-9 28-41

g, e, it, f — all 20-30
working groups, community assembly (consensus) — shared income
association
free spirit, understanding of the world as the unity of spirit and matter
music, all arts, wellness, healing, meditation, catering, agriculture, horticulture, carpentry, handicraft, eco-friendly house construction, alternative energies, alternative school...
ecological, self-managed, anarchistic, feminist, non-violent, spiritual, chaotic, inner growth, meditation, yoga, tantra, sexuality, free love, organic diet, vegetarian diet, vegan diet, adult education, pedagogical, free school, seminar-house, peace-work, animal protection, environmental protection, self-sufficiency, permaculture, organic agriculture, conventional agriculture, organic gardening, conventional gardening, forestry, alternative energy, new technologies, ecological construction, humour, wellness, Ayurveda, ecological construction, carpentry, literature, poetry, music, film, painting, multimedia, video, photo, sound

☺

Mirapuri, the European City of Peace and Futureman in Italy, as well as its first satellite, Miravillage in Germany, have been created for people who like the Integral Yoga of Sri Aurobindo and his life-partner Mira Alfassa and the ideals of human unity and spiritual consciousness growth. Mirapuri and Miravillage are environments where people can practice the Integral Yoga and live and work together to realize its ideals in daily life. Everyone of goodwill who likes Sri Aurobindo and Mira Alfassa and their Integral Yoga can join Mirapuri and Miravillage. If you want to live in Mirapuri or Miravillage you can rent a house or apartment and contribute financially or with work to the growth of Mirapuri. Land and housing properties of Mirapuri and Miravillage belong to the Mirapuri Association. The growth and respect of individuality is understood as the basis and flowering of real communal life.
Each individual follows his/her own spiritual and material evolution according to his/her aspirations and inclinations in consent with the community. For a documentary about Mirapuri see #4890 on eurotopia.directory/booklinks

🌍 Nextcultureinfo.org, Mirapuri
☺ knowledge of Sri Aurobindo and Mira Alfassa and their integral yoga, as well as goodwill and genuine cooperation

Porto 15

Via del Porto, 15 | 40122 Bologna

info@porto15.it | porto15.it, Facebook: "Porto15"

2017 | 8 | 37

- i, s, Arabic, Romanian, Russian
- all who are concerned (consensus)
- some common financing
- rented
- self-managed, Collaborative

Porto 15 is one of the first cohousing communes in Italy promoted by public initiative. The project has been realized by a partnership between the municipality and private citizens that had gathered around a collection of ideas: to implement innovate housing models, to support the youth's autonomy, to revalue public real estate and to realise projects of social significance.

The project allowed the refurbishment of an historical building located in the heart of Bologna, divided into 18 apartments and several common spaces. It aims for young people under the age of 35 to experience a new form of collaborative living.

A training was undertaken by all suitable candidates and the community was eventually defined through self-selection. The group of inhabitants decided the distribution of the apartments and the designation of the common areas. A common "charta of values" including the vision and the mission of the new community was defined. Core values are: empowering diversity and sharing everyone's best practices, places, ideas, activities, relationships, needs, skills and experiences. The main goal of the community is to create social value thanks to proximity and interaction with the urban neighborhood.

- Italian Co-Housing Network
- new cohabitants are supposed to be under 35 at the time of contract signature

Photo: Combo/BoCiack Project

Progetto Gaia Terra We are planning a sustainable future!

Via Petrarca, 45 | 33061 Rivignano Teor (Udine) +39-(0)392 060 72 90
info@progettogaiaterra.com www.progettogaiaterra.com

2018 | 3 ha (near Udine) | 5 % | ♀1 ♂4 | 5

e, i | all of us

all of us (we practice the sociocracy model)

limited common financing for food and small technical expenses

property of one member

sustainable living

building with natural material and gardening

ecological, self-managed, organic diet, vegan diet, adult education, seminar-house, self-sufficiency, permaculture, organic farming, forestry, ecological construction, accessible to the disabled, zero waste

Gaia Terra is a dream come true! Gaia Terra was born from the will of a single person, now it is a shared dream. We are a group of heterogeneous people who through dialogue, confrontation and the method of consensus are growing in a community with a common goal. Gaia Terra is a container of projects, a place where it is possible to experiment and to create alternative possibilities to inhabit the planet in a low-impact way:

- practice sustainable agriculture
- build with natural materials
- educate to be critical of the things we buy
- live in a group to support each other
- feed consciously
- relate to the territory and protect the surrounding environment
- develop cultural projects
- practice hospitality and cultural diversity.

The building we are in was an industrial building until 2015, thus it is huge! We are restoring it with natural materials using the soil of our field, wood and straw.
We have a vegetable garden and a food forest which we are planning to expand in the future. We have a camping site in the woods and other forest space by the river Stella, where we swim in the summer. We aim to become a model for simple living in harmony with all the other creatures of the universe!

→See also description of network "**RIVE**"

food coop, sustainable living network

☺ economically independent, emotionally stable

Reggioli

loc. Reggioli 34 | 53013 Gaiole in Chianti (SI) ☎+39-393-413 15 82
info@reggioli.org www.reggioli.org

2012 126 ha (near Gaiole in Chianti) 25 % ♀2 ♂2 3 7

i, e, f, g, sp
all who are concerned (in the circle with consensus)
property of one member
organic agriculture, sustainable farm tourism, Waldorf education, Action and Talking Circles
ecological, self-managed, non-violent, inner growth, organic diet, vegetarian diet, vegan diet, adult education, seminar-house, environmental protection, permaculture, organic agriculture, organic gardening, organic farming, forestry, alternative energy, ecological construction, construction, humour, seminars, ecovillage, community building, ecological communication, circle culture, arts, handicraft, concerts, music, forum, volunteers, tuscany, bio-lake, geodesic dome, sweat lodges

☺

This is a great place to exchange culture, language, knowledge, wisdom and stories. We embrace diversity not only in our vegetable garden and enforce personal development. Moreover we are big friends of the circle culture and respectful communication to strengthen our community and the individuals living here. On our farm we mainly produce delicious ethical organic olive oil. Moreover we have an eco-farmstay. Our aim is to attract conscious people who want to be part of a ecological change and are willing to contribute to our dream of creating a self-sustainable eco-village while taking responsibility.

→See also description of network **"RIVE"**

Roncadella Ecovillage Under Construction

Via Walter Manzotti 23 | 42122 Reggio Emilia ☎+39-(0)345-466 72 55
sarabigi.via@gmail.com
Facebook: "Roncadella Ecovillaggio in costruzione"

2015 (near Parma, Bologna) 30 % ♀2 ♂3 1 6

i, e
the group circle (assent)
currently 2, changing
shared economy
rented
community, personal growth, ecology, sustainability, territorial networking
ecological, self-managed, non-violent, spiritual, adult education, pedagogical, peace-work, one-world development, physically handicapped persons, permaculture, organic gardening, organic farming, alternative energy, ecological construction, accessible to the disabled, handicraft, facilitation

☺

Roncadella is an intentional community whose main goal is the self-realization of the human being. We have chosen to live right here and now as we dream of seeing the world. We live in a common sense of belonging, aware of the co-responsibility we have towards each other and towards Mother Earth which hosts us.

We choose to give ourselves time, following its flow without trying to possess it; a time to be aware of the needs of each other, of conflicts and efforts. We grow through listening, respecting and caring about ourselves, the community, other human beings and Mother Earth. We maintain the utopia of changing the world with a broader horizon and without getting tired of the effort of this change in our everyday life.
Our project is built on various main areas: Individual, Community, Ecology, Economy, Spirituality and Celebration, Social and Territory. We are currently looking for a new location where to move in 2019, still in this area of Italy.

→See also description of network "**RIVE**"
☺ process to get to know each others

SanMaPa

Podere Pian di Cuccio - Localita Santa Caterina | 58053 Roccalbegna
☎+39-340-357 46 48
post@sanmapa.net www.sanmapa.net

2010 | 70 ha (near Grosseto) | ♀1 ♂2 1 4

g, e | not permanently inhabited yet
household till
still property of individuals, should become an association/ trust
free spirit, clear thinking, holistic
fostering of nature, permaculture, care of visiting groups, building of overtone instruments and earth houses
ecological, self-managed, non-violent, spiritual, inner growth, meditation, yoga, tantra, vegetarian diet, vegan diet, adult education, free school, seminar-house, peace-work, animal protection, permaculture, organic gardening, forestry, alternative energy, new technologies, construction, humour, non-smoking, overtone music, body painting, sweat lodge, vision quest, retreat, Tipi, earth houses

☺ ()

Ecology is not the central issue to us, because we feel it is a given! For example, we were never connected to the electricity network and heat all houses exclusively with wood from our own forest in very effective Finnish soapstone stoves. The essential thing that connects us in SanMaPa is the human-to-human interaction – in other words: How can we mutually support each other in growth and exhaustion of our full consciousness potential! Likewise the topics HEALTH & REAL TIMES HUMAN QUALITY OF LIFE are very important in SanMaPa - based on our wonderful, powerful natural country up here in the mountains of the Maremma with many hundreds of hectares of wild land - an OASE - embedded in vastness and Silence - with the best, living water - spicy, clean air and unloaded soil, as well as FREEDOM of mobile radio radiation, as well as very consciously living friends and neighbors. A true JEWEL in the heart of Europe - which is complemented by the sounds and songs of nature.
Our main concern - as a "service to fellow human beings on Mother Earth" - is the provision of retreat and cleaning times through fasting, meditations, raw vegetable diets, cooperation, also intended for members of other community projects in periods from one week up to several months. From these encounters we will develop a solid core group living here all year and work in harmony with the natural rhythms, in simplicity - just "minimalist" - with plenty of time to wonder and dream - in real paradisiacal quality of life!

Tempo di vivere Ecovillaggio e Associazione

Loc. Piccoli von Calenzano, Bettola | 29021 Piacenza
+39-(0)523-182 00 21, -329-0218941 (mobil)
info@tempodivivere.it
www.tempodivivere.it

2011 3000 m² (near Mailand (40 km)) 10 % ♀4 ♂2 2 8

i, e all

the resident group for decision concerning the project, but we include sociocratic circles in some areas (consensus and consent method, it depends on subject)

total shared economy rented

courses, vegetable garden for internal sustainability

ecological, self-managed, non-violent, inner growth, adult education, free school, self-sufficiency, organic agriculture

Tempo di Vivere is an intentional community, born in 2011, from a small group of people who decided to realize their dream: the birth of a commune based on relationship, personal growth, collaboration and the sharing of our talents, properties, skills and visions. Our values are sustainability, ecology, reciprocity, peace, harmony and respect of any living beings. We follow the Circle Way - an inspiring way of living and sharing our feelings. Since 2015, every summer we organize the Circle Way Summer Camp, a great chance for a new way of listening to everybody and to share ourselves.
Nowadays the group is formed by 8 residents (6 adults and 2 children) and a small group of people who are approaching and are trying the community way of living. Our passions and interests are expanding; we are engaged in many activities: courses about personal growth, communication, sociocracy, wellness, self-handling of fermented products and hygiene products, homeschooling, care of our vegetable garden and disseminate our life choice and project to the people of our region.
Since November 2017 we moved to a suggestive place in the mountains by Piacenza, 800m above sea level, next to the wonderful Perino Waterfalls, and ValTrebbia. Now we live in the middle of the nature, close to the bush and far from pollution and city noises.

→See also description of network "**RIVE**"

Financial Ethic Network

common values, long period (two years at least) to get to know the project and the people

Utopiaggia

Villa Piaggia 21 | 05010 Montegabbione (TR) ☎+39-347-348 78 05
info@utopiaggia.eu, beatrix_ebeling@gmx.de www.utopiaggia.eu

1982 100 ha (near Perugia) 20 % ♀8 ♂8 2 18

g, e, it 4-5 4-5

everyone in common (consensus) limited community financing

common property

intentional, egalitarian community

sheep, cheese making, pottery workshop, dyeing workshop, olives

ecological, self-managed, left-wing political understanding, non-violent, environmental protection, self-sufficiency, organic agriculture, organic gardening, organic farming, animal friendly husbandry, forestry, alternative energy, handicrafts

☺ ()

On a large piece of land in the hills South of Lago Trasimeno in Central Italy, we rear sheep and cultivate olives. We cover a small part of our needs through self-sufficiency. We finance our daily needs from a common fund. Some gainful employment takes place on the grounds itself (lambs, sheep cheese, dyed silk, ceramics, guests) and outside (environmental consultation, restoration of houses, school). The adult members living on the grounds are between 30 and 80 years old; with a medium age around 60. We are looking for new members who ideally should be younger than the majority of us. Visitors who would like to live with us for an extended period are welcome, but must contact us beforehand. Important decisions are made by the plenary meeting on a consensus basis. In recent years we have increasingly focused our activities on the region: we run a small café for the local area, work with the community council in developing an environmental education centre and organise conferences on new rural development models.

→See also description of network **"RIVE"**

☺ probationary year, membership contribution is set by agreement

Villa Vrindavana Hare Krishna Temple

Via Scopeti 106/108 | 50026 San Casciano Val di Pesa (FI) ☎+39-230 26 961
info@villavrindavana.org, reception.villa.vrindavana@gmail.com
www.radiokrishna.com

1979 100 ha (near Florenz) 30

i, e almost all

- one part of the group (majority vote) — limited community financing
- society property
- Hare Krishna Community
- religious services, self-awareness through Bhakti Yoga
- ecological, self-managed, non-violent, spiritual, meditation, yoga, vegetarian diet, animal protection, conventional agriculture, conventional gardening

☺

Villa Vrindavana community is located in the heart of the Chianti hills in Tuscany. It is a beautiful 100 hectares property with a breathtaking view and olive groves, an organic vegetable garden, an orchard, with protected cows, a Tuscan traditional garden and woods surrounding a Renaissance villa with a vegetarian center, an art gallery and an ashram for guests, all centered around a Vedic Temple.
Villa Vrindavana is a project of the International Society for Krishna Consciousness (ISKCON), which belongs to the oldest monotheistic Vedic tradition whose philosophical teachings are based on the great classical Sanskrit texts of India, such as the Bhagavad-Gita and the Bhagavat Purana. Our commitment is to create a place for peace and collaboration where one can learn and cultivate the ancient knowledge of the science of devotion and love for God according to the sublime principles of bhakti-yoga. It's an authentic and lasting alternative that brings one closer to knowing the true self; giving the harmony, balance, satisfaction and joy that thus contributes to the expansion of peace on the world.

→See also description of network "**ISKCON**"

Zonca Ecovillaggio

Frazione Zonca VB | 28843 Montescheno — ☎+49-(0)170-778 53 61
amicidizonca@gmail.com — zonca.info

1100 — (near Domodossola) — 50 % — ♀10 ♂10 6 20

- g, e, i — none — none
- democratic vote (based on the common interest) — private ownership
- no
- gardening, goats, chickens, building
- ecological, self-managed, anarchistic, feminist, non-violent, organic diet, environmental protection, permaculture, organic agriculture, organic gardening, animal friendly husbandry, forestry, alternative energy, ecological construction, construction, humour

☺ ()

Zonca is not a community but a neighbourhood emulating the organic social structure of life in the past. Homes are in private ownership with the intention to create collectively owned spaces for work and recreation.

→See also description of network "**GEN Europe**"

☺ love nature and love people

Lithuania see page 426

Luxembourg one planned community, see page 471f

Netherlands

Find 29 existing communities from the Netherlands in the following section, plus 7 planned ones (pages 472ff), 2 Camphill communities (page 514) and 15 that were published in the last edition of eurotopia and of which 10 probably still exist (page 501).
There is one community network with three names: It's known as "GEN Netherlands", but also as "Ecodorpen-netwerk" and as "Vereniging Ecodorp Nederland" (page 527). Another Dutch community network is "Vereniging Gemeenschappelijk Wonen" (page 540), and it publishes a huge list of "Woongemeenschappen" on http://gemeenschappelijkwonen.nl/adressen-woongemeenschappen, which we have NOT contacted individually. It hasn't really become clear to us how the terms "C.W." (Centraal Wonen) and "Woongroep" - such groups make up the most part of that list - relate to our own definitions of community. We are looking forward to your feedback about this! Gelukkig zoeken in gemeenschappen.

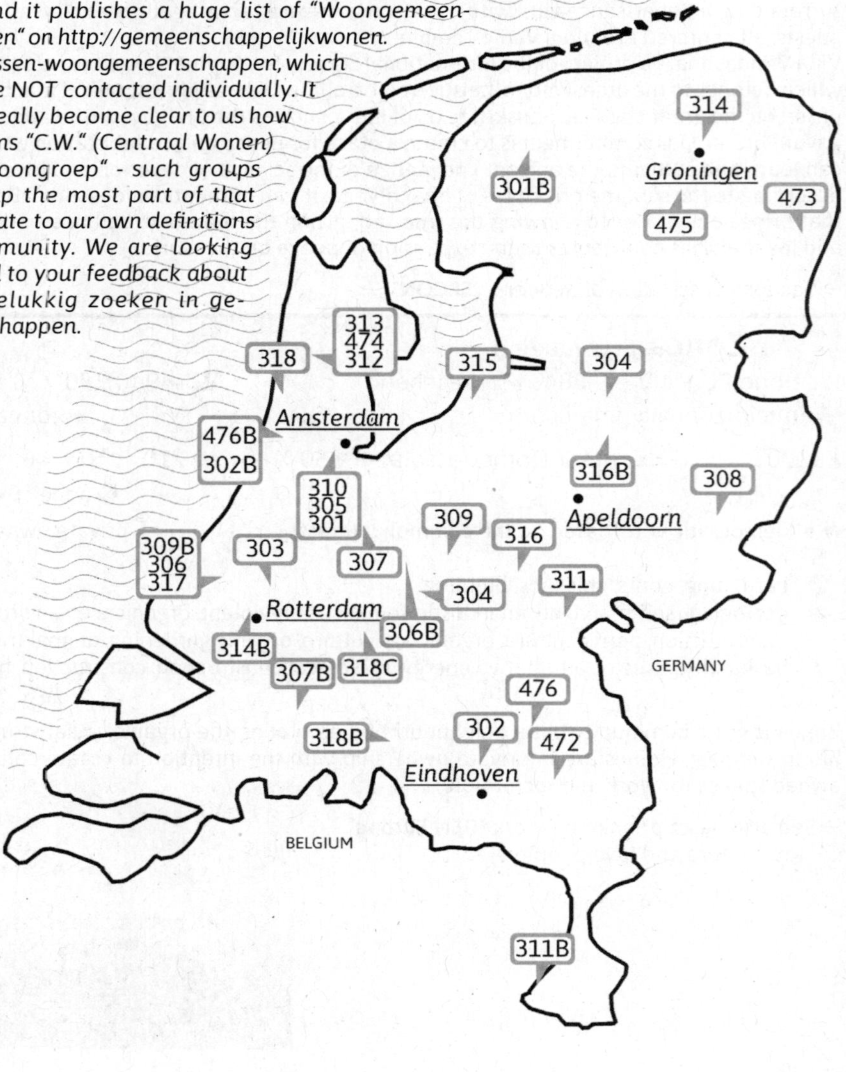

Amsterdam Catholic Worker Jeannette Noël House

Dantestraat 202 | 1102 ZR Amsterdam ☎+31-(0)20-699 89 96
noelhuis@antenna.nl www.antenna.nl/noelhuis

1988 1 % ♀10 ♂10 5 25

- Dutch, e
- all volunteers
- core group (sort of by consensus)
- we rely 99% on donations
- rented
- christianarchy & pacifism
- offering hospitality to undocumented migrants, holding vigils at the deportation prison & other protests, gardening
- ecological, self-managed, anarchistic, left-wing political understanding, feminist, non-violent, Christian, ecumenical, vegetarian diet, peace-work, environmental protection, organic gardening, humour

☺

In Matthew 25, Jesus says "what you do to the least of your brothers and sisters, you do to me". Nowadays, those people who are recognized as illegal by the society, are certainly among the least. So we try to live together and offer hospitality to 11-18 undocumented migrants and their children. We also regularly hold vigils and protests for refugee rights, against (nuclear) weapons and climate change. We have use of a piece of land where we grow some fruit trees and berry bushes and enjoy campfires with the children. In the vegetable garden at the house we grow vegetables. Short visiting guests are welcome by prior agreement and they dont have to work or pay.

→See also description of network "**European Catholic Worker**"

- KNR Kleine Gemeenschappen
- ☺ they have to speak Dutch and like/tolerate Jesus & pacifism & Christianarchy

C.W. Circa

Schoolstraat 9F | 8911 BH Leeuwarden ☎+31-(0)6-204 144 63
a.arnold@chello.nl

1988 50-80 m² ♀6 ♂2 1 9

- Dutch, e
- the member meeting (democratic)
- rented
- no
- very different
- self-managed, humour

☺

We have 7 rented apartments in one house or block. In that block we also share one big accommodation (80 m²) where we have common activities and where the children can play. Also we share the garden, a guest room and launderette. As habitants we form a society that rents the building from the owner and we have rules about using the common property. Our philosophy is very practical: Each of us lives in his or her own apartment, we share some parts of the building and we do have some common activities. Members can form relations that mean more than only those of neighbors, but this is not a necessary condition. So it is a very workable combination of community and individuality. Most

Netherlands

people are single parents of one or more children. Christmas, New Year's Eve and other special days we celebrate together and invite friends from the outside.
The plan of the building is laid out in such a way that through the day, people meet each other naturally. For example, we often meet each other at the washing machine or in the garden. The children enjoy this way of living very much. They are a little bit like brothers and sisters and they do many things together.

→See also description of network **"LVCW"**
☺ it depends

C.W. Lismortel

Tourslaan 22 | 5627 KX Eindhoven
info@cwlismortel.nl www.cwlismortel.nl

1983 | 3 ha | ♀43 ♂32 35 110

Dutch, e, g | 5 or 7
everyone in common (majority vote/consensus) | individual financing
rented/leased

☺

We live in 60 houses grouped in 10 clusters (6 family houses with a common hall, living room, kitchen, washing machine room and a big garden around each cluster). Each cluster has some flexibility in the way it is built, so that houses can be split or otherwise rearranged: there is a good possibility for people living on their own, with or without children. The main goal of Centraal Wonen is to maintain good contact with the people who live around you and some responsibility for each other and our project. The living groups (the 'clusters') choose their new members themselves. Our buildings obey energy-saving principles: the houses are more insulated than 'normal', earth warmth is used to heat the houses.
15 people share three cars in an organisation, several other cars are also shared. Some of the other activit ies in the project are: a paint club, a 'walking club', our own computer network. Centraal Wonen Lismortel has a community building for social (a weekly bar-evening and weekly coffee hours, parties) and other activities (paint and yoga courses, etc).
(Description from 2009, the other data is was updated for this edition)

→See also description of network
"GEN Nederland/Ecodorpennetwerk/Vereniging Ecodorp Nederland"
☺ should support our form of community life

C.W. Romolenpolder

Venkelstraat 25 | 2034 MT Haarlem +31-(0)6 238 466 85
bestuur@cwrom.nl www.cwrom.nl

1992 | ca. 3200 m² | ♀39 ♂16 11 66

Dutch, e, g
the residents association (usually majority vote)
monthly membership fee, used for common expenses | rent
cohousing
self-managed, non-violent, organic gardening, humour, non-smoking

☺

We are a co-housing project with 46 flats that have 2 to 5 rooms each. Our members are singles, families and couples of all ages: the youngest is 1 and the oldest is 88 years old. We share common spaces, laundry machines and bicycle storage.
Although we are in a city environment, we have a very nice large shared garden. In the summer life happens outside as much as possible. In the garden we grow small fruits and flowers and we have many trees. Our garden is friendly for insects and birds, although our cats are not always. We have a guest room and are always happy to welcome guest from other projects.

→See also description of network **"LVCW"**
☺ we look for people who like to share

Centraal Wonen Zevenkamp Rotterdam

August Vermeijlenpad 40 | 3069 WT Rotterdam ☎+31-(0)6-228 197 16
pr@cw7.nl www.cw7.nl

1982 0 ♀37 ♂23 15 75

g, e, f, Dutch 1
all residents (majority) membership fees of the residents

rented
cultural creatives
living, sustainability, neighbourly help, quality of life
self-managed, non-violent, meditation, yoga, environmental protection, permaculture, alternative energy, humour, creative activities, chores, eating together, insect-friendly vegetable garden, gardening

☺

Living in CW7: CW7 is the most beautiful village in Rotterdam. If you walk through the lush gardens, you can hear chickens clucking, neighbourly chats and the wind in the tall trees. People help each other to set up solar collectors, give each other advice on water consumption or energy and maintain the insect-friendly kitchen garden together. At the first sign of sunshine we get together for communal outdoor picnics with tapas and barbecue equipment. About 75 residents between the ages of 0 and 80 have set up a beautiful and safe environment at the city's edge. There are 60 houses with eight communal rooms, three indoor gardens, a café, guest rooms and an orchard. The people who live here know each other and their neighbourhood. They are people who like to help each other out if necessary. CW7 is a particularly brilliant home for children. We love nature and art and try to accommodate different lifestyles.

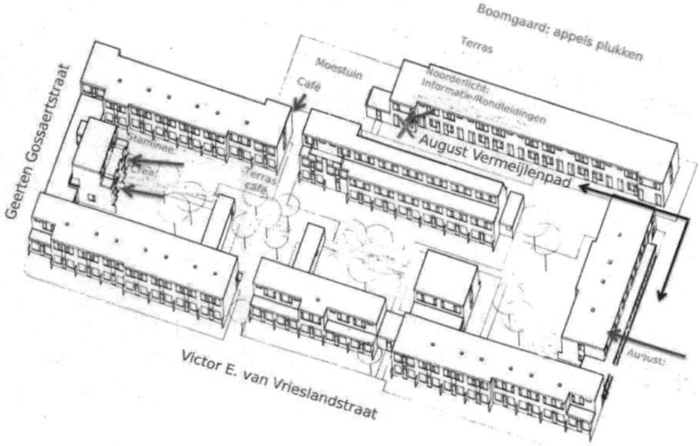

LVCW (landelijke vereniging centraal wonen; engl.: National Co-Housing Association)
☺ cultural creatives

CoHousing Group Kwarteel

Anna Blamanweg 22 | 4103 WJ Culemborg ☎+31-(0)345-513 388
j.hanhart@kwarteel.nl www.kwarteel.nl

1998 0.5 ha (near Utrecht) ♀23 ♂10 0 33

- e, Dutch
- owner gatherings, both about real estate and social matters (absolute majority after informal discussion)
- individual financing
- apartment owners are co-owners of collective provisions
- no
- senior cohousing
- ecological, self-managed, alternative energy, construction, accessible to the disabled, private woodwork shop

Het Kwarteel is an initiative project for seniors. It is based on the assumption that collective provisions like a common garden, meeting room, launderette, guest room etc. is preferred for elderly who are open for good neighborhood. There are no social obligations. The formula has worked in that sense, that up until now the community has been self-financing and -governing. See also www.kwarteel.nl.

LVGO

De Bongerd

Backermarke 55 | 8016 LJ Zwolle ☎+31-(0)38-465 21 42
mmwzdebongerd@gmail.com mmwzdebongerd.nl

1997 0.7 ha 1 % ♀54 ♂22 15 76

- Dutch, e, g
- general meeting of the members (i.e. all inhabitants), preferably by consensus - if necessary by (qualified) majority
- individual financing
- 19 of 38 homes are rented, the remaining ones are privately owned by inhabitants
- ecological way of living
- environmental protection, organic gardening, alternative energy, ecological construction, "noaberschap" (good neighborship)

☺

De Bongerd is a green, attractive, diverse project, as ecologically built as possible. A neighborhood where people feel at home. With a place for different ages, forms of families, ideas and opinions. Different homes for different needs, with the chance to stay in the neighborhood when your housing needs change. The inhabitants maintain the neighborhood mainly by themselves. Living in De Bongerd does not mean being watched by your neighbors, but getting support from them when you need it. Taking up responsibility for the common green, working together, participation in one of the working groups (new: entering the energy transition) are part of our life.

→See also description of network
"GEN Nederland/Ecodorpennetwerk/Vereniging Ecodorp Nederland"
Stichting MMWZ
agree to principles and practices

De Groene Gemeenschap

Barkasstraat 7A | 1086VJ Amsterdam
info@degroenegemeenschap.org degroenegemeenschap.org

2011 10 6

g, e, f, sp, Dutch
ind. financing & collective solidary financing of community spending
collectively rented
non-spiritual, non-religious
ecological, self-managed, anarchistic, left-wing political understanding, feminist, non-violent, organic diet, vegetarian diet, vegan diet, seminar-house, permaculture, organic agriculture, organic gardening, organic farming, alternative energy, non-smoking

De Groene Gemeenschap (The Green Community) is a project on IJburg, the newest archipelago of artificial islands that have been added to Amsterdam in the last decade. It is one of the few bits of the city that is above (the present) sea-level. The area is on the edge of the city, far enough from the famous tourism and close enough to the open fields.
Space for sustainability and solidarity: the project consists of 6 individual apartments and communal spaces for 6 people that form a close community. It shares a neighbourhood living room with neighbours who host a variety of activities with the inhabitants. By working together the inhabitants and neighbours promote social values and a practical sense of community.
Impact on the immediate surroundings consists of organizing and participating in various neighbourhood networks, such as a vegetable garden, a local food coop, a library and a musicians meetup.
The future sees the group hopefully collectively buying the apartments and continueing to rent them together in a network model that takes its example in the German "Mietshäuser Syndikat" and is called Vrijcoop. Owning is the new sharing.

→See also description of network **"LVCW"**
Vrijcoop, Mietshäuser Syndikat (www.syndikat.org/en)

De Oude Nieuwelaan

Nieuwelaan 48 | 2611 RS Delft

wonenopdenieuwelaan@gmail.com www.nieuwelaan.nl

1993 ♀16 ♂14 0 30

Dutch, e

everyone (majority vote) membership fee

property of all

living together

self-managed

We are co-housing project in the center of Delft. We own six townhouses built in 1912 which are split into four groups, called: the nerds, the scum, the old people and the Kenaus. We share a beautiful garden in which we chill in summer. In winter we visit each others kitchens. Our focus is on living together, affordable and with a diverse mix of people. We do not share a common political ideology, diet, practice or religion. The Nieuwelaan is not a place to start a family, so we have a rule that members have to move out when they wish to do so.
Everybody contributes to the community by maintaining the garden or the house, organizing concerts or our yearly party, being active in a committee, or cooking on Sundays when we have a diner-table. Apart from the obligatory 4 hours of work per month and occasional meetings, everybody is free to spend as much or as little time as they want with their house mates.
The row houses were squatted in 1981. 12 years later the squatters organized, founded an association and ultimately bought the houses they lived in and started renovating them. Now we live nicely in the self-renovated houses, including solar panels on the roof, for which we pay a monthly contribution.

☺ applying when there is an empty room

Ecowijk Mandora

Robiniahout 15 | 3991 MX Houten +31-(0)6-486 774 16

ecowijkhouten@gmail.com www.ecowijkmandora.nl

2016 1 ha 10 % 30 100

Dutch, e

all community members (consensus) limited common financing

mostly properties, two rented homes

sustainability, good neighbourhood

ecological, self-managed, non-violent, organic gardening, alternative energy, new technologies, ecological construction, accessible to the disabled

Our ambitions for sustainability: 1. ecological (garden, materials for the houses), 2. financial (investment in houses conform the market, relative contribution to shared assets), 3. technical-sophisticated installations for comfort and minimal energy use, use of clean energy and 4. social (helping each other, working together in the garden, having fun during social programs etc.)

local food-coop

Emmaus Haarzuilens

Eikstraat 14 | 3455 SJ Haarzuilens ☎+31-(0)30-677 15 40
haarzuilens@emmaus-utrecht.nl www.emmaus-utrecht.nl

1966 1 ha (near Utrecht) ♀4 ♂10 14

e, Dutch all of us all of us

shared income rented

justice!

recycling, circular economy

Emmaus Haarzuilens is a living and working community that has been providing accommodation for more than 50 years to people, from homeless people to world improvers.

The residents and volunteers work together in our recycling company. Not only goods but also people get a second chance here. Through the sale of recycled goods we are financially independent and able to support small-scale projects at home and abroad.

→See also description of networks
"Emmaus, GEN Nederland/Ecodorpennetwerk/Vereniging Ecodorp Nederland"

Emmaus in Langeweg and Breda

Kloosterlaan 6 | 4772 RA Langeweg ☎+31-(0)168-324 524
info@emmaus-langeweg.nl www.emmaus-langeweg.nl

1971 (near Breda, Rotterdam) 3 % ♀4 ♂9 1 14

Dutch, e, (g, f) 12-13 members, 6 salaried workers, 80 volunteers

part of the group, called the core-group (partly consensus, partly board)

shared income foundation

humanistic

second-hand shop and support for homeless

☺

Once more! These words are symbolic for the activities of Emmaus in Langeweg. Emmaus houses people who are in trouble, especially the homeless. Emmaus gives them the opportunity to rebuild their lives. Meanwhile Emmaus gives chances to idealistic people to put their ideas into practice. Finally, Emmaus combats the economy of excessive consumption and returns value to discarded objects. Emmaus collects cast off goods and sells them. In this way, Emmaus is financially independent. Part of the income is donated to projects, both in developing countries or nearby.

The community in Langeweg houses about 10 persons. All community members work in the second-hand shop, collecting, sorting and selling goods. They are attending the housekeeping together. Community members receive food and board, insurance, pocket money, travel expenses and items from the shop. Aside from them there are over 25 volunteers and several paid employees.

Emmaus Langeweg is situated in a former monastery in a small village in the South of the Netherlands. A stay in the community gives you the chance to put your ideas into practice and discover your talents. We want to live, work and share together. We choose solidarity.

→See also description of network "**Emmaus**"
☺ enthusiasm for common work, working and living, probationary period required

Erve Veldink

Eibergseweg 42 | 7273 SR Haarlo ☎+31-(0)545-261 448
info@erveveldink.nl erveveldink.nl

1981 0.8 ha (near Zutphen) 20 % ♀3 ♂2 5

Dutch, e, g 2 2

all or the owner, depends on the issue (consensus) individual financing

property of one member

focus on consciousness, self-awareness

guesthouses, therapy, retreats, consciousness

ecological, self-managed, left-wing political understanding, non-violent, spiritual, inner growth, meditation, organic diet, vegetarian diet, seminar-house, environmental protection, organic gardening, animal friendly husbandry, alternative energy, consciousness, cranio-sacral and massage therapy, breathing, sound, Qigong

☺

Our farmhouse is inhabited by a small community, based on the principles of living in consciousness and mutual connection from the heart.
Groups and individuals are very welcome, both as guests as well as volunteer workers. A former stable house has been transformed into a modern group house. We want to offer a place and create a bedding for guests and residents, to live in attention, to relax, to recharge, to reflect, to come to oneself and in connection with each other. With that we also want to pass on something of what we receive so abundantly.
We feel that in life it is about growth to consciousness. Nature gives us peace and healing. By choosing this, taking the time and getting away from fixed patterns, there is room for awareness and health, inspiration and clear insights, and life can become simple again. Various treatments contribute to this.

☺ yes, noted on our website

Evangelical Community Elim

Bergweg 7 | 3941 RA Doorn ☎+31-(0)343-417 064
info@elim.nl www.elim.nl

1975 | 1 ha (near Utrecht) | ♀6 ♂6 1 13

- Dutch, e — 5
- everyone in common (consensus) — shared economy (shared income and capital)
- foundation
- Protestant/Evangelical/Christian
- design, missionary work, church, parish contacts
- self-managed, non-violent, Christian, accessible to the disabled, non-smoking

The Elim community provides for their own livelihood through full-time and part-time jobs, and consists right now of 12 adults and 1 child. Most of the income comes from the running of a technical illustration and design bureau, study-bible project, and part-time jobs outside the community. Community members live with a common purse, from which all costs are paid: rent, cars, holidays, pocket money, etc. The evangelical Elim church emerged in 1985 from the community's circle of friends. This church has 3 meetings each week. We try, both in the community and in the church, to live with God in our everyday life. In addition, there are ongoing contacts with Christians in Third World countries – mainly India - both with Western missionaries and native believers. Elim initiated a community movement of what has grown to become 35 Dutch communities. The main focus is now our intercultural church house, where many refugees come and find Jesus. We hope that the way we lead our lives will help people discover that it is possible to live in peace and happiness through the life-giving and -transforming power of Jesus Christ, our Lord and Saviour. Come and visit us!

Christian Community Network

Geworteld wonen Warmoestuin

Warmoestuin 24 | 2286XZ Rijswijk ☎+31-(0)644 584 582
https://www.warmoestuin.nl/contact/ www.geworteldwonen.nl

2016 | 1 ha | 20 % (of fruits and vegetables) | ♀44 ♂44 38 130

- g, e, f, Dutch — 50
- every household (47 in total) has a vote (majority vote, sometimes 2/3, consensus is always prefered)
- limited common financing — municipal grounds are property of all
- accept everybody in his/her value, humanism
- maintaining the municipal grounds with vegetable garden and orchard
- ecological, left-wing political understanding, non-violent, ecumenical, permaculture, organic agriculture, organic gardening, alternative energy, new technologies, ecological construction, humour, "village-like" neighbourhood, social activities, own initiative, nature minded

☺

We are a neighbourhood in the town of Rijswijk. We have a large municipal garden, which we maintain together. There is a vegetable garden, an orchard, a jetty at the water, flower gardens, seats and playing grounds for the kids. This summer we will add a chicken pen so we also have our own eggs.

Netherlands

Everyone has his/her own house (build partly from reused materials) and life, but we also regularly celebrate or organise a social activity together. Because we are maintaining the gardens together, everyone knows and takes care of each other. We don't always agree on things, but we respect each other.

☺ being able to buy a house in the project (if one is for sale)

GWL-terrein koepelvereniging GWL terrein

Waterkersweg 316 | 1051 PK Amsterdam ☎+31-(0)614 155 460
info@gwl-terrein.nl www.gwl-terrein.nl

1996 6 ha 0.01 % 1400

Dutch, e

koepelvereniging GWL-terrein (umbrella organisation for house owners and residents)(consensus and/or by majority vote)

individual financing | owned and rented apartments

ecological, organic gardening, ecological construction, car free

The GWL site is not a community, but a small eco-district near the centre of Amsterdam on the former municipal water supply site and has been inhabited since 1997/1998. There are 600 houses, 50% rented apartments and 50% private homes. Some of the buildings have a green roof, but our current goal is to green all the roofs and install solar collectors.
The inner courtyard is about 2/3 unpaved, so that the rainwater is directly absorbed into the ground. The water from the paved parts is not led into the sewer, but into the canal, which has an overflow to the Haarlemmer canal. There are 85 vegetable gardens, which are maintained by the residents. The hedges offer nesting places for birds and small animals. There are 60 fruit trees, mainly apple and pear. The area is car-free, as can be seen from the number of bicycles parked on the area and outside.
On the site there is an umbrella association of the residents, which has set itself the goal of preserving the green character of the site. A neighbourhood manager guards the public areas of the rented apartments, provides the residents with information and preserves the car-free charm of the complex. Together with the residents, he also maintains the green areas on the site.

→See also description of network
"GEN Nederland/Ecodorpennetwerk/Vereniging Ecodorp Nederland"

Iewan

Laauwikstraat 106 | 6663 CM Nijmegen ☎+31-(0)6-425 295 39
info@iewan.nl iewan.nl

2015 2220 m² <1 % ♀28 ♂15 8 51

e, g, Dutch 4

everyone (consensus)

indiv. financing with limited common financing for shared spaces and self-management

rented

no

educational

ecological, self-managed, permaculture, alternative energy, new technologies, ecological construction, accessible to the disabled, food cooperative, event venue, coworking space, fossil free

☺

In Iewan three core values of living are shared:

- sustainable and ecological
- collective and social
- open and educational

Iewan is a social housing project initiated by the residents themselves, and it attains to make ecological living possible for people with a low income. Typical for Iewan are the communal areas and shared facilities, as sharing room and resources is considered to be a good strategy to lower the individual ecological footprint of the residents. Through educational activities and advice Iewan tries to be a living example of future proof living, which is both socially and ecologically sustainable.

Woningbouwvereniging Gelderland,
neighbourhood association for self-management of public green areas

☺ yearly income of a household below 36.798 EUR, as it is a social housing project. Tell us about the way you think about living communal and sustainable, so we know you share our core values.

Landbouwbelang Culturele Vrijplaats

Biesenwal 3 | 6211 AD Maastricht ☎+31-(0)628 877 644
info@landbouwbelang.org http://www.landbouwbelang.org

Landbouwbelang houses and attracts a diverse community of inhabitants, artists and other users, volunteers and visitors. This community is what makes the Landbouwbelang an innovative, ever-changing hotspot, operating in the heart of Maastricht. We are constantly adapting to the social and cultural climate in the world around us in order to connect people. Landbouwbelang has joined, collaborated, talked, reflected and worked together with all kind of organisations, artist, politicians, bands, squats, free zones, galleries and more.

The building is officially owned by the municipality of Maastricht, but since 2002 it is being used by artists, musicians, programmers, designers and students who made it their home.
Behind the big yellow doors, the cultural freezone Landbouwbelang always runs full speed thanks to the inhabitants, engaged volunteers and donations from its visitors. With their activities they rebel against the "throwaway society" and stimulate awareness in eating and drinking, consumer behavior and spending leisure time. From beehives on the roof to a freeshop in the basement, every room of the Landbouwbelang has given birth to a fascinating initiative.
Since 2002 we provide the city of Maastricht with countless concerts, live DJ sets, performances, visuals, biological drinks and more. One of the biggest exposition spaces in Maastricht. A living and working community.

MW2 project De Vuurplaats

Mens- en Milieuvriendelijk Wonen and Werken
Keizerfazant 46 | 1704 WL Heerhugowaard ☎+31-(0)72-574 46 07
info@de-vuurplaats.nl www.de-vuurplaats.nl

1994 3000 m² (near Alkmaar, Amsterdam) 10 % ♀13 ♂5 7 25

e, Dutch 2
all who are concerned (consensus (sociocratic)) limited common financing
rented
environmental
cohousing
ecological, self-managed, non-violent, spiritual, inner growth, meditation, organic diet, vegetarian diet, permaculture, organic agriculture, organic gardening, alternative energy, ecological construction, accessible to the disabled, no cell phones/wireless, non-smoking, GPS/UMTS shielding

☺

MW2 de Vuurplaats was founded in 2004 by a group of people to share more than just being neighbours. MW2 stands for "Mens- en Milieuvriendelijk Wonen and Werken" which means that we want to live and work (there are some working facilities in the building) with mindful care for and with each other (people) and the environment.
The founding group, of whom four still are residents, made sure that the place was built in an ecological way and with an architecture that provides many possibilities to meet each other and organise common activities.
Besides many common facilities (living room/kitchen, multifunctional room, workroom, laundry room, etc) there is a big shared garden. Since everyone has her/his own house, there is also much privacy. Decisions are made based on sociocratic mechanisms.
We hold introductory tours 2-3 times a year (e.g. January, May, September). See our website (in Dutch).

→See also description of network "**LVCW**"

☆ **Osho Humaniversity** School for Masters

Dr. Wiardi Beckmanlaan 4 | 1931 BW Egmond aan Zee ☎ +31-(0)72-506 41 14
info@humaniversity.nl www.humaniversity.com

1978 ca. 2 ha (near Alkmaar) ♀25 ♂20 45

e everybody

our directors (together with final decision power by our president)

we are a non profit organization owned

working with people to create a better world; Osho, therapy, addiction therapy, therapist trainings, spirituality, meditation, natural healing, personal development and growth

therapy, therapist training, natural healing, administration, garden, household

self-managed, non-violent, hierarchical, spiritual, inner growth, meditation, vegetarian diet, adult education, pedagogical, peace-work, one-world development, organic gardening, construction, humour, humaniversity therapy and therapy trainings, martial arts, wellness and natural healing, music studio

☺ ( )

If you have never done the "Tourist Program", you have missed a great opportunity to find out what your human potential is and how to become more deeply fulfilled. The "Tourist Program" is probably the most powerful and fastest process of personal transformation.
The Osho Humaniversity is an international community where personal growth, treatment of people in crisis and therapist trainings are happening simultaneously. Its style is highly innovative and its therapists are passionate rather than detached. Psychology students, spiritual seekers, housewives, unemployed labourers, businessmen, addicts, teachers – are all coming to be in the same basic program, learning how to relate to others in a healthy way and how to be themselves. Osho Humaniversity is a social awareness school where you can get an emotional education. It sees the main causes of human misery in isolation and in the suppression of one's true feelings. Authentic love and friendship is therefore considered to be the highest goal a human being can achieve.
The techniques of all major psychological schools and of the Zen and Sufi masters are used to make this the most exciting learning experience in your life. You will emerge from it emotionally purified, powerful, aware and with a new purpose in life.
We also offer a wide range of weekend workshops on many different topics of personal growth and development. You are welcome to visit us.

☺ participate in our basic therapeutic program, the Tourist Program

Oude RKZ

Emmastraat 15 | 9722 EW Groningen
+31-(0)6-239 452 35
info@orkz.net
orkz.net

1979 — ca. 10 — 250

- Dutch, e, g, f, Polish — approx. 150
- different groups, and on important issues we all come together in a general assembly (majority vote)
- individual financing
- theatre, cinema, bar, restaurant, art studios, garden, and several other groups related to the building
- ecological, self-managed, left-wing political understanding, organic gardening, alternative energy, new technologies, construction, humour, accessible to the disabled, arts, music

☺

The Oude Rooms Katholieke Ziekenhuis (ORKZ) was originally a hospital. In 1979, when a new hospital was built, the ORKZ should have been knocked down in order to build luxury apartments. However, different groups united and squatted the building. It was the Netherlands biggest squat at that time until it was legalised in 1985 (so no longer considered a squat by everyone). We are a community of roughly 250 inhabitants, and the group consists of many different types of people. Like the old hospital, our community is organised in wards (or corridors). Each corridor houses a sub-community with their own culture and ways of living together and dividing tasks. As an individual you pay rent to the organisation that you're part of as a renter. You have a vote in the general assemblies to vote where the money goes. As a member of the community you're supposed to do two hours of volunteering per week. This ranges from helping in the kitchen, to working in maintenance, to gardening or taking care of administrative issues for example. We have our own traditions like counting down to New Year on the roof, the New Year's Bingo, our anniversary on the 3rd of September with games, food and party and the city's best Halloween party.

☆ Poortgebouw Vereniging Poortgebouw

Stieltjesstraat 38 | 3071 JX Rotterdam
office.poortgebouw@gmail.com
www.poortgebouw.org

1980 — 1 — 28

- e, Dutch
- the whole group through general house meeting, the members of the board for formalities (consensus or vote)
- individual financing — rented
- cultural platform, performative arts, wood workshop, dark room
- self-managed, anarchistic, left-wing political understanding, feminist, non-violent, chaotic, vegan diet, self-sufficiency, humour, arts, circus, international, political, debate, venue, cafe, bar, events, bands

☺ (✋✋)

In the Poortgebouw ("gate house") resides a community association, where people live together and work together. This is a key element of the group. This group consists of people

of various ages, from different fields of expertise, studies and nationalities.
What connects them is the will to live together as a collective within a society where individualism seems to increase. The national monument (and gatehouse that is still in service) on the Kop van Zuid gives a residence and freedom to have a personal working and living space as well as sharing communal spaces and commodities. For over 35 years the association (vereniging) Poortgebouw has been able to provide a platform to non-commercial art, music, and debates.
In addition, there are social activities such as an eetcafé (vegan meals) and a Give Away Shop. Ever since the building has been squatted in 1980, and after the signing of the rent contract four years later, the association has been able to remain autonomous and realise its wishes...

☺ active members wanted
(active in workgroups and living community)

Sidhadorp Harmonisch Wonen

Rivierenlaan 230 | 8226 LH Lelystad ☎ +31-(0)320-218 077
info@harmonischwonen.nl harmonischwonen.nl, sidhadorp.nl

1985 150 550

- Dutch, g, e
- Transcendental Meditation in a group programme
- Transcendental Meditation
- ecological, spiritual, meditation, vegetarian diet, adult education, peace-work, alternative energy, ecological construction, inner growth

☺

In 1979 Maharishi called on the Dutch sidhas to create an ideal village, so that the effect of the daily group practice of the TM-Sidhi programme could be demonstrated in this Land of Wholeness. The Ideal Village was started in 1985 in an area which offers the most space and the cleanest environment: the new province, Flevoland (reclaimed land). On the outskirts of Lelystad, a city of 60,000 inhabitants, the Dutch sidhas were allowed to build their own suburb, according to their own principles. The Ideal Village has grown steadily and it now includes 300 homes of different sizes, a beautiful dome, a junior school, a vegetarian restaurant, business- and office- accommodation, a Maharishi Ayurveda clinic and much more.
There are about 400 adults and 150 children living in the village. Since Maharishi gave the Dutch sidha village an European status, we have a steady stream of sidha immigrants coming. The Ideal Village has its own property developer and housing agency, which rents out the various houses and business accommodations. All the houses are built using environmentally-friendly materials and are well insulated. There are two-roomed apartments, two- and three- bedroomed houses (all for rent) and also privately owned houses.

☺ practice of Transcendental Meditation

Studentenvrijstaat Droevendaal

Droevendaalsesteeg 95 | 6708PS Wageningen   +31-(0)317-417 961
droef.barrak95@gmail.com   www.droevendaal.nl

(near Arnheim, Nijmegen) 10 %   ♀110 ♂95 200

- expert groups, all of us (advice, consensus, majority vote)
- indiv. financing with limited common financing (e.g. vehicles or food)
- Student housing management company
- respect nature, vegetarian food (not a dominant rule), sharing goods and showing solidarity between us
- gardening, cooking, christmas-spring ceremonies, parties, taking care of our chickens, cats and dogs (literally and metaphorically), studying
- arts, music, dance, parties, cooking and exchange of recipes, campfires, protests, cycling

We are a student housing community that lays focus on freedom in thinking and creativity, there are no rules for this in Droevendaal. People live in small houses of six rooms and they share a kitchen-living room and bathrooms. Usually they cook and eat together, share the tasks of different activities such as gardening, management of the green space, parties and discussions around different topics. Issues like sustainability, human rights, activism, green way of living and alternative economy have always been and are still hot topics in Dorevendaal.

Structure and size (as well as the changing number of members) make Droev a really dynamic society. Our community doesn't have a specific goal, otherwise we could say that we try to set the boundaries of anarchy. We live in an environment giving us freedom and in which we learn to live responsibly with it. Droev has the goal to make its members happy and creative in order to be passionate in their lives. There is no age limit, so everyone who is studying or working in the University Wageningen is able to live in Droevendaal. The town of Wageningen is characterized by its multicultural student society and we in Drovendaal are appreciating and celebrating this diversity.
We are a student community; whatever we do, we do it by ourselves. Only services such as water, electricity, and waste management are provided to us by Idealis (student housing company) and local authorities.

☺ bachelor/master/PhD-student, researcher or university employee; be active and show interest for environmental issues, an alternative way of living and study close to nature

Vereniging Aardehuis Oost Nederland

Earthships and strawbale houses in Olst, The Netherlands
Rietgors 26 | 8121JZ Olst
email@aardehuis.nl   www.aardehuis.nl

2014   (near Deventer) 5 %   ♀22 ♂21 28 71

e, Dutch   several from home

- collectively (collective buildings and grounds) and privately (houses)
- ecologigal
- permaculture, tours for groups on sustainable building.
- self-managed, non-violent, no gas, own water source, solar panels, reed bed filter for sewage, composttoilets

It is our intention to build, work, live in our homes and in our lives in harmony with nature, in constructive relationship with one another and as a source of inspiration to the world around us. This goal is to be realized through the construction of an ecological housing project of self-sufficient Earth-houses whereby all aspects of sustainability are taken into consideration and balanced with each forming a harmonious whole.

→See also description of network
"GEN Nederland/Ecodorpennetwerk/Vereniging Ecodorp Nederland"

Woongroep De Grote Leeuw

Maerten Trompstraat 2 - H | 2628 RD Delft
stofberg@boomdelft.nl

1981 house: 262 m², garden: 135 m² ♀2 ♂3 0 5

- Dutch
- all who are concerned (majority and sometimes consensus)
- individual financing / limited common financing for food
- property of association, all of us are members
- no
- we live and eat together, we maintain the house and garden
- self-managed, alternative energy, non-smoking, media: internet, television

We live with 5 adults (age about 30 - 65 years) in a large new house nearby the centre of Delft. Our house has a communal live-in kitchen, 2 roof gardens, a bathroom, a visitors room and 5 large rooms, one for each of us. We eat often together in the evening and friends are welcome. Sometimes we do something together such as going to the cinema.
At this moment we do (as work) something with ecological architecture, amusement parks, computers, plastic in oceans and graphic design. The house has 10 m² solar panels for electricity, 5 m² solar collector for hot water and a shower drain water heat recovery unit.

Netherlands

Woonvereniging Heemshof

Oosterweg 2 C | 1968 KN Heemskerk
heemshof@gmail.com heemshof.nl

1974 4.5 ha 10 % 3 14
e, sp, f, Dutch 14
all who are concerned (in a meeting) limited common financing
ownership
none

Heemshof is a former dairy farm. In the 1970s the farm closed down and the farm building was renovated to house six apartment-like homes and shared facilities such as a communal kitchen and spare rooms. Since 2011 the pastures are in transition to become an ecological reserve. Old and young live together here now. The responsibility of maintaining the building and its surroundings which include an orchard, a vegetable garden and the ecological reserve is shared.

→See also description of network "**LVCW**"

Woonvereniging S32

Stationsstraat 32 | 5038ED Tilburg +31-(0)651 608 543
info@s32.nl s32.nl

1986 ♀2 ♂2 1 5
Dutch, g, e
group (consensus) limited common financing
property of all
self-managed, shared housing

A big house in the center of Tilburg with a beautiful garden, where we live with a small number of people (5-6). The main reason for existence of the project is to provide housing for the members and to keep the building as a residential building for people who want to share a part of the household.

Zilt

Sint Hubertuslaan 25 | 4205 ZA Gorinchem +31-(0)183-640 50
info@leefgemeenschapzilt.nl www.leefgemeenschapzilt.nl

1988 ♀14 ♂9 7 23
e, Dutch all of us have jobs outside the community
only volunteer tasks for maintaining the property
a core team of eight adults (always unanimously) individual financing
rented
Christian community
we offer long term stay (up to one year) to people with psychosocial problems
self-managed, Christian

Zilt is a Christian community. The core team consists of three families and two singles. The community offers five guest rooms for rent to people with minor psychosocial problems. Usually guests stay for a few months up to a year for a time-out or a new start.

Norway

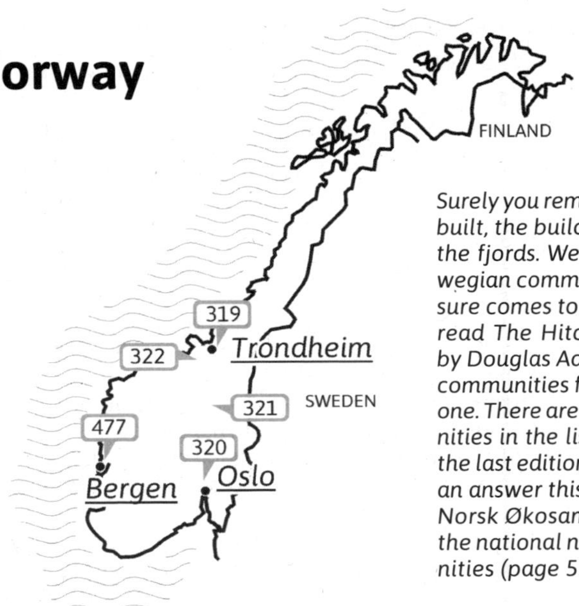

Surely you remember that after the Earth was built, the builders won a price for designing the fjords. We are not certain that the Norwegian communities know about that, but it sure comes to mind to anyone who has ever read The Hitchhikers Guide to the Galaxy by Douglas Adams. We present four existing communities from Norway and one planned one. There are 6 additional Camphill communities in the list on page 513f and one from the last edition of which we haven't received an answer this time (page 502).
Norsk Økosamfunnsforening/GEN Norge is the national network of intentional communities (page 534).

Norway

Camphill Rotvoll

Rotvoll Allé 1 | 7053 Ranheim +47-73-826 850
velkommen@camphill-rotvoll.no rotvoll.camphill.no/hjem/

1989 (near Trondheim) 45 % ♀8 ♂7 5 ca. 35

Norwegian, g,e, Swedish, Suomi, Danish, Dutch ca. 35 ca. 35
council group and steering committee (consensus) income sharing
partly rented, partly property of Camphill foundation
anthroposophy
Workshops (juice production, bakery, weaving mill, canteen, gardening) and community houses (housekeeping)
ecological, non-violent, anthroposophic, organic diet, mentally/psychologically handicapped persons, physically handicapped persons, organic agriculture, organic gardening, organic farming, accessible to the disabled, social therapy, culture

Camphill Rotvoll is the smallest and perhaps most "untypical" Camphill community in Northern/Western Europe. We are not a village community in the traditional sense, but are centrally located in a housing estate on the outskirts of Trondheim. Our community and agriculture are comparatively small, but the workshops are very productive, especially the juice factory, where various juices, jams, dried fruits and applesauce are produced and delivered to many restaurants in the city. In the three house communities, the employees live at different stages of their lives, alone, in pairs or with their families, annual trainees, people in need of care and visitors, organise everyday life and celebrations together and support each other in a practical and human way.

→See also description of network "**Camphill**"

Camphill Solborg

Solborgveien 21 | 3520 Jevnaker ☎+47-(0)321 324 80
solborg@camphill.no solborg.camphill.no/en/home

1975 | 75 ha (near Hønefoss, Jevnaker, Oslo) | ♀30 ♂14 6 47

Norwegian, e, g — all adults
village council and leader group (consensus) — varies

foundation
anthroposophy
bio-dynamic farming and gardening, forest, baking, weavery, housekeeping, woodworking
ecological, left-wing political understanding, non-violent, spiritual, Christian, anthroposophic, inner growth, organic diet, adult education, pedagogical, free school, mentally/psychologically handicapped persons, physically handicapped persons, environmental protection, self-sufficiency, permaculture, organic agriculture, organic gardening, organic farming, animal friendly husbandry, forestry, alternative energy, accessible to the disabled

☺

Solborg Camphill Village in Norway has a bio-dynamic farm, extensive vegetable gardens, a bakery, a weavery, and a large forest for timber and firewood.
Most of us live in large extended families, co-workers (long term people with their families, and young temporary volunteers) and villagers (mentally handicapped or otherwise in need of help) are sharing our lives, our meals, our living rooms and bathrooms. There may be as many as fifteen people gathered around a dining table every day. Each house has its own budget, and is run more or less autonomously by a couple of responsible co-workers. In the morning and the afternoon everyone goes to work, in a variety of workplaces. Each person contributes to the running of the village according to their capability. Within this sphere no money changes hands, and work is seen to be something that is freely given within the fellowship.
Solborg is in Southern Norway, an hour from Oslo, on the edge of an extensive wilderness area. Within the village we have a Steiner school with over 100 pupils.
We are open for volunteers and long term co-workers who would be interested in sharing their lives with us.

→See also description of network "**Camphill**"
☺ enjoying work, healthy, happy

Gaia Eco Village Okogrend i Alvdal

Folldalsveien 6010 | 2560 Alvdal ☎+47-919 021 50
post@gaiasenter.no www.gaiasenter.no

2012 1000 m² common house + more houses (near Oslo, Trondheim)
♀6 ♂5 1 12

- e, sp, Norwegian, Swedish, Icelandic ♲ 15, but only 1 (the cook) is employed yet
- all the residents, our friends and our shareholders
- the boards in each company, and the annual Great Shareholder Meeting
- individual financing — we own the property together
- green and spiritual values; building a new future
- designing healthy homes and lives, grow and cook healthy and tasty food, creating local businesses
- ecological, self-managed, non-violent, spiritual, ecumenical, esoteric/New Age, anthroposophic, inner growth, meditation, yoga, adult education, pedagogical, seminar-house, peace-work, one-world development, animal protection, environmental protection, self-sufficiency, permaculture, organic agriculture, organic gardening, organic farming, animal friendly husbandry, forestry, alternative energy, new technologies, ecological construction, construction, humour, accessible to the disabled, healthy living, healthy homes, spiritual travelling

☺

Gaiasenter: Our mission is to reach for a good life, happy days, working together, growing and making healthy and tasty food. Creating the lives we want at the same time that we create jobs and income to the community using the nature, the forest, the mountains and the river nearby.

Gaia Homes: Our mission is to design and build healthy and sustainable homes and live together in respect and understanding with the nature: A good health for the residents and a good health for the planet.

Gaia Travel: Our mission is to take people to the places where their soul calls for them. We arrange spiritual journeys to sacred energy places all over the world to expand our horizon and understanding.

→See also description of network **"GEN Europe"**
☺ financial, to be a shareholder

Storbuan Oekosamfunn

Fjeldheimveien 14 | 7332 Løkken Verk ☎+47-(0)911 746 51
oko-by@online.no storbuanokosamfunn.com

2016 20 ha, near Orkanger (39 km), Trondheim (73 km) 10 %
♀4 ♂4 2 10

Norwegian, Danish, e
plenum (consensus) individual financing
rented
so far ecological farming, permaculture, sustainability
ecological farming; providing accommodation for retreats, groups and privat tourism; social work with children.
ecological, non-violent, spiritual, inner growth, self-sufficiency, permaculture, organic agriculture, organic gardening, organic farming, animal friendly husbandry, social work, arts

☺

We are located at an old house with several houses/apartments and a bigger common house for renting. We have a barn and stables with sheep, cows and chickens, fields for grass and a small vegetable garden.
We want to build a community of sustainability and a common vision, based on values alternating to the consumer society, open to the surrounding society, particularly with other similar groups of ecological, permacultural and spiritual worldview.
We are in the process of developing the ecovillage in terms of decision-making and organizing the farm to be community-based. We seek including-processes as a base for making decisions that respect both the individual and the community.

☺ ecological living, economical self-sufficiency, participate in the community

Poland see pages 427f

Portugal

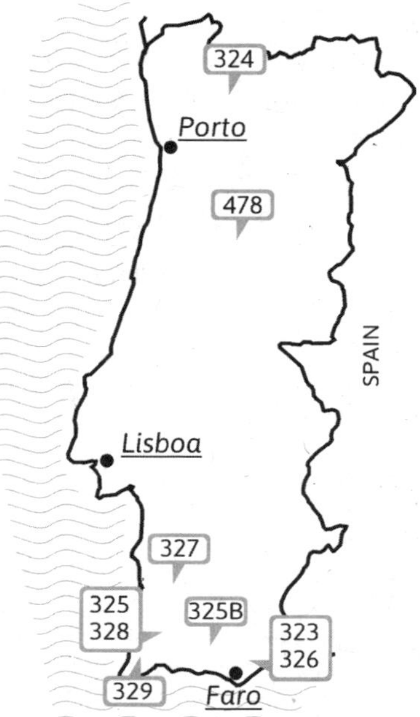

We present 9 existing intentional communities in Portugal, plus one from the last edition on page 502.

ARTLIFE Algarve Quinta Artlife

Quinta Verde / 489 V | 8700-124 Moncarapacho/Olhao ☎+351-917 248 948
✉quintaartlife@gmail.com

1996 | 0.5 ha (near Faro) | 5 % | ♀5 ♂3 1 | 9

g, e, f, Portuguese | everybody
everybody (consensus) | common supply, common projects

rented

household, garden, care of the elderly, projects, computer work

non-violent, sexuality, adult education, pedagogical, humour, art, art therapy, social research

We are a small community with a long history. We lived at Friedrichshof, Austria in the community of Otto Muehl (from 1974 to 1990). We are now seven adults who are over 50 years old and there are still two of our children living with us (13 and 19 years). We organize our household together (cooking, shopping, trips, cleaning, gardening etc...) and eat together every evening.
We meet regularly to discuss topics and projects, trying to decide by consensus and strive for open and humorous communication. We seek a balance between individual, family and social needs, art is important to us.
We do: Painting, music, photography, films, dance, literature, performances - PantomimiC.
Further topics: Communal research, alternative forms of life.
We are not looking for new members, but visitors are welcome (working participation and/or financial contribution by arrangement)!

☺ ensues upon personal encounter

Ecoaldea Vegetariana Espiral

Associaçao Cultural e Ecologista de Portugal
Quinta da Corga, lugar de Vilela de Riodouro | 4960-434 Cabeceiras de Basto
ecoaldeavegetariana@gmail.com
www.ecoaldeavegetariana.wordpress.com

2012 12 ha ♀6 4 12

g, e, f, sp., Catalan, Portuguese all

assembly of all who have definitely been accepted (consensus in the assembly)

individual and shared economy, all common costs are financed from the revenues of the ecovillage

spirituality (no religion), vegan nutrition, health, self-sufficiency, communitary life and social themes, free education, ecological practise, no alcohol, no drugs, electrical devices and mobile phones only in sleeping rooms

gardening; building paths, walls and infrastructure

ecological, self-managed, left-wing political understanding, feminist, non-violent, spiritual, inner growth, meditation, yoga, organic diet, vegan diet, pedagogical, free school, peace-work, one-world development, animal protection, environmental protection, self-sufficiency, permaculture, organic agriculture, organic gardening, alternative energy, ecological construction, construction, humour, no cell phones/wireless, non-smoking, vegan nutrition, no WIFI

☺ ()

Ecovillage with free education for vegetarian persons.
The space where the tribe is settled is in the North of Portugal. There are 12 ha of land, 4 half-refurbished buildings, one large ruin and 2 small ruins of old mills. The property has a beautiful crystalline river and pools for swimming. There are many trees, some very large and also magical space for serenity, meditation and silence. The land is in a sloping mountainous area with flat terraces ideal for kitchen gardens. It is on the bottom of some big mountains, protecting the ecovillage from the northern winds. It is 10 minutes walk from the nearest village. It's also only 20 minutes drive from a larger town (Cabeceiras de Basto).
Everyone lives in a private space (home) and shares community spaces. All activities are based on the principles and designs of permaculture. Community gardens are cultivated. The goal is to be increasingly self-sufficient with our nutritional needs. Shared free education is done where children can grow from respect to individual development and their real needs.
During the whole stay or visit inside or outside the ecovillage, please: no drugs, no alcohol, no coffee, no mate, no TV. Mobile phones and electronic devices only inside the sleeping rooms, internet only outside the ecovillage. We live mostly adapt to what nature gives us. We minimize our dependence on money.

→See also description of networks **"RIE, GEN Europe"**

☺ contribution during the integration process or visits and definitive contribution for the definitive integration as member with full rights, vegan nutrition, spirituality and ecology

La Belle Verte

Aguçadeira, Sitio de Palheiros, CCI 13-J | 8550-322 Monchique
+351-282 913 056
twg@aliceadsl.fr
www.la-belle-verte.weebly.com

2011 2 ha (near Monchique) 30-40 % 1 2 3

f, g, Portuguese, e 3 3

all concerned together (consensus)

individual financing, either common or individual shared

The project values are inspired by the French movie "La Belle Verte" by Coline Serreau.

different construction activities: mainly earth buildings in different techniques, landscaping, fruit trees and vegetable garden, beekeeping, home-schooling of the children, kitchen-work and housekeeping

ecological, self-managed, non-violent, inner growth, organic diet, vegetarian diet, pedagogical, environmental protection, self-sufficiency, permaculture, organic agriculture, organic gardening, organic farming, alternative energy, ecological construction, no cell phones/wireless, non-smoking, voluntary simplicity, emotional authenticity, sensitive view on addictions, consistent ecology, practical solidarity, sustainable lifestyle

☺ ()

La Belle Verte is a beautiful place, located 500m high in the mountains Serra de Monchique of Southern Portugal, 45 minutes drive from Faro Airport in the Algarve. Our land is nestled within a gorgeous rural and secluded valley. We are a family of 2 adults (German and French) and our 3 children.

We have a natural spring water river flowing through our land on its meandering journey to the Atlantic Ocean. We grow as much of our own food as we can and have our own solar power plant for electricity and water for aqua-showers. We are seeking a self-sustainable environment that will also appeal to like-minded humans of all ages and both genders to come and join us. We offer you the chance to come here and also feel nature and the real earth with us. You can stay as long or as short as you "feel" it with us .... We are humane, welcoming and authentic. We have a simple philosophy here: Be Real ... Be You ... and Live.

→See also description of network "**WWOOF**"

Transition Town movement Monchique

☺ try-out periods, no particular financial requirements

Naturalmente – Retreat Center

Alte-Loulé | 8100 Loulé
Naturalmente.Retreat.Center@Gmail.com
Naturalmente-Retreat-Center.Blogspot.com

2012 (near Alte-Loulé) 15 %

e, Portuguese

ecological, self-managed, non-violent, spiritual, anthroposophic, meditation, yoga, tantra, organic diet, vegetarian diet, vegan diet, free school, environmental protection, self-sufficiency, permaculture, organic agriculture, organic gardening, forestry, alternative energy, ecological construction, no cell phones/wireless, non-smoking

☺ (✋✋)

Naturalmente is a Holistic Center focused on a natural way of living where we invite everyone that feels the call for a deep connection with nature and evolution, expansion of consciousness and elevation of our vibration and awakening our essence, and our full potential, evolving and becoming aware of our interconnection/union with all-that-is. Evolving in harmony with the whole universe and consciously contributing for the co-creation of a World of Love.
We receive guests and volunteers who feel in tune with our principles (the 12 Core Petals), see the conditions on our site (page - Volunteers and Visitors).

Editor's note: Please note the article about the Anastasia movement on pages 4fff.

☺ yes (please check website - Visits - Volunteers)

Quinta Alvorada do Ouro Umbria

| 8800-166 Tavira ☎+32-(0)487-209 304
✉victoria.ratledge@gmail.com
www.facebook.com/groups/alvoradadoouro/?fref=ts

2009 — 7500 m² (near Tavira, Faro) — ♀2 ♂2 0 4

e, Dutch, f, Portuguese — 4

all who are concerned (majority vote and/or consensus)

individual financing (for the beginning) — single owner

meditation, conscious living, permaculture, sustainable building, art, music & healing...

ecological, self-managed, non-violent, pedagogical, seminar-house, peace-work, one-world development, spiritual, Buddhist, inner growth, meditation, yoga, tantra, anthroposophic, organic diet, vegetarian diet, environmental protection, self-sufficiency, permaculture, organic gardening, alternative energy, new technologies, ecological construction, adult training, vegan raw food, lots of love and humor

☺ ✋✋

Quinta Alvorada de Ouro offers a space in pristine nature, where city people (alongside others) can mentally and physically relax and detox – where everyone can enter a realm of supported healing and open themselves to the present consciousness behind the chattering streams of thought. Withdraw into tranquillity or participate in community projects. There's a lot to DO, alongside BEING (in awareness): Permaculture-vegetable-garden, solar building projects, cooking and non-cooking, food and herb processing, house building. There is space for 12 people at any given time, long- and short-term visitors, brothers and sisters, children young and old. You can run workshops on a myriad of subjects...: wide and open opportunities to express our hearts' desires... music, dance, art – all forms of expression! A learning field for living and breathing in harmonious unity.

www.gaia.com

☺ sharing our vision, no tobacco- or alcohol-drugs in the area - we get high from loving each other!

Tamera Heilungsbiotop 1

Monte do Cerro | 7630-392 Reliquias
office@tamera.org
+351-283 635 306
www.tamera.org

1995 150 ha (near Portugal, Odemira) 40 %

♀80 ♂60 30 170

g, e, Portuguese all permanent members all

the responsible members (consensus of the responsible people)

communal economy

property of two associations whose members are the community members

ecologically and socially sustainable, truth in love, transparency in community

building a sustainable village, ecological regeneration, Love School, Defending the Sacred (global peace work), spiritual research, art, political network

ecological, self-managed, left-wing political understanding, non-violent, spiritual, inner growth, sexuality, free love, organic diet, vegetarian diet, vegan diet, adult education, pedagogical, free school, seminar-house, peace-work, animal protection, environmental protection, self-sufficiency, permaculture, organic gardening, forestry, alternative energy, new technologies, ecological construction, humour, our skilled worker's group has vacancies for: translators (Engl., Port.), IT, electricians, plumbers, carpenters, car mechanics.

The "Healing Biotope 1 Tamera" in Southern Portugal is an international training and experimental site for the development of Peace Research Villages and Healing Biotopes worldwide. Approximately 170 people live, work and study in Tamera. Tamera's aim is to develop an example of a model for a nonviolent co-existence of people and between people and nature. The main tasks of Tamera are: the peace training of young people within the Terra Nova School, the building of an autonomous village called "Solar Village" which produces its own food and solar energy, and global networking under the name of GRACE and DEFEND THE SACRED.
In 1995, Tamera was founded by the sociologist and psychoanalyst Dieter Duhm, the theologian and peace ambassador Sabine Lichtenfels and the physicist and musician Charly Rainer Ehrenpreis. On a site comprising of 332 acres more than 20,000 trees were planted, houses and workshops built, gardens based on principles of permaculture and experimental biotopes were created.
Tamera started a training programme for adolescents and young adults and a community of people with different professions and from different age groups. People who are interested are welcome to participate in the annual Summer University and numerous seminars and courses.

→See also description of network **"GEN Europe"**

URI, Global Campus, Terra Nova school

study and participation in the basic semester, agreement with main goals of research project, work contribution, taking financial co-responsibility

Another photo on page 6

Valley of Peace

Casa do Pomar Velho 36 A | 8550-286 Monchique ☎+351-2-829 118 10
wolfgang@peacefulvalleys.com peacefulvalleys.com

2007 2.5 ha (near Lagos, Faro) 70 % ♀2 ♂1 1 4

4

- g, e, Portuguese
- all together (consensus)
- community income
- property of individuals
- yoga spiritually
- agriculture, making goat cheese, bread, clay construction
- self-managed, left-wing political understanding, non-violent, spiritual, Buddhist, Taoist, anthroposophic, inner growth, meditation, yoga, tantra, vegetarian diet, animal protection, environmental protection, self-sufficiency, permaculture, organic agriculture, organic farming, alternative energy, ecological construction, singing bowls

"Learning is finding out what you already know. Doing is demonstrating that you know it. Teaching is reminding others that they know just as well as you. We are all learners, doers, teachers" - Richard Bach ("Illusions")

Peaceful Valley is a place where limitations of "I know" and "I think" dissolve into a pure existence of "I feel" and "I am". It is a place where every soul can grow with inner light by uncovering, accepting and loving unconditionally who we really are.
In our Peaceful Valley, we are engaged in healthy living with regard to the main aspects of our existence - physical, emotional, mental and spiritual. Every day we dedicate ourselves to organic farming, learning and teaching permaculture, sustainable living, superadobe (clay bag) building, vegetarian ayurvedic cooking, living consciously and enjoying the Now.
Our mission is to live in harmony with ourselves and nature. Over the years, with all our love and faith, we have been creating a special place to help visitors to wake up, restore purity of mind and body, open up for a more conscious way of living, light up the path to inner resources of love to life and reconnect with the true self and the universe.

- community garden
- spiritual, new members must fund their earth building themselves

Vegan Hills Vegoa

Sito Vale Marreiros E Castelo | 8650-401 Pedralva

hello@vegoa.org vegoa.org/

2015 103 ha (near Lagos) 20 % ♀15 ♂15 10 40

g, e, f, sp, Dutch

all members (consensus or majority vote (when voting is requested))

individual financing

property of the Vegoa Association (property of all members)

be vegan and focused on environmental sustainability

caring for and improving our land

ecological, self-managed, left-wing political understanding, non-violent, vegan diet, animal protection, environmental protection, self-sufficiency, permaculture, organic agriculture, organic gardening, forestry, alternative energy, new technologies, ecological construction

Vegan Hills is the first project of Vegoa, a non-profit organisation focused on the promotion of veganism and Vegan Hills is the first project of Vegoa, a non-profit organisation focused on the promotion of veganism and sustainability. We are attempting to create a vegan community in the Algarve, Portugal, and imagine a land full of food, flowers, and trees. A place of small and infinite footpaths and natural buildings. We want it to be an example of a cruelty-free world, where people help each other and respect nature.
Members live completely off-grid in low-impact housing, such as yurts, camper vans, and wooden cabins, spread over 103 hectares. Dwellings are distant enough for privacy and independence, yet close enough for sharing stories and helping hands.
There are no leaders here. Each voice is heard as loudly as the next in our direct democracy. Our members come from a variety of places (Germany, UK, Netherlands, USA, etc.) and backgrounds, but all are united in creating an ethically-sound environment.
Action speaks louder than words! Would you like to join us? :-)

☺ live vegan and be financially independent

Romania see page 429

Russian Federation

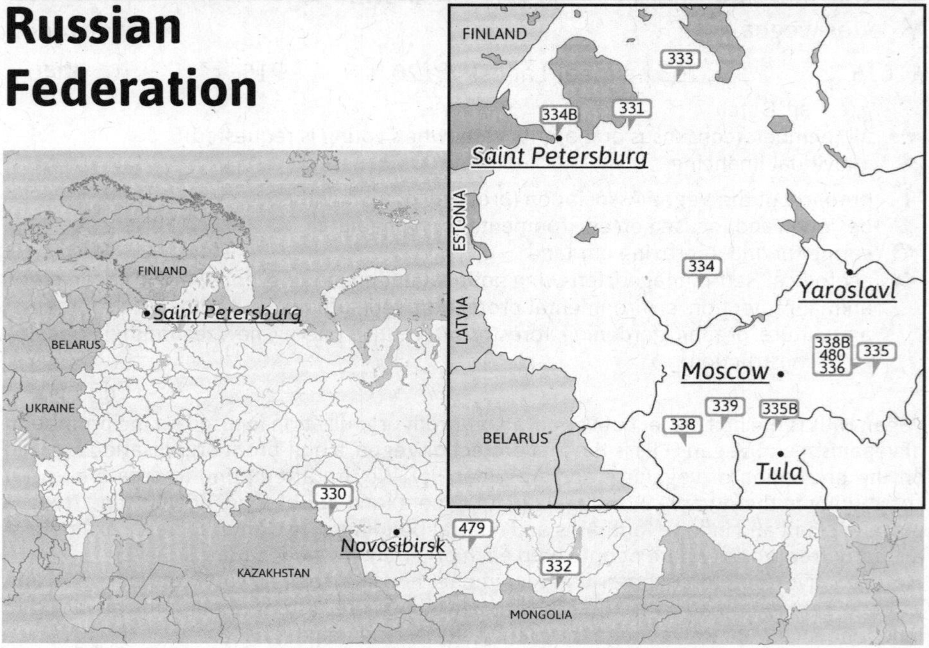

From Russia, we can present 11 existing intentional communities and 2 planned ones (page 479f). Some of them seem rather wild, from the eurotopia editing desk perspective: Huge masses of land, a culture with strong traditional influences and addresses in alien fonts. Also, some values are weighted quite differently from the culture where eurotopia comes from. Much is about family rather than about individuals, and then there is the Anastasia idea, wich seems to make perfect sense for many Russians but is difficult to grasp for others (see our article on pages 49ff and our note on the bottom of page 337).
Apart from Svetlana Village (next page) there is another "official Camphill-partner" (address on page 514) and four addresses (three of them probably still existing) from our last edition (page 503). On page 535f you'll find REEN, a Russian Network of intentional communities. Thanks Google translator for knowing how to deal with Cyrillic script...

AzGrad - New Russian Village Novaja russkaja derevnja

Ul. Pushkina 137 | 644046 Omsk ☎ +7-(0)-913-628-23-71
nikitenko@azgrad.ru www.azgrad.ru

2012 (near Omsk) 80 % ♀6 ♂6 2 14

e, Russian 14 3
a group of 12 people (consensus) individual financing
property of each family
agriculture, animal husbandry, beekeeping, workmanship, education, production of ecological organic food

ecological, self-managed, leader/guru, inner growth, adult education, pedagogical, seminar-house, peace-work, environmental protection, self-sufficiency, permaculture, organic agriculture, organic gardening, organic farming, animal friendly husbandry, forestry, alternative energy, new technologies, ecological construction, construction, workmanship: pottery, woodwork, straw production; education: kindergartens, school, Academy of Knowledge, Abilities and Skill; health and sports: herbalism, gymkhana

☺ (✋✋)

At all times, people built their world. One of the models of the world is the village. It lives by its own rules, offering its inhabitants a place for spiritual and physical being. No village arises by chance: Its birth is the fruit of collective energy. Therefore, we are building it. We are people united by faith in ourselves, the ability and desire to create, the ability to use new technologies and international experience.
AzGrad is not to produce ready-made citizens for a system that can no longer find solutions for the problems it has created. It is to promote the manifestation and appearance of men and women who are able to consciously create themselves on a true basis, at the same time creating a new world in which they have chosen to live with everyone.
The project has absorbed the best and, most importantly, the most viable of the already established and functioning settlements, starting with cottage villages and peasant farms and ending with eco-settlements and ancestral estates. Representatives of science and culture, health and education, entrepreneurs and farmers, people of different ages and professions and their children worked on the project. The project is located near the town Omsk and is constructed with the involvement of specialists from Italy, Austria and Taiwan.
People who want to change their lifestyle are uniting - and are ready to build their and their children's future and living space. *Photo on page 8*

Camphill Village Svetlana Karl-König Trust

Alexina Village, Volchov | 187439 Leningrad County ☎+7-(0)921 9821 335
dsvet1@yandex.ru www.camphillsvetlana.org

1992 57 ha (near St. Petersburg) 50 % ♀13 ♂12 35

Russian, e, g — all
one part of the group (consensus)
shared economy (shared income and capital)
Karl König-Trust in St. Petersburg
anthroposophy, social therapy as developed by Karl König
farm, garden, household, crafts, biodynamic agriculture
ecological, self-managed, Christian, anthroposophic, organic diet, mentally/psychologically handicapped persons, self-sufficiency, organic agriculture, organic gardening, organic farming, alternative energy, new technologies, ecological construction, biological and appropriate animal keeping, woodwork, dolls, felting, inner growth

☺ ✋✋

Svetlana Village was founded by a Russian "pioneer family" with the support of a large Camphill community in Norway. The large piece of fertile land, located along a bend of the River Syass, was empty except for some meadowland - since then a lot of work has been invested. Four large houses and a farm with cows, goats, sheep and chickens have been built, a well with plenty of fresh water was dug, and a large garden was planted. The first

handicapped person, or "resident", arrived before the well was finished. He and his mother knocked on the door and asked whether he could live here.
The village life which we lead is entirely different from the prison-like institutions in which the patients are given nothing besides drugs to keep them quiet. Many people come to us from all over for different lengths of time. Visitors quickly find themselves integrated into important work, help in the family kitchens where we cook and make cheese, or work with the animals on the farm, drive the tractors or dedicate themselves to a construction site.
Building any village is exciting - this special place in Russia gives the whole project additional unique and fulfilling dimensions. A place which was once a part of a giant Soviet collective - which still bears the scars of monstrous machines - is once again being worked by human beings and is giving life to a small human community.

→See also description of network "**Camphill**"
☺ desire to work alongside handicapped adults

Ecopolis Tiberkul Community of Vissarion

A6

c/o Förderverein Ökopolis, Hansenstraße 21 | 53859 Niederkassel
☎+49-(0)228-450 864
info@vissarion.info
www.vissarion.info

1995 | 1000 ha (near Abakan) | 90 % | 2500

Russian, (g, e) — almost all
- the village community (majority vote or consensus, depending on situation)
- shared financing of common projects (e.g., school, kindergarten, meeting building, workshops, horses) by tributing 30 % of income
- houses are usually the property of the inhabitants
- Vissarion's teachings are the basis. People of all denominations are welcome.
- farming, school, handicraft, art, horses, liturgy
- self-managed, leader/guru, spiritual, ecumenical, inner growth, vegetarian diet, free school, self-sufficiency, non-smoking

☺

We live in the midst of untouched woods, mountains, clean rivers and lakes, more than 2000 people, distributed among different villages, in the picturesque scenery at the foot of the Sajan mountains in the South-Siberian taiga. During the hot summer we grow our food for the whole year, and collect, in addition, the fruits of the taiga. By building workshops we prepare for an independent, non-industrial life form.
We educate our children to look at everything in nature as alive, and we promote their creativity and imagination. In our schools we teach, among other things: arts and crafts, dance and music, history of cultural achievements instead of history of wars.
Most notable about our community is the special form of spiritual development which is based on the "Last Testament" by Vissarion. Our members should really live charity as already described in the New Testament, that means e.g. to never have negative feelings towards others, but always and under all circumstances to give the warmth of their souls.
We are happy to cooperate with other communities, also training periods are possible. Everybody is invited to visit us and to settle here.

Editor's note: the community's contact data are

Zerkov Poslednego Zaveta (Church Of The Last Testament)
Russia,
662925 Petropavlovka, Kolkhoznaya Str. 9

Information Center: Tel +7 39136 23594

The German friends' association (adress above in the address field) is, however, better equipped to answer German and English inquiries.

☺ interest in our lifestyle

Ecovillage Grishino

ul. Rechnaya 12, Grishino, Podporozskii rayon | 187780 Leningradskaya oblast
☎+7-(0)921-340 53 03
vasudeva@bk.ru
vk.com/ecovillage_grishino

1993 | 6000 m² (near Podporozhye) | ♀4 ♂6 4 14

Russian, e | all

one part of the group (consensus) | limited community financing

property of individual community members

spiritual workshops, garden, construction

ecological, self-managed, non-violent, spiritual, inner growth, meditation, organic diet, vegetarian diet, seminar-house, peace-work, environmental protection, organic agriculture, organic gardening

☺

In the North West of Russia, in the historic village of Grishino (300 km north-east of St. Petersburg), at the confluence of two rivers, we have built up and lived in an ecological village for over twenty years. We came here from different cities to live on the land in harmony with each other and with nature. We all have different world-conceptions, but we are learning to accept each other and to co-create our community in common.
Fourteen of us live in Grishino all year round. We have several traditional Russian-style community and private houses. We grow what we eat in our vegetable gardens, and gather lots of herbs from a large meadow of wildflowers. Our ecovillage is surrounded by natural forests, where we gather berries and mushrooms, and in which there live many kinds of wild birds and animals, like beavers, moose, bears, hares, foxes, wolves, wildcats and many others. We continue our project on ecological forest management.
We aim to continue the cultural traditions of our ancestors, learning the folk arts, their songs and dances, woodcraft, ceramics and village architecture; and we are endeavouring to create a family school for our children. Each summer we hold seminars for adults and children where one can participate in the spiritual life and work of our community and merge with the spirit of Russian nature.

Editor's note: Please note the article about the Anastasia movement on pages 49ff and the note on the bottom of page 337.

→See also description of networks "**GEN Europe, REEN (Russian Ecovillage Network)**"

Russia

Ecovillage Shiram

Borovoe village, House 18 | 172063 Tver Region, Torzhoksky District

+79-201 707 981

holos@inbox.ru shiram.daism.ru, da.dzin.org

2009 15 ha (near Torghok (50 km), Moskau/Moscow (220 km))

Russian, e

Ecovillage Shiram was created in 2009 by a group of enthusiasts who were looking for a place for spiritual practices and internal work.
We have a beautiful piece of land near a river and a wild forest. Today Shiram consists of a territory of about 15 hectares on which three residential houses, two temporary houses, a rabbit farm and a chicken farm have been built. Evenmore, there is a small sheep herd, and a fostering garden for rare plants. We also installed geodesic domes (of Fuller), an aerial trail, two labyrinths and mirrors inspired by the work of A. Kozyrev. Currently, two people live in an ecological settlement permanently, about ten people a month, sometimes up to fifty people attend training programs. There are seminars on holotropic breathing, transformation games, seminars on spiritual growth and meditation.
The initial Shiram group was founded 1990 and consisted mainly of psychologists, PhDs and doctors of science who were and still are actively engaged in scientific and practical activities in the fields of self-development and nowadays participate in scientific conferences events all over the world. We have many interesting projects, we are happy about all volunteers and new residents in the ecovillage Shiram, who would, like us, strive for self-development and self-improvement.

EcoVillage Yasnoe

190000 Leningrad region

eds.bychcow@gmail.com vk.com/ecoyasnoe

2007 (near St. Petersburg) 20 % ♀8 ♂8 4 20

e, Russian

all who live here permanently (consensus) limited common financing

ownership and other forms

ecology, space of love

permaculture, milk farm

ecological, self-managed, non-violent, adult education, seminar-house, animal protection, environmental protection, permaculture, organic agriculture, organic farming, animal friendly husbandry, new technologies, ecological construction, humour, non-smoking, cows, fly, art, music, stove, water system

We are looking for practical ways of co-creation and harmonious life on Earth. Our settlement is in the Vyborg district of the Leningrad region, 100 km from St. Petersburg. There is a cow dairy farm and an apiary for our beekeeping. We welcome all creative guests! We welcome new neighbours and families, who are ready to live in nature.

→See also description of network **"REEN (Russian Ecovillage Network)"**

Hospitable House in the Forest Домик на Природе

Bronnickij pereulok, d.2, kv.28. | 109202 Moscow ☎+7-(0)926-863 94 09
houseinthewild@gmail.com hut.shestero.info/en

2008 0.5 ha (near Gus-Hrustalnyj) 15 % ♂2 2

Russian, e 2

one person (after consulting the permanent members)

property of one member

healthy lifestyle, internationalism, humanistic values

gathering

ecological, self-managed, ecumenical, vegetarian diet, adult education, seminar-house, permaculture, organic agriculture, organic gardening, forestry, alternative energy, new technologies, ecological construction, non-smoking, nature, forest, mushrooms, berries, herbs

☺

The "Hospitable House in the Forest" was established as a free space for long-term inhabitants and short-term guests. The main idea is to develop a healthy community of friendly people with positive core values (ecological, healthy lifestyle, openness, self-development). This is a place for spiritual, psychological and physical recovery and development through contact with nature, fresh food, simple life and temporal relative isolation from the outer world. It was also planned as a resource spot, a meeting point which hosts various small-scale training events, programmes, exchanges of knowledge as well as putting some of them into life. This place is good for open-air sport activities and for picking gifts of nature such as mushrooms, berries, wild herbs, etc.
We are not a living community yet. The house is usually open/inhabited between May and October. Eventually, we are going to turn ourselves into a small-scale living ecological community. The house is situated in the low-populated area of Central Russia, just 200 km to the east from Moscow in a tiny village surrounded by pine forests.
The house is open for guests FOR FREE. But note that it is a total non-smoking, alcohol- and drugs-free zone. Please, feel free to contact and ask questions.

Baltic Ecovillages Network

☺ no smoking, no alcohol, not any drug addictions

Insight

near village Laptevo, Zaoksky distrikt | 301015 Tula region
☎+7-495-644-62-72
info@mir-insight.ru mir-insight.ru

2015 79 ha (near Zaokskiy) 40 % ♀6 ♂9 3 18

e, Russian everyone 15

collectively (consensus) shared economy (shared income and capital)

group of 4 owners

Gaudiya-Vaishnavism

agriculture, eco-products, training workshops

ecological, self-managed, anarchistic, left-wing political understanding, non-violent, spiritual, meditation, yoga, organic diet, vegetarian diet, adult education, free school,

Russia

seminar-house, peace-work, one-world development, animal protection, environmental protection, self-sufficiency, permaculture, organic agriculture, conventional agriculture, organic gardening, conventional gardening, organic farming, animal friendly husbandry, conventional farming, alternative energy, construction, humour, non-smoking

☺ (✋✋)

INSIGHT team members realise a rural community in the form of labor fraternity, intentional community, philosophically based on the INSIGHT almanac materials. Our community organizational model is founded on universal spiritual values of the Krestovozdvizhenskoe (Holy Cross) labor fraternity of Nikolai N. Neplyuyev. Instructions of Founder Acharia of the International Society For Krishna Consciousness, Om Vishnupad Paramahamsa Parivrajakacharya Ashtottara-shata Shri Srimad His Divine Grace Srila A. C. Bhaktivedanta Swami Prabhupada, are the spiritual foundation of the community.

The INSIGHT team is the community of leadership, therefore it mainly suits goal-oriented, strong, psychologically mature devotees, who aspire to practice pure devotional service to Lord Krishna. The community members have chosen their priorities in their lives. They are stable on the spiritual way. They are ready for active, creative, encouraged practice of devotional service among like-minded persons. They take responsibility for the whole team and consider it their eternal spiritual family.

☺ attitude of Krestovozdvizhenskoe labour fraternity by N.N.Nepluev

Kind Land (Dobraya Zemlya)

Dobraya Zemlya, near v. Studencovo, Sudogodskiy district, Vl601363

☎ +7-926-323-04-60

✉ dmitry.solovyev@list.ru; dobraya-zemlya@yandex.ru 💻 dobrayazemlya.ru

2010 400 ha (near Vladimir, Russian Federation) 20 %

♀80 ♂70 30 180

e,f, Russian, Japanese 10

individual financing — private ownership

vegetarian

ecological, self-managed, spiritual, esoteric/New Age, vegetarian diet, seminar-house, animal protection, environmental protection, self-sufficiency, organic agriculture, organic gardening, ecological construction, non-smoking, vegetarian, spiritual, happy country life, woodwork

☺ (✋✋)

Kind Land is a community center of neighbouring "Anastasia" kins domains and traditional villages and the center of social & business activities. There are about 300 families-members in these settlements and about 80 families living all the year round. But we are not a solid community, we are mostly independent.
Developing our social and cultural project, we learn to live outside towns/in the country anew, we want to find a harmonious way of coexistence of nature and man. But we are not fugitives from city, and we do not deny the greatest achievements of the human spirit and mind. We understand that nature must be ennobled by human culture, human creativity, human labor. We are only at the beginning of the road and we do not have ready-made recipes for success. We are open to cooperation and we are looking for like-minded people.

Our activities: Children's camps, permanent kindergarten, hobby groups, pedagogy seminars, Kind Land Eco Fest (every summer), Forum Sustainable Development of Rural Territories, feasts and concerts, sport feasts.

→See also description of networks "**GEN Europe, REEN (Russian Ecovillage Network)**"
☺ ideological and financial requirements

Editor's note: The way we understand it, Kind Land is a settlement surrounded by Anastasia-style kin domains, and all are on good terms with each other. Indeed, our article on pages 49ff is based not so much on our knowledge of Russian kin domains, which is very limited, but more on what we know of the Anastasia movement in Western Europa/Germany.

Kitezh Children's Community Orion Children's Community

Kaluzhskaya oblast | 249650 Baryatinsky ☎+7-08454-232 24
info@ecologia.org.uk, info@kitezh.org
www.kitezh.org, www.ecologia.org.uk

1992 | 100 ha (near Kaluga, Smolensk) | ♀13 ♂13 39 65

all
everyone in common (consensus) | shared economy (shared income and capital)
common property
pedagogy, school
ecological, self-managed, spiritual, pedagogical, self-sufficiency, ecological construction, orphans

☺ ()

Kitezh Children's Community in Kaluga Region, Russia is a unique example of family-based care and education for orphaned and homeless children in a rural ecovillage community. The families adopt children, raise them together and educate them in the Kitezh school. Therapeutic Education is the main focus at Kitezh. With this experience of secure family and community life the children can recover from the trauma of earlier years to become open, loving, valuable and contributing members of society. Despite harsh economic conditions, since 1992 this dedicated group has built two foster family villages, Kitezh and Orion, which accommodate 25 adults and 40 children. Over 100 children have benefitted from living at Kitezh since then.
Self-sufficiency and an ecological lifestyle is consistent with traditional rural Russian values. Kitezh and Orion have natural biological wastewater treatment plants and each village has a farm, where they grow vegetables and raise chickens, goats and cows. Volunteers, professionals and students from all over the world come to help with building, working in the gardens, on the farm and with the children.
Ecologia Youth Trust, a Scottish charity based at the Findhorn Community in Scotland, has supported Kitezh Children's Community since it began, providing professional training, organizing a volunteers programme and youth exchanges with other ecovillages and raising funds to build both villages. For information contact: www.ecologia.org.uk (Ecologia Youth Trust).

☺ interest in child care

Lyubodar Kin Domain Anastasia-Siedlung

Nahe des Dorfes Dobryatino, Distrikt Melenkovsky | Region Vladimir
☎+7-903-129-27-83
poselenie@lubodar.ru www.lubodar.ru

The "Lyubodar" kin domains' settlement was founded in 2008 by readers of the book "Zvenyashie kedry Rossii" ("Ringing Cedars of Russia") by Vladimir Megre. The settlement is located in the Vladimir area, 330 kilometers away from Moscow. Its size is 178 hectares which is divided into 75 kin domains, 1.4 hectares each. There is also a shared territory of six hectares for organizing gatherings, celebrations or welcoming guests. Up to this day, there are 18 families in the settlement. We were united by the dream of earth's beautiful future, described by Anastasia, the heroine of Megre's books.
A kin domain is a piece of land by the size no less than a hectare, where a family establishes the space of love and works on making themselves, their surroundings and the world better.

Each domain will have its garden, house, bathhouse, lake, forest, well and green fence. We are heading towards self-sustainment. Several families organized the guild and are involved in organic production: Cold-pressed oil, wholesome sweets, chocolate, natural beauticraft, healing ointments, herbs, willowherb, etc.
We keep a healthy lifestyle and tend to interact with the land, world, our neighbors and our families consciously. The family is one of the main values from our point of view.
We support natural parenting, natural childbirth and domestic education.
Here our kinship will live on, this is our small motherland, and we want to pass it on to our children!

Editor's note: Please note the article about the Anastasia movement in this book.

☆ VSCHAST'Ye VSchast'e

Potapovo village, house 1 | 249868 Kaluga region, Dzerzhinsky district
☎ +7-925-355-2502
ecofet2018@gmail.com
vk.com/vschasty, https://tinyurl.com/eurotopia8751 (search for "ВСчастье")

2018

We formed after the non-profit event "7th Festival of Ecotechnology" in August 2018. After working together we, several families, decided to continue to develop the territory. It's placed in the birch forest of the Kaluga region, bordering the settlement Milenki and the national park "Ugra". Formally, it is private property, collectively managed by a project team of three families.
We have a large summer shed for general classes, 3 tents, a kitchen, a covered dining room for 150 people, a permacultural garden, 2 warm houses for living, several compost toilets, a dome shed, a stage, a storage room and two summer houses. We organize parties on national holidays, round dances and sober weddings. Our goal is to create a space for reloading, relaxing and inspiring visiting guests. We strive to live in abundance, developing our territory, creating a balance of spiritual and material.
We also do training seminars, create platforms for the communication of practitioners, invite experts for sharing practical experience of environmental technologies applicable in the own household. Sustainability lies in the fact that we combine different types of activities: tourism, permaculture, enlightenment, green building and the project approach.
In the community, we actively apply permaculture and permacultural principles. We use various forms of barter, self-organisation and are engaged in crafts, creativity. We teach people how to fully experience the rest of the soul, peace of mind, a sense of harmony between space and yourself. Step by step, we are developing a local economy by increasing the cooperation with our settler neighbors. We encourage barter without forgetting to improve the quality of their services and products.
We are happy about helping hands. Our place is suitable for the practice and research work of students, graduate students and scientists.

Slovakia, Slovenia see pages 430f

Spain

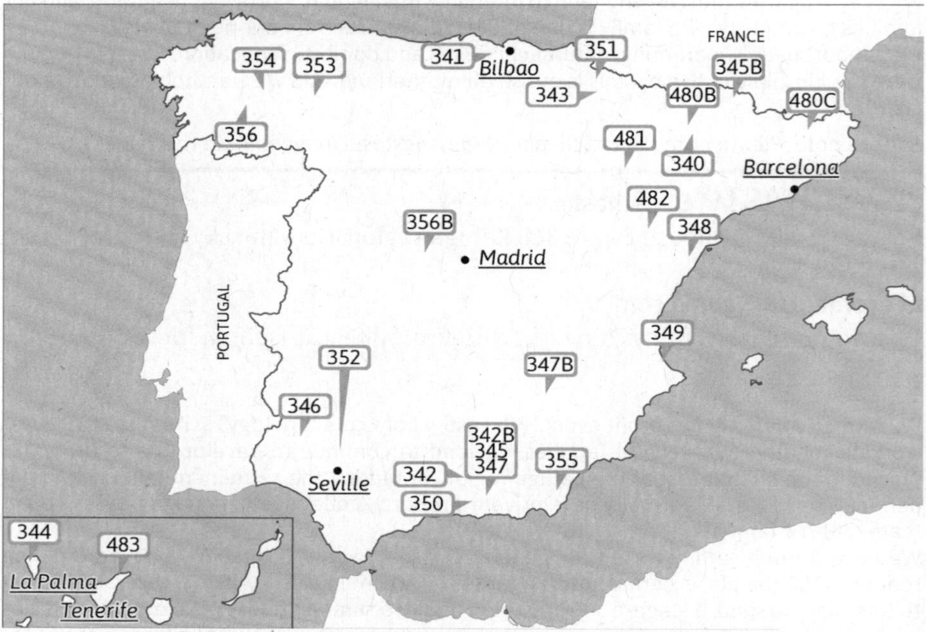

26 intentional communities from Spain, 5 of them still in planning stage (Page 480ff). 8 communities were presented in the last edition and are listed in this book on page 503f. Red Ibérica de Ecoaldeas (page) is the national ecovillage network of Spain (page 535).

Aguinaliu Asociación de Desarrollo Sostenible

C. /Única, 80 | 22588 Aguinaliu - GRAUS-HUESCA, ARAGON
+34-(9)625 780 714, info@aguinaliu.es www.aguinaliu.es

2002 (near Lleida, Saragossa, Barcelona) 12

We have been preserving and reviving the village of Agiunaliu for more than eight years now. Our organization is cooperating with a cultural association and a work cooperative. We see ourselves as a group of humans that support each other through direct mutual help and exchange.
The cooperative (5 houses) consists of members who share their houses and manage the surrounding countryside - or have an employment there. About half of the village community spends a large part of the year working in the city to gradually invest in reconstruction.
In future, the association will integrate the new neighbours (with 6 further houses) in the upper part of the village and accommodate friends, visitors and all further users of the association's offers or seminars. In the summer, we often welcome volunteers (maximum 2

at the same time) when there is money for building/renovation projects and for board and lodging of the volunteers.
The community's motivation is education, therapy and the learning and teaching of the alternative rural life. This includes all activities from before birth until after death: health, social relationships, professions and work. In Agiunaliu we combine all this with criteria towards environmental protection.

☆ Amalurra Artzentales

amalurra@amalurra.com | 48879 Artzentales, Bizkaia ☎+34-946 109 540
amalurracomunidad@gmail.com www.amalurra.org

1993 10 ha (near Bilbao) 20 % ♀15 ♂10 5 30

- sp, f, e, Basque
- whole group in a circle (sociocratic consent)
- indiv. financing and common shares/financing of own association
- property of all
- spiritual, non-religious approach towards personal transformation: we are focused on connecting with the sacred aspect in every expression of life
- inner work (shadow work), outer work: gardening, rural hostel and hotel, restaurant

☺

Amalurra ("Mother Earth" in Basque) was born in the 90s in the Basque Country. Its founder, Irene Goikolea, envisioned it as a platform for transgenerational healing, consciousness transformation and care and well-being for people and nature. A space for coexistence where individual gifts had a social projection and the healing processes undertaken at different levels - physical, emotional, mental and spiritual - could contribute to healing unsolved transgenerational issues we project on other people. Our vision pillars are: the recovery of our ancestral bond with Mother Earth, shadow work, awakening to the deep feminine and awakening to the ancestral wisdom of the culture to which the community belongs.
Living in community was a family initiative, together with individuals who resonated with the project. Our commitment was to explore new ways of manifesting our highest potential to create a collective field of inspiring influence.
Amalurra materialized after a process of inner work focused on promoting a sustainable coexistence at all levels (ecological, economic, social and cultural), through awakening to the sacred or essential aspect of life: the quality that includes every polarity in order to face current challenges in an innovative way, thus contributing to a culture that prioritizes collective well-being.

→See also description of networks **"GEN Europe, RIE"**

- There are 3 Amalurra communities. Other networks: Amma, Kolore guztietako basoak
- ☺ first having a previous experience as a volunteer; being committed to do a process of transformation work; becoming aligned with the community values

Amalurra Granada

Calle Calzada (Caparacena) | 18290 Granada
+34-9-625 136 348
amparifs@gmail.com
amalurra.eus

1996 | 4000 m² | 5-30 % | ♀11 ♂8 9 28

sp
2 4

all community members (consensus)
some common financing

property of all

none

work of conscience, process work and self knowledge

ecological, self-managed, left-wing political understanding, feminist, spiritual, inner growth, adult education, peace-work, permaculture, organic gardening, integration of our traumatic collective memories

The Amalurra Granada community emerged after 7 years of inner work and group cohesion, facilitated and inspired by Irene Goikolea Uriarte, founder of the Amalurra community in the Basque Country. The Andalusian group that make up the current community decided to use the same name (Mother Earth), because we share the same energy and foundational information as the Basque community. These two are our essential objectives:

- To consider the community and its different scenarios as opportunities to make the unconscious conscious. This is so because we understand that owning both that large portion of personal and collective unconsciousness, as well as the feelings and emotions that are imprisoned in it, is a true energy revolution and a transformation tool that brings responsibility and commitment to those who use it in any action they undertake in favor of life.
- As participants of the community project, to become united and support each other in the building process of the inner community within each one of us, by learning from the reflection we receive from the external community.

In short, we would like our community endeavour to contribute to a more caring, free and responsible social paradigm, and that our path is useful for future generations.

→See also description of network "**RIE**"

There are 3 Amalurra communities. Other networks: ESPIRAL, Red Iberica de Luz

Amoraleza

Apdo 3 | 18420 Lanjaron
+34-6-440 375 11 (mobil)
amoraleza.net@gmail.com
www.amoraleza.net

2013 | 4 ha (near Granada) | 10 % | ♀3 ♂3 4 10

e, f, sp, Portuguese
4 6

the core members (consensus)
limited common financing, some shared income

property of 2 members

shamanistic

sacred plant ceremonies, yoga, permaculture

ecological, self-managed, spiritual, yoga, organic diet, vegetarian diet, permaculture, non-smoking, sacred plants: nutrition

☺ ( )

Amoraleza is a sacred space, where everyone can experience an expansion of the heart and the consciousness, within a community whose goal is to live in symbiosis with Nature. Our objective is to create a functional model for an alternative lifestyle to meet our fundamental needs, while following our dreams and without compromising our integrity. We hope that this model can help and inspire others who are looking to step out of the "system", live more in harmony with nature and follow what their hearts truly desire. We aim to provide a space to give and to receive all that is good, joyful and prosperous: a home, as well as a place for creative expression, information and skill sharing and a genuine connection with others. Our main missions are to share our work with sacred plants and conscious nutrition.

☺ ideological, financial

☆ Arterra Bizimodu

Calle Abajo 1 | 31480 Artieda (Navarra)
arterrabizimodu@gmail.com arterrabizimodu.org

2014 8000 m² (near Pamplona) 9 34

sp, Basque, e 10

sociocratic management and coordination circles (sociocracy)

individual financing rented

facilitation, agroecology, transition

ecological, self-managed, feminist, organic diet, vegetarian diet, environmental protection, organic agriculture, organic gardening

☺ ()

What started in 2013 as a small group of people determined to find a better way of living and educating their children has flourished to become Navarra's first co-housing community. A mix of around 40-50 residents, volunteers and visitors make up our day-to-day mosaic. We recognise and celebrate the differences in our community members, each of us living in our own apartment while sharing common spaces and enjoying lunch together in our shared kitchen every day and working together in tasks and management. Once a month we come together to share as a whole community and do auzolan - community work.
Through our 'Centro de Encuentros', we run courses that train people in various areas directed towards the transition to a better world. The range of other projects in Arterra represents the diversity of passions and skills found in the community - entrepreneurial initiatives such as our bakery, a CSA, a brewery, a recording studio, and a magazine, among others. Still more people work outside the community and contribute through cleaning, cooking, waste management, caring and parenting, accounting, communications, maintenance...

→See also description of network "**GEN Europe**"

Iberian Ecovillage Network, REAS Navarra (solidarity economy network)

☺ visiting and integration phase for mutual approval, a small financial contribution

AUTarcaMatricultura - Autonomous Permaculture Academy

Camino La Ermita 26 | 38780 Tijarafe +34-92249-0215
autarca@matricultura.org www.matricultura.org

2007 3.5 ha (near Los LLanos de Aridane/La Palma) 90 %
♀2 ♂2 2 6

g, e, f, sp, Swiss German 4 4
all of us together: including: Earth, fire, water, air, all living beings (consensus)
economy of community: “not to grow not to go”, matriarchal subsistence farming, gift economy
property of some
matriarchal values, eco-feminism, earth-democracy
reforestation with forest gardens, repair of water and mineral cycles, creation of humus and terra preta, conscious reintegration of wo&men into living nature, education: permaculture courses, permaculture for kids, visitor days
ecological, self-managed, feminist, non-violent, spiritual, inner growth, tantra, organic diet, vegetarian diet, vegan diet, adult education, animal protection, self-sufficiency, permaculture, organic agriculture, alternative energy, ecological construction, humour, no cell phones/wireless, non-smoking, solidary permaculture, godparenthoods for trees, curriculum for alternative practitioner for mother earth, integral landscape healing: retention ponds, biodynamic and holistic energetical methods, food forest systems, alternative beekeeping, earth-spirituality: deep ecology, matriarchal basic values: body-centered psychotherapy, bioenergetics, rebirthing

Theory and practice of the creation of life-friendly and autonomous forest-gardens.
The step-out-pioneer recognizes the exponentially growing destruction of Mother Earth and the oppression of her/his creation as a consequence of the patriarchal world view, and therefore she/he steps out deliberately. The step-out-pioneer says no to the chaos of war, economic and social injustice, nuclear, genetically engineered and electromagnetic contamination, abuse and exploitation, logo-driven perception and ego-centred lifestyle.
The step-in-pioneer puts life in the centre of her/his work and her/his activities. She/he encourages vividness with heart and creativity, the joy and the health for all beings; she/he feels that she is part of the whole, part of Mother Earth. The step-in-pioneer walks the path towards life-centred livelihood focused on the origin, she/he creates powerful visions of a healthy Mother Earth and she/he assumes full responsibility for her/his actions. In order to become more uncompromising and more life loving there are changes to plan and to realize in all areas of life. See also “AUTarcaMatricultura” on page 560.

La Palma Transition Initiative, local La Palma (ecological) networks
☺ put life into the centre of activity and put own life in service of Mother Earth

Badulina Community

Tienda Ecológica, Mercado Municipal, Puesto 13, c/correo | 18400 Orgiva
+34-633 630 472, info@badulina.org www.badulina.org

2017 9 ha (near Orgiva) 30 % ♀4 ♂3 3 12

e, Dutch, g, Polish, sp not yet all
members of association (majority vote)
loan for limited common financing, common purse for food
property of association
farming, building infrastructure, self sustainability, raising children
ecological, self-managed, non-violent, inner growth, self-sufficiency, permaculture, organic agriculture, organic gardening, organic farming, alternative energy, new technologies, ecological construction, construction, healthy lifestyle

We are Badulina, a young, consciously living community in the south of Spain in the development phase. We like to live with heart, hand and mind. We are from all age groups and cultivate this land for our children and their children, so that they grow up in abundance and harmony. We like to live healthy and open-minded lives, strive for independence and at the same time enjoy the advantages of modern technologies. We are like settlers on a new land, working on our infrastructure, our organizational structure and our communication skills. The first families started to settle and are now developing ways to grow. We are looking for families and individuals who want to settle and build their own homes in the countryside. We need carpenters, builders and of course the people who can cook and carry good energy and joy. The joining members must participate financially, in order to become co-owners of the project. The place is about 9 hectares in size, a whole hill with two river beds as a natural barrier. On the land there is currently a beautiful house in good condition with electricity, water, internet, solar heating and other basic luxury. We have a 6 m community yurt as a common room and some nice private yurts and domes spread around us.

EcoAlpujarra
families or individuals with the financial capacity to buy into the project

Comunidad de San Feliu · the slow village ·

c/Unic s/n | 25514 Senterada +34-632 483 870
info@reguard.online www.reguard.online

2014 (near Pobla de Segur) 10 % ♀0 ♂5 0 5

sp, e, f, g 3
village council (consensual decision making process) individual financing
individual properties
rebirth of the village
ecological, self-managed, non-violent, spiritual, inner growth, meditation, organic diet, peace-work, self-sufficiency, permaculture, organic gardening, organic farming,

Spain

alternative energy, new technologies, construction, humour, monthly sweat lodge ceremony, meditative archery, tiny houses, old and new yurts, individual spirituality, artist's residences, spiritual retreats, carfree village, bushcraft, glutenfree enclave

☺

A new beginning: the millennia-old terraced hills around our formerly abandoned village have been nearly totally reclaimed by nature. A place full of contrasts: extraordinary mountain wilderness meets high tension lines, busy tourist road, industrial cattle farms - we never lose sight of the toll modern civilization takes. A gentle restart is our aim and we are looking for people who want to share their knowledge and skills. The priorities are the creation of living spaces (jurts, tiny houses, tree house etc.) for artist residencies and spiritual retreats as well as replanting the terraced gardens. Every month around full moon we celebrate a temazcal with a mexican shaman.

☺ we would like to get to know you well, if you are interested in our village

El Calabacino

21340 Alájar (Huelva)
elcalabacino@gmail.com aldeaelcalabacino.org

♀44 ♂40 30 114

sp, e, f, g, Dutch 35

individual financing most members own their land

ecological, self-managed, anarchistic, left-wing political understanding, feminist, non-violent, spiritual, esoteric/New Age, chaotic, inner growth, meditation, yoga, alternative energy, ecological construction, construction, humour, resilience, respect, helping each other, celebration of live, big family of families, donkey love

We are not a specific intentional community. El Calabacino is an old abandoned village and in the early eighties the first families came to live and to raise their children in the mountains. Over the years more and more people searching for a more natural life settled down.
The whole village runs with solar energy. There's nearly no car coming up and we are working with donkeys to bring up heavy things. We have some activities to make communal money which we invest in maintenance of paths and the commune house. But here are also many people who just want to live the way they like to, without organizing themselves with anybody. There's no accommodation for visitors, because the commune place is really small. And also there is no space for new people, because nearly all the bits of land and houses are occupied. There is no way to just come by. However, some people have a Workaway or WWOOF site so they do the exchange on a private way. If somebody wants to visit it is absolutely necessary that they write to us beforehand to our e-mail address.
People living around here do music, theatre, garden, bioconstruction, education, agriculture, natural therapies, handcraft, building, working on the campside, and many other things. The spoken language is Spanish, there are some people speaking English, French, German and Dutch.
Our commune work is usually about maintaining and fixing our paths and stonewalls and about taking care of our communal spaces. We love to come together and share music and laughs and give a hand to each other if needed.

→See also description of networks **"RIE, GEN Europe"**

El Morreon

Buzón Benizalte 6 | 18400 Órgiva
solarmichel@hotmail.com, tomtomjana@web.de

1990 60 % 20 50 (Sommer) - 150 (Winter)

g, e, sp — it changes
God (meetings strictly forbidden) — great! we are all poor!

ownership, rented and occupied places
multiculty
surviving, circus, theater, music, healing, alternative technology
ecological, self-managed, anarchistic, left-wing political understanding, feminist, non-violent, spiritual, esoteric/New Age, chaotic, inner growth, meditation, yoga, free love, organic diet, vegetarian diet, vegan diet, macrobiotic diet, adult education, pedagogical, free school, peace-work, mentally/psychologically handicapped persons, physically handicapped persons, animal protection, environmental protection, self-sufficiency, permaculture, organic agriculture, conventional agriculture, organic gardening, conventional gardening, organic farming, conventional farming, forestry, alternative energy, new technologies, ecological construction, construction, humour, no cell phones/wireless, circus, music, handicraft, therapy, yurt and dome builder, solar energy and mobility, and other... People

☺

We are living in a valley 3 km from Órgiva, which has been known for years as a hippie village. People are living together in different forms and projects. There is no fixed structure or central organisation. Some of us have tap water from town, and there is solar electricity or generators. There are two rivers, one is dry in summer. Only some land is irrigable in summer.

☺ You should be able to take care of yourself.

El Pardal Vida Holística

El Pardal, carril Peña Corva 1 | 23300 Villacarrillo-Jaén ☎+34-646 484 131
ecopardal@yahoo.es
www.facebook.com/groups/227559473966911, pardal36-es1.webnode.es/

1981 27 ha (near Villacarrillo, Ubeda, Granada) 30-50 %
♀0-2 ♂3 0-3 3-12

g, e, sp, f — all present, usually 7-12
everyone in common (consensus) — limited community financing

property of individual community members
all people are brothers in spirit and should work and learn to help one another
horses, goats, food production, reforestation, improvement of the premises
ecological, self-managed, left-wing political understanding, non-violent, spiritual, Christian, ecumenical, inner growth, meditation, yoga, organic diet, vegetarian diet, adult education, pedagogical, environmental protection, self-sufficiency, permaculture, organic gardening, organic farming, animal friendly husbandry, alternative energy, new technologies, ecological construction, humour, work with children from orphanage, deep massage, spiritual focussing, natural medicine, handicraft: construction and furniture, music instruments

Spain

☺

The over 100 years old farmhouse that we renovated (natural stone walls, grass roof) is situated at 1500 m altitude, surrounded by nature, inside of Spain's biggest natural park.
We try to live as self-sufficient as possible, with four sources on the land: sun & wind power, ecological cultivation of veggies, fruits and herbs, some animals and natural medicine. I (Wolfgang) want to make an effort to make this world a better place for all living beings, to help people change their life and to subsequently improve their health. I wanted to learn all that will allow me to help people in every phase of life, and learned a lot about inner work (techniques of humanistic psychology), natural medicine, acupuncture, yoga, Shiatsu, meditation etc.
We recycle/reuse what we can, respecting and adapting the norms of permaculture. Part of the land is used as pasture, we plant many trees and keep some of the 27 hectares simply protected to enjoy what nature can do.
No neighbour lives nearer than one hour walking or driving, the nearest villages are at a distance of 2-4 hrs walking or 2-3 hrs by car.

→See also description of network "**WWOOF**"
Workaway, HelpX, Helpstay, "neighbours" in the parc
☺ good will and mutual understanding

Editor's note: We have received several reports of visitors to Er Pardal who claimed that El Pardal is in fact not a community, and that conditions were much harsher than expected - in some aspects more difficult than at other places they have visited. El Pardal emails say that they do consider themselves a community. Please form your own opinions.

Familia Feliz

Calle Las Parras, 19 | 12570 Cervera del Maestre ☎+49-(0)175-24 68 955
welcome@familiafeliz.eu www.familiafeliz.eu

2013 (near Vinaros) 25 % ♀2 ♂1 0 3

g, e, f, sp, Catalan 1 full-time, 2 part-time
all who are concerned (consensus) shared economy on voluntary basis
property of different community members
art & culture; ecology; sustainability
ecological, self-managed, non-violent, inner growth, free school, seminar-house, environmental protection, organic agriculture, organic gardening, animal friendly husbandry, humour, art: painting, sculpture, installation, social sculpture, artist residence; garden: dye garden, poultry keeping; general pursuit of happiness

☺

Familia Feliz is a decentralized living community with shared economy. We are EUROPEANS, talking different languages, and we have different skills and backgrounds based on our socialization. We think it is "contemporary" to develop various places in Europe as home units. Presently we are at home in Spain and Germany. Moving from one to another home and being multinational enables us to take most advantage of the European idea and culture. Our community life is based on shared economy, decisions by consensus, veto options for

everyone and solidarity with each member. We respect and develop private property and focus on life in peace and freedom. One characteristic of Familia Feliz is the strong link between art, culture and nature.
The community clearly separates between ownership or property and possession. Usufruct of all goes into the community, the property remains with the person.
If any member leaves the community upon her/his own request or by exclusion, all property returns with all rights and obligations to this particular member. This favours autonomous individuals who remain free in their decision to become actively involved in the community and do not fear any disadvantage when leaving.
Besides the Spanish location there is a second base in Windeck, Germany near Cologne/Bonn.

☆ **Interneland/Innerland** Harmonysession

Appartado de Correos 141, San Rafael | 07816 Islas Baleares
☎+34-671 946 512 (mobile)
anetanet2@yahoo.com
www.harmonysession.com

1970 (near Barcelona) 50 % ♀800 ♂600 0 1400

g, e, f, i, sp, Polish, Greek, Portuguese, Hebrew, Japanese and more — some
for material work some, for spiritual work all
the corresponding council - each one belongs to a council of choice (through talks and votes)
individual financing — private
religious: no, spiritual/cosmic: yes
our collective existence serves to a spiritual and cosmic awakening of humankind through a contact with higher consciousness and in form of meditations, exercises and energy exchange
ecological, non-violent, leader/guru, spiritual, inner growth, meditation, yoga, organic diet, vegetarian diet, vegan diet, peace-work, organic agriculture, organic gardening, new technologies, humour, non-smoking, cosmic contact, divine connection

☺

We are a community of friends who are living a life in respect and communion with higher, energetic, cosmic and spiritual levels of consciousness. We have international communities in Spain, Brazil and Argentina as well as centers in 30 countries and we also offer virtual communities for online meetings and meditations on our websites. We are working for inner growth which allows us to control our energies and put limits in our life.
We are preparing to reunite ourselves through meditations, seminaries and energy work with our original, cosmic dimension and a reunion with the circuits of the universal transmission of the united planets. Thanks to a permanent contact with higher levels of consciousness we are trying since 50 years to hold a brotherly and interdimensional energy.
Thanks to this connection, we achieved a happier, harmonious and fulfilled life, and accomplished to mostly free ourselves from material slavery and the global tumult of the macro society.
We would like to meet other people who feel such an inner search for truth, goodness and beauty within themselves, who through an inner understanding, respectfully and with a pure heart want to take a step into a life in harmony with harmonious, universal connectedness.
This initiatory movement was found in the 70's. We have since then participants all over the world and offer people the possibility to integrate themselves into a community, or to themselves create a small community or to take part with us in internet meetings.

We are, so to say, an energetic family, which is in contact with a spiritual family, that has often manifested through precise messages in the sky and who is leading us to the return into our immortal figure.
The pictures on page 9 are from our virtual community, which can be visited with an avatar from home. We meet there online weekly with others from all over the world.

☺ respectful and harmonious living together/search for truth/being vegetarian. No drugs, alcohol and tobacco.

☆ La Bolina

Calle Escuela 6 | Saleres El Valle de Lecrín, Granada
info@labolina.org www.labolina.org

(near Málaga)

La Bolina is a collective of people from Europe, West Africa, South America and the Middle East who live and work together in the rural village of Saleres, Valle de Lecrín, Granada, Spain. We have diverse identities, legal statuses and professions. What brings us together is a shared vision around integration, sustainability and regeneration. Our aim is to integrate refugees and migrants in rural environments, through the creation of sustainable livelihoods based on agroecology, permaculture and sustainable commerce.
We provide trainings, develop micro-enterprises and sustainable livelihoods, set up housing cooperatives and coordinate cross cultural integration to transform multilevel challenges into genuine opportunities.

Our values are:
LIVING COMPLEXITY: a holistic approach – adapting and responding to the complexity of living systems.
TAKING ACTION: a commitment to taking courageous, responsible and positive action.
CREATIVITY: to think, do and act with creativity.
RESPECT: to meet all humans as humans and embody the values of equality, love and respect.
REGENERATION: to embed ecological regeneration in all that we do, acknowledging the non-separateness of ourselves and nature.

Lakabe

Valle de Arce | 31438 Nafarroa - Euskadi

+34-948-392 002

lakabe.org

1980 550 ha (near Pamplona (Navarra)) ♀11 ♂9 7 20

Eusquera, Catalan, f, sp all

weekly meeting, circles, general assemblies (consensus)

shared economy (shared income and capital) occupants

non-violent action, political activity in region

farming, organic bakery, organic animal husbandry, village reconstruction, self-sufficiency, "amorro energetico"

ecological, self-managed, anarchistic, left-wing political understanding, feminist, non-violent, environmental protection, self-sufficiency, permaculture, organic agriculture, organic farming, forestry, alternative energy, ecological construction, construction, no cell phones/wireless, holistic management of land, water management

Lakabe is a village that was occupied in 1980. It developed from the antimilitarist movement. Its goal is to put philosophical principles of active non-violence into practise. We foresee the development of a small society based on various social parameters such as equality, justice, anti-patriarchy, feminism...
Thus from the very first moment, Lakabe witnessed the development of a community. Decisions are made at the assembly, the earth and the livestock are treated on an ecological basis, the forest and the surroundings are cared for. We are conducting experiments with alternative energy forms. At the moment we are self-sufficient and use sunlight as well as water and wind power (along with a generator for emergencies). We bake ecological full-grain bread for our own needs and to earn money for our village. We have rebuilt ten houses and the church with stone and wood according to organic construction principles and in the traditional style of the region.
We have shared economy; all resources belong to everybody. There is no private land. We remain an occupied village but the government of Navarra is not particularly interested in resolving this issue. We are now focusing a lot of the holistic management of the land.
Please do not come to visit us without contacting us first. We do not host spontaneous visitors.

→See also description of network "**RIE**"

IFFACE, Red de Pueblos Okupados, CEPAIN, Local Cooperatives

☺ First come to our open doors days, then follow the integration process of 2 years. consensus principle, life in a rural commune

Spain

Los Portales

Finca Los Portales | 41230 Castilblanco de los Arroyos ☎+34-690 845 928
info@losportales.net www.losportales.net

1984 200 ha (near Sevilla) 75 % 40

sp, f, e 8 35

different circles in a sociocratic structure (the decisions in every area are made by the corresponding circle and sometimes, need to be confirmed by the general circle)

property is owned by a limited company of the shareholders of which are community members

Jungian psychology

organic agriculture, horticulture, bread making, goat cheese making, olive oil production, hosting courses and workshops, hosting visitors, maintaining and developing renewable energy systems including homemade wind turbines, solar panels

ecological, self-managed, non-violent, spiritual, inner growth, meditation, yoga, organic diet, vegetarian diet, adult education, pedagogical, seminar-house, peace-work, environmental protection, self-sufficiency, permaculture, organic agriculture, organic gardening, organic farming, animal friendly husbandry, forestry, alternative energy, new technologies, ecological construction, humour, accessible to the disabled, dreamwork, process work, family and systemic constellations, community building, ZEGG-style-forum, jungian therapy, ESC (European Solidarity Corps), deep demoracy, human permaculture, music, dancing, singing, sociocracy, natural remedies and cosmetics

☺

This is a space to learn, grow and care for the human spirit and the planet.
Los Portales is an intentional community that practices innovative ways of life in different areas including organic agriculture, holistic education, deep psychology, empowering and power sharing systems of governance, conflict resolution, natural medicines, art, clean energies, an economy based on sharing, and above all, the empowerment of each individual, continually advancing towards greater sustainability and self-sufficiency.
Founded in Andalucía in 1984 (although born in 1978 in Brussels), the inspiration of the founders was nurtured by Jungian psychology and dreamwork as a way to know and transform our fears and pains into power, and to access our hidden talents.
We have a firm commitment to personal and global development and harmony through respect for oneself, each other and our environment. We conceive the community not as an end in itself, but as a means of creation and a path of transformation. Our mission is constantly evolving to respond to the global need for change at this very special time in the history of humanity.
After 34 years of existence, Los Portales is a well-established, stable and broad project that has developed in many directions and integrated the main dimensions of sustainability.

→See also description of networks **"RIE, GEN Europe, WWOOF"**

The Fellowship of Intentional Communities, Red Andaluza de Semillas, local trade networks

☺ financial participation of 200 EUR/month, commitment to inner work using the tools we apply in the community, participation in an Experience Week, two retreats annually and participation in the daily work of the community

Matavenero y Poibueno Pueblo Ecológico Europeo

Lista de correos | 24300 Bembibre, Leon ☎+34-987-190 652
joernickes@hotmail.de, contacto@matavenero.org www.matavenero.org

1989 3000 ha (near Bembibre/Astorga/Ponferrada/Leon)
♀20 ♂30 18 68

g, sp., e 8
everyone in common (consensus) limited community financing and individual
restored and administered as official community
no
organic gardening, wood gathering, preserving village centre, pedagogy, hospitality
ecological, self-managed, anarchistic, spiritual, esoteric/New Age, chaotic, inner growth, meditation, yoga, sexuality, organic diet, vegetarian diet, adult education, pedagogical, free school, environmental protection, self-sufficiency, permaculture, organic agriculture, organic gardening, organic farming, alternative energy, ecological construction, humour, music, drumming, jewelry, wood, chestnuts, sewing, sport: pushing barrows up the hills, hippie soccer, painting, dancing, fiestas, juggling, watching movies, games; village-gossip; international, sweat lodge, solar electricity
☺ ()

Matavenero and Poibueno are two sister villages at an altitude of ca. 1,000 m in the mountains of North Western Spain. Before our first "pioneers" arrived in 1989 the villages had been abandoned. Destroyed by a fire, a core of four buildings survived, which we repaired

for the community's infrastructure (bakery, school, shop, alcohol-free bar, communal workshop, meeting and banquet hall, guest space). Settlement began in tepees and tents, then we built huts and wooden houses. We grew to 100 people from around 10 countries. It later declined and is now on the rise once more. The basic idea - partially influenced by the Rainbow Movement - is ecological living and working close to and in harmony with nature - close to and in co-operation with like-minded people - as well as the development of our own pedagogy. Since 1993 we have been formally recognised as a village with our own administration. We fund community projects such as the water system etc. from our shared purse. We have a free school with a Spanish teacher, an on and off bakery and a co-op store, professional musicians and a lot more. Big animals are welcome by prior notice, but there are already more than enough cats and dogs. It's hard to earn money here and only a minimalist economy is possible. We practice a lot of recycling and bartering. Partly we live kind of neo-hippie-tribal-culture.
The village can only be reached on foot or on horseback. We generate electricity with solar collectors and grow our own organic vegetables. Thanks to birth-experienced women and men, more than 50 children have been born here. We practice alternative medicine as much as possible (homeopathy, herbs, Bach flower remedies, raw vegetables and fruits, but also allopathic first aid etc.).
The best time for visiting is May-September. Spanish skills are desirable. Financial independence is necessary. Pioneer-spirit remains important.

→See also description of network "**RIE**"
☺ one-year stay (exceptions are possible!)
+ cooperation + acceptance by the council

Proyecto O Couso

LUGAR O COUSO, S/N | 27620 SAMOS ☎+34-982 157 684
info@dharana.org http://proyectocouso.org

2014 (near Samos) 20 % ♀3 ♂7 10

e, sp, Catalan, Galician 10 10
members of the foundation (consensus) limited common financing
foundation
no ideological principles
education, agriculture
ecological, self-managed, non-violent, spiritual, esoteric/New Age, inner growth, meditation, vegetarian diet, vegan diet, pedagogical, free school, peace-work, one-world development, animal protection, environmental protection, self-sufficiency, permaculture, organic agriculture, organic gardening, organic farming, alternative energy, new technologies, ecological construction, humour, non-smoking

We are creating an integral and open community, where the values of the New Ethical Culture are put into practice. We have been working for some time on mutual support and cooperation and especially on the universal values that help to improve our world. All this under our main motto: "leave what you can, take what you need".

☺ no conditions

Sunseed Desert Technology

Los Molinos de Río Aguas, buzón 2 | 04270 Sorbas, Almería +34-950-525 770
sunseed@sunseed.org.uk www.sunseed.org.uk

1986 32 ha (near Almería) ♀9 ♂9 18

e, sp 15-35

community, staff, trustees
(communal meetings, meetings of staff team, meeting of trustees)

individual financing The Sunseed Trust (GB)

low tech solutions to climate change

low impact education and training for young people

ecological, self-managed, organic diet, vegetarian diet, vegan diet, environmental protection, self-sufficiency, permaculture, organic agriculture, organic gardening, organic farming, alternative energy, ecological construction, low impact, off-grid, appropriate technology, sustainability

☺ ()

Sunseed Desert Technology is the Spanish Project of The Sunseed Trust. We are a group up to 35 people, 10 full-time staff members and short and long-term volunteers who stay from 2 weeks to one year and make a financial contribution of 90-120 EUR/week to cover their food and accommodation (the amount depends on the length of stay).

Sunseed Desert Technology aims to develop, demonstrate and communicate accessible, low-tech solutions to climate change. We live in a low impact way, practicing off-grid community living, growing some of our own food, composting, reusing and recycling our waste as far as possible and making full use of solar energy and other appropriate technologies.

Volunteers are welcome throughout the year to work in all our departments: appropriate technology, organic gardening, drylands management, sustainable living, eco-maintenance and communication & education. All members of the group share the cooking and cleaning duties. For more information on joining us, please see our website or contact us.

→See also description of networks "**GEN Europe, RIE**"

REPESEI

☺ weekly donation for room and board: 90-120 Euros

Tanquián Finca Ecológica Tanquián

Tanquián s/n | 27437 Deade, Pantón, Lugo, Galicia ☎+34-672 650 728
emmytanquian@gmail.com tanquian.es

1993 5.5 ha (near Santiago de Compostela) 60 % ♀1 1
1-10

- g, e, sp
- all who are concerned (together)
- individual financing, hopefully shared economy in the future
- private property
- no other than a strong desire to stop harming the planet
- self-sufficiency, veg, jam and juice production, workshops and retreats
- ecological, self-managed, left-wing political understanding, non-violent, yoga, organic diet, vegetarian diet, seminar-house, self-sufficiency, permaculture, organic agriculture, animal friendly husbandry, forestry, ecological construction, humour, permaculture, yoga, self-sufficiency, ecoturism

☺

Tanquián is a project which was started in 1993 by Paul Baker and Emmely Föhring. The initial ideas were: to raise the three kids in contact with nature, to become self-sufficient and to live in a community. The farm is 5.5 ha big and holds a mix of woodland and meadows. Our main crops are fruit trees and bushes and the big housegarden. In the past we kept all kind of animals, but right now we have only a horse and a donkey. Since 2008 we have run a conserve-kitchen, where we produce jams and juices from our own harvest.
During the years various families and many volunteers lived with us. Right now we are looking for qualified members, especially handy wo/men and gardeners. We can offer food, lodging and share of profit.

→See also description of network "**WWOOF**"
Workaway
☺ only skilled craftspersons or experienced gardeners (best female), secretary

☆ **Valle Despierta**

El Hoyo de Pinares (Tierra de Pinares) | Ávila
monicabioconstruccion@gmail.com

2003 15

We are a small community where we live together in different families. Our projects are focused on life in the countryside. We share a natural environment in Tierra de Pinares, surrounded by vines and wooded mountains of pines, oaks, cade juniper...
We develop different activities. We work in the recovery of traditional trades such as carpentry, beekeeping, natural wine making, bio-construction, renewable energies, music, gardening, recovery of abandoned vineyards... We collect rainwater and have dry toilets and solar energy for electricity.
We organize open days in the solstices and equinoxes. We also organize courses and workshops on bio-construction, voluntary work and learning days on family beekeeping, wine making, yoga, retreats... and days of coexistence.
We are united by our love for nature and we have a "principle" that is essential for the proper functioning of coexistence: mutual respect.

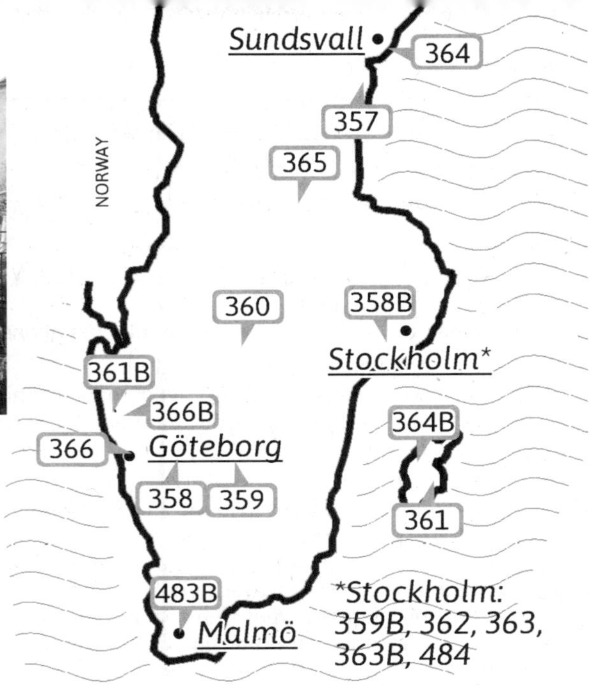

Sweden

We can present 18 intentional communities from Sweden, two of them planned (pages 483f). Additionally, there are three Swedish Camphill communities in the list on page 514 and five addresses from the last edition that haven't gotten back to us or have ceased on page 504. There is one community network called Kollektivhus NU (page 529f) and a national GEN branch called Ekobyarnas Riksorganisation which has not returned our calls, but can be found at ekobyar.se in the internet. There's actually a mentioning of Bullerby on page 363 (Ängsviksgården, photo above)...

Bålarna

Bålarna 238 | 82951 Bergsjö — +46-(0)652-712 69

1991 — 43 ha (near Sundsvall) — 10 % — ♀4 ♂5 0 9

Swedish — 8

all who are concerned (consensus) — individual financing

property of all

forest harvesting

ecological, permaculture, organic gardening, organic farming, forestry, alternative energy, ecological construction, accessible to the disabled

We started this ecological village 28 years ago and we still live here. All children had moved out, and now they are returning with their partners/children to develop the village further. Some years ago, we invested in solar cells and we are still very satisfied with them. Even a new bakery house has been built from timber from our own forest.

Njord

Sweden

Bogemenskapsföreningen Tunnan

Tunnlandsgatan 9 | 50744 Borås ☎+46-(0)73-443 44 36
ordforande@tunnan.org www.tunnan.org

1987 ♀18 ♂17 10 45

Swedish, e, Finnish, Arabic, sp, Bosnian, f, Dutch, Danish, Russian, Norwegian, Japanese, Thai, Congolese

community board, advised by all members (consensus)

common financing as part of rent — rented

communal meals, gardening, maintenance of house and premises

ecological, self-managed, environmental protection, organic gardening, accessible to the disabled

☺

We are a co-housing project in Borås, close to Gothenburg. We live in a house with six floors, with 27 flats ranging from 30-120 m^2. Every resident is a member and we decide about activities and resolve issues together during communal meetings.

We strive towards having an even mix of residents in the house: age-, gender- and culture-wise and with a mix of family types. The goal is to provide a friendly living atmosphere where all residents feel safe, welcome and part of the house community. We have an ecological profile and strive to make eco-friendly choices in all activities.

What makes Tunnan special is our communal areas with dining room, large kitchen, playrooms for kids, sauna, gym and training area, pub, art studio, music studio, wood working shop, sewing studio and guest accommodation as well as a large garden with patio, greenhouse and gardening area. We have communal dinners and Swedish "fika" (with coffee and fikabröd) 2-3 times per week. We also celebrate different holidays from our range of nationalities together.

If you happen to pass by, don't hesitate to contact us for a visit or stay a few nights in our guestroom!

→See also description of network **"Kollektivhus NU"**

☺ become a resident

Charlottendals gård

153 90 Järna ☎+46-(0)70-173 86 17, -(0)8-551 700 97, -(0)73-954 86 36
peter.hagerrot@telia.com, merle.hagerrot@telia.com charlottendal.se

1995 35 ha (near Järna, Södertälje) 20 % ♀7 ♂5 5 17

Swedish, e, g, f, Estonian — 4

mix of private and housing board (working on sociocracy) — private

farm, kindergarten

ecological, inner growth, yoga, environmental protection, organic agriculture, organic gardening, forestry, alternative energy, ecological construction

☺ (✋✋)

Charlottendal is an old settlement with a hill overlooking the "garn", once a sailable sea fjord that gave Järna its name. The existing residential building – today a women's collective – was built at the turn of the last century as part of the "National Movement Against Emigration to America", providing farmsteads for people to sustain themselves on a small size farm. Nowadays, the small farm was not big enough to generate profit as a farm. 35 hectares were too much to be used only as a residency, though. This situation created an opportunity to use the property in new ways, to grow food, to cooperate between neighbors, to enhance quality of life.
Recently Charlottendal has been going through a new phase of development. We have built six wooden residential houses and a Waldorf kindergarten all with sustainable building materials and waste management. Also the community's first yurt was built at the end of 2016, and it has been followed by two more yurts and some trailers. The first group of long-term volunteers joined Charlottendal in 2018 as part of the European Voluntary Service programme, living in a newly constructed log house
Charlottendal regularly hosts events, ceremonies and courses, among which the annual "One Year in Transition" course that's partly held at the farm.

→See also description of network "**ERO/GEN Sweden**"
transition network

Ecovillage Smeden

Korgebogvägen 67 | 55308 Jönköping ☎+46-(0)70-819 95 87
pia_larsson@swipnet.se www.ekobyn.se

1994 4800 m² (24 x 200 m²) (near Jönköping) ♀26 ♂27 20 73

- mainly property of one member
- traditional vegetable gardening for a few inhabitants, hens and chickens
- self-managed, organic gardening, alternative energy, ecological construction

☺

The former ecovillage initiative on ca. 4 ha now turned into individually owned homes for 24 households. Good and friendly living for both people and the environment.

☺ You have to buy a semi-detached house to live in the village.

Färdknäppen Kollektiv

a housing project for people in the second half of life

Fatburgsgatan 29B | 11854 Stockholm ☎+46-(0)8-669 17 89
kerstin@karnekull.se www.fardknappen.se

1993 ♀38 ♂19 0 57

- Swedish, g, e, f, sp, Norwegian, Finnish 5
- all at common meetings, five times a year (majority vote)
- limited common financing rented
- collaborative housing for people in the second half of life

self-managed, adult education, humour, accessible to the disabled, non-smoking, second half of life, self-administered groups for cooking and cleaning, culture, gardening

☺

Färdknäppen is a publicly owned apartment house located on Södermalm in Stockholm. Here we are developing a community in our daily lives, which makes life easier and more fun when the "togetherness" of family life and working life diminishes as we grow older.

- We have cooking groups that prepare communal meals on weekdays.
- We take care of our house and garden ourselves.
- We learn from one another while carrying out our daily chores.
- We can help each other keeping up our spirits if – or when – our strength declines.

→See also description of network "**Kollektivhus NU**"
☺ being interested in collaborative housing

Goda händer ekoby

Råbäck 101 | 693 94 Åtorp ☎+46-(0)586-731 060, -785 078
godahander@gmail.com www.godahander.nu

2011 12 ha (near Laxå) 10 % ♀2 ♂2 4 8

Swedish, e, Arabic, g, f 1
all who are concerned (consensus) limited common financing
property of all
none
create an ecovillage; social aspects, ecology, economy; practical work is often cultivation, construction work, forestry and work with common property or housing
ecological, self-managed, organic diet, organic gardening, forestry, ecological construction, community building

☺ ()

Goda Händer Ekoby is an ecovillage that strives for a sustainable lifestyle where we take care of local resources and respect the limitations of nature. We work with practical issues such as building ecologically and cultivating our food. The ecovillage is also a lifestyle project, where we want to create a community which promotes sustainable values and life quality. Instead of stress and materialism we want more time for nurturing our relationships and realizing our dreams. We also wish to build up a local market where we can cooperate with our neighbors and keep the chain from producer to consumer as short as possible.

We value our community. We regard relations and cooperation as the most important components in our ecovillage. We eat dinner together twice a week. Several times a month we have meetings where we decide upon big and small issues. One weekend every month and one week in the summer we have communal working days. Even though we are separate households we also share social life; for example invite each other to birthday parties, go swimming on the beach together or just work side by side in the garden.

→See also description of network "**ERO/GEN Sweden**"
☺ passionate about ecologically sustainable lifestyles and community building

☆ Gotlands Djurfristad Stiftelsen Väderkvarnen

Garde Smiss 154 | 62363 Ljugarn ☎+46-(0)70-486 05 59
stift.vaderkvarnen@gmail.com gotlandsdjurfristad.se

1980 100 ha (near Visby) 50 % ♀3 ♂2 0

Swedish, e 2
all who are concerned (consensus) limited common financing
foundation
vegan, nature-free-spiritual love
eco-construction, organic gardening; animal sanctuary
ecological, self-managed, anarchistic, left-wing political understanding, feminist, non-violent, spiritual, Buddhist, inner growth, meditation, yoga, tantra, sexuality, free love, organic diet, vegan diet, pedagogical, peace-work, one-world development, animal protection, environmental protection, self-sufficiency, permaculture, organic agriculture, organic gardening, organic farming, animal friendly husbandry, forestry, alternative energy, ecological construction, humour, non-smoking

☺

We live in the countryside 35 km from the city Visby on the eco-community-island Gotland. Seven years ago we started an animal sanctuary on the farm with 150 animals that we have saved! We have 4 dogs, 4 cats, 4 horses, 4 mountain-bulls, 6 pigs, 100 sheep and 30 hens on the farm. We hope to stop all animal-industry as soon as possible!
Also some wild animals are hanging around in our forest and have a free place without hunting on our land! A part of our land is a nature reservation with some unique species. Gotland is a very beautiful and magical island and a very exciting place to visit! Love to all!

Kollektivkontakten Sweden
☺ vegan, no drugs, peace minded, self-sufficient, living love

☆ Hästekasen farm

Permaculture, small scale farming and natural building
Hästekasvägen 6 | 45196 Uddevalla ☎+46-(0)72-151 22 44
info@hastekasen.se www.hastekasen.se

2010 11 ha 25 % ♀3 ♂3 6

e, Swedish 1.5 2-6
investors of money, time, skills, commitment (depends on level of question; long or short term)
courses, outside work, food production, grants
rented by association and also owned by private owners
documentary "There is no tomorrow"
natural building, permaculture gardening, courses, youthwork
ecological, hierarchical, yoga, vegetarian diet, pedagogical, environmental protection, self-sufficiency, permaculture, organic gardening, organic farming, forestry, ecological construction, no cell phones/wireless, non-smoking, adventure, story telling

☺ ()

Sweden

Hästekasen Farm is a collective, an association and a small scale permaculture farm. Depending on season a bunch of people lives or visits here (many during summer, few during winter). We are aiming for a self sustainable traditional homestead. In the future we hope that more of the inhabitants of this small valley will be able to work on the site and make their living here, off the land and it's resources.
We use ecological permaculture principles. The association's aim is to learn about and spread knowledge and skills regarding nature, ecosystems, small scale food production and natural building of small structures. Solving basic needs about: water, food, sewage, recycle, compost, warmth, shelter etc. Our activities also include health and personal development and adventure in nature.
We are situated 1.5 hours north of Gothenburg, close to a fjord that leads to the north sea.

→See also description of network **"GEN Europe"**
- collapsology
- ☺ like-minded in regards to work ethics, practical work, no addictions, no nicotine, no drugs, no industrial meat, low internet use

Kollektivet Världshuset Vita Hästen

Ringarstigen 3 | 16858 Bromma ☎+46-(0)837 81 17
info@kollektivetvitahasten.com www.kollektivetvitahasten.com

1970 | 200 m² (near Stockholm) | ♀5 ♂5 10

- Swedish, e, g, sp
- everyone in common (discussions and consensus)
- limited community financing — foundation
- vegetarian, community, solidarity, equality
- social life
- ecological, self-managed, left-wing political understanding, feminist, non-violent, inner growth, meditation, yoga, free love, organic diet, vegan diet, peace-work, one-world development, animal protection, environmental protection, self-sufficiency, permaculture, organic agriculture, organic gardening, humour, art, great parties, music

☺

The Kollektivet Världshuset Vita Hästen ("White Horse Collective") is a living community, constituted by the wishes and needs of its members.
We have all agreed on a set of principles that we follow:

respect for life, respect for each other, respect for the house, sustainability.

We honor our roots as a creative and engaged community. We encourage both community individual spiritual, creative, physical and political development. We want to cultivate openness and stand for gender, ethnic, social, interspecies and global equality.
By living together with shared values we are an alternative to how people traditionally live in the cities. We aim to be an affordable household so that people can have more freedom to follow their passions and can distance themselves from the traditional financial constraints that might affect their work/life balance. As a community we encourage awareness and participation in our political, social, and creative environment regarding issues that are in line with this mission statement.

☺ veganism

Kollektivhus Angsviksgarden (Ängsviksgården)

Ängsviksvägen 10b | 13950 Värmdö
info@angsviksgarden.se www.angsviksgarden.se

(near Stockholm) ♀12 ♂8 12 32

Swedish, e
all who are concerned (majority vote)
no
conventional gardening, shared meals, extended neighbourhood

none none
individual financing

Ängsviksgården is a community outside of Stockholm, close to the archipelago. We are a cohousing community with shared meals in our community rooms and a big garden that we cultivate together. The collective activities depend on the people living here. Right now we enjoy having chickens and using our new floating sauna on the lake. The beautiful woods and the closeness to the water/sea combined with excellent daycare and school nearby makes us feel like we live in "Bullerby" :-). *Photo on page 357*

→See also description of network "**Kollektivhus NU**"
☺ people have to buy an apartment (bostadsrätt)

Kollektivhuset Södra Station c/o Sofia Rickberg

Magnus Ladulåsgatan 15–21 | 11865 Stockholm ☎+46 (0)739 810 959
sofia.rickberg@gmail.com sodrastation.org

1987 ♀ca. 70 ♂ca. 60 ca. 40 ca. 170

e, g, f, sp
some members work from home in their apartments
the board and the house-meeting (majority vote, some by qualified majority)
Individual for apartments, common purse for shared food Monday-Thursday dinners
individual property
eating dinners together (Monday-Thursday), organising a bar with music, sewing, dancing, yoga
non-religious, non-political community

☺

We are a co-house community located in a calm spot of inner city Stockholm. We all own our apartments but share spaces where we can dine Monday-Thursday, play and work out, have pub evenings, play music and pingpong, paint, sew and woodwork. We welcome external members for dinner/cooking teams and together we cook approx. 80 portions of food. There are vegan/vegetarian alternatives served every day. We also welcome planned study visits from international groups but we cannot offer accomodation, only dinner at self-cost price.

→See also description of network "**Kollektivhus NU**"
☺ buy an apartment in the buildings

Sweden

Rumpans by

Pråmviken 128 | 86296 Njurunda ☎+46-(0)70-399 88 68
anarkab@telia.com

1968 18 ha (near Sundsvall) ♀25-40 ♂25-40 40
90-120

e, Swedish 3
the whole village (consensus) individual financing
29 plots, owned by private persons
social activities, maintenance, new projects
ecological, self-managed, organic gardening, conventional gardening, animal friendly husbandry, forestry, alternative energy, ecological construction

☺

The settlement is on 18 ha of land, based on a social vision. Individual holiday and dwelling houses belong to the individual families: small-scale wastewater treatment, own well, own harbour with boathouses and a sauna, eco-cycle adapted infrastructure, windmill, firewood from our own forest, local sawmill, farming, cattles. Project started 1966. All buildings are built by the owners themselves by using local materials. There is a book in Swedish describing the village during 50 years and an article at eurotopia.directory/the-rumpan-village

Suderbyn Permaculture Ecovillage

Toftavägen 211, Västerhejde | 62261 Visby, Gotland ☎+46 (0)76 218 73 69
info@suderbyn.se suderbyn.se

2008 5 ha (near Visby (8 km)) 20 % 1 25

e, sp, f, g, Swedish, Russian, Ukrainian, Dutch, Romanian 3 20
community together, in extraordinary questions: Cooperative board, NGO council (consent)
paid food and rent, publicly funded NGO projects, donations
property of all cooperative members
voluntary simplicity, permaculture, diversity, non-violence
non-formal education, international projects, DIY renewable energy, gardening, eco-building, experiments, community development, outreach
ecological, self-managed, feminist, non-violent, chaotic, inner growth, meditation, yoga, sexuality, organic diet, vegan diet, adult education, pedagogical, seminar-house, peace-work, environmental protection, self-sufficiency, permaculture, organic agriculture, organic gardening, organic farming, alternative energy, new technologies, ecological construction, construction, humour, cultural diversity, experiment, social laboratory

☺ ( )

Suderbyn Ecovillage is an intentional community of open-minded people from various countries and cultures, located on the island Gotland in the middle of the Baltic Sea. All of us who live here are building the place - not only physically but also socially.
Our goal is to live in a way that creates a prosperous living environment while minimizing our environmental footprint. In creating Suderbyn Ecovillage together, we aim to live close to nature and achieve a more sustainable lifestyle, striving for self-sufficiency in food production and renewable energy, prioritizing ecological and local materials.

We question patterns of consumption, social structures and cultural stereotypes. We experiment in different areas with innovation and traditional knowledge with the attempt to create a model of healthy and happy society, enjoying the diversity of people. We work with educational projects on the local and international levels. We regenerate the connections between people and between people and nature.

→See also description of networks "**GEN Europe, ERO/GEN Sweden**"

ECOLISE, BEN (Baltic Ecovillage Network), SCI-Sweden (Service Civil International), Transition Gotland, Permaculture Sweden

☺ two weeks test living

Tersen

Tegelvägen 18 A | 79135 Falun ☎+46-(0)706 651 002

john.fletcher@tersen.se www.tersen.se

2005 2600 m² individual plus 400 m² common rooms

♀37 ♂18 0 55

Swedish, e, g, f

no paid workers

house meetings open to all (majority vote (if needed))

partly individual financing, mostly bank loan

property of all

no

community life, house management, garden

self-managed

☺

The Tersen cohousing community offers an environment which actively encourages co-operation and friendship between the inhabitants. We get to know each other. We learn from and with our neighbours and develop new joint interests. We develop space for living active lives.

We own our house and help one another in managing and developing Tersen. Monthly 'house meetings' offer the opportunity to share ideas and needs in our home. We cook and eat Friday dinners together. If you enjoy being a member in a group taking care of for example our garden, our house, our finances or being a member of our board of directors, you would be likely to enjoy Tersen.

Each inhabitant has his/her own apartment with a door that can be closed for privacy. At the same time, Tersen offers access to a large restaurant kitchen and dining room, a TV room, a library, a carpentry shop, a textile room, a gym and a sauna bath – and guest rooms for family and friends. Spontaneous activities flourish.

Our security is based upon caring for each other. Beyond that, local community social and medical services are fully available.

We share resources such as a car pool, shared bikes, computers, newspapers etc.

→See also description of network "**Kollektivhus NU**"

Kollektivhus.NU, Sweden

☺ > 40 years, no children

The Fig Tree – A Jona House Resistance and Catholic Worker Community c/o Per Herngren

Sandeslätt 54 | 424 36 Angered
+46-(0)708 877 211
konfliktingripande@gmail.com
ickevald.net/catholicworker/english/

1999 (near Göteborg) 5 % ♀3 ♂3 6

Swedish, e, Arabic 1 3

working groups (consensus and feminist meeting methods)

rented and property

nonviolence, civil disobedience, queer-feminism

nonviolence training, housing for refugees, organize locally

self-managed, anarchistic, left-wing political understanding, feminist, Christian, ecumenical, anthroposophic, adult education, pedagogical, peace-work, one-world development, alternative energy, ecological construction

☺

The Fig Tree Community is an ecumenical nonviolent resistance community located in Hammarkullen, Gothenburg, Falköping (Sweden) and Cairo (Egypt). Hammarkullen (Hammer Hill) is one of Sweden's most ethnically diverse neighborhoods, with people from 130 nations and 70 languages. The Community is part of a molecule of communities in Hammarkullen. Different communities organize resistance, soup kitchens, reflections, parties. Rather than thinking the Fig Tree Community as a whole it works like a rhizome and a molecule: becoming, connecting, departing, moving, resting. We have been focused on nonviolence training in Europe, Kurdistan, Kyrgyzstan, Turkey and Iraq, a soup kitchen on Fridays in Hammarkullen and we have built a house for three women from Romania who were working in Sweden as beggars. Our community is inspired by the tradition of Jona House in Baltimore, and the Catholic Worker movement.

→See also description of network "**European Catholic Worker**"

☺ working with nonviolence, civil disobedience or the local community

Utsikten ekoby Lilla Krossekärr

Krossekärr 705 | 473 94 Henån
+46-(0)705-479 945 (Rickard), +46-(0)708-7905 (Marie)
marie@fotoochform.se
www.utsikten-ekoby.se

2001 (near Uddevalla) 10 % ♀15 ♂10 11 36

Swedish, e, g, f

individual financing

the houses are privately owned

main goal is inspirational

ecological

Life in the eco-village in Lilla Krossekärr is characterized by a mixture of the private and the collective. Each house is unique in terms of both form and material, but the houses together form a whole through a harmonizing color range.
The village's operations are organized in a board and working groups. The groups do much of the ongoing work such as forest management, land management and renovation. One concrete result is the common agricultural land, divided into small cultivation lots. The barn, the recycling house, the tool shed and the hives are other collective spaces. We are now a part of a forestry garden projekt.

Switzerland

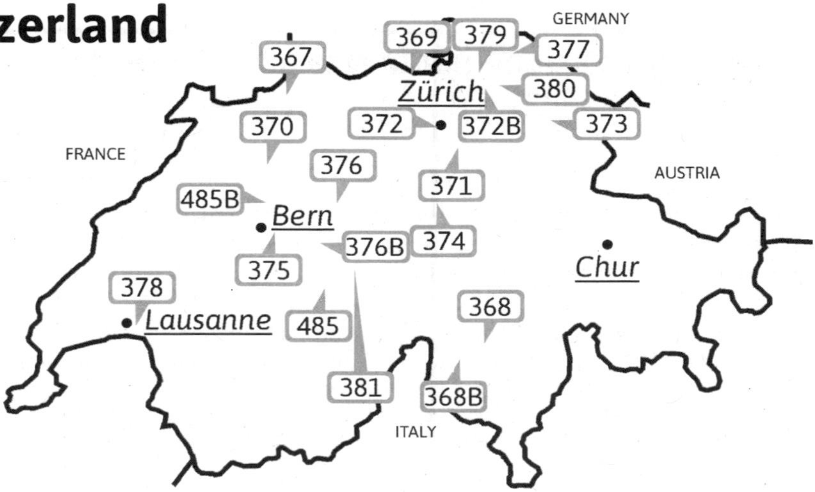

In the last edition of eurotopia, we published 8 intentional communities from Switzerland - this time it's 20. 13 of those are new to eurotopia, two of the 20 are still planned (page 485). Four communities from the last edition are listed on page 504. There's Common Life Network (page 519) as well as GEN Suisse at gen-suisse.ch as Swiss community networks (don't mix up with gensuisse.ch, which seems to be a GMO lobby site...)

☆ Arche im Nauen

Unterdorfstr. 18 | 4143 Dornach ☎+41-(0)61-599 70 20
arche@imnauen.ch www.imnauen.ch

1985 (near Basel) ♀12 ♂12 24

g 90 %

everyone together/some decisions made by association board (majority vote)

individual financing and state subsidies — association property

Christian, ecumenical

living and working together in house communities (incl. assisting 'mentally disabled' residents)

spiritual, Christian, ecumenical, inner growth, pedagogical, mentally/psychologically handicapped persons

☺

The Ark Community Im Nauen opened its first house in Hochwald (SO) near Basel in November 1985. In September 1991 it moved into the second house in Dornach. Since 2014 all flat-sharing communities and the workshop have been in Dornach. During the day, various handicrafts are made in our workshop.
As a community we would like to live with each other on the basis of the Gospel with our abilities and limits. We always welcome and look for people who would like to live and work in our community; there is the possibility to complete a professional internship or to perform a civilian alternative service. The main areas of work are the shaping of life with cognitively impaired people in flat-sharing communities, nursing care, household and workshop.

→See also description of network **"L'Arche (Jean Vanier)"**

☺ a previous visit to get acquainted, later discussions

Photo on page 12

Ces - Fondazione per la rinascita di Chiesso

Monte di Cés (Chiesso)
Ces, c.p. 15 | 6747 Chironico (Ticino) +41-(0)91-865 14 14
info@cesnet.ch www.cesnet.ch

1972 10 ha (near Bellinzona) <10 % ♀2 ♂2 0 4

g, i, e, f 4 4 (from May to October)
all who are concerned (consensus) limited community financing
Fondazione (foundation)
visitors, gardening, construction, landscape conservation, farming
ecological, self-managed, left-wing political understanding, non-violent, organic diet, adult education, seminar-house, environmental protection, self-sufficiency, organic agriculture, organic gardening, organic farming, animal friendly husbandry, forestry, alternative energy, ecological construction, group work

CES is a formerly abandoned mountain village on a high plateau in Northern Ticino (1450 m above sea level). Ces is far away from the traffic chaos and can only be reached on foot, two hours uphill. Already on the way up one encounters a varied landscape. In Ces itself the naturally cultivated plateau with its waterfall, moor areas and rare flowers is striking.
Ces consists of some 25 houses and barns and a chapel. One third of the houses belong to us, the self-administered "Foundation for the Reconstruction of Ces". Since 1972 we have been dedicating ourselves as a collective to the preservation of the village and its surroundings in their original form and to the accessibility of Ces for alternative communal forms of living and working. In winter Ces is quiet, in summer we cultivate our organic garden, care for our guests and for the animals, make hay and chat; we renovate, play and discuss.
We consciously choose to live simply, and work with our hands. Since we want to live as independently as possible, we produce our own electricity (with a small hydroelectric plant and a solar energy system) and live - where possible - from the produce of our organic gardens. Ces is remote but not isolated: it is intended to reach out and demonstrate alternatives. Each year a large circle of friends and guests from many countries come to visit. If you would like to visit or work here, please make sure to give us a call beforehand.

→See also description of network "**GEN Europe**"
La Ragnatela
ecological consciousness, social competence, agreement with community's goals

Cooperativa Pianta Monda Menzonio

Pianted | CH-6692 Menzonio +41-(0)91-754 18 53, -(0)794 176 721
www.piantamonda.ch

1993 3 ha (near Locarno (35 km)) ♀1-2 ♂3-4 0 4-6

g, e, it, f all, incl. externals
everyone in common (majority vote (2/3 majority))
limited common financing & gift-/giving-economy
co-operative & rented/leased
communal living/learning in harmony and creating a new holistic culture; reviving a beautiful abandoned mountain valley

- gardening, building, wood-cutting, tree-clearing, hay-making, caring for guests, living together, infrastructures, landscape management
- ecological, self-managed, non-violent, inner growth, organic diet, peace-work, one-world development, environmental protection, self-sufficiency, permaculture, organic agriculture, organic gardening, organic farming, alternative energy, ecological construction, humour, no cell phones/wireless, non-smoking, social works: camps for adolescents, crafts, media: own village-newspaper

☺

Slowly and persistently, the community of Pianta Monda is restoring the hamlet of Pianted with its 30 terraces and 10 buildings to create a small eco-village, 15 minutes above the village Menzonio at 850 m in the upper Maggia valley. In addition to the practical work in the mountainous terrain, learning of living-together and practicing independence is of central importance. Our team consists of 2-5 long-term members and is supported by short-term helpers, external co-operative members, a large circle of friends and even workcamps from all over the world.
Organic vegetables, herbs, berries and fruit are cultivated on the former terraces to maximize self-sufficiency. The rustici (=tradit. granite-buildings) are ecologically refurbished for the residents, groups, school classes and our agriturismo guest farm, which - together with the gardens - covers our basic needs. In addition to all this, we are dedicated to the care and preservation of our sunny terrain, the care for our decent infrastructure, seed duplication, soil improvement and biodiversity.
Autarky is important to us; we are combining old traditions with modern technologies: compost toilets with urine separation, solar energy, panels for 12V power, parabolic stoves, etc. We are one of the rare offline zones, non-profit, tax-exempt and mortgage-free.
We work hard and live simple and original, voluntary, drug-free and without salary. We are cooking and heating with wood and solar power and are using almost no machines. Our eco-village is open to all interested guests and offers free food and lodging for volunteers (when staying longer than seven days!).

- Bergheimat
- ☺ several weeks trial period (please pre-register!), motivation, idealism, team ability, joie de vivre

☆ Familiengemeinschaft Inanna As Life is meant.

Propsteiweg 24/25 | 5463 Wislikofen ☎+41-(0)565 253 093
inanna-family@posteo.ch

2018 (near Zürich (25 km)) ♀6 ♂5 5 16

- d, e, f, sp, Portuguese
- all members together (consensus) — shared economy
- rented
- ecological, non-violent, spiritual, inner growth, meditation, yoga, organic diet, vegetarian diet, vegan diet, pedagogical, peace-work, organic gardening, alternative energy, new technologies, ecological construction, humour

After several years of community building, we have been living together since autumn 2018 as a family community. We live in an old homestead, consisting of three houses, which were biologically renovated by the landlord (clay and straw insulation, rainwater harvest, solar panels, etc.).

The Community dares to have a fully shared economy, meaning sharing all our financial and material possessions and all the spaces. We want the money to stimulate dynamic and life-sustaining processes and not further strengthen the rating systems or interpersonal boundaries. We invite each other to develop our potential professionally. We are looking for groundbreaking ways to sustainable forms of abundance.
We relate to each other in an authentic, mindful, conscious, and benevolent way to nourish each others inner basis of self-acceptance, compassion, trust, and security. We value the diversity of individuals and encourage each other to personal responsibility and self-development. We create a child-friendly environment for our children with natural playgrounds and their own rooms, where we lovingly accompany them in their adult and learning area.
We have no spiritual, political or religious dogma. Each of us recognizes in his own way a spiritual reality greater than our own. Together we discover, explore and celebrate old and new rituals. We like to celebrate and enjoy the fullness of life.

Bio für Alle, GEN-Suisse

Gemeinschaft Kirschblüte

4574 Lüsslingen-Nennigkofen ☎+41-(0)76 510 02 71, +41-(0)76 568 24 25
juhit@gmx.ch, joshuan@live.de, mail@hof-zur-kirschbluete.ch
www.kirschbaumbluete.ch, www.samuel-widmer.ch

1995 (near Solothurn) 70 % ♀72 ♂46 77 195

g, e, f, sp, i varies all, not on a daily basis, rather on the "field"
all concerned (reaching a consensus in a joint conversation)
usually individual financing, joint financing within the Agricultural Cooperative and the Rössli Community Farm

individually, collectively or partially collectively
own businesses or employed in the surroundings
ecological, self-managed, left-wing political understanding, spiritual, inner growth, meditation, tantra, sexuality, free love, organic diet, vegetarian diet, vegan diet, adult education, pedagogical, seminar-house, peace-work, one-world development, mentally/psychologically handicapped persons, environmental protection, self-sufficiency, organic agriculture, organic gardening, forestry, alternative energy, humour, love, self-knowledge, psychotherapy, psycholysis, loving children, new history

What brought us together is, above all else, our interest in self-knowledge. It was through our interest in self-knowledge alongside our joyful engagement in Tantra that we gathered around Samuel Widmer Nicolet (died 18.1.2017) and Danièle Nicolet Widmer, who form the crystallization point for our community. As result of the therapeutic work and seminar activities of Samuel Widmer Nicolet and Danièle Nicolet Widmer, our community is embedded in a large and widely distributed network of community consisting of hundreds, if not thousands of people, in which it simultaneously forms the centre.
We are all motivated by the great questions of life: love, closeness, relationship, unbreakable bonds between people, liberated sexuality, education, capacity for happiness, enlighten-

ment... What does it really mean to be human? How can life be lived with dignity in a field of community, in which love and compassion are not bound by the limits of the couple relationship, but are allowed to develop further? Can love at the physical level also be completely free? Which conditioning, attained through education, society and the history of mankind to date, needs to be transcended? Our passion is to create a field in which it is not fear of loss and greed for possessions, but love that may unfold its power unhindered. In which we experiment and that is our joy.
In contrast to other communities, we see ourselves as a very airy community, like cherry blossom petals in the wind, who want to commit ourselves to few structures and no ideology. We reject as hostile to community all forms of authority and hierarchy that do not result as a matter of course from natural power relationships. From our origins in psychotherapy and self-knowledge, we devote ourselves to a deep spirituality that nevertheless works in the world without being politically oriented.

Editors note: This community and its founder have been discussed controversially in the past. Please do your own research and, as for any other community, take full responsibility for any approximation.

🌍 WorldWide Magic Movement
☺ New "members" usually come through existing friendships. You should want to live with us and be able to support yourself and your family. Financial and physical engagement in our wide range of current projects. *Photo on page 12*

☆ Gemeinschaft Unter Ort

Alte Landstr. 133 | 8804 Au ZH ☎+41-(0)79-735 08 28
✉claudiokol@yahoo.de

2016 200 m² (near Wädenswil (4 km), Zürich (20 km)) 2 %
♀4 ♂6 2 12

g 1
everybody (consensus) self-developed "game" for internal common economy
rented
non dogmatic
community building processus, gardening
ecological, self-managed, organic diet, environmental protection, permaculture

We live together in a big old farmhouse in the agglomeration of Zurich. The path from a student shared flat to a community was taken in 2016. The decision to join GEN was sealed with an orientation according to its four pillars (ecology, economy, social affairs and ideology). At the same time, we began working on an innovative solution for the research on common households. We had enough of cleaning plans and discussions about who cooks and when, and wished for a system in which everyone could do what they would like best or care about most. The individual stakes (time and/or money) are then matched against each other via an internal currency. With the birth of the first child in 2017, the family issue has become topical ... and dealing with the common areas has become a new challenge.
The second freshly baked family procured in 2018 a nearby, energy-efficient Tiny House for additional space. The community lives ... and if it does not die, then we still explore living together today.

→See also description of network "**GEN-Suisse**"

Genossenschaft Kalkbreite

Kalkbreitestr. 2 | 8003 Zürich
+41-(0)43 317 17 22
mail@kalkbreite.net
kalkbreite.net

2014 | 6350 m² | ♀145 ♂115 69 260

g | 3 10

general council (consensus) | solidarity fund

building right

2000-watt-society

volunteer work

ecological, self-managed, left-wing political understanding, non-violent, environmental protection, alternative energy, new technologies, ecological construction, construction, participative

☺

In 2006, a handful of district residents and professionals began to materialize their vision of a sustainable and innovative development of the Kalkbreite. One year later, the group was established as the broadly supported and well-networked Kalkbreite Cooperative, which was granted the 6350 m² site by the City of Zurich under building code. After many participatory processes, the vision was condensed into a project that mastered the complex local conditions and at the same time transformed the requirements of a socially and ecologically pioneering urban living space into contemporary architecture. The estate was opened in 2014.
Thanks to its location, its commercial enterprises and restaurants, its innovative living concept and excellent public transport connections, Kalkbreite has quickly developed into a lively district center.

Cooperative Suisse, Association Neugass, KOVI, wbg-zh

☺ member of the cooperative

Giesserei

by Gesewo (cooperative for self-managed living)
Ida-Sträuli-Str. 79 | 8404 Winterthur
+41-(0)52 511 83 00
info@giesserei-gesewo.ch
giesserei-gesewo.ch

2013 | 1.1 ha | 1 %

♀ca. 120 ♂ca. 120 ca. 70 ca. 310

g, (e, f, i) | some

general assembly of the tenants association (majority vote)

individual finaczing: Tenants have to pay 10% of the investment capital for their apartment

owned by a coop (Gesewo), the apartments and shops are rented out to members of the coop only

none

self-managed, ecological apartment complex with shops

ecological, self-managed, internal labour contribution obligatory for maintenance, administration, cultural & social activities

The multigenerational housing project Giesserei is a complex of 156 apartments and 14 business units (restaurant, a public library, art galleries, a center for music etc.). There are many communal rooms like a large assembly hall for internal and public events, workshops and guest rooms. The age distribution of the inhabitants is very close to the one of the Swiss population. We have members from 12 nations with 12 different mother tongues.
For the self-management of our community every member is obliged to work 30 hours per year. Practical tasks like the up-keeping and cleaning of the premises, gardening, and administrative and organizational tasks for cultural and social events offer opportunities for everybody to actively participate in our community.
The buildings meat the Minergie P Eco standard of the Swiss Association of Engineers and Architects and we are a so called car-free project with only 30 underground parking spaces including visitors parking and racks for 600 bicycles.

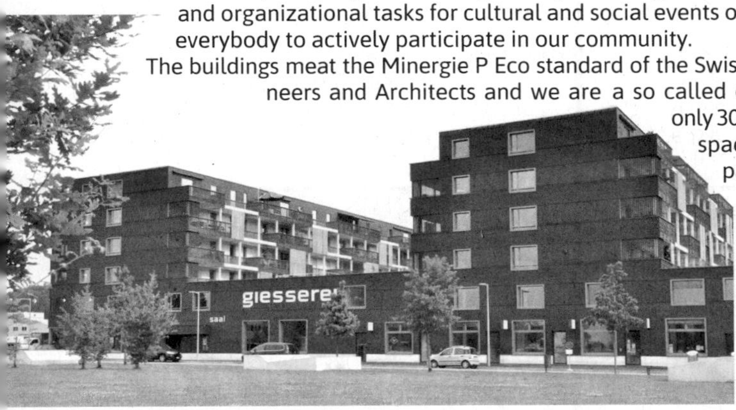

🌍 Swiss Building Co-op Association "Wohnen Schweiz"
☺ in vacant apartments only (new tenants)

Herzfeld Sennrüti

Sennrütistr. 23 | 9113 Degersheim
☎ +41-(0)71-511 25 68
✉ networking@oekodorf.ch
💻 www.sennrueti.ch

2006 | 1.2 ha (near St. Gallen) | 10 % | ♀32 | ♂13 | 25-29 | ca. 72

Swiss German, g, e, f, Dutch, sp, i,
all concerned together (by attunement)
approx. 4 | approx. 20
individually, partly in common

owned by all (cooperative)

garden, workshop and guests area, green building techniques, gathering place

ecological, self-managed, non-violent, spiritual, inner growth, meditation, yoga, organic diet, vegetarian diet, vegan diet, adult education, pedagogical, free school, peace-work, environmental protection, self-sufficiency, permaculture, organic agriculture, organic gardening, alternative energy, ecological construction, humour, non-smoking, "a new WIR" - culture, consciousness development, therapy, singing, dance, music, art and culture, hospitality, time-out care

☺

We honor the diversity of each individual, encouraging personal development and inner growing. We aspire to each individual finding and growing into their full potential.
In order to create and maintain healthy relationships in increased consciousness we meet regularly. We use sharing circles, shadow work in triads, the Transformation Game, integral community training (Forum) and ho'oponopono.
The integration of our children and youth is a crucial ongoing process. By growing up together there is a shared paternal feeling, resulting in collective nurturing of the younger generation. Children can easily relate to other adults in the community and some choose

their 'own grandparents'. We run integration days with lots of fun and games.
We installed thermal and photovoltaic panels, which means we have more than enough to power the entire building with solar energy. Furthermore, rainwater is collected from the roofs and used for flushing toilets, washing machines and in the garden. Organic waste is collected and composted or used for making charcoal for the permaculture garden. We grow fruits and vegetables and are lucky to have organic farms nearby that can fulfill our additional needs all year round. We try to do all practical work that is needed with care and love. We often celebrate life, all good things and ourselves.

→See also description of network **"GEN Europe"**
GEN Suisse, World Peace Prayer Society, some people: Talent, ARTABANA, Holon
☺ appreciation of the existing, YES to our common grounds, willingness to personal development, active participation and involvement in the community, membership in the cooperation

☆ Hofgemeinschaft Gruebisbalm

Grubisbalm 1 | 6354 Vitznau ☎+41-(0)41-397 20 98
www.biohof-grubisbalm.ch

1994 20 ha (near Luzern) 70 % ♀3 ♂4 3 10-12
Swiss-German, g, f, e 6-10
for daily concerns (work and well-being): together (consensus for daily concerns, large weekly meeting and round of well-being)
common income rented
ecological
organic farming, cowshed, garden, wood, household, social work
ecological, non-violent, spiritual, Christian, Buddhist, Taoist, meditation, yoga, organic diet, animal protection, environmental protection, self-sufficiency, organic agriculture, organic gardening, organic farming, animal friendly husbandry, forestry, alternative energy, humour, ecological, self-governing, non-violent, spiritual, meditation, organic food, mentally/psychically handicapped: 3 places for people with psychological problems, environmental protection, organic agriculture, organic horticulture, organic livestock farming, alternative energy, crafts: spinning, embroidery, felting (from time to time), art: painting, other: drumming

☺

Grubisbalm lies above Vitznau on the southern side of the Rigi at 850 m above sea level and offers a wonderful view of Lake Lucerne, the Bürgenstock and the magnificent mountains of Central Switzerland. The place is remote, quiet and can only be reached by train or on foot. In the car-free surroundings, relaxing hiking trails invite you to discover the Rigi. A waterfall flows along the edge of a farm. Forests, meadows, caves and waterfalls are suitable for outdoor work. On the farm "Grubisbalm" people live and work in a politically and religiously independent community. It is important to us that there is room for cultural exchange at Grubisbalm. We are also socially committed to young people and adults. Solidarity, openness and commitment enable the individual development of the individual. The community does not see itself as an island, but as part of the public. On the organic farm, local products are offered for course cooking and direct sales.

☺ interest in organic farming, care, living and working together

☆ Humanushaus Beitenwil Camphill Humanus-Haus

Postfach 55 | 3113 Rubigen ☎+41-(0)31-838 11 11
info@humanus-haus.ch www.humanus-haus.ch

1973 5 ha (near Bern) 60 % ♀70 ♂60 3 150

g, e, f 60

organised in regulations (different from case to case; goal: unanimous)

contract by Federal Social Security Agency foundation

anthroposophy

caring for people with a mental disability

ecological, self-managed, hierarchical, spiritual, anthroposophic, inner growth, organic diet, adult education, pedagogical, mentally/psychologically handicapped persons, physically handicapped persons, environmental protection, self-sufficiency, organic agriculture, organic gardening, organic farming, animal friendly husbandry, forestry, alternative energy, ecological construction, humour, therapy: painting, music, language, eurythmics, massage, handicraft: wood, weaving, candle-making, instrument building, herb workshop

The Humanus House, a Camphill village community, is situated between Rubigen and Worb - more precisely in Beitenwil - in an idyllic location in the green countryside, embedded in a hilly landscape and surrounded by forest.
As the largest employer in the municipality of Rubigen, the Humanus-House offers 90 living and working places for young people and adults with a disability (including various internal trainings) and 18 external workplaces for people from the surrounding area. Some of the professional employees live on the premises with their families.
In addition to the main building (Paracelsus House) - a 300-year-old listed estate - the settlement includes several residential buildings, a weaving mill, a swimming pool, a shop and a café (where the products of the Humanus House are sold), a carpenter's workshop, a bakery and a lyre workshop. Communal work areas are kitchen, housekeeping, laundry, construction and maintenance. The Humanus-House offers a wide range of therapeutic treatments.
The Humanus-House Foundation was founded in 1973 by Hans and Johanna Spalinger, who headed the community until 1989. In the meantime two "remote stations" were established: a biodynamically managed farm with 18 ha of land and 4 ha of forest as well as a musical instrument workshop. For the future, we are working on a conceptual extension, which combines the advantages of the grown village infrastructure with the increasing need for independence and autonomy.
Public lectures, concerts and theatre performances take place regularly in the Humanus-House auditorium. Everyone interested is cordially invited!

→ See also description of network "**Camphill**"

vahs, INSOS, SOCIALBERN

☺ training in social fields

Jurtendorf

Unter Niespel | 6156 Luthern Bad ☎ +41-(0)41 978 01 93, +41-(0)79 307 53 28
info@jurtendorf.ch jurtendorf.ch

2003 5 ha (near Luzern) 20-30 % ♀4 ♂3 3 10

g, f, e, (sp, i) 4 3-7

after discussions with all, the owners decide (talk)

courses and seminars, guest house, gardening individual owners

open spirituality

working on the land, cooking for guests, garden, maintaining the yurts

ecological, self-managed, left-wing political understanding, non-violent, hierarchical, leader/guru, spiritual, inner growth, organic gardening, construction, humour, preferably undogmatic

We live in a little valley surrounded by woods. Our private rooms are mostly in yurts, in the main house we have an apartment for use for all who live here permanently.
We work with compost toilets and a water purification that works with plants. From spring to autumn we try to grow our own veggies for us and all the guests that come and go. Apart of our own yurts there are a lot of yurts for guests and seminars. We have a summer season with a lot of coming and going in the village and a winter season with less guests. This is also the time when we build yurts for selling. These two activities are the main pillars of our income for living and making investments in the project. During the summer-season life is mainly outside, we also cook in a outdoor-kitchen, with fire.

To us, a loving coexistence with humans but also with the land and all the entities that were here long before we came, is very important. We do not have a defined religious or spiritual direction that we practice together regularly... or maybe yes? Love?

Pro Luthertal, Willisau Tourismus, GEN Suisse

Lebensgemeinschaft MonteBasso

Niederberg 78 | 3537 Eggiwil ☎ +41-(0)76 244 94 99
zuGast@MonteBasso.ch MonteBasso.ch

2010 (near Langnau im Emmental) ♀3 ♂1 2 6

g, e all of us work primarily outside the community

all persons involved together (no rules defined)

joint financing of household cash, partly common economy

building and ground in the property of individuals, garden rented

without dogmas

subsistence: consulting, coaching, theatre, projects

anarchistic, non-violent, spiritual, inner growth, meditation, sexuality, free love, seminar-house, peace-work, workshops for self organisation, polyamory, magic

☺ ✋✋

MonteBasso is a free-spiritual, humanist-humanistic community project that is not committed to any concrete ideological tradition. It is a biotope for conscious, life-affirming and best possible personal development and distinguishes itself from social conditions in need of overhaul: "connect and transform" instead of "divide and rule". We see our community as a kind of independent living being that - like any other form of life - functions as a self-organizing system without hierarchies, gurus or programs. We seek to understand and internalize the life principle of self-organization. Community also means to accept and support the group dynamics and the process-oriented character in the community system. Attempt and error are welcomed. Since all community members have been independently socialized somewhere, it is a special task to see and promote community development as a connecting process. All activities that strengthen the bonds between members are welcome. Open events and festivities are community highlights. Spirituality for us is a mindset that strives for awareness, reflects experiences, seeks resonance and meaning in all encounters and events, pursues heartfelt concerns, shares joy and dreams. MonteBasso sees itself as a neo-tribalistic community in which the members can permanently root themselves and provide their children with a rich field of experience. MonteBasso strives for the greatest possible economic autonomy and is expressly committed to the philosophy of permaculture. (from the website)

☺ no rules defined

Schloss Glarisegg

Schloss Glarisegg | 8266 Steckborn

welcome@schloss-glarisegg.ch gemeinschaft.schloss-glarisegg.ch

2003 5 ha (near Konstanz) 5 % ♀18 ♂16 22 56

5 11

- g, e, f
- all community members (consensus minus 10 %)
- combined individual and shared economy
- our corporation Schloss Glarisegg
- community building according to Scott Peck, social forum, transparent communication, spirituality rooted in everyday life
- guests and seminar centre, garden, construction sites, therapy centre, creative workshop, instrument making
- ecological, self-managed, left-wing political understanding, non-violent, spiritual, inner growth, meditation, yoga, organic diet, adult education, seminar-house, peacework, self-sufficiency, permaculture, organic gardening, alternative energy, new technologies, ecological construction, humour, various therapy methods, instrument making, crafts, music, art

☺ ✋✋

Schloss Glarisegg – "A place for encounter and consciousness" – is a place to learn and study community in all its dimensions. The focus is on communication with oneself and others. Our common practise is singing, silent times, gathering as a circle in our weekly sharings, a daily morning circle, yearly conventions, monthly decision making meetings, and various other events like our festival, Thanksgiving Day, working in the garden and feedback circles. For this we have integrated Community Building (Scott Peck inspired), ZEGG Forum, Possibility Management and various other forms of communication into our community culture. We

love to experiment with new forms of encounter.
We are hosting a seminar center, a school with forest kindergarten, a permaculture garden, a small food coop, a workshop for musical instruments, an artists workshop and a bistro. Many of us are working in the therapeutic field.
Our property with several buildings is placed between a wild forest and grasslands with a demeter farm, a stream with waterfalls and a proper beach at Lake Constance. We are living and house keeping individually as families, in single households and shared flats. Our vocation and living with our children makes a big part of our daily life.
We can offer three vacancies for young adults to support us in the garden and the school financed by the European Voluntary Service. At the beginning of each year we are organizing the five week Ecovillage Design Education workshop. It allows a big international group to live in community "on trial" and get to know the tools to build ecovillages and communities. From this event the first sister communities have already emerged.

→See also description of network "**GEN Europe**"
GEN Suisse, Gemeinwohlökonomie
☺ one year trial time, acquisition of a 5450 CHF share, self-responsibility, desire to design a new culture

Smala Ecovillages

La Smala tribe & institute for ecovillage living & social entrepreneurship
Rue de la Gare 17 | 1091 Grandvaux (Bourg-en-Lavaux) ☎ +41-(0)21-311 30 47
info@lasmala.org www.lasmala.org

1993 (near Lausanne) 50 % ♀20 ♂20 11 51
f, e, sp, i, Portuguese 50 % 50 %
all members with more than 3-5 years experience within our community (Meritocratically and wiki-alike: digitally driven organic multi-methods with permanent updates and full traceability. Sometimes we laugh, let things flow and that's it!)
cooperative financing cooperative ownership, non profit real estate
social entrepreneurship, creative souls
social entrepreneurship, incubation of ecovillages
ecological, self-managed, left-wing political understanding, non-violent, Taoist, anthroposophic, inner growth, yoga, organic diet, adult education, pedagogical, self-sufficiency, permaculture, organic gardening, alternative energy, ecological construction, humour, accessible to the disabled, social entrepreneurship
☺ ( )

Ever since 1993, the Smala team has tackled the challenge of incubating 40 short and mid-term creative eco-homes, eco-life learning centres, with 12 to 30 cohabitants and co-workers in each location, in the Lausanne area. These temporary hives had a duration of 1 to 10 years, and have been developed in partnership and/or supported by local and

federal administrations for housing, social services, culture, academic research, adult training, NGOs and socio-cultural centres. Since 2013, we bought land, built and now have various little long term ecovillage-alike intentional communities, where we co-live and co-work. We also support organizing TEDx Geneva and offer an ecovillage living training programme (named Smala CoCo), a gateway for newcomers.
These activities have made La Smala institute & communities resilient to societal changes and demonstrated its ability to manage complex governance situations.

Smala
Tout un art de vivre

→See also description of network "**GEN Europe**"

Smala Ecovillages, ENOLL (European Network of Living Lab), Chamber of social economy, Alternatiba and the network of beneficiaries of the 10% discount national eco-supermarket (yeah!)

☺ case per case, trial period

Verein Holzlabor

Thurtalstr. 30 | 8478 Thalheim an der Thur
labor@xylem.ch www.xylem.ch

2007 2.4 ha (near Winterthur) 20 % ♀5 ♂5 4 21

g, e, f, (sp) 3 7

association and board (consensus) household budget, association budget

association property, lease

association work, fruit cultivation, landscape gardening, domestic work, small animal husbandry, carpentry, events, organisation

ecological, self-managed, anarchistic, left-wing political understanding, non-violent, chaotic, inner growth, organic diet, vegetarian diet, vegan diet, environmental protection, organic agriculture, organic gardening, ecological construction, humour, creative, sustainability, collective responsibility, locality, compost toilet, craftsmanship, carpentry + wagon building, landscape gardening, beer brewery, music, 4 generations, 2 houses, wagon dwellings, events (culture, celebrations, information event), high stem fruit garden, lunch table, alternative medicine, farm shop

☺ ()

We are a motley living and working community. Together we use, cultivate and design this rural habitat in the old buildings and surroundings, with planting areas, tree meadows, animal enclosures and hedges.
The place offers living, working and toddling space for young and old, for people with different professions, interests or projects. The living and working spaces as well as the surroundings are designed and maintained by all those who live and work here with their shared resources (cooperation and co-financing).
In constant change, we develop our values together, but also take individual needs and interests into consideration. We are a place of inspiration and joie de vivre.

☺ energetic, with heart → according to statutes

Yamagishi Verein

Schneitbergstr. 12 | 8523 Hagenbuch ☎+41-(0)52-364 15 28
info@yamagishi.ch www.yamagishi.ch, www.tokkoh.ch, www.tokkoh.nz

1983 40 ha (near Winterthur) ♀5 ♂6 2 13

g, Japanese, e — all
everyone in common (through Kensan) — shared income
common property
agriculture, food production, direct marketing
self-managed, inner growth, seminar-house, peace-work, environmental protection, animal friendly husbandry, alternative energy, humour, humanity, natural charity, Kensan = communication method; to recognise the love that is in everything, one-body-life

☺ ()

Yamagishi, Miozo was a peasant from Japan who died in 1961 and made a close study of (human) nature. His goal was a stable world community which makes it possible for everyone, including our children, to be happy.
Starting from the fact that all beings and all existences are inextricably linked with one another – I flourish when everything else flourishes – this principle equally applies to marriage, the family, the workplace... With the knowledge that every person in this world wants to live in peace and satisfaction, Yamagishi developed a model for society, a guide on "how to construct the world revolution". We believe that the happiness of all people begins in each of our everyday lives and only each small individual like you and me can make it happen. This has been practised with "Kensan" for over 40 years in the 80 Yamagishi villages around the world. There are no bosses or gurus, there are no rights or responsibilities. Instead, everyone does what he can and when he wants. Everything is free, as in nature. Let us build such a society together. We believe we have found a practical, viable path and heartily invite you to join us.

→See also description of network "**GEN Europe**"
Yamagishi
☺ desire to help build a happy society and participate in the Tokkoh, the Yamagishi seminar

Zentrum der Einheit Schweibenalp

Schweibenalp | 3855 Brienz ☎ +41-(0)33-952 20 00
info@schweibenalp.ch
www.schweibenalp.ch, www.alpine-permakultur.ch

1982 20 ha (near Interlaken, Bern, Luzern) 20 % ♀14 ♂13 27

g, e, f, i 27 27

concerning community, seminar center, permaculture project: teams, core group (consensus of those present)

unitary pay for staff, board and lodging for volunteers, donation, long – term loan, fundraising

property of Foundation Schweibenalp

day-to-day spirituality, open to all spiritual orientations

kitchen, housekeeping, outer area, permaculture, construction, management, reception, seminar organisation

ecological, self-managed, non-violent, spiritual, inner growth, meditation, organic diet, vegetarian diet, vegan diet, adult education, seminar-house, peace-work, environmental protection, self-sufficiency, permaculture, alternative energy, new technologies, ecological construction, construction, biodiversity, economy, volunteering, internal growth, self-awareness

For the first 25 years, our community has served as an ashram and as a seminar center for alternative disciplines, for interreligious and intercultural peace work. Since 2008 we are broadening our awareness in focusing on a balanced development of sustainable ecology, fair economy, authentic transparent interpersonal relationships and universal spirituality. We are members of GEN and happy to be host of Green Phoenix.
As a COMMUNITY we run the seminar center, the gardens and the temple, in short: The Center of Unity. Volunteers are welcome to stay from one month up to three, in individual cases longer.
The platform GREEN PHOENIX, which we founded, has since become the "think, feel and consciousness tank" of the "community of communities". Each European Community is invited to join our annual meetings.
Our ALPINE PERMACULTURE PROJECT at 1100 m altitude is now the largest of its kind in Switzerland and attracts many interested parties to seminars, guided tours and open days. Likewise, we watch with appreciation the growing number of teachers, instructors and their participants, who support Schweibenalp by staying here... and loving it. Our UNIVERSAL TEMPLE is always open to people to come to meditate, to gain power, to... enjoy.
WELCOME!

→See also description of network "**GEN Europe**"

GEN Suisse, Global Campus, Green Phoenix, Babaji Network

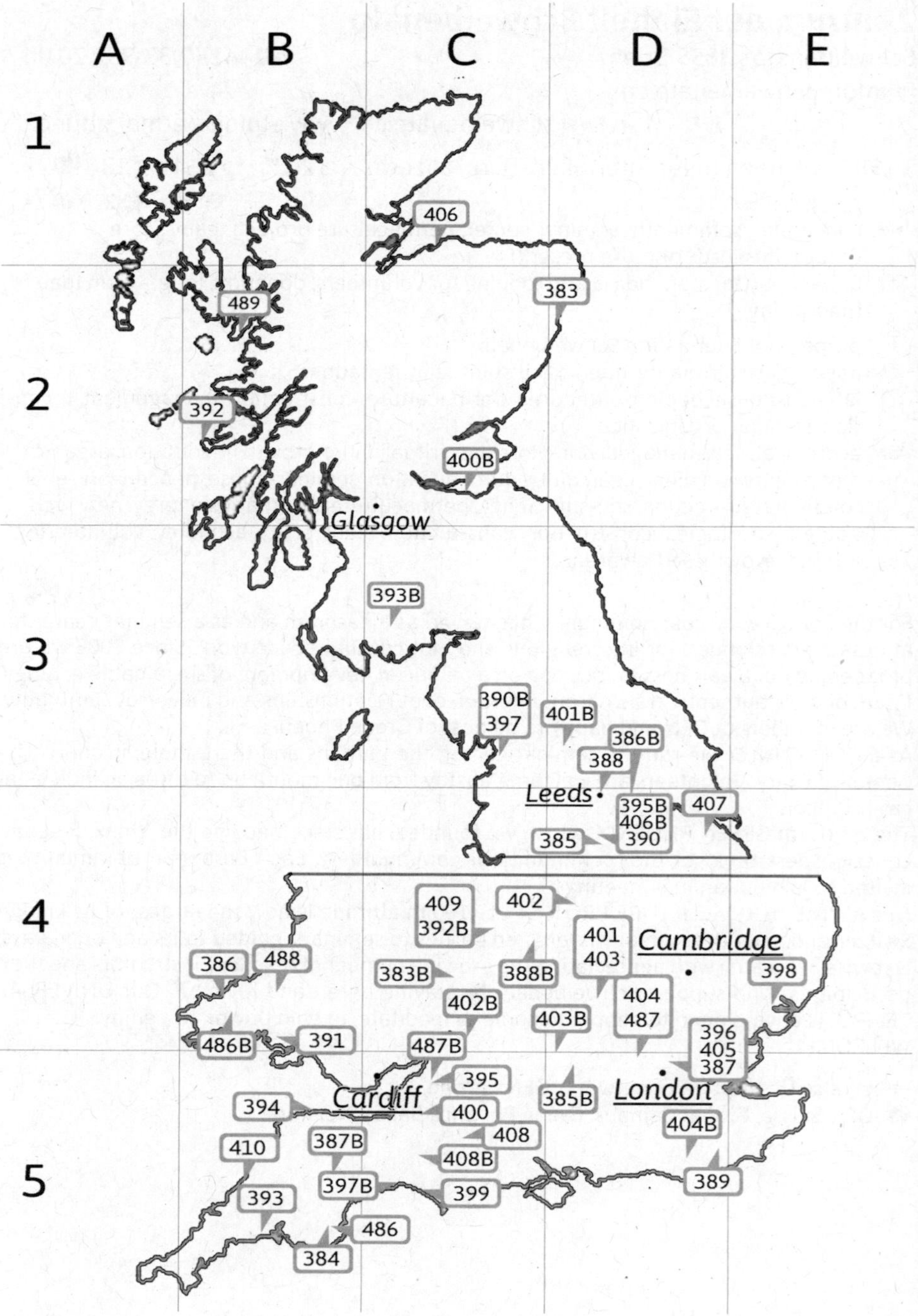
A
B
C
D
E
1
2
3
4
5
406
383
489
392
400B
Glasgow
393B
390B
397
401B
386B
388
Leeds
395B
406B
390
407
385
409
392B
402
401
403
Cambridge
386
488
383B
388B
398
404
487
402B
403B
486B
391
487B
396
405
387
395
Cardiff
394
385B
London
400
408
404B
410
387B
408B
389
397B
399
393
486
384

United Kingdom

A lot of intentional communities in England, Wales and Scotland, which make up the United Kingdom - at least when we edited this book, they did. The UK is home to 45 existing communities and 6 in planning stage (p.486ff). With another 38 Camphill communities (see page 514ff) it takes second place in the number of communities of the countries of Europe. Many addresses are co-housing projects that have, for unknown reasons, not declared themselves as such. There's another 22 communities that were described in detail in the last edition of eurotopia (page 505ff) and a community network called Radical Routes (see page 534). Also, Diggers and Dreamers (see bibliography) publish their own directory of communities in the UK - we're on good terms with them and value their work greatly.

Beannachar Camphill Community

D2

South Deeside Road, Banchory–Devenick | Aberdeen AB12 5YL

+44-(0)1224-869 250

office@beannachar.org — www.beannachar.co.uk

1978 | 11 ha (near Aberdeen) | 50 % | ♀28 ♂16 | 4 | 44

e — all

all who are concerned (consensus)

shared income for meeting the needs of coworkers — Camphill Foundation

anthroposophical, christian, open

persons with mental disabilities

ecological, self-managed, non-violent, spiritual, Christian, ecumenical, anthroposophic, inner growth, meditation, organic diet, vegetarian diet, vegan diet, adult education, mentally/psychologically handicapped persons, physically handicapped persons, environmental protection, self-sufficiency, organic agriculture, organic gardening, organic farming, alternative energy, social work, therapies, art, sports, crafts

We are an integrated, therapeutic community providing care, meaningful work and a rich social and cultural life for young adults with learning disabilities.

→See also description of network "**Camphill**"

Biodynamic horticulture

☺ fluent English

Birchwood Hall Storridge

D4

Malvern | Worcestershire WR13 5EZ

+44-(0)1886-884 203

info@birchwoodhall.org.uk — www.birchwoodhall.org.uk

1970 | 3,5 ha (near Malvern) | 25 % | ♀8 ♂8 | 2 | 18

e — none — none

all members (consensus) — limited community financing

common property

green, left, feminist

external work (multiple roles)

ecological, self-managed, left-wing political understanding, feminist, non-violent, organic gardening, humour, volleyball, various media

☺ ✋

Birchwood Hall is a housing co-op with two houses (Hall & Coachhouse), set in 3.6 ha of garden. There is the Victorian hall with lots of rooms, and a converted Coach House which is nearing the end of a conversion build. We have solar panels to generate some of our electricity and grow some of our own food, although we aren't full self-sufficient by any means. Our children go to local schools. We have a range of full and part time occupations, in a variety of jobs. We live here for ethical and practical reasons. We like each other and enjoy sharing the house. Our joint politics are leftish, greenish and feminist, with enough variation for some rowdy teatime discussions! We eat together in the evening, taking it in turns to cook, trying to accommodate likes and dislikes. Some of us are vegetarian with occasional meat dishes for the omnivores! Meetings are once a week, alternating business and social meetings. We don't have many defined expectations of each other but trust that everyone will add to the community and pull their weight in some way. The garden is both beautiful and wooded, but takes a lot of effort to maintain!
The key to our success is stability. We are looking for new members and are enthusiastic to hear from interested people out there.

Diggers and Dreamers
☺ green, left, feminist

Bowden House Community Bowden House

B5

Totnes | Devon TQ9 7PW

info@bowdenhouse.co.uk — bowdenhouse.co.uk

2006 | (near Exeter) | 5 % | ♀16 ♂16 11 43

e, g, sp, f — a few
whole community (consensus minus two) — individual financing
private and shared ownership
none
mixed mostly off site
ecological, self-managed, left-wing political understanding, feminist, non-violent, Buddhist, inner growth, meditation, organic gardening, alternative energy, humour

☺ ✋

We are a group of families and individuals developing conscious, authentic and eco-mindful living. We enjoy a culture of singing, meditating, working, gardening, celebrating and learning together, as well as hosting events for the wider community. Some things that inspire us:
Developing our resonance with each other and our surroundings - singing together - authentic communication - coming together to share our feelings and challenges, and resolve any conflict - living in a beautiful place in touch with the spirit of the land - growing some of our food - developing skills for sustainable living - the alchemy of doing things together - what we discover through our journey together and how this helps us grow - Respect for individual spirituality and different approaches to life.

☺ financial and ideological

Brambles Housing Co-op 80-82 Andover Street

Pitsmoor | Sheffield S3 9EH ☎+44-(0)114-2797164
brambles@riseup.net
diggersanddreamers.org.uk/communities/existing/brambles-housing-co-op

1992

e, sp, f, i

all members

all members (consensus in monthly meetings)

co-operatively owned

Brambles is a home for an eclectic mix of people and urban wildlife and a stop-over point for many more. We come from a mix of genders and social backgrounds with a wide range of skills to share and between us explore and promote co-operative ways of living.
The co-op itself has been running for over 10 years and has housed over 40 people during this time with members staying from months to many years.
We share two neighbouring houses, vegetable patches, fruit trees, wildlife garden and frog pond in a vibrant multicultural but economically poor area of Sheffield. Brambles provides its members with a secure, supportive environment from which we both work together and do our own stuff too. The co-op acts as a focal point for resources and information and as a free meeting space for local groups.
We are working on plans to open as a resource centre and radical library for use by the local community. Brambles is ecological and anarchic in focus with no rotas and very few, but fair boundaries and a lot of learning under our collective belts.
Solar water heating is being installed, communal meals generally happen and the recycling and compost systems are pretty funky and efficient.

→See also description of network "**Radical Routes**"

Braziers Park School of Integrative Social Research

D5

Braziers Park, Ipsden | Wallingford OX10 6AN ☎+44-(0)1491-680 221
admin@braziers.org.uk www.braziers.org.uk

1950 22 ha (near Oxford/Reading/Wallingford) ♀8 ♂10 1 19

e, g, f

12

committee of Management (consensus)

individual financing

In trust

secular

hospitality, maintenance, estate work

ecological, self-managed, left-wing political understanding, non-violent, adult education, permaculture, organic gardening, alternative energy, construction, education

☺

Braziers Park is a community, a residential college and an architectural treasure hidden deep in the south Oxfordshire countryside in an Area of Outstanding Natural Beauty. It was founded in 1950 as an educational trust, and is a continuing experiment about the advantages and problems of living in a group. Community members share responsibility for running the mansion house (a Grade II* listed building), 55 acres of land, organic kitchen garden and livestock, facilitating courses and organising events assisted by visiting volunteers from around the world.

☺ vacancies and info on website

Brithdir Mawr Community

B4

Cilgwyn Road, Newport | Pembrokeshire, Wales SA42 0QJ

☎+44-(0)1239-820 164

visit@brithdirmawr.co.uk brithdirmawr.co.uk

1994 32 ha 75 % ♀4 ♂4 3 11

Welsh, e

nearly all earn their money outside

everyone in common (consensus)

limited community financing

ecology, simplicity

garden, coppicing, farm animals, alternative technology, preserving

ecological, self-managed, organic diet, adult education, environmental protection, self-sufficiency, permaculture, organic gardening, organic farming, animal friendly husbandry, alternative energy, ecological construction, humour, rural handicraft, "greenwood skills"

Brithdir Mawr is a land-based community endeavouring to live as sustainably and simply as possible. We grow most of our own organic vegetables; chickens and ducks provide eggs and our goats provide milk. Our power needs are met entirely by renewable resources from the land and the weather. We welcome volunteers for a week at a time to share our work, in return for food and accommodation. Please email a brief introductory letter with some idea of your availability and skills.

☺ interest in resource-conserving construction, enthusiasm and tolerance

Brotherhood Church

D4

Stapleton | Pontefract WF8 3DF

☎+44-(0)1977-620 381

enquiries@thebrotherhoodchurch.org www.thebrotherhoodchurch.org

1898 4 ha (near Leeds, Doncaster) 60 % ♀2 ♂2 4

e

most of us

all who are concerned (consensus)

limited common financing

property of all

environmental, pacifist, low impact

self-managed, anarchistic, left-wing political understanding, feminist, non-violent, organic diet, vegetarian diet, vegan diet, peace-work, animal protection, environmental protection, self-sufficiency, permaculture, organic agriculture, organic gardening, organic farming, animal friendly husbandry, forestry, alternative energy, ecological construction

We strongly object to consumerism and greed. We are pacifists and hold strong views on human and animal rights and have anarchist leanings. A main concern at present is climate change. We plant a lot of trees. We do not participate in many aspects of modern life and refuse to be brainwashed by the mainstream media. We feel there is a better way to live and are trying to work towards it. We are low-impact and consider our effect on the environment and the natural world in everything we do, but this does not mean we are against all modern technology. We make part of our living from the land but individuals can also pursue their own interests or careers separately, whether on the land or in another way.

Our name comes from the original settlers who were linked to the Quaker movement. Although we have common sympathies with the Quakers, we are not a religious community. We are not conspiracy theorists. We are not strict vegans, but most of our main meals are vegan. We have some rescue hens, who lay some eggs, and we are part of a network which is trying to re-introduce native bees.

☺ ideological

Copper Lane Cohousing

D5

Copper Lane | London N16 9NG

copperlanecohousing@gmail.com copperlanecohousing.wordpress.com

2014 5 % ♀6 ♂4 3 13

- e, sp, Danish
- everyone (consensus, majority vote in last resort)
- individual and limited common financing
- leaseholds, members jointly own freehold
- ecological, self-managed, feminist, non-violent, anthroposophic, pedagogical

We are a group of six mixed households living in a small cohousing project at Copper Lane in the Stoke Newington area of North London, completed in 2014. We have shared gardens and a hall/workshop space that we make available to other local community groups. Visitors very welcome - contact us in advance - but we don't have space to accommodate you at the moment.

Corani Housing and Land Co-op Ltd.

B5

1 Taw Vale Terrace Station Road | Crediton, Devon EX17 3BU

info@corani.org, herthat@gmail.com www.corani.org

1978 1 6

- e
- owned by the co-op
- climbing

We are some friends living together in a large Georgian house on the edge of a small town in Devon, very close to the rail station, with gardens front, side and rear; fields nearby.
We enjoy eating and cooking together every day, sometimes with veg from the garden, eggs from the chickens and honey from the bees. Other things we enjoy doing are climbing on the sea cliffs around Devon and further afield. We are around 6 adults, one baby/toddler and one dog. Corani is not currently seeking new members but this could easily change.
Visiting/passing through is possible by prior arrangement.

Cornerstone Housing Co-op

D4

16 Sholebroke Avenue, Chapeltown | Leeds, West Yorkshire LS7 3HB

☎+44-(0)113-262 1534 cornerstone@cornerstonehousing.org.uk

www.cornerstonehousing.org.uk

1993 2 x ca, 2000 m² 2 % ♀10 ♂5 1 17

e, f, g, sp (basic), Serbian, Urdu, mandarin (basic)

all who are concerned (consensus) limited common financing

housing co-operative

no

(eco) direct action, repurposing food waste, feminist direct action, economic inequality campaigning, queer film, permaculture, solidarity economy

ecological, self-managed, anarchistic, feminist, chaotic, vegetarian diet, vegan diet, animal protection, environmental protection, permaculture, organic gardening, humour, non-smoking

☺ ( )

Cornerstone has two large, shambolic Victorian houses in Chapeltown, an ethnically diverse part of inner-city Leeds, Yorkshire (slowly being gentrified, though). Both houses need a lot of maintenance, but have large gardens front and back, where the odd co-op member does their best to create beautiful and productive landscapes. Each house has space for eight members. The political focus shifts as membership changes, one of our topics was/is to explore aspects of privilege, workplace organising, community organising, ecological education and cycle projects.
One house has an accessible bathroom, although there are 3 steps to get into the house. A printing company (Footprint Worker's Co-op), operates from the cellars of one house.

→See also description of network "**Radical Routes**"

Radical Routes

☺ ecologically aware, actively working for radical social change

☆ Coventry Peace House

C4

311 Stoney Stanton Road | Coventry CV6 5DS ☎+44-(0)2476-664 616

info@covpeacehouse.org.uk www.covpeacehouse.org.uk

1999 320 m² 1-2 % 7

everyone in common (consensus) limited community financing

co-operative

peace and environment

refugee support, working with people on the margins of society, bicycle recycling

ecological, self-managed, left-wing political understanding, non-violent, vegetarian diet, peace-work, environmental protection, organic gardening, social work: working with people on the margins of society, solidarity with refugees

☺

Coventry Peace House is based near the centre of Coventry. It consists of a row of 6 terraced houses. One house is lived in by refugees, and the remaining 5 are being converted into a community resource and housing co-op.

The main ethos is to live simple green lifestyles and to work actively for peace. We currently run projects for asylum seekers based in Coventry, and for disengaged young people. We will soon open a bike workshop, where we will be recycling bikes.
We are members of the Radical Routes network of co-ops, which means that we have made a commitment to work for radical social change.
Coventry peace house needs new members. If you like the idea of living in a co-op and want to work towards peace in any way, then we want to hear from you.

→See also description of network "**Radical Routes**"
🌍 Radical Routes
☺ participation in the life of the co-op and also working towards social change (e..g on peace or environmental issues etc.)

☆ Darvell Bruderhof

D5

Brightling Road, Robertsbridge | East Sussex TN32 5DR ☎ +44-(0)1580-883 300
✉ darvell@bruderhof.com 💻 www.bruderhof.com

1971 | 120 ha (near Hastings, London) | ♀100 | ♂100 | 100 | 300

g, e, f, sp, Korean | 180

everyone in common (consensus) | shared economy (shared income and capital)

common property

Christianity, non-violent

woodworking, kitchen, construction

non-violent, Christian, humour

We are a Christian intentional community, which belongs to the Bruderhof Movement. This movement consists of more than 2700 people living in over twenty settlements on four continents. We are a covenant of families and singles who want to follow Jesus without compromise and in the spirit of the first church in Jerusalem. We gladly renounce private property and share everything with each other.
Darvell is one of the larger communities of the Bruderhof Movement. The community includes a small factory where we make furniture, a school with a kindergarten, a laundry, a construction team to maintain the buildings, and many other work areas. None of us have a career and every job has equal status.
Living and working in the community gives us the opportunity to follow Jesus with our whole lives, to help each other and to accept help.

→See also description of network "**Bruderhof**" with a photo of Darvell Bruderhof, page 510
☺ just live with us

Fireside Housing Co-op 55-63 Melrose Road

D4

Sheffield | Sheffield S3 9DN ☎+44-(0)114-272 59 08

fireside@blueyonder.co.uk

www.diggersanddreamers.org.uk/communities/existing/fireside-housing-co-op

1995 ♀6 ♂4 10

e

everyone in common (consensus) — limited community financing

common property

living

ecological, self-managed, non-violent, vegetarian diet, organic gardening, alternative energy, ecological construction

☺

Fireside has a row of five terraced Victorian houses in Burngreave, Sheffield. We've knocked our long and skinny gardens into one, so there's plenty of room for us to play, grow organic veg, have bonfires and parties. The gardens back onto a Victorian cemetery so there is lots of green space and wildlife around even though we're in an inner-city area. There's also an adventure playground at the end of the road.
The age range of the 10 adults living here is staggered from 20 to over 60 years old. We live mostly as five separate households, but we have knocked a connecting door between numbers 55 & 57 and numbers 59 & 61 have a shared kitchen area. We eat communally on a fairly regular basis, particularly when having a meeting, or a workday on the houses or gardens, and some households eat together more frequently.
A few years ago we created a highly insulated 2-storey extension across the backs of 55-59, and since then we've had a second solar thermal system put in, so 4 houses now have solar hot water. We also have PV panels across the south facing roofs.
Burngreave is a quite deprived inner-city area but very multicultural and vibrant. Many of the co-op members are involved with local community groups.

→See also description of network "**Radical Routes**"

Forgebank

C3

9 Forgebank Walk, Halton | Lancaster LA2 6FD ☎+44-(0)782 561 0342

info@lancastercohousing.org.uk lancastercohousing.org.uk

2012 2 ha (near Blackpool (35 km), Preston (40 km))

♀36 ♂20 11 67

e — volunteer work only

all (consensus) — leasehold, freehold, privately rented

ecological, self-managed

☺

Our multi-award-winning eco-cohousing-community at Forgebank consists of private homes, community facilities, workshops/offices/studios and shared outdoor space. We are around 56 adults and 11 children. We eat some meals together, make our decisions by consensus, and enjoy meeting our neighbours in our lovely pedestrianised street and common house. Our cosy and comfortable homes meet "passive house" and "code for sustainable homes"

(level 6) standards, and we benefit from renewable technologies (solar, biomass and hydroelectricity). Living at Forgebank also enables us to live a lower-impact lifestyle in many ways e.g. through our travel plan and car club, cooperative food store, shared meals and other shared resources.
The site is just outside Lancaster on the outskirts of the village of Halton right next to the river Lune, and there are 35 leasehold and 6 freehold homes, shared buildings, Halton Mill (a low carbon public work and events space), and riverside woodland habitat. We aspire to be a cutting-edge example of sustainable 'eco'design, for both living and working, with close links to the local community.

☺ they need a home (rented or leasehold) and to sign up to agreements

Glyn Abbey

B4

Glyn Abbey | Llanelli SA15 5TL
☎+44-(0)1554-811 234
kinesiologyinwales@yahoo.co.uk

1976 (near Llanelli) ♀9 ♂11 5 25

- e
- trust with shares
- no, vaguely green/vague
- various

Begun in 1976, the place has developed organically ever since. We seem to just enjoy living here and have no ethos other than to look after the place and ourselves. We host a Mayday celebration the 2nd Saturday in May every year.

☺ no conditions

Isle of Erraid Community Findhorn Foundation

B2

Fionnphort | Isle of Mull, Argyll PA66 6BN ☎+44-(0)1681-700 384
erraid@live.co.uk www.erraid.com

1978 256 ha ♀2 ♂3 0 5

e, g, f — all of us

the whole group (consensus) — shared income

private property leased to us in exchange for looking after the buildings and the land

inner listening, co-creation with nature, service to the world

gardening, chopping wood, looking after the animals, cooking, cleaning

ecological, self-managed, non-violent, spiritual, inner growth, meditation, organic diet, organic gardening, organic farming, humour, organic farming, vegetable gardening, boating, candle-making, holding guests, sacred singing, celebrating Celtic Festivals

☺

Erraid is a small Hebridean island in Scotland, in view of Iona. The island is one square mile in size and our small community lives and works on the land. We live in old light house keepers' cottages, which we share with our guests. We work in the gardens, heat the cottages and our water with wood-burning stoves, drink rain water and live a life very close to the elements. What we are doing on a day to day basis is often determined by the sea, the tides, the moon and the wind.
One of our main purposes on the island is to hold guests in the island's ancient energy and allow them to experience and explore its magnificent landscape and wildlife, our unique lifestyle and community life. Guests come throughout the year for a week or even months to live with us and join in our activities: cooking, chopping wood, working in the gardens, helping with the animals, making candles, meditating and singing sacred songs in our beautiful sanctuary and celebrating the turning of the seasons in our Celtic Festival weeks.

Findhorn Foundation

☺ going through our membership process

Jesus Fellowship

C4

The Anchorage | Birmingham B20 2PJ ☎+44-(0)121-551 18 85
barney.taylor@gmail.com

1983 0 % ♀4 ♂2 2 8

e — all voluntary

everyone (consultation)

at present most share all income and capital but it is changing

owned by Jesus Fellowship Community Trust

Christian

local to the house

☺ ( )

We are an intentional Christian community seeking to live a lifestyle which is based on Act 2 and Acts 4. Our aim is to love and serve one another and the wider community in the love of Jesus.

→See also description of network "**New Creation Christian Community**"

☺ it can vary, must be sympathetic to Christianity

Keveral Farm

B5

Keveral Farm | Looe, Cornwall PL13 1PA ☎+44-(0)1503-250 936
keveralfarm@yahoo.co.uk www.keveral.org

1973 12 ha (near Plymouth) 10 ♀3 ♂4 0 7

e, i. f, Portuguese 4 7

everyone present at the co-op meetings (aim for consensus, but if not possible then majority vote)

individual financing housing co-operative

organic land management

organic horticulture, orchard and soft fruit, trees, campsite

ecological, self-managed, permaculture, organic agriculture, organic gardening, organic farming, forestry

☺

Keveral Farm has been a community since 1973, when a group from London moved here to try to live self-sufficiently while helping people with drug-related problems.
Today the farmhouse and farm-buildings are owned by our housing co-op, and our workers co-op oversees the management of the land, which is certified organic.
Ten adults plus one child, at this time, live in the main house, a barn conversion, a wooden cabin, and 3 static caravans. We live quite independently, getting on with our various projects, but coming together in the hub of the community, the farmhouse, for bi-weekly meetings and occasional communal meals.
We run a veg box business that started in 1997, which delivers to about 100 households each week. We have workshops where some members do craft work such as green woodwork, carpentry and stone carving. There is also a small campsite and a Visitor's Barn which is for volunteers, events and courses. In recent years we have held an annual Open Day event to welcome local people to come and get to know the farm and have a good old knees-up together! In the past we have also held a number of permaculture design courses here.
The land consists of vegetable plots, polytunnels, a campsite, an orchard, soft fruit, woodland, and pasture. These are rented out to community members, who either grow for themselves or sell commercially, to the veg box and beyond. Other business activities include tree-work and firewood, apple juice and cider, growing mushrooms on logs. Those that work off the farm predominantly work on other land-based projects in the area.
We have our own spring, a derelict water wheel, an ancient track through the woods to the sea (15 minutes walk), and a picnic spot overlooking the sea. Some people call this place paradise.

→See also description of networks "**Radical Routes, WWOOF**"

Land Workers Alliance

☺ participate in the work and management of the community *Photo on page 7*

Laurieston Hall Housing Co-operative

C3

Laurieston Hall, Scotland | Castle Douglas DG7 2NB ☎+44-(0)1644-450 682
enquirylh@gmail.com
www.diggersanddreamers.org.uk/communities/existing/laurieston-hall

1972 56 ha (near Dumfries) 85 % ♀7 ♂7 0 14

many 5 all adults

everyone in common (consensus) | limited community financing

- common property
- holiday centre, farm, woodland management
- self-managed, feminist, non-violent, self-sufficiency, conventional gardening, animal friendly husbandry, conventional farming, forestry, alternative energy, ecological construction, humour, sauna, card games, handicrafts: felt-making, woodworking, jewellery making, photography, volleyball

☺ ()

We live around a large country house: individually, as traditional family units or in the living group - in the main house or in cottages. Each adult pays rent to the housing co-op and is domestically and economically independent. We enjoy a high degree of co-operation and caring. Much of what we eat is home-produced. There is a garden, a shop, and a workers co-op providing some income. We also have cows, calves, hens, bees and pigs (three quarters of us eat meat). We have wood stoves for space heating and our own hydroelectric scheme which supports the lighting, power and hot water. Part of what we call Workshare (2-3 days each week) is taking a share of responsibility by being on committees. Additionally everyone is expected to attend the weekly meetings and wood-collecting trips, and to take part in Maintenance Weeks each year. Visitors are invited to join in the building maintenance, gardening, land work and domestic work on Maintenance Weeks - the times when we function as one big group. Except for personal guests, we're not able to invite people to stay at other times. We are not actively seeking new members: people who want to join us should discuss the possibility with friends they make among residents during Maintenance Weeks.
In summer 2019 we're taking in a group of 7 adults (5 women, 2 men) and 4 children.

Lee Abbey Devon Community

B5

Lynton | North Devon EX35 6JJ

www.leeabbeydevon.org.uk | www.leeabbey.org.uk

(near Barnstaple)

We seek to glorify God in our love for Christ, for one another and for all those who visit us.

☺ ()

Do you want to be part of a community of Christians from 20 nations around the world, who encourage each other to put Christ at the centre and to live out the gospel in down-to-earth ways? Lee Abbey is home to a Christian Community that hosts retreats, holidays and conferences on an estate on the dramatic North Devon coast. Individuals, couples, families, small groups and whole churches come to be transformed and renewed by God. Being in Community is about living, working, serving and worshipping closely together, being open to being known for who we are. It's challenging but great fun, too, and if you let it, it can be life-changing.
We are looking for people with a personal faith in Christ and a desire to deepen that faith by surrendering more of their lives to Him. You need to be prepared to give of yourself to our guests and fellow Community members. Fitness, health and energy are pretty essential too! Lee Abbey is committed to safeguarding the welfare of children and adults and expect all who join Community to share this commitment.

The Lee Abbey Network

☆ Lee Abbey Knowle West

C5

Lee Abbey Small Missional Communities/Household Communities

Filwood Broadway | Bristol BS4 1JN ☎+44-(0)117-239 01 47

KnowleWestSMC@leeabbey.org.uk www.leeabbey.org.uk/SMC

1992 800 m² 2 ♀3 ♂3 2 8

e 3

all community members with support from trustees (consensus)

individual financing with some elements of common financing

owned by the registered charity

Christian

mission and ministry, supporting the local churches, social action

Christian, ecumenical

☺

The Lee Abbey community was established at the invitation of the local churches, with whom they continue to work in close partnership. The community house is home to a non-denominational multi-generational community of Christians called to serve in the local area, individually and collectively. Living under one roof, they adopt a rhythm of prayer and shared meals. They worship at a local church and are engaged in many different facets of mission and ministry.

Lee Abbey family of communities, Knowle West Churches Together

☺ Christian faith, time to invest in local area

☆ Lifespan Community Collective Townhead Collective

D4

1-19 Townhead Nr Dunford Bridge | Sheffield S364TG ☎+44-(0)1226-762 359

lifespancommunitycollective@yahoo.co.uk www.lifespancommunity.co.uk

1975 1,2 ha (near Holmfirth, Huddersfield) 10 % ♀7 ♂7 3 17

e

general meeting of all members (consensus or majority voting)

property of the cooperative

housing cooperative

ecological, self-managed, organic gardening, alternative energy

☺

Lifespan Community Collective (aka Townhead Collective) is made up of around 15 adults and 7 kids. We have 19 old railway cottages on 3 acres of land in rural Yorkshire. Lifespan was formed in the early seventies as a rural community and has been through different stages of communal living. At present, it is a housing co-op whose main aim is to provide housing for its members. We have our own personal space, as well as communal areas comprising of rooms for visitors, living room, kitchen, bathroom and games room. We try to live with as little impact on the environment as possible. We are off grid and produce our own electricity from solar and wind and compost our toilet waste. Some organic fruit and vegetables are grown in the gardens and help is always appreciated.
If you are interested in visiting, please contact us via e-mail and we will let you know the date of the next work weekend. We hold around six work weekends throughout the year and

all visitors are welcome. Work weekends are a good opportunity to meet members, have communal meals and get to know how the place works. We are open to new members and occasionally have vacancies; the first meeting preferably being at a work weekend.
We have a down to earth vibe and DIY culture. Public transport is available but minimal.

→See also description of network "**Radical Routes**"

London Catholic Worker Guiseppe Conlon House

D5

49 Mattison Road | London N4 1BG ☎+44-(0)20-8348-8212
londoncatholicworker@yahoo.co.uk www.londoncatholicworker.org

2010 ♀2 ♂1 3

e — all of us full time, plus volunteers

live-in community and live-in volunteers (consensus)

live-in members get board & lodging — rent free lease

Christian, ecumenical, personalist anarchist, pacifist, intersectionality

house of hospitality for destitute migrants, soup kitchen, anti-war and anti-borders, creative civil disobedience

self-managed, anarchistic, left-wing political understanding, feminist, non-violent, Christian, ecumenical, peace-work, organic gardening

☺

The Catholic Worker Movement was founded in New York in 1933, initially as a radical newspaper and then communities offering hospitality to the poor and marginalised. Today there are over 200 such houses and farms worldwide and each is autonomous.
The London Catholic Worker/Giuseppe Conlon House is an intentional Christian community living together with homeless asylum seekers and undocumented migrants. We aim to live simply, in solidarity with the most vulnerable members of society, and in resistance to the violence of wars, capitalism, borders and other forms of oppression.
Community members work full time running the house and our projects. Our work involves cleaning, cooking, maintenance, admin, emotional support, meetings, and gardening.
We are currently involved in creative non-violent actions against the arms industry, deportations, nuclear weapons and for environmental justice. We have daily prayers, we publish a small newspaper, host events and participate in our local churches and community organisations.
Our Christian faith is at the heart of our life and work. However, we welcome members and volunteers of all faiths and none. We aim to be an inclusive and diverse community welcoming people of all gender identities and sexual orientations.

→See also description of network "**European Catholic Worker**"

☺ sympathetic with or open to Christian pacifist anarchist views

☆ Lunesdale Community

C3

78 Beech Road | Lancaster LA2 6QH ☎ +44-(0)1524 888 588
community@lunesdale.community lunesdale.community/

2015 (near Lancaster) ♀2 ♂2 2 6

- e, Finnish
- core members (through discussion and agreement in meetings.)
- shared costs of living — owned
- ecumenical, Christian, contemplative, sustainability
- shared life with meals and household tasks as well as twice daily common prayer and Christian learning.
- ecological, self-managed, non-violent, spiritual, Christian, ecumenical, inner growth, meditation, vegetarian diet, humour

☺

We live together with a sense of vocation for Christian community living and a commitment to a regular prayer life. We seek to grow as human beings and in our knowledge of God and support each other through making ourselves available and pursuing a common life, with space to spend time on our own as well. We live in a beautiful village just outside Lancaster in a large house and share meals and tasks to make life together easier.

☺ our members commit to sharing the contemplative prayer life of the community.

Monkton Wyld Court

C5

Elsdons Lane, Monkton Wyld, Nr Charmouth | Dorset DT6 6DQ
☎ +44-(0)1297-560 342
info@monktonwyldcourt.org www.monktonwyldcourt.co.uk

1982 4,5 ha (near Exeter, Bournemouth) 30 % ♀5 ♂7 5 17

- e — all of us
- everyone in common (consensus) — common till
- trust
- education and sustainability
- gardening, farmwork, catering, housekeeping
- ecological, self-managed, non-violent, organic diet, vegetarian diet, vegan diet, adult education, seminar-house, environmental protection, self-sufficiency, permaculture, organic agriculture, organic gardening, conventional gardening, organic farming, animal friendly husbandry, forestry, alternative energy, ecological construction, humour

☺

Monkton Wyld Court is an educational charity run by a resident community of about a dozen adults, plus children and animals. The 11-acre grounds include a 13-bedroom Victorian rectory, a converted stable block and other outbuildings, a large organic vegetable garden, terraced lawns, fields, woods and stream. Facilities include a library, art hut, workshop and piano room.
Living at Monkton is a full-time commitment. Each morning we meet together to talk and plan the day. Cooking, cleaning, etc. is organised by rota, but each resident is also responsible for specific areas of work, from gardening to admin. We generate income through a programme of educational courses and other events.

United Kingdom

We have weekly business meetings where we make decisions by consensus. We are committed to making continual improvements to our levels of ecological sustainability, and usefulness to the wider community as an educational resource.
We welcome volunteers to work with us for two weeks initially. Volunteers often return for longer visits, and sometimes end up living here.

Old Hall Community

E4

East Bergholt | Suffolk CO7 6TG
+44-(0)1206-298 294
secretary@oldhall.org.uk
www.oldhall.org.uk

1974 30 ha (near Ipswich and Colchester) 80 %
♀25 ♂25 10 55 - 60

- e, g, f, sp
- everyone
- all members (consensus)
- limited common financing
- property of members
- commitment, consideration, communication, co-operation
- self sufficient organic food production and consumption
- ecological, self-managed, left-wing political understanding, feminist, non-violent, spiritual, Christian, Buddhist, esoteric/New Age, chaotic, inner growth, meditation, yoga, organic diet, peace-work, environmental protection, self-sufficiency, permaculture, organic agriculture, organic gardening, organic farming, animal friendly husbandry, alternative energy, humour, accessible to the disabled, non-smoking

☺

We are a group of families and individuals who have chosen to live together in a large old building in rural Suffolk. There are approximately 60 people living here, though that can vary from year to year. A number of the founder members from 1974 still live here and, although people come and go, we are a fairly settled community. We see ourselves as part of the wider community rather than separated from it.
We own 30 hectares, which we try to farm organically. We have vegetable gardens, orchards, woodlands, soft fruit, and a large lawn and play area, plus pastures and arable farmland. We grow wheat for our own bread, oats and other crops for animal fodder. We have a flock of Lincoln Longwool sheep and Redpoll dairy cows which are milked by hand. We make our own butter, cheese and yoghurt. We are largely self-sufficient in meat and some vegetables, although we do buy in whole foods, fish and other foodstuffs. Some of us are vegetarian. We also have chickens, geese, bees, pigs and cats. What livestock we have at any particular time depends on who is willing to manage them. We have a 'no dogs' policy.

→See also description of network **"WWOOF"**

- we have hosted PBI, IVS, Transition Town, WWOOF small conferences and meetings
- ☺ egalitarian green ideology, a capital investment

Othona Community, West Dorset

C5

The Othona Community, Coast Road | Bridport DT6 4RN
+44-(0)130 889 71 30
recruitment@othona-bb.org.uk   www.othonawestdorset.org.uk

1965   2,5 ha (near Exeter, Bournemouth)   5 %   ♀2-4   ♂3-5   0   7

e, f   7

all residents contribute, manager and trustees have ultimate responsibility

owned by The Othona Community charity

open and inclusive community, welcome people of all faiths and none, ecumenical

we host residential events for adults and families - see website

ecological, self-managed, hierarchical, spiritual, Christian, ecumenical, inner growth, meditation, environmental protection, organic gardening, alternative energy, humour, accessible to the disabled, arts, crafts, music, diversity, mindfulness

Othona Community, West Dorset is run by a small residential team who provide an experience of living in community for visitors staying in our 14-room community house, as they attend programmed events throughout the year.
We see Othona as "a place to be real together". While our roots are in the Christian heritage, we are open to all based on mutual respect and a sense of the everyday sacred. In approaching spirituality we are light-hearted, down to earth and unafraid to face change. Othona particularly suits people who are on a journey to discover their spiritual path or find a way to evolve within Christian faith. Some of us on the team see ourselves as Christian and some do not, or acknowledge other faith traditions as a guiding presence. Within the team and visitors we affirm diversity among people in faith, spirituality, age, ability, race, background and sexual orientation.
Othona is distinguished by the way visitors can join in a whole rhythm of life from gardening to leading chapel services. Singing, music, silence, candlelight and mindfulness are part of our life. Events during the school holidays are usually child friendly.
Our stone house is set in 6 acres on the West Dorset coast and can accommodate 30 visitors. Located above the first natural World Heritage Site, Chesil Beach: walking, bird-watching and beachcombing are popular activities.

- Othona
- all residents have paid roles - see website

United Kingdom

Pendragon House

C5

Butleigh Road | Glastonbury BA6 8AQ
+44-(0)1458 830 135
roger@pendragonhouse.org
pendragonhouse.org

2013 2000 m² 5 % ♀2 ♂2 2 6

e, f
1
all (meetings)
community fund
property of one member
sustainable living
maintaining the community
ecological, self-managed, left-wing political understanding, non-violent, inner growth, organic diet, organic gardening, alternative energy, humour, non-smoking, fun

Pendragon House is located close to the heart of Glastonbury. The house has rooms and suites for residents and guests. We live together as family, sharing in one anothers lives. Of course we enjoy privacy whenever needed, but we also try make an effort to contribute of our energy to the community we live in.
Our gardens include an orchard, vegetable patch and lawns. We all share a big kitchen and normally there is at least one communal meal a day. The living room is big enough to also be used for other things, it has table tennis, a grand piano, movie projector and sound system. All rooms have south facing windows and the sun keeps the house naturally warm. We try to be fossil fuel free, there is no gas or oil, in the winter months a 45 kW biomass boiler heats the house, and individual wood stoves. We have a large solar system and put as much electricity back into the grid as we use. Sunday is Funday all day.

Ploughshare Housing Co-op

C2

Edinburgh

1984 at least 1 (see page 410) 8

e
the whole group (mostly consensus, some majority)
limited common financing
ecological, self-managed, non-violent, vegetarian diet, non-smoking

We have no real shared ideology, though we hold vaguely similar attitudes in terms of environment, social justice, politics etcetera. The co-op was originally set up to accommodate people on low income in a traditionally wealthy area. Ploughshare Housing Co-op is based in a Victorian terraced house with small gardens to front and rear. It is in a quiet street in the heart of Bruntsfield, close to the Meadows and a short walk from the centre of Edinburgh. Members of our Co-op, eight in total at capacity plus one cat, are both tenants and directors of the limited company that owns the building - we have no landlord. We also have no right to the proceeds if the co-op is wound-up. We have a organic box scheme, an ethical milkman and bulk-buy food for communal evening vegetarian meals from Green City.

Redfield Community

D4

Buckingham Road | Winslow MK183LZ ☎+44-(0)1296-713 661
info@redfieldcommunity.org.uk
www.redfieldcommunity.org.uk/index.html

1978 | 7 ha (near Aylesbury, Milton Keynes) | ♀4 ♂7 6 17

e, g, Czech, sp — none
all decide in consensus (fortnightly meeting)

individual/limited community financing — housing co-operative
environmental and ecological focus
property maintenance, food production, living together
ecological, self-managed, non-violent, environmental protection, permaculture, organic gardening, animal friendly husbandry, alternative energy, ecological construction, humour, football, volleyball

☺

We are a community living together in a large country house in Buckinghamshire. The buildings are situated in an area of 7 ha with gardens, fields, newly planted and older trees. Redfield was established in 1978 and has developed progressively towards closer community. We live together as a single household, eat together and share the work. We try to be self-reliant, producing our own fruit, vegetables and meat by using organic techniques. We make decisions through consensus at our weekly meeting. One meeting a month has no agenda and offers the possibility to examine and celebrate different aspects of our communal lives. Spending so much time together places many private and common demands on us. Coping with them can be worthwhile and inspiring. We have found that the result of this way of life is just as great and positive as the effort we put into it.

☆ Scargill Community Scargill House

D3

Skipton House, Kettlewell | Skipton, North Yorkshire BD23 5HU
☎+44-(0)1756-760 500
admin@scargillmovement.org www.scargillmovement.org

1958 | 37 ha (near Skipton) | hardly | ♀19 ♂15 1 35

g, e, f, i, Hungarian, some African languages — 34 34
leadership team and council, members give input at chapter meetings
community members receive board and lodgings plus a monthly allowance

owned by a charity
Christian
retreat centre for guests
self-managed, non-violent, Christian, ecumenical, environmental protection, alternative energy, new technologies, humour

☺ ( )

We have a thriving, multinational Community from many different Christian denominations offering warmth, welcome and hospitality to all. Prayer, food and laughter seem to be the main hallmarks of our life together, with Jesus at the centre. Together we run a busy programme of events including quiet retreats, exuberant family holidays, teaching from some amazing visiting speakers and our own team, musical, creative and craft-based events, walking weeks, environmental and working holidays, house parties celebrating Easter and New Year and much more besides. In between our programmed events we host many church weekends, youth events and gatherings large and small, as well as individuals who make themselves at home with us for a few days.

☺ living Christian faith

Scraptoft Housing Co-op

D4

105 Scraptoft Lane | Leicester LE5 2FE ☎ +44-(0)116-282 79 29

scraptofthcl@gmail.com

2015 | 10 % | ♀4 ♂1 0 | 5

1 2

everyone (consensus) | individual financing

housing co-operative

growing organically, living cooperatively

self-managed, left-wing political understanding, organic diet, mentally/psychologically handicapped persons, environmental protection, permaculture, organic gardening, alternative energy, non-smoking, Aspergers

☺

We are still settling in as a group with two main focuses. Creating secure housing for people with Aspergers, and living lightly with interest in social justice in various forms. We have space to expand, but have learned that this needs to be slow.

☺ being familiar with autism

Springhill Cohousing Community

C4

Springhill | Stroud, Gloucestershire GL5 1TN ☎ +44-(0)453-766 466

info@springhillcohousing.com www.springhillcohousing.com

2003 | 1 ha | ♀40 ♂40 13 | 93

all volunteers

monthly meeting and sub groups (consensus using coloured cards)

each households contributes according to size

long leases given by the community for each house

consensus decision making

cohousing

ecological, self-managed, non-violent, organic diet, vegetarian diet, vegan diet

☺

Cohousing and consensus decision making. Ecological, many vegans.

Cohousing Network

☺ buy in or rent and agree to principles of Cohousing

St. Michael's Priory

D4

Newport Road, Willen | Milton Keynes, Buckinghamshire MK15 9AA

☎ +44-(0)1908-242 190

library@stmichaelspriory.org.uk www.stmichaelspriory.org.uk

1973 (near Luton, Northampton, Oxford) ♀2 ♂3 5

e 3

provincial & prior and professed and trustees (majority vote)

limited common financing

owned by our order, The Society of the Sacred Mission

Christian ecumenical, interfaith

retreats, hospitality, library, peace work

ecological, non-violent, spiritual, Christian, ecumenical, meditation, peace-work, organic gardening, accessible to the disabled, non-smoking, social work: asylum seekers, refugees & homeless, respite care / retreats, art: music, some drama, clay & art work, other: workshops on various themes relating to our spiritual program

St. Michael's Priory is a house of the Society of the Sacred Mission (SSM), an Anglican religious order. The first SSM brothers came here in 1973, and the women and men who comprise today's community continue to live the values of SSM's founder, Herbert Kelly.
The life of the Priory is centered around the daily rhythm of prayer and work. St. Michael's is a place of peace, encounter and welcome; and our community life provides a space where all may meet as one, in Christ. We seek to share our life with all who come here and we welcome all around our common table. Strangers soon become friends, and those new to our life will find space to pray, somewhere to enjoy the silence of the gardens or enjoy our food.
The religious community at the Priory finds God in contemplation, action and prayer, and in mission and ministry. St. Michael's is a sign of peace and hope, not only for our town and region, but to a world which is broken and which seeks renewal and new life.

Christian communities

financial and ideological

The Abbey Sutton Courtenay

C4

The Green, Sutton Courtenay | Abingdon OX14 4AF ☎ +44-(0)1235-847 401

admin@theabbey.uk.com www.theabbey.uk.com

1980 2 ha (near Oxford) ♀4 ♂1 0 5

e, f, i 4

members and trustees (consensus) charitable trust

holistic spirituality, universal Christian root, open to all faiths and those with no faith

education and retreat centre

self-managed, spiritual, Christian, ecumenical, Buddhist, Taoist, esoteric/New Age, chaotic, inner growth, meditation, vegetarian diet, organic gardening

The Abbey is a beautiful and historic 13th Century house. It is set round a peaceful courtyard, surrounded by informal gardens. It is a place which encourages the growth of a holistic

spiritual life. This requires nourishing our relationships with the divine, our true selves, the Earth and the world-wide community.
Our own community seeks to be inclusive: we welcome all who are searching, whether or not they find themselves within a particular faith. The source and inspiration of the ongoing work here has been the ecumenical Christian tradition. The emphasis is on the mystical expression of this tradition, which draws us to interfaith exploration and the incarnating of a universal vision.

☺ spiritual seeking outlook, willingness to be open and inclusive

The Catholic Worker Farm Lynsters Farm

D5

Old Uxbridge Road | Rickmansworth WD3 9XJ
☎+44-(0)1923-777 201, +44-(0)7983-477 819
thecatholicworkerfarm@yahoo.co.uk www.thecatholicworkerfarm.org

2006 1,4 ha (near London) 85 % ♀17 ♂3 7 27

e 4 8

8 core community members (consensus (99% of the time))

limited common funding rented at 2 locations

Christian/Catholic

we live with the poor in community; we do the Works of Mercy, resistance to war and grow organic vegetables

ecological, self-managed, anarchistic, left-wing political understanding, non-violent, spiritual, Christian, inner growth, vegetarian diet, work, resistance, community

☺

Our primary reason for existence is to serve and love the poor and, in the process, to discover that we too are poor and are loved. We take care of 18 homeless women and children who are without papers. These have suffered greatly.
We offer legal direction, psychotherapy, food and accommodation, fun and love. The women help to maintain the houses. We in the Core Community work a five day week, gardening and doing repairs, selling bread and looking after the needs of our women and children.
Live-in volunteers receive food, lodgings and a small amount of pocket money.

→See also description of network "**European Catholic Worker**"
☺ We ask that new members defer to the Catholic Worker tradition.

The Community Project at Laughton Lodge

D5

17 Laughton Lodge, Laughton | East Sussex BN8 6BY ☎+44-(0)1323-815 731
mel@laughtonlodge.org www.laughtonlodge.org

1999 (near Lewes, East Sussex) 2 % ♀24 ♂22 24 70

e, (f) about 50 % work from home/community owned office

everyone (consensus)
individual financing; shared financial responsibility for maintenance of land and common areas
properties owned by members; shared land and common house
self-managed

Laughton Lodge is a rural, intergenerational cohousing community. Our overall aim is to create a cooperative environment which transcends the isolation many families experience living in modern cities. We have no rigid philosophy and most of us work outside the community but we aim to create a community in which the whole is greater than the sum of its parts - offering opportunities to all members which they would not otherwise have had. This ranges from the freedom for children to range over our large site and play together in an "old fashioned way" to the opportunity to share in the keeping of animals which need too much space or work for a single family, the opportunity to share resources like cars or tools, share skills, and join with others in creating new enterprises. We benefit from the range of skills - for example the artists amongst us share their work, those with skills in gardening or animal husbandry or event organisation provide other benefits. And there are enough of us to have a good party.

UK Cohousing Network (headquarters)
usually have to be able to buy a house

The Drive Housing Co-op

D5

8 The Drive | London E17 3BW
+44-(0)20852-15433
enquiries@thedrive.coop
www.thedrive.coop

♀5 ♂5 10

e, f
all members present in meetings (consensus)
limited common financing
fully mutual cooperative
sustainability
public sector, charity sector, coaching and tuition, therapy
ecological, self-managed, anarchistic, left-wing political understanding, feminist, organic diet, vegetarian diet, environmental protection, permaculture, organic gardening, humour

The Drive Housing co-operative is an 11-bed, fully mutual member-run organisation in Walthamstow, London. We aim to provide an alternative and better way of being together, as a guard against the isolating forces of modern capitalist existence in London. Whilst we do not have a shared spirituality, we work to ensure that the coop provides a non-discriminatory place to call home and is sensitive to the psycho-social needs of its members and the wider community.

London federation of housing coops
committed to community living

The Findhorn Foundation

C1

Findhorn Foundation, The Park | Forres IV363TZ — +44-(0)1309-690 605
pr@findhorn.org — www.findhorn.org

1962 — 10 ha (near Inverness) — ♀80 ♂60 25 144

e, g, f, i, sp, Japanese — 50 140
everyone in common (consensus) — individual financing

Findhorn Foundation
spiritual, ecological
education, gardening, kitchen, workshops
self-managed, non-violent, spiritual, inner growth, meditation, organic diet, vegetarian diet, adult education, peace-work, physically handicapped persons, self-sufficiency, permaculture, organic gardening, organic farming, forestry, alternative energy, new technologies, ecological construction, humour, accessible to the disabled, ecovillage, non-traditional healing, handicrafts, art

☺

The Findhorn Foundation and Community is a spiritual community and holistic learning centre - an experiential search into new ways of living. It is a dynamic experiment where everyday life is guided by the voice of spirit, where we work in co-creation with nature and take inspired action towards our vision of a better world. It is at the heart of the largest intentional community in the UK.
We share our learning and way of life in experimental workshops, conferences and events. This provides experiences and tools for people to relate to each other and to the world in new ways. We recognise that the future of humankind and the natural world depends upon a worldview that is compassionate and interconnected. The Foundation regularly hosts well-known presenters such as Eckhart Tolle, Caroline Myss and Robert Holden. Through its work to build a sustainable lifestyle, Findhorn's Ecovillage Project has contributed significantly to the growing ecovillage movement worldwide. The community's home and energy footprint is 22 percent of the UK national average.
The Findhorn Experience Week is a practical experience of how to apply sustainable values in daily life. We welcome guests to become part of the ongoing exploration.

→See also description of network "**GEN Europe**"
UNESCO
☺ experience week, living in community programme, living education apprentice programme

The Open House Project Sheffield

D4

The Long Barn, Bracken Hill | Sheffield S351RD — +44-(0)780 316 81 96
openhouseprojectsheffield@gmail.com — openhouseproject.co.uk

2012 — 5 % — ♀5 ♂4 8 17

e, f
the members/directors (consensus/majority voting) — individual financing

freehold by the co-housing company, dwellings are leasehold
developing farm buildings into dwellings; managing the site, field, growing area and orchard; undertaking informal communal events

ecological, self-managed, organic gardening, alternative energy, new technologies, ecological construction, construction

☺ (✋✋)

We are a group of people working within a historical site to establish individual homes and shared spaces for people sympathetic to our ideals of mutual support and respect – for each other, our buildings, and the environment. Aims:

- providing a range of homes, for families or individuals, that are suitable and comfortable for people at any stage in their life
- being sensitive to our environmental impact
- developing useful and rewarding links with the wider community
- homes for purchase or possibly rent
- disabled access wherever possible, single floor living in some units
- safe environment for children
- regular meetings for discussion, decision making, and sharing ideas, was well as regular communal meals and participation in activities
- caring for communal areas, gardening, managing livestock provision of communal space and facilities
- appreciation of the contributions made by each one, from the very old to the very young
- working towards Low-Impact Sustainable Settlement status in areas such as energy production and use, waste disposal, use of vehicles, tool sharing.

UK Cohousing Network, Sheffield Cohousing Network

☺ financial buy-in for full member status + approval of the group

The Well House

D4

10 High Street | Market Rasen LN7 6TX

+44-(0)-147 285 19 40

gytribe1@gmail.com

2017 (near Caistor) ♀2 ♂3 1 6

e, (sp, f)

collective (meeting)

some common financing

owned by 2 members

none

ecological, self-managed, anarchistic, left-wing political understanding, non-violent, inner growth, meditation, vegetarian diet, vegan diet, environmental protection, self-sufficiency, permaculture, organic gardening, new technologies, humour

☺ (✋✋)

A shared living environment/house with a desire to grow and to expand the number of people living in our community. We want to increase our self-sufficiency but currently still work and pay bills as per the usual way of living. The house is owned by one couple but not in a landlord way. Contribution in terms of cleaning and running the house is expected. We do have standards we expect from community members. We have an allotment.
We are interested to find people who want to live together and actively contribute, not just be tenants. This includes being proactive and joining in and using initiative.
Ideally we want people to live here, loving and caring for the community as if it was their home, finding new lifetime friends.

☺ cooperation, contribution

United Kingdom

Threshold Centre Cohousing Community

C5

Cole Street Farm, Cole Street Lane | Gillingham, Dorset SP8 5JQ

+44-(0)1747-821 929

info@thresholdcentre.org.uk www.thresholdcentre.org.uk

2005 8000 m² (near Bristol) ♀13 ♂7 0 20

all contribute

all (consensus) self-managed community finances (CIC)

freehold owned by Threshold CIC, leaseholds by Housing Association and private owners

organic gardening, sustainability, spirituality

ecological, self-managed, non-violent, spiritual, meditation, organic diet, vegetarian diet, permaculture, organic gardening, humour

☺

The Threshold Centre at Cole Street Farm is a unique, pioneering cohousing community and sustainable education centre, which aims to demonstrate a way of life, which is more green, affordable, and neighbourly. Cohousing combines small, self-contained homes with shared facilities, which are a key part of this.

The Centre runs regular courses on cohousing and other subjects. Having operated on a pilot basis since late 2004, the Threshold Centre won planning permission and grant funding in 2008 to create a cohousing scheme at Cole Street Farm. It consists of fourteen dwellings, seven of which are affordable rent and shared ownership with a local housing association - a first in the short history of cohousing in this country. There is also a common house with shared facilities and guest rooms, green energy systems and a community market garden. The Threshold Centre shows how to achieve major cuts in the three main aspects of domestic carbon emissions: home energy, food and travel.

Our Vision is to work towards something greater than our own needs, and to tread as lightly as possible on the earth.

UK Cohousing Network

☺ an interest in living sustainably on the planet

Tinkers Bubble

C5

Little Norton, Norton sub Hamdon | Stoke-sub-Hamdon, Somerset TA14 6TF

tinkersbubble@riseup.net www.tinkersbubble.org

1994 16 ha (near Yeovil) 75 % ♀3 ♂4 1 8

e, sp 7 7

everyone (consensus) limited common financing

property of all

environmentalist

forestry, vegetables, apple juice

ecological, self-managed, non-violent, organic diet, pedagogical, environmental protection, self-sufficiency, permaculture, organic agriculture, organic gardening, organic farming, animal friendly husbandry, alternative energy, ecological construction

☺

Tinkers Bubble is a small woodland community which uses environmentally sound methods of working the land without fossil fuels.
We have planning permission for self-built houses on the condition that we make a living from the land. We make our monetary incomes mainly through forestry, apple work and gardening. As a result we're money poor but otherwise rich! We manage about 28 acres of Douglas fir, larch, and mixed broadleaf woodland using hand tools, horses, and a wood-fired steam-powered sawmill. We press apple juice for sale, grow most of our own vegetables, keep cows, chickens and bees. We make loads of preserves, including cider and cheese. We have solar powered 12 V electricity, spring water on tap, and use compost toilets. We burn wood for cooking, heating, and for hot water in the bath-house. Most people wash their clothes by hand. We eat little meat (mostly game or reared here), and try to cater for all diets. Though some of us would consider ourselves to be spiritual, we have no shared spirituality. Life is lived mostly outdoors, so it's cold in the winter, but we live on the top of a steep hill, so there's plenty of chances to get warm!

☺ must be passionate about working the land and sustainability *Photo on page 15*

☆ Torch Housing Co-op 10 Richmond Road

C4

Hockley, Birmingham | West Midlands B18 5NH ☎ +44-(0)121-554 42 56
jantorch@aol.com www.radicalroutes.org.uk

1993 | 1 house. 2 gardens | ♀2 ♂2 0 4

e, f, g,
everyone in common (consensus then voting) | individual financing
cooperative
being happy
environmental/green
ecological, self-managed, anarchistic, left-wing political understanding, chaotic, humour

☺ ()

Torch Housing Co-op is in the Hockley area of inner-city North-West Birmingham. We have been around quite a long time and have seen many changes. We believe in being happy, having good friends, getting along with each other and being open to new ideas, but sadly are no longer as active in our passions. We welcome working visitors by prior arrangement only.

Radical Routes
☺ getting along with other people, sense of humour

Trelay Cohousing Community

B5

St Gennys, Bude, Cornwall | Bude EX23 0NJ
info@trelay.org
www.trelay.org

2007 11 ha 20 % ♀12 ♂10 11 33

e varies

all of us (consensus)

individual capital, mutual home ownership, shared mortgage

owned by a workers cooperative (non-profit company)

community and connection

local food, local energy

ecological, self-managed, non-violent, spiritual, Christian, Buddhist, esoteric/New Age, meditation, yoga, organic diet, adult education, environmental protection, self-sufficiency, permaculture, organic agriculture, organic gardening, organic farming, animal friendly husbandry, forestry, alternative energy, new technologies, ecological construction, humour, accessible to the disabled, benign capitalism; sustainable enterprises; synergistic activities using waste from one as a resource for another; diversity; freedom with respect

☺ ( )

Trelay Cohousing Community is a vibrant community of twenty-two adults and eleven children, established in 2007 at Trelay Farm. The farm has thirty-two acres of land and is near Bude in North Cornwall, UK, two miles from the Atlantic Ocean.

Trelay is a rural intentional community. We help solve problems of affordable housing and social isolation. We aim to provide a rewarding and satisfying lifestyle to each of our members. We work hard but we also have fun.

Trelay Covenant:
We aim to care for each other, our home, our animals, our plants and our land.
We are committed to supporting each other through all challenges and difficulties.
We believe in a healthy balance between individual and community needs.
We are committed to learning how to trust, communicate and learn from each other.
We believe in arriving at consensual decisions through listening to everyone.
We wish to play a positive part within the wider community of North Cornwall.
We aim to make Trelay a good example of how to live more lightly upon the Earth.

Cat (i.e. picture of cat):
Sent to us by
Ploughshare Housing Co-op, page 400

Ukraine see page 434f

Eastern Europe

Sorry for the alphabetical mixup at the end of the existing communities' descriptions - that's due to the shared map. We can present not more than four communities from each of the following countries, which is too bad, of course. We will reward any national effort to bring more communities into our directory with an own map in the next edition, ok? We know that you can do it: At least Israel, home of the kibbutzim, could have plenty of contributions. Anyway, please refer to the table on page 52f for detailed numbers to the following countries and to GEN Europe (page 523ff) for networking needs. There are also Camphill communities (510ff), and networks like Emmaus or L'Arche might be active as well.

Bosnia-Herzegovina

EP Zelenkovac

Zelenkovac bb | 70266 Podrašnica ☎+387-(0)50-278 649
zelenkovac@blic.net www.zelenkovac.org

1998 2 ha ♀203 ♂282 58 485

e, Russian 5 10
leader and board
self-sustainability of the project, donations the association
no
environment, culture, tourism
ecological

☺ ( )

Coming to Zelenkovac is simply unique: an alternative place that started as an art gallery in 1985 in the middle of a mountainous forest. Nowadays, our small eco-village has grown and in each year we welcome more and more travelers, tourists and artists from all nationalities who meet the locals that have been supporting us from the beginning. Our environment looks like a movie scene: tall pine trees, a little river with many small bridges and wooden buildings that combine ethnographic tradition and artistic creativity. It is also a commitment to promote alternative ways of life and ecology, to support cultural events in a rural and wild area and to reinforce cultures and religions mixing in a country that has been in war a few years ago.

Balkan Ecovillage Network
☺ no conditions

Bulgaria

Spiritual Rainbow Community

Wtora 1 | 6584 Odrintsi ☎+359-(0)361-617 91
odrintsi@spirebo.com www.spirebo.com

2006 80 ha (forest, pasture and farmland) (near Ivaylovgrad) 90 %
♀9 ♂10 6 25

g, e, f, Bulgarian all all

all together/founders of the community (according to energetic criteria)
common economy — property and lease
own free philosophy
according to the season, everything that belongs to a self-sufficient life
ecological, self-managed, leader/guru, inner growth, meditation, organic diet, environmental protection, self-sufficiency, permaculture, organic agriculture, organic gardening, organic farming, animal friendly husbandry, ecological construction, non-smoking, shepherd farm for goats/sheep/cattle and water buffalo, own herd of horses, beekeeping, self-reflection, spiritual teacher/counselor, wool processing, weaving, textile craft, woodworking, herbal medicine, metalworking, loam construction

☺ (✋✋)

"What you have built up there is like another world, a real paradise, just incredible!". This is a response we received from visitors time and again.

Our day starts early with a common morning meditation, to tune in to the essential, the spiritual. We try to preserve the feelings of meditation in everyday life and to perceive the energies in all natural things. To live as naturally as possible is important to us, because we are convinced that the distractions of civilization have deceived us in previous years, so that we no longer knew who we really were.

We strive for absolute self-sufficiency in our community, not just in nutrition. We have already implemented: cheese dairy, bakery, grain mill, oil press, wine press and brewery. In addition, we operate joinery, blacksmith, locksmithery, textile manufacture, tailoring, leather workshop and a herbal pharmacy.

We have renovated some houses with clay, in the way that we feel comfortable in them. There is no electricity in our homes, we cook and heat with wood. We spend most of the day outside - in a beautiful natural landscape with a view of the Rhodope foothills: We take care for the herds of our animals, our gardens and fields, and we often collect gifts of nature.

In 2019 a new documentary has been published about us at https://youtu.be/OWYGA-ABHhBM (German language, Bulgarian subtitles). More Bulgarian language clips at spirebo.com/clips.htm and a German language book at https://amzn.to/2C0SA9N (for Amazon Kindle reader).

no
☺ individually

☆ Xperiment Let's reinvent our lives together

7 Juriy Venelin str. | 6000 Stara Zagora — +359 (0)878 539 752
thexperiment2019@gmail.com — thexperiment.info

2018 — 3000 m² (near Gurkovo) — ♀1 ♂3 4

e, Bulgarian, Russian

core members team (majority vote)
currently financed by team members, will self-support in future
one member, due to be transferred to NGO in spring 2019
first and main focus is "no division of any kind"
cultural and social center, combined with sustainable life
ecological, self-managed, non-violent, meditation, yoga, seminar-house, animal protection, environmental protection, self-sufficiency, permaculture, organic agriculture, conventional agriculture, organic gardening, conventional gardening, organic farming, conventional farming, alternative energy, new technologies, humour, social, art-centre

☺

This project is aimed at achieving a new sustainable life as a society of humans in all of its aspects. In its core it is an open platform for anyone, who desires to practically change their paradigm.
The specific goals of the Xperiment are aimed at all areas of life, including:

- Creating a shared space for sustainable life out of the urban areas in harmony with nature and ourselves for people of all ages, background or family status.
- Creating a steady micro-society, aimed at closing as many energy cycles as possible (i.e. food, water, electricity, heat etc.) by combining traditional knowledge with all technologies available today.
- Creating a dynamic social space, clean of prejudice and division, which is being constantly evolved by its members regarding its values, principles and structure.
- Creating an open cultural space, aimed at nurturing the free flow and exchange of inspiration between people from all cultures and artistic backgrounds.
- Creating a local society, aimed at shared development and implementation of social, spiritual and educational practices.

Combining all of the above into a single living system, which translates into creating a whole new life environment. During summer 2018 around 200-220 volunteers worked with us.

☺ sharing our views in general. More details on the website

Croatia

Eko Centar Latinovac

Latinovac 11 | 34350 Latinovac-Caglin ☎ +385-(0)958 525 498
ekocentarlatinovac@gmail.com www.ekocentarlatinovac.hr

1995 5 ha (near Pozega, Dakovo) 10 % ♀2 ♂2 1 5

e, d, Croatian 3
the group (majority vote) individual financing
rented, property of different members
permaculture, education, social entrepreneurship
ecological, non-violent, vegetarian diet, seminar-house, environmental protection, self-sufficiency, permaculture, organic agriculture, organic gardening, ecological construction, humour

☺

Our community/NGO is called Eko Centar Latinovac. It was founded in Latinovac, Croatia, in 1995 and received its legal status three years later in 1998. Through its existence, Eko Centar Latinovac had many and varying goals, from peacebuilding and personal growth to environmental activism and raising awareness. In the last five years, our activities are mostly based on permaculture, development of local community, work with youth and volunteerism. Throughout the year, we are hosting several international volunteer groups. All activities are based upon an educational and practical part, responding to the principles of lifelong learning. The goal our organization pursues is to contribute to the community development, to connect and network with local people, to create possibilities for a better life and future and to provide a higher quality of living for all generations.

☺ ideological

Czech Republic

Osho Shangri La Osho Meditation Center

Lažany 49 | 36452 Žlutice — +420-(0)353-395 310
shangrilaomc@gmail.com — www.shangrila.cz

1995 — 6 ha (near Carlsbad) — 20 % — ♀6 ♂7 4 17

Czech, e, g — 6

one part of the group (consensus)

individual and limited community financing

property of individual community members

Osho

gardening, cooking

ecological, self-managed, spiritual, meditation, organic diet, vegetarian diet, pedagogical, seminar-house, environmental protection, permaculture, organic agriculture, organic gardening, alternative energy, new technologies, ecological construction, therapy

The Shangri La OMC is a place of lively togetherness for people who love silence and untouched nature. The centre is located in a wide, wild, silent and natural landscape near Carlsbad in the Czech Republic, 50 km from the German border. There are daily meditations and a wide range of offers for meditation and therapy. Our spiritual leader is Osho. Members of the community share seven houses. An additional house hosts meditation- and seminar room, kitchen, dining hall as well as the guest area. All dwellings are within generous gardens with old trees, ponds and fields.

Osho

☺ meditation experience and/or interest in meditation/Osho

Photo on page 11

Czech Republic

PadmaFarma Jakub Green

Belá u Jevícka 125 | 569 43 Belá u Jevícka +420-(0)737 550 156
eko.renata@seznam.cz www.padmafarma.cz

1999 (near Jevícko) ♀2 ♂1 3 6

e no

ecological, self-managed, non-violent, spiritual, Christian, ecumenical, Buddhist, Taoist, esoteric/New Age, meditation, yoga, tantra, organic diet, vegetarian diet, vegan diet, macrobiotic diet, adult education, pedagogical

PadmaFarm in Belá u Jevícka is a precious place where people are seeking for alternatives to a contemporary lifestyle based around consumption, which we feel is developing a little too fast. We have been rebuilding (from 1999) an old ruin of a farm into a centre devoted to ecology, art and meditation. That is here for all of you:

- People interested in ecology and who have a concerned attitude towards the world
- Families who are aware that we are responsible for what we leave behind for our children
- All those who not only talk about good things, but also do them
- All those who are willing to give up their egocentric thinking, open up and co-operate with others and
- people searching for their own way and for whom spiritual development has become a natural part of their lives.

PadmaFarm consist of people, animals, buildings, meadows, fields, garden, pond and forest. The forest, altogether five hectares, is not commercially used because the trees are too young. The older forest (one hectare) was used for a log chalet and farm buildings. We raise our stock of cows, sheep and goats on five hectares of meadows. The overall capacity of the PadmaFarm (when all the work is completed) will be seventy people; so far it is designed to hold fifty people.

☺ no drugs

Estonia

Lilleoru Lilleoru training center and conscious community

Taevasmaa tee 2, Aruvalla, Rae vald | 75320 Harjumaa +372-5649-9199
ave.oit@lilleoru.ee www.lilleoru.ee

1993 (near Tallinn) 15-40 % ♀30 ♂16 14 50

e, g, f, Estonian, Finnish, Russian 6 7

different ways: small teams, meeting of all members (mostly majority vote in general meeting; consensus)

limited common financing, community owned business property of all

holistic view on human nature and practices for mastering one's inner world

adult and child education

ecological, self-managed, non-violent, leader/guru, spiritual, inner growth, meditation, free school, yoga, organic diet, vegetarian diet, adult education, seminar-house, permaculture, organic gardening, forestry, ecological construction, kriya yoga

☺ (hands icon)

Lilleoru is a training center and conscious community in Estonia, established 25 years ago. The heart of our activities is to support people's consciousness change and development by introducing methods for mastering one's inner world, orientating quickly in changing situations, being creative and innovative. The core method is the Art of Conscious Change, practiced also by all members of the community.
For more than a decade Lilleoru has also been active in promoting ecologically sustainable way of life, through cooperation with partners in Estonia and abroad. Lilleoru permaculture garden is considered to be largest in Estonia. Both vegetable garden and herb garden are certified organic. We're also employing sustainable forest management techniques to manage Lilleoru forest.
The new main building of Lilleoru centre is The School of Practical Awareness – built mostly with the help of volunteers – is the largest reed-bale building in Europe, and the first public building of its kind in the world. The opening of the house was in Autumn 2018 and also Lilleoru Elementary Scool started its work here. Lilleoru is also an Ashram, in Babaji's Kriya Yoga and Haidakhan Babaji spiritual lineages. The project was started by an inner growth study group of about 20 people, who gathered around the well-known Estonian spiritual teacher Ingvar Villido. At the moment, 60 persons live at the community premises and SkyEarth village. Lilleoru is managed and owned by Lilleoru NGO that has close to 150 active members. Many members carry regular voluntary tasks and this is mainly how the project has been built up and brought forward. Some people are employed by our organization.
We have 20 hectares of land with a unique Flower of Life park and we have a future vision for the premises. *See photo on page 14*

→See also description of network "**GEN Europe**"
GEN-Estonia, Estonian Coalition for Sustainable Development

Small Footprint

Ökokeskuse | 78237 Mõisamaa
info@vaikejalajalg.ee vaikejalajalg.ee

2014 33 ha (near Märjamaa, Tallinn (75 km)) 5 % ♀7 ♂3 9 19

Estonian, e 1 2
big circle (all adult members) (sociocracy)
common purse for food and housekeeping co-owned
Gaia education principles
trainings, gardening, agriculture, beekeeping, various
ecological, self-managed, non-violent, spiritual, chaotic, inner growth, vegetarian diet, seminar-house, environmental protection, permaculture, organic agriculture, organic gardening, organic farming, mixed individual diets, vegetarian in common areas

☺ (hands icon)

Small Footprint Community is an eco-community founded at the end of 2012 and since 2014 it is located in Mõisamaa, Rapla County, Estonia. Small Footprint is a member of the Network of Estonian eco-communities. The founding members of the community were brought together by the summer gatherings of the Network of Estonian Eco-communities, Ecovillage Design Education (Gaia Education) course in 2010 in Lilleoru, community building training Rising Together in 2014 in Esna.

We in Small Footprint believe that the creation of new and sustainable life is possible through practical everyday life experience, by living and working in an intentional community. The members of Small Footprint are connected by a common understanding of the sustainable operation of community through all four aspects – economic, ecologic, social and worldview.

Colour drawing on page 10

GEN Estonia

Greece

Estia Agios Nikolaos Life-sharing community

Keramou | 33052 Galaxidi ☎+30-(0)22650-420 62
evs.estia@gmail.com www.estia-agios-nikolaos.eu

1993 11.3 ha (near Amfissa) 10 % ♀15 ♂18 1 34

Greek, e, sp, f
all and five external partners
those responsible in the community, managing committee and supervision board (consensus)
community ownership
respect for people and the environment
making jewellery, pottery workshop, gardening
ecological, non-violent, mentally/psychologically handicapped persons, environmental protection, organic agriculture, organic gardening, animal friendly husbandry, accessible to the disabled, pottery, handicapped, disabled

☺

Estia Agios Nikolaos is a community of people with and without developmental disabilities, in which all members live, work and spend their free time together in small family structures. At the moment there are three functioning houses, each one accommodating a "family", as well as two workshops, a conference hall and a small coffee shop on private land of 28 acres. Each house hosts 6 to 8 co-workers (people with developmental disabilities) who live together with the house parents (staff) and volunteers. There is also supporting staff from the local community, who help with household tasks and run activities such as art therapy, physiotherapy, gymnastics and music.
In the mornings, all residents are involved in the work of the community, in pottery workshops, making jewellery, embroidery, cultivating the farm and household work. The afternoons are devoted to music, exercising, practicing musical instruments, walking, and - in the summertime - swimming. The weekends are dedicated to artistic and spiritual activities, such as small trips, musical events, theatre and participation in local celebrations.
Estia Agios Nikolaos is situated in Galaxidi, next to the sea and not far from Mt. Parnassos. There are buses running regularly to the towns of Itea and Amfissa (the province capital). The capital of Greece, Athens, is 200 km away.

☺ willingness to share life in community

Free and Real The Telaithrion Project

Aghios | 34300 Edipsou ☎ +30-(0)222 603 16 96
info@freeandreal.org freeandreal.org

2010 15 ha (near Istiaias Edipsou) 50-60 % ♀6 ♂5 0 11

f, e, Greek 11
website: https://en.freeandreal.org/research/
shared economy property of the project
website: https://en.freeandreal.org/thevision/
ecobuilding, permaculture, DIY, technology
ecological, self-managed, non-violent, vegan diet, adult education, seminar-house, one-world development, environmental protection, self-sufficiency, permaculture, organic agriculture, organic gardening, organic farming, forestry, alternative energy, new technologies, ecological construction, construction

☺ ( )

We have made it our life's journey to research, create and implement an efficient social structure in a prototype, sustainable eco-community on the island of Crete, Greece, which can be imitated and applied in any part of Greece, and why not the world, to work as a research school on self-sufficiency and sustainability. This social structure will be founded on inalienable human rights and an economy defined by strategic use of technology, available resources, practical knowledge and hard facts. Selfless service (not exchange) and an eagerness for development and new sustainable solutions will be the driving force of our project. During this time we encourage and actively support like-minded people and groups in their effort to research and/or realize their own productive and sustainable ideas, structures and communities. This way, our community will serve as a school for anyone who wishes to learn about self-sufficiency and sustainability.

☺ ideological *Photo above is a composition*

Kalamos Island Eco-community

Kalamos Lefkados | 31081 Kalamos ☎ +30-(0)697 344 21 92
contact@terrasylvestris.org www.terrasylvestris.org

2015 30 % ♀2 ♂1 3

Greek, e, sp, Portuguese, f 3
as a non-governmental organization, we have a board of directors (majority vote sometimes, consensus sometimes, it depends)
individual financing, common financing, donations, income generating activities (e.g. seminars)
We believe in conserving nature.
development of income generating alternatives to destructive practices, a more sustainable way of life in general

Greece

ecological, self-managed, non-violent, spiritual, inner growth, meditation, organic diet, adult education, seminar-house, animal protection, environmental protection, self-sufficiency, permaculture, organic agriculture, organic gardening, organic farming, animal friendly husbandry, ecological construction, construction, accessible to the disabled, activis, scientific

☺ (✋✋)

We started this project as a part of the Kalamos and Kastos sustainable development project of our non-governmental non-profit organization.
The goal is to transform a part of the main island of Kalamos into productive permaculture areas, populate them with ecologically conscious people and develop income, generating alternatives to currently environmentally destructive practices on land and at sea. The ground we work with (or plan to work with) are lands which are abandoned but are given to us by local people in written formal agreements to manage. These are lands which would never contribute to wildlife conservation if abandoned or managed in another way. Examples are former abandoned sites full of junk in the middle of the main village, agricultural areas far from ecologically sensitive areas like forests or unmanaged grasslands, abandoned houses with their large front garden and other similar types.
People from all over the world come to help us, learn and some will hopefully stay longer.

→See also description of network "**GEN Europe**"

Transition Network

☺ volunteering to come and help us out, with time integration in our community; membership by relevant ethos and character that is needed

Sarakiniko Alternative Life

P.O. Box 12 | 28300 Vathy/Ithaki ☎+30-(0)697 863 72 77
villawunderbar@gmx.de

1979 | 50 ha (near Vathy) | 5 % | ♀30 ♂30 | 60

g, e | 20

one group votes (decision by majority) | limited common financing

property of the community

caring for the nature around us, maintenance of the rudimental infrastructure

self-managed, non-violent, animal protection, environmental protection, alternative energy, humour, holiday destination, arts: painting, music etc.

☺

The Sarakiniko peninsula on Ithaki (Island in the Ionian Sea - Greece) was bought in 1979 by approximately 200 Germans who were looking for an alternative way of life.
Today we are only 60 members and at least half of them don't live permanently on Sarakiniko. Many singles, some couples, and families live in a friendly neighborhood together.
We have solar-energy for our electricity. And we love our wonderful piece of land and respect nature.
Visitors are welcome, preferably by agreement.

☺ financial deposit,
some time to get to know each other

Hungary

Gyürüfü Ecovillage (Gyurufu)

Gyurufu 13 | 7683 Dinnyeberki — +36-(0)620-665 59 51
kili@gyurufu.hu — www.gyurufu.hu

1991 | 300 ha (near Pécs) | 50 % | ♀8 ♂9 | 5 | 22

e, g — 10

one part of the group - legally elected bodies (majority vote)
individual financing
property of a foundation/property of the local municipality
applied ecology
farming, ecotourism, computer science
ecological, self-managed, environmental protection, self-sufficiency, permaculture, organic agriculture, organic gardening, animal friendly husbandry, forestry, alternative energy, ecological construction, construction

Gyurufu is an emerging full-fledged ecovillage based on the permaculture principle and other methods of ecological planning. It is designed for a 1100 ha outstanding watershed partly owned by the Gyürüfü Foundation. The rest is either privately owned land or state forest. The village was started from scratch; no road, no electricity, no infrastructure were present in a formerly populated area which was abandoned during the communist era. The community has set up a master plan for the area which contains the main rules for activities as well as housing, transport, agriculture, water management and forestation goals. Building permits have been granted for family houses and agricultural facilities since 1996. Available land is held in common, officially owned by the Gyurufu Foundation and the municipality of Ibafa, respectively. Plots are sold to the house owners. Upon completion, the village may not hold more than 300 permanent residents as determined by the ecological carrying capacity of the area. Appropriate building technologies include mud bricks, rammed earth buildings, composting toilets, reed bed systems and rainwater harvesting as well as a host of other methods. Gyurufu might well be the first ecovillage where ecological constraints are integrated into the state administration in form of the master plan, the nature conservation requirements and other resolutions of the cognisant authorities. Beside 10-15 residential houses, the village accommodates a community house, a camping area and several businesses, a goat farm, an integrated agroforestry project, horse riding facilities, and an information technology company based on the internet. Plans for the future include a village school, an afforestation project, an aquaculture pond and other water management schemes.
(Description from 2013, other data has been updated for this directory edition)

→See also description of network "**GEN Europe**"
Hungarian Network of Living Villages (MÉH)/GEN Hungary
☺ newcomers should be able to build their own houses
and provide themselves with a livelihood

Krisna Völgy

Gauranga tér 1 | 8699 Somogyvámos
+36-(0)30-377 1530
info@okovolgy.hu
www.krisnavolgy.hu

1992 | 200 ha (near Kaposvar) | 20 % | ♀70 ♂90 | 28 | 188

Hungarian, g, e | 95 %

the board of directors (majority vote) | limited community financing

common property

Hare Krishna

education, Vedic culture, ecological farming

ecological, self-managed, non-violent, hierarchical, leader/guru, spiritual, inner growth, meditation, yoga, organic diet, vegetarian diet, adult education, seminar-house, peace-work, animal protection, environmental protection, self-sufficiency, organic agriculture, organic gardening, organic farming, animal friendly husbandry, forestry, alternative energy, ecological construction, non-smoking, social work (distributing food), bakery, pottery, candle workshop, art, dance, painting, Indian music, non-violence, media: magazine, therapies currently developing, Krishna consciousness, cow protection

Krisna Völgy ("Krishna Valley"), established in 1992, is a religious, non-profit organization in Hungary. It is Europe's largest and most well organized ecological community*. Our farm is a spiritual center that attracts many pilgrims yearly as well as an ecological farm community that focuses on self-sufficiency and environmentally friendly agriculture and way of life. Our temple is part of ISKCON, its uniqueness and beauty make it a national attraction. We provide training for both local and international practitioners in the science of Bhakti Yoga and Vedic culture.

In our agricultural fields, we strive to cultivate all the community needs. Our cow protection project houses more than 50 herd members saved from slaughter. The oxen help us cultivating our land as we strive to be environmentally friendly, using as less machinery as possible. Furthermore, becoming self-sufficient means not being part of the national grid, our required electricity is produced by solar panels, our water is supplied by wells and the processing of wastewater is done with a reed-bed sewage system without the use of any chemicals or electricity.

In short, becoming self-sufficient is not merely necessary to help our degenerating environment but also to help us in our own spiritual practice. We believe that living in a peaceful, God-centered environment, not harming any living entity around us is crucial for any spiritual practice. We strive to share this knowledge with others by our example as well as our various workshops, courses and programs.

→See also description of networks **"ISKCON, GEN"**

Hungarian Ecovillage Network

☺ religious, no meat, no drugs, no "illegal" sex, no loafing

**editor's note: We are glad that this distinction has been made without our participation. The process could easily have gotten ugly...*

Nyim Eco-Community

Rákóczi utca 4. | 8612 Nyim
info@felkelonap.hu www.felkelonap.hu

2010 26 ha (near Siófok) 20 % ♀6 ♂6 5 12

g, e, Hungarian 3 8
whole group (consensus)
individual financing and cooperative (shared capital and income)
private property of members
ecological values, humanistic psychology
community enterprises, community development, permaculture
ecological, self-managed, non-violent, inner growth, meditation, vegetarian diet, adult education, pedagogical, permaculture, organic gardening, organic farming, ecological construction, humour

☺ ( )

Our community lives in a small village called Nyim, near Lake Balaton. We have a community garden, orchard, bee-keeping and a community owned book manufacture. We also cooperate with other local NGOs and ethical businesses, organise short courses and workshops (ecological, self-care, communication).
We also own 26 hectares of land with hundreds of trees we planted 6 years ago. The long term goal of the community is to create an ecovillage on this land.

Permaculture network, Transition movement, GEN Hungary
☺ willing to cooperate, communication culture, self-responsibility

Israel

Kibbutz Mishol

Carmel 25 | 17661 Nazareth Illit
anton@communa.org.il

♀38 ♂37 75 150

e, Hebrew 98%
shared economy rented
education
ecological, self-managed, anarchistic, left-wing political understanding, feminist, non-violent, adult education, pedagogical, seminar-house, peace-work, permaculture, organic gardening

Kibbutz Mishol is the largest urban kibbutz in Israel. Established 18 years ago, Mishol is a community of educators situated in the city of Nazareth Illit in the north of Israel. Practicing full income sharing, the Kibbutz consists of 75 adults and 75 children who live together in an eight-storey building in a deprived neighbourhood in the periphery of the country.
The community has a rich cultural life, choosing to celebrate weekly, yearly and life-cycle

events together. Other facets of the community include the practice of consensus decision making, communal study, a car-pool and various committees that deal with all aspects of life. However, where this community comes into its own is in the societal missions which its members take upon themselves. Work teams made up of kibbutz members spend their day-to-day in our own educational projects set up and run through our NGO.

Kibbutz Neot Semadar

88860 Eilot ☎ +972-(0)8 635 81 11
Volunteers@neot-semadar.com www.neot-semadar.com

1989 4000 m² (near Eilat) 50 % ♀52 ♂34 47 86

e, Hebrew most of us

all members as general assembly (consensus on important issues, majority on day to day minor issues)

shared economy

We focus on learning by ourselves and observing ourselves in a relationship. We live, work and learn together.

agriculture, organic food production, operating our shop, construction

ecological, self-managed, non-violent, inner growth, meditation, vegetarian diet, free school, seminar-house, peace-work, environmental protection, self-sufficiency, permaculture, organic agriculture, organic gardening, organic farming, animal friendly husbandry, alternative energy, ecological construction

☺

Kibbutz Neot Semadar is located in the Southern Negev Desert of Israel, about 60 km north of Eilat. The Kibbutz was established in 1989 by a group of friends interested in creating a learning community that focuses on cooperation and creativity in daily life. As individuals in a dynamic community, we explore what it means to live together as co-learners, observing ourselves in everyday activities and relationships. Our interest in relationship extends to the environment through organic farming, desert architecture, water recycling, The Art Center, winery, and more.
Today there are 90 adult members (families and singles), 70 children of various ages, and about 50 volunteers of all ages. These volunteers, from Israel and from abroad, come to Neot Semadar to find a fresh and independent perspective on life.

☺ minimum 2 years candidacy

Kibbuz Lotan

Kibbutz Lotan | 88855 D.N. Chevel Eilot ☎+972-(0)54-979 9 030
alex@kibbutzlotan.com, mmlivni@gmail.com, info@kibbutzlotan.com
www.kibbutzlotan.com

1983 | 72 ha (near Eilat) | 10 % | ♀30 ♂30 | 50 | 160

e, Hebrew, sp, g, f, Dutch | 30

everyone in common (majority vote)

individual income, high community tax, community infrastructure investment from cooperatively owned enterprises, safety net for special needs.

common property

reform Judaism

dairy, tourism, ecology, date plantation

ecological, self-managed, left-wing political understanding, non-violent, hierarchical, seminar-house, environmental protection, permaculture, organic agriculture, conventional agriculture, organic gardening, organic farming, alternative energy, new technologies, ecological construction, accessible to the disabled, adult education, hands-on education, desert conservation

☺ ( )

Kibbutz Lotan is located in the Southern Arava desert, 55 km North of the city of Eilat. Population varies - about 160.
We offer opportunities to learn about and interact with our community via our Desert Inn guest accommodations and with our Center for Creative Ecology (CfCE) - Permaculture education and research institute's interactive educational programs.
The Lotan economy is based on a dairy barn (275 head), date plantation (9000 trees), CfCE and guest accommodations, partnerships with other kibbutzim in field crops and maritime agriculture. Many members work outside the kibbutz, in the regional services or the city of Eilat. Every member is responsible for their income, and everyone pays to support a wide range of community services such as cultural events, youth education, and the communal dining hall. The communal budget is comprised of funds from members and residents as well as a portion of profits from the communally owned businesses.
Kibbutz Lotan belongs to the Israel Movement for Progressive Judaism - Reform movement. Our approach to Judaism is liberal, creative, egalitarian, and non-coercive.
Kibbutz Lotan strives to relate to the fragile desert ecosystem in its daily life and development plans and is part of GEN – Global Ecovillage Network Europe. In Israel, we are pioneers in the use of waste materials, alternative building technologies and desert organic agriculture. The CfCE is a training hub for groups visiting from South American, Asia and Africa, and hosts peace-building workshops for Israelis, Palestinians and European participants.
Please visit our website www.kibbutzlotan.com for more information and contact us at info@kibbutzlotan.com to arrange a visit.

→See also description of network **"GEN Europe"**

Kibbutz movement

☺ ages 25-40

Israel

Lithuania

Ropëja Community

Ropëjos st. | 56438 Paparèiai ☎+370-695 58 91, +370-601 419 86
audrone.neniskiene@gmail.com
facebook page: #6894 at eurotopia.directory/booklinks

2000 | 30 ha (near Kaunas) | 30 % | ♀7 ♂3 4 10

e, f, Russian, Norwegian | 6 3

all who are concerned (consensus) | some common financing

property of all

Christian

husbandry, traditional crafts, beekeeping, clay construction

ecological, self-managed, non-violent, Christian, adult education, environmental protection, self-sufficiency, permaculture, organic agriculture, organic gardening, organic farming, animal friendly husbandry, ecological construction, humour, wool felting, clay construction, woodwork, art

Our community is located between two nature reserves, within a historical-archaeological park. Because its members "matured" separately, there is less "communism" and more "individualism". However, we are united by our understanding and philosophy concerning nature, ecology, self-reliance, arts and traditions. And since our lands are adjacent, we naturally form a small eco-island in the sea of conventional farming.
Currently we keep bees and raise sheep, making felt boots from their wool. We also collect medicinal plants and engage in horseback riding. Construction from clay and woodwork are also among the crafts practiced in our community, because most of our male residents have some experience with these works. Hence, clay construction is noteworthy, since we have to construct the buildings in our community with very limited resources and mostly by ourselves. Until recently our main focus was self-sufficiency, but since our situation is improving, we will soon be able to make our products available to a wider market and also practise more of the traditional crafts, such as blacksmithing or pottery.

☺ hard-working, intelligent, visionary

Photo:
Bhrugu Aranya Ecovillage
(next page)

Poland

Bhrugu Aranya Ecovillage Fundacja Terapia Homa

Wysoka 151 | 34-240 Jordanow, Malopolska +48-(0)502-347-898
info@homatherapypoland.org www.homatherapypoland.org

1995 5 ha (near Krakow) 50-75 % ♀6 ♂6 4 14

e, Polish, sp, g, f all adults in varying capacity most of us
all who are concerned (general agreement)
individual family financing, community business
individually owned homes and shared land
We practice the science of Homa Therapy for healing, nourishing and energizing the atmosphere and consequently the environment. We are a universal spiritual community.
working in the Foundation to promote Agnihotra and Homa Therapy internationally, then: gardening, microgreen production, bee keeping, fruit orchard, farming, cows, cottage industries
ecological, self-managed, non-violent, spiritual, inner growth, meditation, yoga, organic diet, vegetarian diet, vegan diet, seminar-house, environmental protection, self-sufficiency, permaculture, organic agriculture, organic gardening, organic farming, animal friendly husbandry, alternative energy, new technologies, ecological construction, humour, non-smoking, Agnihotra, Homa Therapy, spiritual ecology, healing, herbs, ecological, educational, Vedic farming, crafts, creative arts, sound therapy, music, medicinal herbs, Mandala Art Therapy, Ayurveda

☺

Bhrugu Aranya is a blossoming international ecovillage. Residents range in age from 5 to 65 and are an eclectic group of artists, healers, writers, film makers, musicians, therapists, gardeners, teachers and herbalists. Workshops are offered in Homa Organic Farming, Herbs for Healing, Mandala Art Therapy, Sound Therapy, Creative Expression and many more.
Our non-profit ecological/educational 'Homa Therapy Foundation' is based at Bhrugu Aranya, situated in the Beskida Mountain region with views of the majestic Tatra Mountains. The atmosphere of this farm is infused with high vibrations from Ayurvedic healing fires that are performed daily. It is this energy that radiates to heal the Earth. We invite you to come and experience transformation in a super-charged, creative, healing environment.
In 2010, the community designed a project to create the Centre of Light to be a place of refuge, solace for those seeking healing and rejuvenation in a purified and harmonious atmosphere. The Centre, a unique double octagon structure, will be built with ecological materials. It is to be a gathering point where visionaries and healers from around the world can come and share their knowledge with other kindred spirits.

Homa Community Network, FIC
☺ interest in Agnihotra and Homa Therapy. All members are vegetarian, no alcohol or drugs allowed.

Realearth Thomas & Claudia Notter

Bachanowo 21 | 16-404 Jeleniewo ☎ +48-(0)87-568 35 35
realearth@interia.pl www.realearth.pl

1993 19 ha (near Suwalki) > 50 % ♀2 ♂3 1 6

g, Polish, e 2 or more

all who are concerned (consensus, selecting the best solution for all of us)

common financing common property

a strongly grounded pure spirituality full of light, beyond all religions

self-sufficiency (gardening, agriculture, animals, cheese production)

ecological, self-managed, spiritual, inner growth, organic diet, environmental protection, self-sufficiency, permaculture, organic agriculture, organic gardening, organic farming, alternative energy, new technologies, ecological construction, humour, non-smoking, flexible social structure, being in balance, joie de vivre, sensual, school of life, cheese production

Since 1993 we have been living in north-eastern Poland, in a rather harmonious community with a great variety of cultivated and wild plants, farm animals, domestic animals and wild animals in the Suwalki Landscape Park.
Due to our 19 ha organic farm with cheese production and direct sale, we have succeeded in creating a good foundation for our growing family. However, the planned and hoped integration of long-term residents from outside could not be realised.
Contrary to our initial expectations, there was little interest in [agricultural] community projects in Poland. After some years we realised that the main reason for this was the Polish social security system. Only family members of farmers and students do not pay additional contributions. If another person joins, it becomes expensive. In our place, here in the northeast of the country, the harsh climate also can be a challenge, which is why we have mainly remained a seasonal community. A place where especially during the summer months, guests and friends came to visit and to collect experiences in self-sufficiency and community living.
Now we face a new challenge: Our children have grown up and the majority of them have moved away. There is space and new opportunities for other people to join us. At the moment we are very busy finding out exactly how we want to continue - we are open for new hints, encounters and ideas in any direction. And we urgently need to update our website, which unfortunately has not been maintained since 2008.

Pólnocno-Wschodnia Ofensywa Spoleczna

☺ we are happy to let you surprise us - if it fits, it fits

Ermitaj Malin

Malin (comuna Nuseni), nr 199 | 427 200 Malin
+40-(0)769 380 968
contact@ermitajmalin.com
ermitajmalin.com

2010 | 8 ha (near Beclean) | 25 % | ♀1 ♂3 0 4

f, e, Romanian
all | all
all who are concerned (consensus)
individual financing & limited common financing
property of all
permaculture
permaculture, gardening, farming, eco-building, community management
ecological, self-managed, non-violent, adult education, pedagogical, seminar-house, environmental protection, self-sufficiency, permaculture, organic gardening, organic farming, animal friendly husbandry, alternative energy, ecological construction, therapy and hiking with donkeys, experimenting community lifestyles, off-grid

Ermitaj Malin is an off-grid eco-place of 8 hectares in a valley of Transylvania, surrounded by streams, woods and pastures. We want it to be a place of living and welcoming, in search of well-being, autonomy and resilience.
The project and the team is not totally stabilised yet (2019), as it is a place of experimentation, we are in constant changing and adjusting! One of our objectives is to promote exchanges and experimentation around permaculture, agro-ecology, and eco-construction, thanks in particular to the welcoming of volunteers and the organization of internships, participative workshops and events. We mean to be an international meeting place, and a place of solidarity, for people wishing to relax, and for those who wish to learn and experience our daily live. We would like to achieve optimal food and energy self-sufficiency, to encourage local exchanges, and learn more about peasant's skills.
We have strong links with the Romanian Association of Permaculture, and also some Belgian roots! We are seeking a way to organize our activity in order to find a financial balance, to enable us to live, and to make this project sustainable.

☺ yes

Slovakia

Ecovillage and Live Ecomuseum Moravia

Petr Skorepa
Vrbovce 844 | 90606 Vrbovce ☎+421-(0)944 293 223
ekovesnice@gmail.com
www.ekovesnice.cz/, ekovesnice.cz/229-projekt-eko-osada-bile-karpaty.html

2012 (near Myjava, Bratislava) ♀5 ♂8 9 23

e

ecological, self-managed, Taoist, organic diet, seminar-house, environmental protection, self-sufficiency, permaculture, organic gardening, organic farming, alternative energy, new technologies, ecological construction, non-smoking, crafts, worskhop, dark therapy, 3D, archery, sports, recycling, arts, museum

☺ ()

We are an ecovillage in Slovakia on the border to Moravia. We built up a Live Ecomuseum with craft, self-sufficiency and recycling. Furthermore, we organize a Natural Building School. We have natural accommodations and a space for dark therapy.

Zaježka

Zaježová 132 | 96263 Pliesovce ☎+421 (0) 915 271 700
sekier@zajezka.sk www.zajezka.sk

1994 77 ha (near Zvolen) ♀50 ♂50 50 150

Slovak, e, g, f, sp 20 - all part-time from 1 to 3 days per week

all who are concerned and interested (consensus of concerned)

partly common, partly rented, partly private

none

workshops, youth informal (experience) education, natural building, waste management, eco-agriculture

ecological, self-managed, adult education, seminar-house, environmental protection, self-sufficiency, permaculture, organic agriculture, animal friendly husbandry, ecological construction, crafts, youth activities, waste management, rural development, community run school and kindergarten, permaculture, alternative buildings

☺

Zaježka is our nickname for this area of old lonely farms in the hills, with the centre in the village Zaježová. Lots of the houses were abandoned in the past. The area was "discovered" by enthusiasts of traditional culture who started to organize weekend craft workshops here in 1990. Later we decided to buy a local farmhouse in Zaježová (named Sekier) where people can stay, try a simple lifestyle and help us with our activities. Usually 3 to 8 young people from several countries live on Sekier for a couple of months.
Later, more and more people started to move here and new houses were built and bought. Now, around 150 community members live here permanently. Many children were born here in the past decade and the community is still growing. The community is still far from being self-sufficient, but it is a good place to learn about a simple lifestyle, sharing and solidarity to all. We meet often for various reasons. Either it is for communal projects such as working

on our communal centre, shopping in our food-bank, meetings in the local school and forest kindergarten, and there are also a lot of informal meetings for being together, singing, dances, discussions, eating, celebrating, meditating, practicing yoga, wall-climbing, volleyball, football and other sports activities.
There are several NGOs operating in this valley. We host ESC volunteers, organize youth exchanges, offer many weekend workshops and summer camps for public each year.

→See also description of network "**GEN Europe**"
☺ for rented houses only - one month trial period

Slovenia

Sunny Hill

Topolovec 29 | 6272 Gracišce ☎+386-(0)402 996 67
manja.vrenko@gmail.com sunnyhills-istria.eu/

2014 1 ha (near Koper) 50 % of vegetables ♀7 ♂5 2 14

e, i, g, Slovenian, Croatian 2 5
all who are concerned (majority vote, striving for consent)
limited common financing (food, heating, electricity and internet)
property of cooperative
no
workshops, youth work, volunteers, self sufficiency, rural development
self-managed, seminar-house, permaculture, organic agriculture, new technologies, ecological construction, humour, Dragon Dreaming

☺

Sunny Hill is a space of cooperation and innovation. We provide a supportive environment for research, education and active practice of sustainable living. We organize events, workshops and gatherings. We also offer mentoring and counseling to emerge sustainable initiatives. Sunny Hill is involved in the Erasmus+ funded Community Learning Incubator Programme for Sustainability (CLIPS).
We are cooperating with competent individuals and organisations and we are a member of GEN (Global Ecovillage Network) and ECOLISE (European network for community-led initiatives on climate change and sustainability).

→See also description of network "**GEN Europe**"
ECOLISE

Slovenia

Turkey

Imece Evi The Farm Natural Life and Ecological Solutions

Turgutlar Köyü, Menemen | 35660 Izmir +90-(0)543-677 58 09
ismailyenigun@gmail.com www.imeceevi.org

2007 2 ha (near Menemen) 30-40 % ♀3 ♂2 2 7

e all adults
all together (consensus) selling the own products and sharing
property of one member
anti-capitalist, anti-militaristic
agriculture, producing detergent, olive oil, vinegar
ecological, self-managed, non-violent, organic diet, environmental protection, self-sufficiency, organic agriculture, organic gardening, organic farming, forestry, ecological construction, non-smoking, solidarity

Imece Evi is located at Dumanlidag, which is a wild and remote place near Izmir. Living far from civilization and high-technology makes us coherent with the rhythm of the mother Earth. We already learned alternative living methods and are convinced that another way of living is possible. Living at Imece requires to be able to cope with nature and not to be addicted to modern city benefits.
Imece Evi, as a civil initiative and an ecologic village project, is targeting to develop a self-contained, sustainable, non-hierarchical and co-operative complex. We prefer not to use industrial products but rather the natural ones that are produced without use of chemical fertilizers or pesticides. In our gardens we have enough fruit and vegetables and we can produce our cleaning products, olive oil, vinegar etc. "Imece" can be translated as "to work collectively". Well, solidarity and sharing are our first priorities and our ethical beliefs make us help the others as much as we can and even sacrifice our own comfort.
We receive every year hundreds of visitors and volunteers from around the world. We expect our volunteers to stay at least two weeks and share their knowledge and abilities with us.

→See also description of networks "**WWOOF, GEN Europe**"
Nükleer Karsiti Platform, Ege Çevre Platformu
☺ be able to live with least technology and civilisation

Kaz Dagi Bugdaygilleri

Kucukkuyu | 17980 Canakkale +90-(0)533-4766461
gunesin@bugday.org

2015 (near Çanakkale) 10 %

e, Turkish 10 10
all who are concerned (consensus) individual financing
ownership
ecological living
ecological, self-managed, non-violent, inner growth, organic diet, vegetarian diet, seminar-house, environmental protection, organic gardening, alternative energy, ecological construction, humour

We don't have only one settlement. People are living in different places, villages and and in the nearest town. Everybody works his/her own business or job. Here many different small social groups are coming together for different aims in different circles. Such as:

- people with children are coming for a school
- some people have more intense relationships and come together for social activities
- other relationships are more loose, as people are coming together for occasional circumstances such as olive picking, on some special celebration dates, community meals etc.

This is a place of transit for many people, and the people who are living here have many connections with some other communities and individuals; therefore this here is a meeting point of ecological minded people in Turkey.

Editor's note: our mail to the postal address above came back as undeliverable

→See also description of network "**GEN Europe**"
local exchange trading systems, bugday.org, Mt. Kaz Protection Association

Refikler

+90-(0)554-896 01 78
refiklerciftligi@gmail.com

5 ha

e, Turkish
all who are concerned (consensus)
shared economy
property of all
agriculture
ecological, social ecology

Refikler is a commune which operates on two different locations in Turkey, aiming to raise awareness towards communal living practices with the vision of a world without hierarchy and domination. The two locations are in Bayramiç, Çanakkale and Fethiye, Mugla - and are called "Zeytinli Ecological Community for Communal Living" and "Refikler Farm". On a land of approximately 5 ha, we have been cultivating medicinal herbs, fruits and vegetables, gathering wild herbs and producing agricultural goods on a minor scale. Though we pay utmost attention to preserve nature, we certainly do not embrace the mystified idea of a naturalistic asceticism or "going-back-to-wilderness." With this in mind, we also collaborate with the local population.
The work intensity is distributed very heterogeneously. Within our perspective of living, we plan our time and labor flexibly but responsibly. We value reading, discussions, and, in general, the times that we spend communally. Our economy is communal. We keep distance from alienated, speculative finances and enrichment that we can't collectivize. We oppose to specialization by division of labor, thus each of us is responsible for each aspect of the work.
We expect that everyone who visits us for volunteering will also spend as much time with us as possible. It is important for us to share meals together, although the kitchen is always open. The toilets, showers, and resting/sleeping shelters are also communal facilities.

Ukraine

☆ Community of family estates Chmyrivka v. Chmyrivka

Novoselyttsya area, Chyhyryn district, Cherkassy region | P.O. 20934

☎ Roman +38-066 4315145, Sergiy +38-095 0949470

2010 (near Chirgirin (25 km), Cherkasy (50 km)) 18 44

Ukrainian, f, e, g

creation of dome furnaces, baking bread at home, beekeeping, gardening and horticulture, milk and cheese, visceral chiropractic, IT, play table tennis and volleyball, needlework

☺ ( )

We aim to live in unity with nature, search for the spiritual path, create our own special eco-space and form a new view on the relationship between people.
The settlement is located in a picturesque are with beautiful hills and ravines framed by forest and located along the small ancient river, with several lakes. This is a historic area - Kholodnyy Yar (Cold ravine), associated with the formation of the state of Ukraine.
In 2019, 11 of our 13 families live here longer than 2 years. 6 children were born in Chmyrovka. There is an experience of home births.
Some families come in the warm season and on weekends, there are several locals who take an active part in public life. Most of the members are vegetarians.
We have aim to create a new large general house in which we wish to organize a school for adults and children, to host guests, to hold joint events, sports competitions. We are creating a green tourism project.
Eco-village is opened only for scheduled events and excursions by prior arrangement. The village does not use alcohol. Tobacco and meat are consumed as little as possible. The community is open to new families, and especially with kids.

☺ From new neighbors we expect hard work, purposefulness, active participation in settlement life and a creative approach to the development of new, neighboring territories.

Ecovillage and permaculture center Zeleni Kruchi

Lisova str. 82 | 09750 Dmytrenky, Kyiv Region ☎ +38-(067) 967 87 93
zelenikruchi@gmail.com zelenikruchi.in.ua

2012 about 10 ha (near Boghuslav, Kiyv region) 50 %

♀9 ♂5 9 23

e, f, Ukrainian, Russian 2-3

meetings (majority of 70 %)

indiv. financing + monthly compulsory fees to the common budget

rented as well as individual and common properties

organic gardening, permaculture, organic farming, nature building, IT

ecological, esoteric/New Age, inner growth, meditation, yoga, organic diet, vegetarian diet, vegan diet, adult education, seminar-house, animal protection, environmental protection, permaculture, organic agriculture, organic gardening, organic farming, animal friendly husbandry, forestry, ecological construction, humour, non-smoking, woodwork, handmade

We are a community of like-minded people who decided to move from cities to the countryside and lead a sustainable way of life. We build our houses with natural materials, grow our own food, lead a healthy lifestyle and spend our free time together. Furthermore, our members constantly develop themselves, organize different studies and build up our permaculture center.

OUR GOALS:

- growing ecological food for our own needs and for sale
- giving ecological festivals and seminars
- developing green tourism and camps for children
- protecting the environment, to increase the biodiversity
- spreading the ideas of permaculture and sustainable development.

Ukrainian Permaculture Centers

☺ ecological way of life

It's not entirely logic to list two communities so far from Europe in an European Communities Directory - but then, they insisted on having European roots and at some point we just accepted their data. Everything happens for a reason, some say...

Bolivia

☆ Terra Universalis

Samaipata ☎+591-(0)736 468 56

✉info@terra-universalis.org 💻www.terra-universalis.org

sp, e, g

ecological, self-managed, non-violent, spiritual, inner growth, meditation, yoga, tantra, sexuality, free love, organic diet, vegetarian diet, vegan diet, adult education, pedagogical, free school, seminar-house, peace-work, one-world development, animal protection, environmental protection, permaculture, organic agriculture, organic gardening, alternative energy, new technologies, ecological construction, construction, humor, self-knowledge, inner growth and deconditioning, sacred architecture and sacred geometry, Adobe building, energy work and healing work, art & culture laboratory, craft, small business, transition town movement in Samaipata, global networking, new wingmaker philosophy and new warrior spirit, work on/with extraterrestrial intelligences, time travel, remote viewing, transhumanism, emergence of a new evolutionary consciousness, orientation towards an emergent behaviour of all group members, cooperation with indigenous people - Munay-Ki, silent retreats and meditation, synergy between individual and common economy, medical work with San Pedro, Ayahuasca

The individual members of the still young community Terra-Universalis live freely in different rented or also in individual houses, as well as in adobe houses that have been acquired together and are based on on holy geometry at the edge of Samaipata village at an altitude of 1650 m in the Bolivian Andes. The climate is regarded as one of the best in the world and practically all types of fruit and vegetables from all continents of the world thrive there. We see ourselves as a free wingmaker community (with regard to the ultimate unfolding of the energy body), which means, among other things, that we mainly intend to develop our inner energy being and the energy body, i.e. to peel out the innermost essence. We work predominantly in mindful self-knowledge with a profound deconditioning work, connected with meditation and sometimes also with medical work, in order to form, in community, a vessel for the new conscious (warrior) spirit. The interaction in the community is not organized by a hierarchical system, but arises organically and responsibly from the conscious state of being alone (without the old ego structures) of all individuals and therefore develops out of the common circumstances of a natural spiritual process of energy and consciousness. Volunteers and guest helpers are welcome. Interested guests, families and/or people who want to get to know the place and the community are also welcome.

New Zealand

Global-Native @ Spirit of Nature Ursus Schwarz

1891 Takaka-Colingwood Higway | 7182 TAKATA, Golden Bay province, Nelson
☎+64-(0)27 352 09 52
uur@gmx.ch www.nuovo-walser.org, www.global-native.org

2002 7 ha 4-10

g, e, (f, i)

trust

the spirit of Mother Earth

learning to connect with your higher selves, how to do permaculture, eco-building/renovation with recycled materials, planting a food forest, vegetable and herb gardens...

☺ (✋✋)

We are "GLOBAL-NATIVE.ORG" an educational eco-awareness group and grassroots activists; we welcome people seeking to learn to connect with their higher selves and with Mother Earth. In order to learn and experience this wider range of topics we ask people to stay for a minimum of 1 month (ideally it would be 6 to 12 months, for planting, maintaining, harvesting etc.) If you are unable to stay for this long, please let us know and we can make other arrangements. We also welcome those with carpentry or artistic talents to help us improve the accommodation and living areas.
We are living on a scenic hillside overlooking the beautiful landscape of Golden Bay, just a 10 min walk from the beach, 5 km to Collingwood and 22 km to Takaka. Easy access by hitchhiking. Living includes a daily natural rhythm, with sharing experiences, singing and playing music together and having talking circle customs. We request volunteers to help contribute towards the food cost (but there is no pressure to give donations). For experienced and suitable long-term volunteers we can offer alternative arrangements.
Our daily meals are eaten together like a family where we get to know each other, discuss the day, sing a song or two, and share our experiences. We are drug, alcohol and smoke-free.

☺ open for all people who like the vision of the Global Natives

Book challenges capitalist mindset

mer Swiss mountain- Ursus Schwartz, aka Bear, hopes his newly hed book will change ets and help save the

tled *How to Survive lism? With the Global e Vision*, it is the ation of 15 years of hought by the author.
us, who lives at pirit of Nature near orpe explains that the native vision concept, anchors his thesis, ted among the native an Hopi tribe. "Ten nd years ago they vision that people are the planet."
asn't until the 1960s e threat was taken ly by western culture hen it was, Ursus got involved. "I was

Paperback writer: Author Ursus Schwarz is launching his new book *How to Survive Capitalism? With the Global Native Vision* at events in Takaka and Collingwood. Photo: Jo Richards.

capitalism. "It works by making profit the most important thing. We need to change

lost the connection to the superconscious. We need to reconnect with it or we will kill

Planned Communities

For the first time in eurotopia history, planned communities get their own chapter. We made that decision before we knew that the number of planned communities had tripled (!) compared to the last edition. It means that in the chapter of existing communities you only find physically existing groups in actual locations. In this chapter, however, there's a place even for mere ideas. It might be especially interesting for people seeking community, because most of the projects presented here seek members. And maybe it's just you that's missing to make one of the dreams of this chapter come true.
If a planned community has a location, you'll find it in the country map of the previous chapter. An "X" in the upper right arrow means that there's no location yet (applies only to Germany, France and the United Kingdom). Czech Republic, Hungary and Turkey: See pages 490f.

Austria

Divine Foundation Divine Village for Body, Mind & Soul

X

Dorfstraße 17 | 7474 Eisenberg an der Pinka ☎+43-(0)699-177 897 89
welcome@divine-foundation.org divine-foundation.org

2011

g, e

live and build together, recycling - natural cycles, working with people and animals, training center, workshop, creation of a "School of life"

ecological, self-managed, non-violent, spiritual, Buddhist, Taoist, anthroposophic, inner growth, meditation, yoga, organic diet, vegetarian diet, adult education, seminar-house, self-sufficiency, permaculture, organic agriculture, organic gardening, alternative energy, new technologies, ecological construction, humour, free energy, seminars, coaching, mediation, research, transformation & healing, health, simply living happily

We are a group of people who want to create a unique place "Divine Village", which is trend-setting and stands for sustainability and self-development.
"Divine Village" is a place of co-operation, communication, spiritual growth, of healing & growth, of teaching and learning, silence & inner encounter, timeout, joy & humour, play & fun, pleasure & life. An open encounter oasis in a peaceful, harmonious atmosphere, between young and old, from heart to heart, a co-operation for joint action, a revealing of the skills and talents of each individual, from the I to the you - from you to we - from we to oneness. We want to connect the people of this Earth to an ecological community, to leave the ecological footprint as small as possible, to touch and accompany people in their true being, to protect nature and to lead a life in line with it, to build and live close to nature, to bundle research related to natural building, living and life, and to combine it with new approaches, to be a focal point for innovative projects, to provide a network for communities & ecovillages worldwide.
Important! 2019: preparation phase, creating an infrastructure, public relations, establishment of a network...
Address = not the place of the project. Latest news can be found on our website.

☺ caring, manually talented, dynamic, creative and responsible people who would like to support the project development in any form - please contact us via phone or mail!

Garten der Generationen Living Well in Solidarity!

Herrengasse 5 | 3130 Herzogenburg
info@gartendergenerationen.net www.gartendergenerationen.net

2007 3 ha (near St. Pölten)

g, e
leading circle (sociocratic consensus) wealth pool for land and buildings
property of the association
no
gardening, construction, administration, child care
ecological, self-managed, non-violent, inner growth, organic diet, adult education, pedagogical, free school, self-sufficiency, permaculture, organic gardening, alternative energy, ecological construction, sociocracy, coexistence and work of all generations, modern subsistence economy

☺ (✋✋)

Our plot is about 3 ha at the edge of Herzogenburg village. A part of the property is dedicated building land, the rest is agricultural. Our community-house is the first building on the property. The planning of the next construction phase is finished. We expect completion of the first buildings in summer 2020. In the medium term, up to 100 people will live and/ or work here. We are happy to hear from interested people who would like to actively participate in the implementation of an ecologically sustainable, life-sustaining and community-minded way of life!

→See also description of network "**GEN Europe**"

☺
trial-period, attending of some events, agreement on our handbook

ISSGRAS Oase des Lebens

Oberhofenweg 28 | 6380 St. Johann in Tirol +43-(0)664-730 225 34
info@issgras.at www.issgras.at

2014 2 ha ♀1 ♂1 3
2-4

g,e,f
circle way (circle way)
no problem with any religion
everything, building, construction, gardening,
ecological, self-managed, non-violent, inner growth, organic diet, vegan diet, free school, environmental protection, self-sufficiency, permaculture, alternative energy, ecological construction, non-smoking, carpentry, all crafts, music, campfire, children, free time

☺ ✋✋

When envisioning a perfect future, we do not see large cities, factories and anonymity, but self-sustainable, small-scale living and a shared living space with other families or groups of people. We have always tried to be as self-sustainable as possible if it comes to our energy or food. Besides this, we have spent years planting thousands of plants in our garden with the intention of creating an enormous biodiversity. Just as plants benefit from diversity, living in a truly sustainable way also comes down to different elements that strengthen each other. Our dream is to have different families living together that share land, supplies, information and skills, to strengthen each other. To receive additional information or to arrange a visit to our living space, please contact us!

☺ member of www.oasendeslebens.com

Editor's note: Please note the article about the Anastasia movement on page 49ff.

✎ midanond

Tiefenbach 12 | 4753 Taiskirchen ☎ +43-(0)650-960 01 07

2018 g, e

The following values, aspirations and beliefs are part of our common vision that will guide our future community life:

- personal development
- vitality and creativity
- shared everyday activities
- (political) relations of the community
- ideational-spiritual basics

Personal development as a value includes the mutual effort to value the person as he/she is in his/her 'authentic being'; this includes systematically working on fear-laden topics and thereby experiencing secureness in a group of like-minded people.
Vitality and creativity is celebrated and expanded through various media, such as celebrating, dancing, singing, playing together. Shared everyday activities on the one hand ensure the functioning of the community, but also promote the sense of community (e.g. through cooking, eating, gardening or economic activities...).
Our vision is to promote good relations between the community and its immediate neighbourhood and other projects with the aim of promoting a peaceful, ecological and sustainable life. In the area of an ideational-spiritual basis the effort is addressed to build a spiritual basis of our community, which should include among other things:

- each individual contribution is better than none
- the pursuit of a balance of trust and control
- the search for a spiritual basic attitude of life, love and growth, as the basis of our thinking and acting in the community

These value areas (and their implementation strategies) are not carved in stone, they will evolve vividly with the needs of the group members. If you feel addressed by our vision, be welcome and contribute!

Naturnah - Miteinander - Einfach - Leben

☎+49-(0)7071-709 94 40, +43-(0)6503-885 858
karuna_mueller@yahoo.de (checked approx. once a month)

2018 (probably bear Villach) ♀1 ♂1 0 2

developing an equal say during the trial period; private decisions = where it does not affect others (consensus)
self-sufficiency all together
naturally and socially orientated, focusing on the relevant
partial/self-sufficiency with own garden & winter storage, wild herbs, social life, household, renovation work, wood

☺

We - Torsten & Karuna (55 & 54 years) - have been living together for nine years. For the Carinthia region (Kärnten) we are looking for people who like us want to live close to nature. We feed mainly on vegetables, wild herbs, sprouts, mushrooms, fruits (95 % vegan). We want to provide ourselves with the necessities. An 'outdoor life garden' is what we want as the centre of our life. As a group we would like to become 8 people. We are looking for singles or couples between 40 - 60 years - without or with children who are no longer living at home. Together we would like to design a common space: In which experienced Earth children live a beneficial contact with each other.
The way we deal with natural beings, e.g. in gardens, meadows and forests also means a lot to us. A bit slower in doing and in encountering everything means the highest quality of life for us. Health & respecting our limits of strength is precious to us! We would like to ask you: How do you live with yourself? Do you give yourself time for self-reflection and reflecting on what has been?
We need a weekly evening for an exchange on well-being and would separate organisational matters. In our small group we don't want to support alcoholics or people with drug problems. The basic needs must be fitting.
Who wants to build and live the above with us?

☺ 2 years trial period

Rainbow shelter place Nomadic base starting 2020

Pleissing (Neustift) | 2083 Pleissing ☎+43-(0)316-340 019 04
living.communal@gmail.com www.flickr.com/photos/benjaminsmit/

2011 600 m² (near Retz) 5 % ♂2

e, g 1
together
common economy one member, but can be shared with others later
sharing skills
ecological, anarchistic, non-violent, inner growth, free love, peace-work, self-sufficiency, permaculture, organic gardening, organic farming, alternative energy, no cell phones/wireless, non-smoking

☺ ( )

This place had some troubles to startup in 2011, I bought the house without knowing anything about community (or cohousing) life! But that is changing from 2020 on, or

earlier if you want to try it.
There is one friend managing the place until 2020, and he likes to have others around (not volunteers but participants who have the same rights, to use the place how they want it).
The people will make this community, nomads who don't have a normal house can share a room here. And share their skills to the other community members.
We'd like to create more of these places, and that is why I was mapping (only digitally visible at the moment with their email addresses) all the other similar projects and more (1000+ connections, in 2019).
Don't hesitate to write us, or send us a message by Whatsapp: +43-(0)-31634001904.
In winter, there is space for 2-4 nomads; 2-4 more could stay there from May - September as summer friends. Like us @ facebook.com/ParticipateLOOSE/
Namaste!

☺ living here 2-4 months a year at least

Wohnprojekt Hasendorf gemeinsam einfach hase

Hasendorf 88 | 3454 Hasendorf
wohnprojekt-hasendorf.at

2015 (near Tulln an der Donau, Krems an der Donau, St. Pölten) 50 %
♀14 ♂10 12 24

g, e, sp, i, Croatian
work groups, sometimes plenary meeting (consensus/consent)
owned by our association
individual occupations and lifes, the common house with garden, the CSA 'Ackerschön'
ecological, self-managed, non-violent, organic diet, environmental protection, self-sufficiency, permaculture, organic agriculture, organic gardening, alternative energy, new technologies, ecological construction

The housing project Hasendorf is an association which plans a community house in Hasendorf, in which we should live as we could all live in the year 2050: joyfully, consciously, reduced.

→See also description of network **"GEN Austria"**
WoGen Wohnprojekte, Initiative Gemeinsam Bauen Wohnen

Belgium

Bijgaardehof: Wijgaard, De Spore & Biotope

9000 Gent
info@cohousingprojects.be bijgaardehof.be

2019 100-200

e, Dutch
together (consensus) property (ownership) - association of co-owners
humour, work-life balance

☺

Bijgaardehof includes three cohousing projects and a public health center.

See photo on page 14

Wijgaard:
We're a group of twenty, thirty, forty and fifty somethings and some kids, born in Belgium and our neighbouring countries. Good in managing businesses, restructuring the world, repairing people and bikes... Love to stand in front of a class or audience or behind cameras. Some of us prefer working behind the screens.
We built 23 apartments and some common spaces. Wijgaard wants to create a perfect life-work balance: next to our common kitchen, shared library, roof terrace, living space, polyvalente space..., we also invest in a coworking-space. A place where it's cosy and spacious, where we can work quietly or inspire each other. We live close to the city center and public transport. We have green, common space, our own private apartments and wonderful neighbors. We find ourselves lucky.
De Spore:
We're a group of fifteen families. We choose cohousing because we enjoy chitchatting with each other, seeing our children play together, be able to help each other doing groceries or babysitting kids. We'd like to believe that this is one of the most sustainable ways to live in a city, having a healthy balance between privacy and community. We share a common kitchen, playroom, living room and a roof terrace.
Biotope:
19 households that find ecology and sustainability of great importance.

Cohousing Gummarushof

Kerkhoflei 49 | 2800 Mechelen
sawomech@gmail.com samenwonenmechelen.wordpress.com

2017 39

- Dutch, e, f, g, sp
- all who are concerned (consensus)
- individual financing
- individual property & property of all
- living together, neighbourhood activities
- ecological, self-managed, non-violent, permaculture, organic gardening, ecological construction, non-smoking, sharing economy

"Gummarushof" is a former school complex that includes an art-deco church and a former rectory bordering on a collective garden. Its residents are mixed in both age and family structure. As such, we want to support and appreciate each other and be open towards neighbours and neighbourhood organisations. In addition, we aim to renovate in the most sustainable

manner possible, building compact living units and complementing them with a large communal kitchen, a common room, a laundrette, a nursery, guestrooms, a library, a workshop, a shared office, etc. Our ambition is also to share tools, materials and vehicles, together with our individual expertise and skills. The "church" can, in time, be rented out for courses, exhibitions, (small) parties or concerts.

Samenhuizen vzw (Flemish/Brussels network organisation)

Cohousing InDruk

Prins Leopoldstraat 60 | 2140 Borgerhout ☎+32-(0)278 08 43

info@cohousingprojects.be www.cohousingprojects.be/indruk

2020 (near Antwerp) 30-50

e, f, Dutch

the whole group (consensus) individual

property of all

ecological, self-managed, alternative energy, new technologies, ecological construction

Cohousing Indruk is situated at Borgerhout close to the heart of Antwerp. It contains 18 apartments with their own kitchen, living spaces, bathroom and terrace. The inhabitants share a large city-garden and a common house with professional kitchen, studio, guest room, co-working space and washroom. The group of owners is a mix of generations, professions and lifestyles. They now (2019) construct the building, at the beginning of 2020 the construction works should be finished and the group should be able to live in their cohousing project.

Cohousing Negenhoek Kortenberg

Engelenstraat | 3070 Kortenberg ☎+32-(0)927 808 43

info@cohousingprojects.be www.cohousingprojects.be/negenhoek

2021 30-50

e, f, Dutch

the whole group (consensus) individual

ownership

ecological, self-managed, alternative energy, humour

☺

Cohousing Negenhoek is situated in the heart of Kortenberg, close to a train station that brings you to Leuven and Brussels within minutes. On the former chicory firm will rise a new cohousing project with 26 households and common spaces like a kitchen, studio, co-working space, washing room, guest room...

A place where people can enjoy their privacy and the company of others when needed. Living in a neighborhood where you know your neighbours and you can enjoy a large garden and a common space for spontaneous contacts.

☺ you'll have to be able to pay and built your own house

planned communities

Denmark

Bofællesskab Kirkebakken Frit Bo

Thygesensvej 15, Krogsbølle | 5450 Otterup ☎+45-214 552 60
kjeldkramsbjerg@gmail.com http://www.krogsboellesognslokalraad.dk/

2018 4 ha (near Otterup)

g, e, Danish 2 8
all together (consensus) individual financing
property of all
no

☺ (✋)

Bofællesskab Kirkebakken "Frit Bo" ("free living") is under construction and therefore many details have not yet been determined. Potential members of the community can help define the principles for participation.
The location is particularly suitable for a community. Among other things, it is possible to use an adjacent area of approximately 50.000 m² for growing vegetables and the like.

Gaias Have

3630 Jægerspris
info@gaias-have.dk gaias-have.dk

2018 (near Nykøbing Sjælland)

e, Danish
Martinus Cosmology
ecological, self-managed, non-violent, spiritual, vegetarian diet, vegan diet, seminar-house, one-world development, organic agriculture, organic gardening, alternative energy, humour

In Gaias Have, we are creating a community of 40 homes plus a community house, located on high grounds near the city of Nykøbing Sjælland, Denmark. Our community is spiritually based on the undogmatic Martinus Cosmology in order to 'ensure' a peaceful and friendly neighborhood.
In our process we are at the point of final negotiations with partners so that we hopefully can start building in late 2019. All buildings will have the same architecture to ensure an aesthetic appearance that will be combined with parklike grounds and 'wild' nature.
Our goals are sustainability, organic and vegetarian food, safety, kindness, privacy, community and other measures that target the 17 UN Sustainable Development Goals.

France

Coogite
C2

32 ter rue de St. Germain | 78260 Acheres
+33-(0)663 240 680
coogiteurs@framalistes.org
www.coogite.org

2019
♀4 ♂5 1 10

f, e
1 1

all concerned inhabitants (consensus)
property of all

eco friendly cottage

environmental protection, ecological construction, participating in community living

☺ ( )

COOGITE is an initiative driven by a community aiming to promote energy transition and its social dimension in order to insure a brighter future for our children in a lasting ecofriendly environment. So the COOGITE initiative aims to grant low to moderate income families access to positive energy and low carbon emission building, while creating a warm and friendly living environment in ecological buildings.

→See also description of network **"Colibris"**

Collectif d'Animation pour l'Habitat Participatif (CAHP), Coordinaction de l'Habitat Participatif

☺ financial engagement (PLS revenue)

Ecolieu Langenberg
E2

Chateau Langenberg | 67160 Wissembourg-Weiler
+33-(0)388 073 519
contact@ecolieu-langenberg.eu
ecolieu-langenberg.eu

2014 20 ha (near Wissembourg, Hagenau, Karlsruhe)

f, g, e

the general assembly of the "SCIC" (French cooperative) (systemic consensus principle)

individual financing
ownership by all (purchase not yet completed)

no

construction and renovation, vegetable growing, horse care

ecological, self-managed, non-violent, organic diet, adult education, free school, seminar-house, peace-work, animal protection, environmental protection, self-sufficiency, permaculture, organic agriculture, organic gardening, alternative energy, new technologies, ecological construction, accessible to the disabled

☺

Our "Ecolieu Langenberg" is a residential and life project that puts humanity and ecological values first.

The estate "Chateau Langenberg" with its 20 hectares of land is a wonderful place to realize our French-German project. Dreamy, quiet and yet not very far away from the small French town of Wissembourg, it is situated directly on the border with Germany. Our ecolieu is for people who want to live together and/or engage in common activities, solidarity, diversity and cross-cultural dialogue, living together with animals, in close contact with nature, a su-

stainable life designed according to the principles of permaculture.

The estate is to be purchased in early 2019. The ecological renovation of the existing building of the estate with common rooms and first apartments will then begin. To the east of the castle, 12 barrier-free low-energy houses (straw bale and clay buildings) will be built. Living space is to be created for the different housing needs and financial possibilities: Houses and small apartments, but also large apartments for shared flats and single rooms...

Furthermore, we're planning: a permaculture agriculture (2 ha) with greenhouse, a permaculture park open to the public, a plant-based purification system, an eco-camping-ground and the continuation of the horse boarding house.

☺ yes, details on personal request

L'Espace des Possibles

D4

Le Riousset | 07150 Vagnas
☎+33-(0)650 709 837
contact@hameaux-legers.org
lespacedespossibles.com

2018 | 10 ha (near Aubenas) | 20-30

- e, f, i, sp
- neighborhoods: taking decision on daily live. The whole federation decides on space-related issues only.
  (consent decision making in SMALL groups <10-15 persons)
- common financing for shared infrastructure (land, community buildings) and individual financing for each house
- individual ownership of houses, common property of land
- depends on each neighbourhood as well
- yet to be decided (likely to be food production, art & co.)
- ecological, self-managed, anarchistic, left-wing political understanding, feminist, non-violent, organic agriculture, organic farming, forestry, alternative energy, new technologies, ecological construction, humour, tiny house, neo-nomadism

☺

L'Espace des Possibles (The Space of Possibilities) is a collective, self-managed and ecological space with individual, mobile housing (or housing which can be easily moved).

It is a federation of largely independent neighborhoods, gathering a limited number of people eager to live together on a daily life and to share time, resources and facilities.

Common rules apply to all members of L'Espace des Possibles, to protect the space and ensure all can live together in peace. L'Espace des Possibles also aims at welcoming businesses and non-profit organizations in network with the local ecosystem and at collaborating with the city council to open facilities to all.

This is an inclusive project open to small budgets (250 EUR monthly contribution for collective spaces) which does not require to lend money to the project. We are open to non-French speaking people and welcome volunteers.

- local networks, partnership with the city council
- ☺ requirements may vary according to each neighbourhood

Quartier&Co

C1

14 Rue Notre Dame | 76200 Dieppe ☎ +33-(0)235 045 998

bocquetmouquet@gmail.com, charlionet.roland@gmail.com

quartierandcodieppe.com

2009 3.5 ha

f, e, sp, g

Our association Quartier&Co is working at the project of reconverting the "site Lebon" into an active eco-district within the town of Dieppe. Ecology, social economy and social cohesion are the values of this eco-neighbourhood whose vocation is to mutualize the various areas, services and equipments, whilst advocating the advantages of urban and rural living. The group of citizens, who created the Quartier&Co association in 2009, intend to continue with the management and running of the area once completed. The originality of the project has enabled Quartier&Co to be recognised as a partner by the local and regional authorities, as well as by the ADEME (Agency of sustainable development and energy control) and the Ministry of Sustainable Development.

The feasibility study was worked at with a group of professional experts and was financed by the local partners and by a crowdfunding. It clarified a first phase of realization: Cooperative housing (100 flats), an international youth hostel (capacity for 150 people - a specialised language centre with organic food restaurant), a medicalised respite home for 16 people with chronic progressive disease and for their carers), an organic grocery store, shared gardens, a management and coordination structure for shared equipments and home services.

UTOPIA

D4

Eglise vieille, Rue de la mouthe | 30460 Lasalle ☎ +33-(0)674 645 934

kaitomarcos@gmail.com

2015 (near Montpellier) ♀5 ♂6 3 11

f, sp

no

adult education, organic agriculture, ecological construction, humour, theatre, cantina, popular education

Collective housing with 4 families and cultural space: theatre, circus, music, cantina, brico café, conferences... In association with the village.

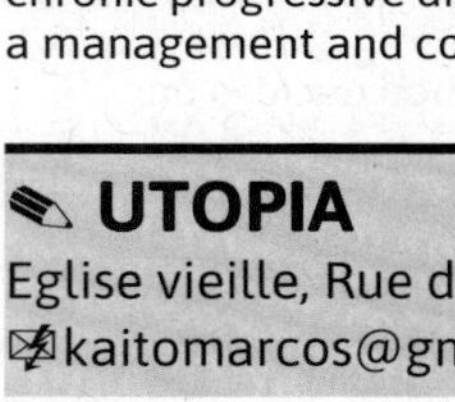

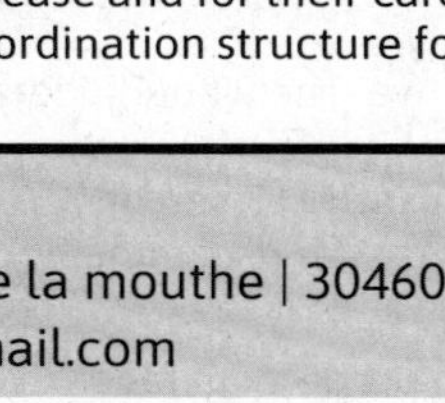

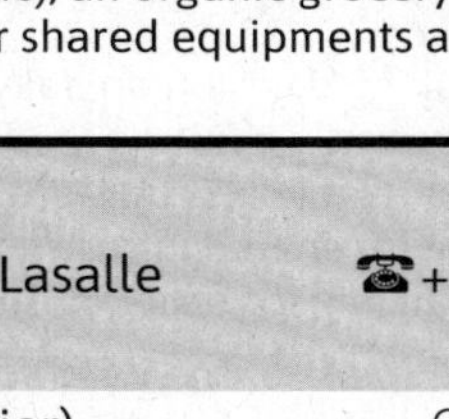

Germany

back2connect [up the top & in the river]

Unterlauchringen | 79787 Lauchringen
☎ +49-(0)176-404 992 80
i-think-so@t-online.de

2006 1 ha, near Zürich (CH), Freiburg (D) 30 % ♀3 ♂3 3 9

g, e, f, i, nl, Swiss German 5 5

depending on relevance (in a way that everyone can live with it)

common and individual 1 person

constructivism, concordance, openness

PR-coaching, team and leadership training and eco architecture and culture

ecological, non-violent, ecumenical, inner growth, meditation, tantra, organic diet, adult education, seminar-house, peace-work, one-world development, permaculture, alternative energy, ecological construction, humour, liberalism, outdoor sports (canyoning, trekking), improvisational theater

☺

Hello! Our community serves as a network. Our members are currently dispersed over distant parts of Europe, so in terms of our local efforts we are still in the set-up phase of finding a place to live. In the long term, we aim to have our own village of huts. We are mostly active in the area of entrepreneurial project constructions in the areas of architecture and landscape planning. We are individuals with skills in economics, business and craftsmanship, and consequently we work as a team within the framework of a foundation. We are also active as business consultants for small and medium-sized businesses in crafts, sports and catering. We are not subject to any kind of political nor fundamentalist ideas. Our ongoing goal is to bring about entrepreneurial projects with a balanced and proactive nature, which ensure that people are mentally fit for the reality of daily life. At the same time, we want to remain modest and open-minded. In principle we are always open to new initiatives, but we want to maintain a healthy balance between sustainability and risk-taking. Truly "making a change" is only possible when we make use of the current economy and work well with good ideas, without having to philander. We meet regularly as a group and currently have at our disposal a house with a vegetable and herb garden, where we can receive guests, but which is not our goal in terms of the land we wish to inhabit. People interested in community are invited to contact us with meaningful e-mails. Cha Dieu ans allegra – may life please us.

TFG - Time for Good

☺ 1-minute-video like an elevator-pitch

Biotopia c/o Öff Öff bei Gasthof Gemauerte Mühle

E5

Alte Lausitzer Str. 2 | 02708 Löbau OT Kittlitz

+49-(0)157-850 656 76, -(0)176-271 102 48

schenker_oeff@yahoo.de

www.global-love.eu

2006 (near Löbau, Dresden) <100 % ♀1 ♂2 0 3

2

- individually and consensus (with different weight of votes) (rooted consensus)
- gift-economy, (family) sharing according to needs
- premises are used in the sense of moral using rights
- givers-(or gift-)movement, working with light and love
- experiment of living like nature-people (wood-men) in wild nature
- ecological, self-managed, anarchistic, feminist, non-violent, spiritual, Christian, ecumenical, Buddhist, Taoist, anthroposophic, inner growth, meditation, yoga, tantra, free love, organic diet, vegetarian diet, vegan diet, adult education, peace-work, animal protection, environmental protection, self-sufficiency, permaculture, organic agriculture, organic gardening, ecological construction, humour, living without luxury, state and money; (existential) survival

☺

The project "Biotopia" (planned as a "biotope for plants, animals and humans") is a radical self-sufficiency-experiment of "Schenkers-(givers- or gift-)Movement" (besides other projects like "social work as a gift" in the "house of hospitality" and normal (permaculture-)gardening and agriculture in the "peace-garden"...): The Biotopia-Project is the attempt to leave not only the state- and money-system, but also the normal civilization, and to live wild-nature-self-sufficiency, on a squatted place of land in a nature-reserve. As "nature- or wood-men in wild-nature":
"If we can live directly from wild nature, we do not need governments or bosses, and we can survive everywhere - and can live the way, we really want..."
The "system-dissident" Öff Öff began the Biotopia-Project in 2006. Meanwhile he is mostly living in the projects "holistic culture-laboratory lilitopia.de" and "real-utopia.de". The whole, including the connection of the different projects, is a continuing experiment of "holistic (r)evolution", 'personally, politically and concerning every aspect of life':
"As much as possible without ideology we must create a system of reason and love!"
As explained, Öff Öff is nowadays touring around between the different community-projects of gift-movement. The situation in Biotopia – also concerning presence of persons – is changing a lot, so it is useful for interested persons to make a telephone-call...

→See also description of networks "**Givers-Movement, European Catholic Worker, L'Arche de Lanza del Vasto**"

- Gandhi-Movement, working with LIGHT and LOVE, international gift economy (moneyless.info)
- ☺ supporting the ideals of the Givers-Movement and the concept of the project

Bunte Lebenswelten

C5

Ludolfshausen 6 | 37133 Friedland
+49-(0)5504-7553
mail@bunte-lebenswelten.de
bunte-lebenswelten.de

2013 (near Göttingen, Witzenhausen, Kassel) 3

g, e 1-6

association members (consent oriented) individual financing

interreligious, also for open atheists

creativity - with humor and/or clowning with all our hearts; permaculture and active climate protection

ecological, self-managed, non-violent, spiritual, inner growth, adult education, peace-work, environmental protection, permaculture, organic agriculture, organic gardening, alternative energy, humour, non-smoking

☺ ()

We founded Bunte Lebenswelten association aiming at community living and working with adults and children. Dwelling and living in an active neighborhood is possible in the village of Friedland-Ludolfshausen at the time. Depending on the further development of the village, distributed urban community living may become an option in the future.
We like guests and work with mostly young people from all over the world. We see communal living and working as our contribution to a committed climate and environmental protection through a reduced ecological footprint.
At the beginning of 2019 we are a small association and in a state of increasing village integration. There is a temporary shared flat in the village. Others may follow. We want to start carefully with the neighbourhood housing project - with families, partnerships, small communities. We organise our educational work in small personal compositions. We are officially approved as a workplace for a voluntary ecological year (FöJ). The aim for our community is to expand our professional commitment in cooperation with Bunte Lebenswelten e. V.

Transition Town, welfare association “Paritätischer Wohlfahrtsverband”, regional association “New Housing”

☺ active engagement in association Bunte Lebenwelten and village

Commonie33 Multi-generational Community

E4

Gorgasring 21 | 13599 Berlin
+49-(0)30-788 940 25
projekt-wir@battes.de
projektwir.jimdo.com

According to Oya magazine, a “commonie” is a “form of society of the commons, of respect and esteem for all that is there.”
“Commonie33” is a project with the aim to found such a community in the green surroundings of Berlin. Some people with the longing for a life close to nature, an attentive coexistence with all generations, animals and plants, for a self-governing farm project, have already joined the group. We are in the process of establishing an association and have started with the concrete planning.

EOS/Schloss Hohenfels Erlebenspädagogik-Institut B10

Wildbachweg 11 | 79117 Freiburg ☎+49-(0)761-600 800
info@schloss-hohenfels.de www.schloss-hohenfels.de

2019 5 ha (near Überlingen) 30 % ♀4 ♂6 0 10

g, e 7 3

all in their areas, central responsibility lies with the board (consensus procedure, holocracy, heterarchy)

individual financing, household budget, associative economy, concept of brotherhood, tripartite wage

property of our nonprofit association

anthroposophy, undogmatic

conference center, gastronomy, garden, building team park, culture

ecological, self-managed, non-violent, spiritual, ecumenical, anthroposophic, inner growth, meditation, organic diet, vegetarian diet, vegan diet, adult education, pedagogical, seminar-house, peace-work, environmental protection, self-sufficiency, organic gardening, alternative energy, new technologies, ecological construction, construction, accessible to the disabled, no cell phones/wireless, non-smoking, team-seminar-venue, social threefolding, holocracy, outdoor education, youth, festivals

☺

EOS-Erlebnispädagogik wants to offer the opportunity to create a modern, yet at the same time innovative conference centre at Schloss Hohenfels that preserves historical and spiritual traditions. A place of encounter can be created here, where the spirit of community, sustainability and peace can revive. In setting up this team conference centre, we attach great importance to the fact that as many employees as possible work on the site and live here at the same time. From the outset, the establishment of an intentional community is an essential goal. In this respect we will cultivate a culture of togetherness from the very beginning, we will continue to cultivate the common vision work, we will hold future conferences, wisdom councils, open space conferences etc. on many weekends in order to develop the spirit of the community. Our real vision is to transform the traditional triple virtues of the members of the different orders (poverty, obedience, chastity) into the present (freedom, equality, fraternity).

EOS has a few other branches

☺ ecological, spiritual orientation

FLOW | Frei leben - organisch wachsen

X

Drahthammerstr. 24 | 92224 Amberg +49-(0)170-411 96 29
info@frei-leben-organisch-wachsen.de
www.frei-leben-organisch-wachsen.de

2016 ♀3 ♂4 1 8

- g, e
- all members
- individual finances with limited shared finances
- planned: corporate property with right of abode for inhabitants
- undogmatic
- planned: seminars, psychotherapy work, vegetable farming, a farm shop, animal keeping
- ecological, self-managed, left-wing political understanding, non-violent, inner growth, meditation, organic diet, environmental protection, self-sufficiency, permaculture, organic gardening, alternative energy, ecological construction, humour, community building workshops, seminars, social projects, shared workshop, farm shop

In August 2016 our first community meeting took place, with the dream to create a place of living and working together. Currently we are seven adults and one child, as well as many interested. Once we are fifteen people, we will start looking for a place for us.
We want to acquire an object close to nature for 20-40 adults plus children in southern Germany with sufficient area for agriculture (field, forest, river, lake) and existing buildings together. Individual living arrangements will be possible, needed renovations and new buildings shall adhere to ecological principles and techniques.
FLOW wants to create a place for growing old and being young. A place for a communal life concept that meets the social needs of individual residents in harmony with nature and the environment. Self-sufficiency through permaculture, old-age provision and communal childcare are some of the ideas we will implement.
We want to live peacefully, non-violent and without dogmas and to cultivate open, respectful relationships. Community education and personal development play a key role at FLOW, everybody has the same rights and responsibilities, with decisions being shared as much as possible. This creates a space in which closeness is experienced, conflicts are articulated and authentic community is cultivated. We try to realise all of this with a good dose of humor and love of life - come over and join in!

☺ acquaintance, come to meetings, consensus of all members, financial contribution

Gemeinschaft Rosenkohl

B8

76xxx Karlsruhe
lesmana@gmx.de gemeinschaft-rosenkohl.de/

2018 ♀1 ♂1 0 2

- g
- all who are affected (consensus)
- shared economy

☺

We are a small community in Karlsruhe, currently consisting of two adults.
We are looking for more people to live together, work together, and love together.

We want to share everyday life, have breakfast together, clean up together, play together, laze around together.
We want to be real, express emotions, express desires, say yes if it's a yes, say no if it's a no.
We want lovingness, tenderness, mentally and physically, to embrace, to listen, to acknowledge.
We want to support each other, by fulfilling the wishes, staying in contact, reflecting, growing.
We want to be connected, be there for each other in case of need and enjoy the benefits together if successful.
We want to be free, just doing what we think is good and right.
We want to grow old together.
We want to start right now.

☺ agree with our way of life

Gemeinschaftliches Wohnen im Hexental

B9

Freiburg — +49-(0)7633-808 828

andreabachstein@wohex.de, info@wohex.de — www.wohex.de

2300 m² (near Freiburg i. Br.) 1 % 3 12

g

all members of the community (consensus) — individually financing

property of cooperative

ecological, inner growth, meditation, yoga, organic diet, adult education, alternative energy, ecological construction, temporary living space for people in crises

☺

We are a group of currently 10 adults and 3 children. In total, we would like to be 15 adults and a few children. We are planning communal and intergenerational living in the Hexental area near Freiburg (Black Forest) on the basis of relationship and conflict abilities, and with the goal of growing together. The core of our communal living is the process of community building, we also offer to introduce interested groups to this process. In addition, we are contemplating a so-called "free space apartment", which would offer a temporary home to people in crisis. As a legal structure, we have opted for a cooperative, and we will most probably build new housing. Interested people with children – including single parents – are welcome to contact us.

☺ people with children, single parents are welcome

Gemeinschaftserfahrung

E4

Müggelseedamm | 10000 Berlin

gemeinschaftserfahrung@t-online.de — gemeinschaftserfahrung.de

2010 150 m² 2 % ♀1 ♂1 0 2

g

together (consensus) — limited common financing

rented

rational - humanistic - atheistic

active participation in the process of community building, self-knowledge, self-enquiry, life in responsibility, commitment, love research ecological, non-violent, spiritual, inner growth, meditation, sexuality, free love, organic diet, peace-work, humour, no cell phones/wireless, non-smoking, research community, awareness/becoming aware

☺ (✋✋)

Right on the outskirts of Berlin, we try to keep our lives simple and to treat nature and our Earth as considerately as we treat ourselves. We are interested in open, truthful encounters and relationships. To us, freedom does not mean arbitrariness, letting oneself go or something like self-realization, but to say "yes" to oneself and one's own path and to unconditionally accept the consequences that this path brings within.
Having time and accepting each each other, listening, supporting, challenging and accompanying each other and sharing gifts and simply being, in deep affection and belonging - this is what characterizes our being together.
All those are invited, who want to start exploring and sharing the issues mentioned above, who are determined in their intention and who are open to allow something new to move in with them. We are not striving for an ideal dream place elsewhere, but are starting to build community right where we are. The available living space offers only limited possibilities for growth. It is thinkable to live together in the closer environment, but to shape the community-building process together. How and where such a process develops is completely open and depends on the energy that grows between and within the participants.

- Liebeskunstnetzwerk Berlin (love art network)
- ☺ own financing, participation in common costs, you are willing to continuously and permanently provide time and energy for the community building process

Gemeinschaftspfad Apfelblüte

D6

Jena ☎+49-(0)151-591 524 25
pjschleicher@gmx.de

2016 ♀3 ♂2 7 12

- g, e, f
- full members (consensus)
- we find wealth in diversity
- freedom, sustainability, self-administered, inner growth, non-violent, humor, living together, living consciously

☺

We cultivate our common and individual understanding of community and create a space for life and personal development for our group. For this purpose, we meet regularly to discuss content and enjoy the time with walks, meals, team work, games and getaways. At our summer picnics we celebrate the contact to former members, interested parties and our circle of supporters. After the current finding, concept and growth phase, we want to find and design a property in the Jena-Weimar-Erfurt area and live our vision of community together there. We are pleased about interest, new members and co-creation.

health - eco village Projekt Ökosiedlungen

X

Postfach 110826 | 60043 Frankfurt ☎ +49-(0)69-739 25 28
an-stiftung@t-online.de

2001 20 ha 20 % ♀4 ♂6 7 17

g, e 2 7

one, later more (consensus) different forms

property of some/all

UN Sustainable Development Goals etc.

fundraising, concepts of ecological architecture, national and international networking

ecological, self-managed, non-violent, inner growth, meditation, yoga, free love, organic diet, vegetarian diet, vegan diet, macrobiotic diet, adult education, peace-work, permaculture, organic agriculture, organic gardening, organic farming, alternative energy, new technologies, ecological construction, humour, non-smoking, political art, left and free, deep ecology, direct democracy, organic security, creative art, educational art, health

☺

The project is in an advanced planning phase. The result shall be a creative, technical and socially innovative "model city" for an ecologically oriented life in a metropolitan area and/or rural area. This includes the integration of living, working, leisure and transport.
Participating are scientific institutions, companies, architectural offices, etc. The experiences shall be scientifically evaluated and passed on to interested parties.
Planned priorities: Political arts and health (including occupational).
Suggestions from other projects are welcome.
Current planning: 50-300 inhabitants/users

☺ competence and creativity in concept realisation, willingness to invest, humour

InGe Wohnprojekt

B7

Individuell wohnen - Gemeinschaftlich leben in Kastellaun

Hauptstr. 18 | 56281 Dörth ☎ +49-(0)6747-948 750
inge-wohnprojekt@gmx.de inge-wohnprojekt.jimdo.com

2018 3800 m² (near Kastellaun) ca. 25

g, (e) 2

all together (consensual) individual financing

our property

no

ecological, non-violent

☺

We are a housing project and want to realize more than a good neighbourhood. More to this you will find on our website www.inge-wohnprojekt.de

Landesarbeitsgemeinschaft Gemeinschaftlich Wohnen Rheinland Pfalz

☺ mutual sympathy, financial participation *Sketch on page 15*

Jagelhof Gemeinschaftsprojekt Prignitz

D3

Jagel, Dorfstr. 8 | 19309 Lanz ☎ +49-(0)38780-12 99 93

mail@jagelhof.de jagelhof.de

2018 13 ha (near Wittenberge an der Elbe, Stendal) 6

g, e

private property

none

preparation of building for restoration

The JAGELHOF is located in the Elbtalaue nature reserve between the Hamburg and Berlin urban agglomerations. In the joint project JAGELHOF, a group of people has come together to shape life and survival in one of Germany's most sparsely populated but also most scenic regions. The aim is to master the challenges of everyday life together and to bring about practical changes that are as close to nature as possible, as well as to give impulses that are needed in an eventful and fast-moving time.
At the core of all considerations is the community idea and the question: How can the good life at the JAGELHOF succeed? How can we be socially responsible for each other, how can we unfold our potential and bring it into harmony with our values? Ecologically sustainable agriculture and a resource-conserving approach to nature are just as important to us as art and culture.
An attractive site with an ambitious concept is still looking for people who want to take action here, who have the courage and entrepreneurial spirit to experiment. This means that they want to work and live together, build and manage themselves.
If this appeals to you, get in touch, come here and join in.

☺ individual examination

Lebensgemeinschaft Ammersee

C9

Dammweg 4 | 82229 Seefeld-Hechendorf ☎ +49-(0)1577-735 89 43

info@FamilienKunstschule.de

www.lebensgemeinschaft-ammersee.de, www.FamilienKunstschule.de

2017 200 m² + garden (near München) 5 % ♀2 ♂2 4

g, e 1

all together (talking-circle, vote, veto is possible)

individual financing rented

open

website, renovation, gardening, cleaning, music, creating culture

ecological, self-managed, non-violent, spiritual, inner growth, meditation, yoga, organic diet, vegetarian diet, adult education, pedagogical, multi-generations, culture, healing, music, art, literature, dance

We are a small, open, easy, humorous community with interest in the common welfare economy and positive, sustainable, fair projects in the world.
We help each other to develop further. Once a week we meet separately for either organizational or personal meetings in a deliberately limited time frame, and then sing, dance or

watch an encouraging, stimulating film about people and projects that make a difference: for example about Christian Felber's Gemeinwohlökonomie, Jane Goodall, ...

☺ ecological, cultural interests

Lebensraum Gut Altfresenburg

C2

Altfresenburg 11 | 23843 Bad Oldesloe OT Altfresenburg
lebensraum@gutaltfresenburg.de www.gutaltfresenburg.de

2017 6 ha (near Bad Oldesloe) ♀7 ♂8 7 22

g, e about 50 planned

groups, partly the whole community, with systemic consensus (subject to change)

common financing in some areas, partially shared economy in some business areas

property of some and of a group

spiritually open

seminars, offices, practices, garden, catering, hotel service, renewable energy

ecological, self-managed, non-violent, spiritual, Christian, ecumenical, Buddhist, esoteric/New Age, anthroposophic, inner growth, meditation, yoga, organic diet, vegetarian diet, vegan diet, adult education, pedagogical, seminar-house, animal protection, environmental protection, self-sufficiency, permaculture, organic gardening, alternative energy, new technologies, ecological construction, construction, humour, accessible to the disabled, health centre, healing, events, guests, focus on alternative energies, art, culture, ideally a kindergarten or free school, respect for humans, animals and plants, history, future, present

☺ ( )

Historic manor in Schleswig-Holstein consisting of six buildings and 6 ha land (including two ponds with more than 1 ha each) and a neo-classical mansion. Unique location in the middle of nature, but only about 2 km from Bad Oldesloe city centre and 30-40 minutes from Hamburg. Lübeck 20-30 min. Ponds and the river Trave invite you to relax on and in the water. Walking paths and a unique nature reserve adjoin the estate.
Property units are in the newly remodeled outbuildings. New buildings wrapped in historical substance. Starting in 2020, these will be developed by several builders' associations. A limited number of rental apartments should be finished by 2025, these are intended for people who want to work on the estate. Seminar and business centre (offices, seminar rooms & practices) will be in the prestigious mansion. Guest apartments for seminar participants, customers and private guests, among others.
Emphasis is on sustainable health, energy and environmental solutions. Community management of the external area, community activities and community focus to make this place a sustainable, as self-sufficient and holistic place as possible. Art, Culture, Health, Energy, Hospitality. All together and everyone for themselves.
Live. Work. Create. Be. Without dogmas, but with an open heart and in self-responsibility.

☺ ability to finance construction, ideally additional share of the community company, sustainability consciousness, spiritual alignment would be nice, ecologically motivated, honest character

Nahe-Miteinander

X

Am Bettenheimer Hof 75 | 55576 Sprendlingen ☎ +49-(0)6701-204 77 23
pfeufer-michael@t-online.de www.nahe-miteinander.de

2017 (near Bad Kreuznach, Mainz) ♀8 ♂8 4 20

g

all of us together (majority decisions at the time)

ecological, non-violent, spiritual, ecumenical, meditation, yoga, environmental protection, organic gardening, ecological construction, humour, accessible to the disabled

☺

Our goal is an ecological, holistic, healthy lifestyle and a spiritually oriented, but not dogmatic, community in which everyone acts in concert. Terms such as mindfulness or alternative healing methods should be known and accepted, and people should be interested in some of these things (people who are still at the beginning are also welcome). Through our own internal group work, seminars and workshops, both internally and externally, we want to initiate a common growth process. Mutual help, support, recognition, mindful treatment of oneself and others should not only be catchwords, but should also be lived.
We want to live with people of different ages and with different ways of living and working, so that each and every one of us can determine how much contact we have with each other. We do not want people with aggressive and radical tendencies of any kind. We accept other people as they are and respect them, regardless of religion, skin color, origin or even just their eating habits. A pillar of our community is spirituality. We welcome this in almost all its many variations and reject only extreme directions.
Everybody can be and live like they want. We accept this and expect everyone to respect those who think and live differently.

☺ none, it has to be a fit on both sides

Niepkuhler Krähennest e.G.

A5

Lüschdonk 20 | 47809 Krefeld ☎ +49-(0)2151-54 66 71
vorstand@n-kn.de www.niepkuhler-kraehennest.de

2014 (near Krefeld) ♀6 ♂4 4 14

g, e

all together (consensus) property of association members / rented

no

ecological, self-managed, spiritual, inner growth, meditation, yoga, mentally/psychologically handicapped persons, physically handicapped persons, environmental protection, permaculture, organic gardening, alternative energy, ecological construction, humour

☺

We are united by the desire to realize a future-oriented form of co-living in the green, natural suburban area of Krefeld. We want to add further people of all ages, genders, nationalities and orientations to our group. Especially families with children (including single parents!) are encouraged to enrich our community. As a result, the cooperative is expected to grow at least twice as high as the current membership. We intend to include people with disabilities

as equal full members in our project. We want to become a meeting place for the neighbourly environment. In addition to the social community, the basic ecological orientation is particularly important to us. In addition to a resource-saving construction method, an energy standard promoted according to KfW-criteria and other environmentally friendly individual measures, we aim to acquire garden and meadow areas in the neighborhood that we manage ourselves or with professional allies (e.g. NABU), thus contributing to a greening of the surrounding area. At the same time, these areas can also be used as an extended play area, for vegetable cultivation and as creative design spaces of the residents.
We are still open for all forms of communication and still try it out. Often the way is already the goal. We get to know each other, because we want to learn real, open and therefore sincere encounters. Our goal is to create a shared home where everyone respects each other so everyone can always feel accepted. We want to work for a successful coexistence of young and old.
As this is printed in 2019, we have an option on our desired property and we expect a commitment soon. We hope to move in in about 2021.

☺ accept our principles

Öko.See.Dorf Co-operative Living Friedrichshafen

B10

Ulrichstraße 7 | 88677 Markdorf ☎ +49-(0)176-228 434 94
mail@oekoseedorf.net www.oekoseedorf.net/

2018 1 ha (near Friedrichshafen) planned: 80-150

g, e
core group incl. the delegated from the labor groups (sociocratically in consent)
shared financing of the coop will be in ownership of the community
no
design-architecture-technology, administration-finances-legal status, community-communication, publicity-education
ecological, self-managed, non-violent, adult education, environmental protection, permaculture, organic agriculture, organic gardening, alternative energy, new technologies, ecological construction, humour, ecologically & regenerative, sociocratic & non-violent, appropriate & fair

☺

The Öko.See.Dorf (= Eco.Lake.Village) Friedrichshafen is a registered residential cooperative prepared for foundation, that wants to provide payable living space for its members, free from speculation, for a social, neighbourly living together across all generations.
We are active people, seeing a chance to take responsibility, developing an ecologic and social living culture, that starts now and has an effect for generations.
We do not simply want to construct and reside, we want to encourage liveliness, sustainability, shape our society consciously and with responsibility.
The represented values are ecologic and regenerative, sociocratic and non-violent, reasonable and fair.
Based on our visions and aims we follow the advancement of freedom, engagement and creativity of the individual. We also expect the willingness to act counscously for public interests.

☺ application procedure, contribute to building our community

Ökodorf Blankensee - Dorfgemeinschaft Lehmhausen

E3

Herzershof 10 | 15328 Küstriner Vorland
+49-(0)1575-662 77 35
info@lehmhausen.de
www.lehmhausen.de

2015 3.5 ha (near Frankfurt (Oder), Gorzów Wielkopolski)
100 % fruits and vegetables ♀40 ♂40 80 160

- g, e, i
- to be determined, income-sharing intended
- to be determined, property of cooperative intended
- currently preliminary planning, crowdfunding
- ecological, self-managed, Christian, ecumenical, organic diet, vegetarian diet, vegan diet, macrobiotic diet, adult education, pedagogical, free school, seminar-house, peace-work, one-world development, mentally/psychologically handicapped persons, physically handicapped persons, animal protection, environmental protection, self-sufficiency, permaculture, organic agriculture, organic gardening, animal friendly husbandry, alternative energy, new technologies, ecological construction, construction, humour, accessible to the disabled, no cell phones/wireless, non-smoking, Crafts, carpentry, straw-building, earth-building, theater, music, children

"It takes a whole village to raise a child." - African saying

The core of the project is the community and the people working in it. We strive for an intercultural and solidary all-generational settlement community in which the residents support and complement each other. Being inclusive, our project addresses people of all ages, religions and cultures.
The members contribute their personal skills and material resources to the community to the extent that is good for them and the community. Everyone in the community can realize their ideas in social, musical, cultural and artistic projects or other areas of competence. For us, this means the connection of family and career and the participation of young and old in community life. Together we want to give children a safe framework to grow. In contrast to the modern, capitalist-oriented zeitgeist, Lehmhausen sees himself enriched by his more mature fellow members and regards older people as an elementary part of the desired community structure.

- Think Camp
- ☺ interested people have to fit the group and share our goals / ideals

Ökodorf Gutleben - einfach gemeinsam leben -

D10

Haushoferstr. 40 | 83358 Seebruck ☎ +49-(0)157-875 566 65

info@oekodorf-gutleben.de www.oekodorf-gutleben.de

2018 g, e, sp

- none
- communal culture, self-sufficiency, ecological building, direct democracy
- ecological, self-managed, non-violent, inner growth, organic diet, adult education, pedagogical, seminar-house, environmental protection, self-sufficiency, permaculture, organic gardening, alternative energy, new technologies, ecological construction, humour

☺

We are a group of 25 active people who are in the discovery phase. We are engaged in community building, are working on location finding and intend to start soon with a project and residential centre in the region of Chiemgau. The core team of the group meets weekly. In addition, some creative groups work regularly on topics such as law and finance.

Pilotprojekt Mehrgenerationensiedlung

X

Freiburger Straße 75 | 77749 Hohberg-Hofweier ☎ +49-(0)7808-913 29 34

info@mehrgenerationensiedlung.org www.mehrgenerationensiedlung.org

2007 ca. 600-700 ha

- g, e, f, i, sp, Polish, Russian
- all members of the project together (consensus, majority vote)
- individual financing, some common financing
- ecological, self-managed, organic diet, adult education, pedagogical, free school, seminar-house, one-world development, self-sufficiency, permaculture, organic agriculture, organic gardening, organic farming, animal friendly husbandry, forestry, alternative energy, new technologies, ecological construction, humour, no cell phones/wireless, social, sustainable, old handcrafts, art, creativity

☺

Our vision: 200 families (children, parents and grandparents) each living on 1ha of land. These "garden farms" with self-sufficient characters are the foundation of the settlement. The centre of the settlement, the "Eco-Valley" industrial zone and roads and paths/roads merge seamlessly into the landscape. At present, land negotiations are being held at three locations south of the river Main. (Hohberg is NOT the location of the settlement. The ultimate location has not yet been determined). Through intelligent energy management and the use of all available regenerative technologies, the settlement is CO_2-neutral. Constant technological development guarantees the sustainability of this project. Springs or wells as well as rainwater ensure sustainable water supply. Water cleaned by plant-based water treatment will flow into the settlement lake and subsequently into the nearest stream or river in a controlled manner. Ecology, biodiversity and local wild plants will promote natural equilibrium. Soil re-naturalisation, mixed crops and species-appropriate animal husbandry according to the guidelines of ecological agriculture (Bioland or Demeter) are basic components of the settlement. A state-accredited alternative school promotes the input of reformed educational concepts which encourage childrens curiosity and joy of learning.

☺ attendance of a visitor's day, joining the project

Sieben Eichen – Mehrgenerationen-Projekt D2

Basisdemokratisches & soziales Wohnprojekt in M.V.
| 17111 Borrentin ☎+49-(0)39952 236 67
siebeneichen@t-online.de www.sieben-eichen.com

(near Rostock, Stettin)

☺

Democratic and social housing project in the German state of 'Mecklenburg-Vorpommern' with houses of different sizes, types and locations between Müritz, Mecklenburg Lake District and Peene up to the area of the North Sea coast: Living and work for singles, couples, families, elderly people and communities.

As a non-profit association and through the development of a network of multi-generational housing projects, education and new cultural sites, we aim to improve the regional living, housing and working environment and make moving to rural areas more attractive.

Our "starter houses" from 2018 for 10, 20 or 30 people each are:

1) 'FAMILIES' house in a large estate with a park
2) 'GUEST' house with 16 holiday apartments + restaurant, right on the Müritz
3) 'Old HORTICULTURE' with 4.5 hectares of natural land (3 houses + 1 new construction area) with picnic area on the Peene
4) 'Great TRAIN STATION for culture' - a house with mixed & urban use between Lake Malchin and Lake Kummerow
5-7) Further houses (farm house/barn & manor complex)

Siedlungsgenossenschaft proWAL eG i.G. A8

Association for the promotion of artistic and sustainable living
Am Bungert 3 | 66625 Nohfelden ☎+49-(0)6852-809 461
info@prowal.org prowal.org

2015 (near St. Wendel, Birkenfeld) ♀18 ♂8 4 30

g, e, f, i

ecological, self-managed, left-wing political understanding, non-violent, spiritual, Christian, anthroposophic, inner growth, meditation, organic diet, adult education, peace-work, mentally/psychologically handicapped persons, physically handicapped persons, animal protection, environmental protection, self-sufficiency, organic agriculture, organic gardening, organic farming, animal friendly husbandry, alternative energy, new technologies, ecological construction, humour, accessible to the disabled

☺

Twenty-five years ago, in the northern region of the Saarland (German state), the seed was planted to start a Waldorf School and build up a community around it.

In the summer of 2015, a group of people took up the founding impulse to found a living, working and living community in Walhausen. It should create a meeting place where people can live, work and grow old in mutual awareness and dignity. Meanwhile, we are 30 people and are in the process of buying a ground of almost 3 hectares in the immediate surrounding of the Waldorf school and the school farm, where we would build our meeting house with a large kitchen, café and small shop area, therapeutics, guest rooms and office space.

With our project we want to strengthen the village character: craft (hairdresser, bicycle repair

shop, bakery, joinery) and a village shop café should be integrated into the village life again. We want to offer services and cultural activities for children, young people, families and old people. Disabled multi-generational homes and living areas for the elderly should be built according to ecological and sustainable criteria. We want to offer people with disabilities, illnesses or disabilities opportunities to engage in meaningful work in our community and to be supported by the community. Even more, there is going to be an organic agriculture farm with vegetables, medicinal plants and herbs, and several small animals (goats, sheep, chickens...).
We want to create a place that healingly extends into the environment and are looking for people who want to help shaping our project, especially in this crucial start-up phase.

Stamm

C5

Im Oberland 22 | 37269 Eschwege/Eltmannshausen ☎ +49-(0)5651-220 814
info@helgebartels.de, kontakt@freie-schule-eschwege.de helgebartels.de

1995 (near Kassel) 10 % ♀5 ♂5 8 18

g, e, f, sp
all (council) individual financing
individual property, rented, etc.
none
gardening, foundation of a free school, construction, various projects
ecological, self-managed, non-violent, inner growth, organic diet, adult education, pedagogical, free school, seminar-house, organic gardening, ecological construction, humour, love, fun, art, party, dance, children

☺

We consider ourselves as a tribe, an original form of community in a modern shape.
What is a tribe? Something like a village community? People who somehow belong together? People who have common ancestors, family? Perhaps also people who feel related in spirit, realize common goals, want to develop together by pushing each other's buttons and being there for each other?
We live in a time in which it has become normal to leave one's hometown and family and to build one's "own life". Nevertheless, many of us are longing for community, for like-minded people - at the latest when we have children of our own and notice that everyday life becomes easier and more beautiful when we can share tasks, worries and joy with other people.
Our idea of a contemporary tribe does not refer to a spatially limited location. Everyone designs their living space as they wish, in their own apartment, their own house, in the village/town of their choice.
We want to connect the people here in this region in the wild middle of Germany with each other. And we would like to invite YOU to become part of this network. Here live some city divers, attracted by abundance and space of this area. We are open for refreshing new tribe members. We are looking for motivated young and no longer young people, families, children, dreamers, doers who believe in a better world through community, in which there is still room for their own desires and visions.

☺ sympathy

Photo on page 41

Strohplatz

A7

pls. inquire via e-mail | 54411 Rorodt

strohplatz@strohplatz.com www.strohplatz.com

2012 3500 m² (near Thalfang) 5 % ♀1 ♂1 0 2

e, Dutch 2

one person (consensus) limited common financing

property of all

We give courses in self-sufficiency, non-violent communication, permaculture and mindfulness.

ecological, non-violent, inner growth, vegetarian diet, adult education, environmental protection, self-sufficiency, permaculture, alternative energy, ecological construction, non-violent communication, mindfulness

We are a small settlement still in building phase. Currently we are giving courses in self-sufficiency, non-violent communication, permaculture and mindfulness in Germany and in the Netherlands. The whole Strohplatz site is being built as self-sufficient as possible as a demonstration site. The house is currently filled with 20 beds as accomodation and the big shed is being used as teaching area and common area. Regarding housing, we are not sure where this adventure ends. We might actually end up as a small community with some families, or something similar. In summer a lot of people visit Strohplatz for short and longer stays.

WIRklicht

x

64678 Lindenfels/Odw.

leben@wirklicht.de, wirklicht.de/kontakt wirklicht.de

solidary

spirituality, nature, nonviolence, deep ecology, sustainability, resilience

WIRklicht wants to become an "emancipatory community" whose goals are the overcoming of individual and social alienation, peace inside and outside, home and happiness as well as a sustainable intact environment. We want to be part of a new global culture, which should enable the survival of all beings permanently. To achieve this, we want to

- develop into a community in which we encounter each other respectfully and want to live an undogmatic, non-exclusive spirituality that makes our connection with everything that is tangible
- lead a life which is marked by appreciation for the planet and all life on it and which gives the following generations the chance of a dignified life. By striving for a modest way of life, we want to curb the consumption of superfluous goods and give ourselves more time for deep and joyful encounters with ourselves, with others and with nature
- to become more resilient against a system based on the illusion that unlimited growth is possible; in which power and wealth are concentrated in the hands of a few; in which our nature, our raw materials and global human rights are plundered and in which the foundations of life are deprived of subsequent generations and nature. This is why it is important for us to develop self-sufficiency and networking with other communities.

Wohngenossenschaft im Schwarzwald

B9

Hegestraße 16/1 | 78609 Tuningen ☎+49-(0)7464-981 77 25
info@schwarzwald-genossenschaft.de
www.schwarzwald-genossenschaft.de

2018 (near Freiburg) 45 % ♀3 ♂4 7

g, e
house community / cooperative members (it has to work for all, consensus)
common property/cooperative
biological, close to nature, a little anarchist, joyful living, low tech
care for the elderly, biological vegetable farming, keeping chickens
self-managed, anarchistic, non-violent, spiritual, esoteric/New Age, anthroposophic, inner growth, meditation, organic diet, adult education, seminar-house, physically handicapped persons, animal protection, environmental protection, self-sufficiency, permaculture, organic gardening, organic farming, animal friendly husbandry, alternative energy, ecological construction, humour, accessible to the disabled, no cell phones/wireless, non-smoking

☺

We are in the planning phase of the Schwarzwald-Genossenschaft ("Black Forest cooperative"). Our aim is to unite many positive people who wish to think and live biologically. Live sustainably and with respect for nature and the environment, e.g. composting toilets, solar power... There will be spaces for tiny houses, they should be self-sufficient.
"Business as usual" is not an option. Each person do their best to make the place they're living in a better one. No WiFi or mobile phone is a requirement for participation. We wish to integrate assisted living for elderly people, to offer and celebrate a life in dignity. Childcare is also part of our vision. Animal keeping, chickens and ducks. Biologically grown vegetables and fruit from our own garden. Homemade food is a big topic, we buy as little as possible. Seminars in the fields of politics, health, biological gardening, crafts etc. You can always find current information on our website.

☺ membership in the cooperative, biological living, mobile free

Ireland

Enriched Earth Ecovillage

Glenballythomas | F45 WK18 Tulsk
EnrichedEarth@gmail.com www.EnrichedEarth.org

(near Boyle)

e, Irish
ecological, inner growth, adult education, pedagogical, peace-work, one-world development, environmental protection, self-sufficiency, permaculture, organic agriculture, organic gardening, alternative energy, new technologies, ecological construction, humour, accessible to the disabled, Celtic Culture

☺

Enriched Earth is working to set up an Educational Ecovillage in North Roscommon and also working to establish a series of Ecovillages across Ireland. We are exploring a clustering approach via acting as an umbrella, in partnership with GEN, for existing Eco projects, businesses and small-holdings. Our Logo is from the Ancient Druidic Tree Alphabet.

ENRICHED EARTH

Italy

Bad Oberhaus

Vertigen 29 | 39020 Partschins ☎+39-339 831 31 89
ewald.thaler@gmail.com

2013 20+ % ♀1 ♂1 0 2

g, i, e 1 1

owner or financially involved persons (by agreement)

has yet to be developed property of one person

no dogmas. Aiming to live more self sufficiently in natural harmony

improvement of self catering and independence; wood; repairs of stonewalls; garden

left-wing political understanding, non-violent, inner growth, organic diet, pedagogical, self-sufficiency, permaculture, organic gardening, animal friendly husbandry, forestry, ecological construction, humour, operation of board saws, transport cableways, cult site

☺

The remote Oberhaus is located in the mountains of South Tyrol, at an altitude of 1200 metres, in the forest. We have 20 hectares of land, a large house and a stable (under construction). The aim is to function as autonomously as possible, independent of money and society. This is partly achieved by private gardens, own sheep, woodworking, guests and tourism. Partly we work in the valley or town. We try to live together with 5-6 persons or 3 families. Since 2010 there have been and still are various temporary communities of helpers (Workaway, Helpx). We intend to be a community where we want to run and support the farm in cooperation. Each one brings his/her own expertise and responsibility. What we share is the joy of nature and the desire to live in mutual harmony and freedom. It is a place with many possibilities, surrounded by forest, without access road, a fertile oasis amidst the best drinking water sources (approx. 20 l/s). The place is remote. You have to walk about 20 minutes.

☺ realistic (not esoteric); nature-loving; motivated for independence; able to be social

Casa del Sole

☎+39-388 063 87 50
casa.del.sole.ne@gmail.com casa-del-sole.wix.com/casadelsole

2002 2 ha (near Genua) ♀1 ♂1 1 3

☺

Casa del Sole was founded in 2002. It was born as a project to become an ashram or a yoga center with agricultural production, later we were looking for members to create a small community farm. In the summer we are three people to live and work together in the sun (so

we are 2 adults and a 10 year old girl in 2019), but the place offers so much more, as there have been many people who have frequented Casa del Sole, also for long periods of time. We would like to build a small but determined community or group of people who rotate or live in the Casa del Sole, collaborating and sharing their lives together.
The place is surrounded by the stunning nature of the hinterland of Liguria just 13 km from the sea. Casa del Sole is located half a hill above a valley called Vall Garibaldo, the weather is always mild, never too hot or cold, the ground is laid out in terraces and exposed to the south, there are two springs of freshest water and a good neighborhood. There is an olive grove (6.000 sqm), a chestnut wood (2.500 sqm) and vegetable garden (3.000 sqm) as well as regular forest and fruit orchards. We live vegan and vegetarian, are non-smokers and practice yoga and meditation at sunrise and sunset. We are looking for conscious people to organize together, possibly sharing our attitudes.

Cohousing Sosozono

Via Ponte Pellegrino, 39/B | 81040 Pontelatone
www.sosozono.com

The final aim of the project is to stop and reverse the degradation of the natural environment and to help building a future in which the community can live in harmony with nature. Here, an extended community of people can live to exchange life, work and social experiences and to form the youth of the future residents.
Our cohousing project will include 3-6 families living together as an eco-sustainable community, managing common spaces and "ecological residences" in a collective way, thus obtaining economic savings and benefits of an ecological and social nature, using only the resources of the place and giving impulses to the global ecological consciousness.
Cohousing is emerging as a strategy of sustainability: on the one hand, participatory planning and sharing of spaces, equipment and resources facilitates socialization and mutuality among individuals, on the other hand this practice, together with other approaches such as the setting up of joint purchasing groups, car sharing or the different services used in common, promote energy savings and reduce the environmental impact of the community.

Community Sicily

Via dei Platani, 12 | 91022 Castelvetrano-Selinunte ☎ +39-388 446 24 92
infoegeria@virgilio.it

Today (2019) our community is existing like a spiritual's researchers group.
At present we're in searching of a site near Selinunte to sart a cohousing project where to share common targets about the environment, the nature and the earth.

Dangri

Dangri | 22010 Peglio ☎ +39-0344-898 17
paulus.joerg@gmx.de

1979 1 ha (near Lugano, Como) 55 % ♀1 1

- g, i, e, (f)
- owned and rented in parts
- open-minded spirituality
- garden, wood, holiday guests, maintenance

ecological, non-violent, spiritual, inner growth, meditation, organic diet, vegetarian diet, self-sufficiency, alternative energy

Dangri is a small, secluded place in the northern Italian mountains above Lake Como at 650 m above sea level, largely free of electric smog. Between large, old chestnut trees there are several natural stone cottages. 6 of them have been developed for living. 100 m below there is a clear mountain river where you can swim in summer. I have lived here since 1979. In the beginning we were a group of 4 adults and 2 children and in summer we went with goats to the alp. The emphasis today is simple life, self-sufficiency (without farm animals), holiday guests and spiritual orientation. There are 2 springs with fresh water, 12-Volt electricity from 2 solar cells and a small water turbine and organic fruit and vegetable gardens. In the village I rent out 2 bright holiday apartments. In summer people come to help for food and lodging and holiday guests who rent one of the natural stone cottages or camp with a view of the river. We like to make campfires with chats, singing and making music. I wish for energetic, open-minded people with an interest in self-sufficiency and inner growth who would like to rent or buy a natural stone house here and build a livelihood for themselves.

I imagine a mixture of neighbourhood and community, mutual support and enrichment, sharing of joyful things like making music, singing, excursions, meditations...
I am looking forward to you – Lisa
(please only call by phone, e-mails are rarely checked!)

☺ active; interest in partial self-sufficiency

Il Cerchio Sacro

Via Roma 3 | 60048 Serra San Quirico (AN) ☎ +39-334-798 6 566
schiastam@gmail.com www.ilcerchiosacro.it

2006 6.5 ha (near Ancona) 1 1

i

one owner

gardening, construction

ecological, spiritual, inner growth, organic diet, vegetarian diet, self-sufficiency, ecological construction, construction

The "cerchio sacro" ("sacred circle") is first and foremost a place where one can return to the wilderness and prepare oneself for a change, where respect is imparted and deep ecology is lived with the aim of reconnecting to Mother Earth. Furthermore, the "cerchio sacro" is a spiritual ecovillage. It is a magical place where everyone can change their life in love. The place is a mystical portal and has been serving as a center for retreats and spiritual healing for years, as an example of sustainable living, as a nature school, center of ecotourism and environmental education. At this place a loosely scattered spiritual ecovillage is to emerge, which is to be built up together with people who fully share these ideas. But it should also become a place for people who want to experiment and prepare for a new lifestyle: alternative, sustainable, self-sustaining and much more...

☺ willingness to volunteer

Lo spazio delle impronte

Via Sopramuro n.66 | 29121 Piacenza ☎+39-348-035 37 84
cohousingtuttipiuuno@libero.it, cohousingimpronte@libero.it
#7857 at eurotopia.directory/booklinks

ecological, feminist, non-violent, animals

The "space of the footprints" is a cohousing project that tries to associate the benefits of co-residence and cooperation with an eventual help in the care of the animals. The initiative is still being planned and cannot yet be visited. The project involves the construction of residential units each with a protected garden, which can be used freely and safely by your pet, and a common space and service. A structure that maintains the individuality of your home and way of life while letting you also find the dimensions of sociality and mutual help.
The initiative, as an alternative housing model to the traditional one, is open to everyone, even to those who do not live with animals. It should lead to a varied and intergenerational community.

Pian del Rosso

località pian del rosso 52 | 19028 Varese Ligure ☎+39-333 913 02 28
piandelrosso@gmail.com https://piandelrosso.wordpress.com

2015 (near La Spezia) >30 % ♀1 ♂1 1 3

e, i, sp 1 2

all who are concerned (in the future will be consensus)

individual finances

property going to be shared in an agricultural society

no dogmas here

permaculture small production, art & craft, culture

ecological, self-managed, spiritual, inner growth, meditation, yoga, vegetarian diet, vegan diet, adult education, animal protection, environmental protection, permaculture, organic agriculture, organic gardening, organic farming, forestry, alternative energy, new technologies, ecological construction, humour, accessible to the disabled, learning, craft, art, media, self growth, beauty

☺

The community will be a cohousing based on shared vision of better sustainability, shared resources, organic food production with no compulsory labor. Important topics are culture, art, craft, personal freedom through reduction of expenses. More info at our website.

☺ yes, people looking for sustainability and ecology, willing to invest for a quote of the place

Urbeto

Via Savonarola 39 | 46034 Borgovirgilio
+39-328-584 45 59
perchiara@alice.it
www.urbeto.it

(near Mantova)

e, i

ecological, non-violent, spiritual, inner growth, pedagogical, seminar-house, peace-work, one-world development, animal protection, environmental protection, organic agriculture, organic gardening, alternative energy, new technologies, ecological construction, humour, accessible to the disabled, Bahá'í inspired

Urbeto is still a project. We would like to create a place where it is possible to overcome difficulties with consultation and respect for our differences, we are inspired for doing that by the Bahá'í teachings and principles.
We would like to use a new kind of architecture using a 3D printer for the private houses and other ecological strategies for the common buildings, also because it is less expensive.
We hope to have many young couples with us to have good energies and children, elderly people as well, with experience and wisdom. We intend to take care of all these people in a spiritual and, if possible, in a material way. We intend to create co-working opportunities for those who are young and a safe place for those who are old. We are plannig to create educational projects from toddlers up to grown-ups as well. We would like to grow organic vegetables and fruit trees, plants and flowers as well. Animals are important to us.

Luxembourg

Adhoc Adhoc habitat participatif s.c.

6 Rue Vauban | 2663 Luxembourg
+352-691 664 823
mail@ad-hoc.lu
ad-hoc.lu

2014 | 3700 m² | 5 % | 60-70

g, e, f, Luxembourgish, Polish | none | none

all of who are concerned ((systemic) consensus)

property of all - it is a housing cooperative and the future building will belong to all the members of the cooperative

none

the main activity of the cooperative is to provide affordable housing to its members - other fields of activities regarding the commercial areas are to be defined

ecological, self-managed, non-violent, anthroposophic, seminar-house, environmental protection, organic gardening, alternative energy, ecological construction, construction, humour, accessible to the disabled

☺

Adhoc is about creating an alternative to the existing housing models in Luxembourg. The main focus lies on collective living/co-housing. The future inhabitants will participate in the process of co-creating the shared common areas as well as the private living spaces. They also share tools and can engage in (neighbourly) community activities.

The project supports the creation of social bonds, and the fostering of exchanges and solidarity between the inhabitants and within the community. Another goal of the project is to create affordable housing through an optimal usage of the available living space, alternative financing models and self-management. In this sense, the project is being run as a non-profit-organization. It furthermore provides social, but also economic, practical and environmental benefits.
The legal form of the project is a (housing) cooperative. A cooperative allows all the members to be equal in terms of voting rights and of participation in the decision-making process. In terms of ownership, the inhabitants are at the same time co-owners of the residential building they live in and tenants of their private space. The collective is currently working on the implementation of its first housing project to be built on Kirchberg (Luxembourg-city).

☺ both financial and ideological - the focus on the ideological part being more important

The Netherlands

Allemansland

Berkenstraat 55 | 5753 GJ Deurne ☎+31-(0)6-228 858 88
welkom@allemans.land allemans.land

2 ha ♀15-20 ♂10-15 20-25 45-60

- Dutch, e
- all people who live in Allemansland (sociocratic (consent-method))
- individual financing
- each one buys its own house, there is a shared community building
- no
- unknown yet
- ecological, self-managed, self-sufficiency, permaculture, organic agriculture, organic gardening, animal friendly husbandry, new technologies, sociocratic, intrinsic motivation

☺

Less house, more living. Living in harmony with nature. A place to enjoy life. Your own idyllic Swedish wooden home, while also being together with others.
Decluttering. Making decisions together. A kitchen garden, a herb garden, a petting zoo, a natural playground for children, people of all ages.
Allemansland is a sociocratic, ecological community in the southeast of the Netherlands. About 20 idyllic Swedish wooden homes, completed with a community home. Allemansland wants to inspire other people and organises all kinds of events. It also supports the possibility of exchange with other projects and wants to provide the opportunity to temporarily experience living in Allemansland.

☺ agree with our vision and be willing to carry it out

Allemansland
samen ecodorp sociocratie

Anastasiadorp

Geraniumstraat 34 | 9581 BP Musselkanaal
+31-(0)15-257 69 97, +31-(0)6-832 832 78 (mobile)
anastasiadorp@gmail.com anastasiadorp.nu

2006 (near Emmen) 120 %

- Dutch, e, g
- all who are concerned (attunement) — shared economy
- the villagers own the initiative themselves
- To create a space of love, as Anastasia teach us about in the books.
- creating kin domains
- ecological, self-managed, non-violent, inner growth, vegetarian diet, free school, one-world development, animal protection, environmental protection, self-sufficiency, permaculture, organic agriculture, organic gardening, forestry, alternative energy, ecological construction, humour, non-smoking, living fence, food forest, water management, space of love, Anastasia

☺

Anno 2004, the Dutch translation of the book Anastasia by Vladimir Megré appears. The content inspires many readers to realize this vision and meet in real life. This included the realization of the kin-domain villages mentioned in the books. In the following years, this idea, supported by a group of young and older people, was taken up by the local authorities in the northeastern Netherlands. Since 2011, the municipality Borger Odoorn basically facilitates the arrival of a new village according to the ideas of Anastasia.
Some of the initiators are now settling north to make all the necessary preparations for founding an Anastasia village. At a distance of ten kilometers from the possible region where the village will emerge, the core group meets in Musselkanaal. The ideology of Anastasia is shared here in theory and practice with the local population and the aspiring residents of the family plots of the village. "Our neighbors are renovating their residential area, and we, as a core group, are invited and involved in reinterpreting public spaces in a natural way."

Editor's note: Please note the article about the Anastasia movement on pages 49ff.

- the ringing cedars international network
- ☺ We require from people to read the book series and let them experience for themselves if it resonates within them.

Coöperatieve Vereniging Ecodorpen Gelderland

info@cveg.nl www.cveg.nl

2015 (near geplant: Nijmegen) ♀30 ♂20 10 60

- Dutch, e
- the general circle (consent according sociocracy)
- limited common financing — rented
- none
- develop two or more ecovillages
- ecological, non-violent, ecumenical, vegetarian diet, adult education, pedagogical, positive health, community care

☺

The Ecovillage association plans to build two-three villages of about 50 living units each. The association has three pillars: care, ecology and community, with awareness in and for all four standard dimensions of European ecovillages. We will be in close relation with the people living around our ecovillage, sharing what we have in space, opportunities and vision. The village will be open for neighbours of all ages and state of health.
Our pillars are:
CARE: we have special focus on neighbours with need for additional care. We have a positive perspective on health: what can you do instead of how ill are you.
ECOLOGY: in our buildings and environment we focus on closing material cycles, using permaculture principles in design of our environment and sharing the ecological principles with builders and financers.
COMMUNITY: working together on a thriving community spirit with many common activities and meals, shared property and equipment, and focus on the needs of the neighbours. The ecovillage vision is also shared with the people living around and in the municipality. We will open our shared spaces with them and in this way promote our values.

→See also description of network
"GEN Nederland/Ecodorpennetwerk/Vereniging Ecodorp Nederland"
☺ agree with the vision and the mission with 3 pillars

Ecodorp Bergen

Groeneweg 14 | 1862 XL Bergen (NH) ☎+31-(0)654 763 157
info@ecodorpbergen.nl www.ecodorpbergen.nl

2013 15 ha 5 % ♀4 ♂5 2 11
Dutch, g, e 4
group (sociocracy) mixed
ownership
ideological, ecological
environmental remediation, organisation
ecological, self-managed, non-violent, anthroposophic, inner growth, meditation, tantra, organic diet, adult education, free school, environmental protection, self-sufficiency, permaculture, organic agriculture, organic gardening, organic farming, alternative energy, new technologies, ecological construction, humour, crafts, healing polluted soil, festivals

☺ ()

We started as a group in 2010 and bought a wonderful plot in May 2013. We have a strong sociocratic organisation and a clear vision. In 2019 our final ground plan shall be designed together with the city council.
We are growing to more practical work and we work with the help of our friends. Since the existing building must be removed and there is ground pollution, we are developing new ways of cleaning. Also we are creating new ways of financing. Being the first, we want to be an example ecovillage in the Netherlands.

→See also description of network
"GEN Nederland/Ecodorpennetwerk/Vereniging Ecodorp Nederland"
Ecodorp Steunstichting
☺ approval of our vision, financial

Ecodorp Noordeland

Oosterduinen D68 | 9331 WE Norg
☎+34-614 915 254
noordeland@gmail.com
ecodorpnoordeland.nl

9.5 ha (near Emmen) ♀6 ♂3 0

Dutch, e, g
all members (9) (sociocratic)
individual financing
property of all
spiritual
cleaning and building
ecological, self-managed, spiritual, inner growth, meditation, organic diet, vegetarian diet, self-sufficiency, permaculture, organic agriculture, organic gardening, alternative energy, new technologies, humour, non-smoking, permaculture, food forest, crafts, social workshop creativity

☺ ()

Our Vision:
We are dreaming of a community where we live in harmony with each other and with the earth. A community that is self-sufficient and strives for a healthy, natural, non-polluting and sustainable lifestyle. A community that gives space to growth of consciousness and personal development, healing and transformation. A community that is aware of the greater whole in which everything is inextricably linked. A place where we enjoy each other's qualities in a loving connection.

- ecologically sustainable
- socially sustainable
- economically sustainable
- spiritually sustainable

☺ ideological and financial

Ecovillage Boekel

Klein Rondeel 103 | 5427 GB Boekel
info@ecodorpboekel.nl www.ecovillageboekel.nl

2016 50 % ♀9 ♂8 9 26

Dutch, e 4

all (consent) common financing

property of the cooperative, we will only build houses for rent

education, living lab for circular economy innovations, sustainable building

ecological, self-managed, inner growth, adult education, environmental protection, self-sufficiency, permaculture, organic agriculture, organic gardening, organic farming, animal friendly husbandry, alternative energy, new technologies, ecological construction, humour, inspiring

Ecovillage Boekel is an inspiring example of sustainable living. In Ecovillage Boekel we will cover our needs in a very sustainable way. By doing so we will contribute to solutions for climate change, environmental pollution, loss of biodiversity and exhaustion of natural resources in close cooperation with our surroundings.
Our goal is a (mostly) self supporting village when it comes to energy, water and a big part of our food. It will be financially supported by the sustainable enterprises that we are projecting. Everything will be done in accordance to our ambitious vision and mission and will be completely fossil free.
Ecovillage Boekel will act as a Living Lab, a real-life test and experimentation environment for a sustainable future. Ecovillagers will be users and producers and will co-create innovation and foster user-driven open innovation.

GEN NL, VrijCoop

stand behind our vision and cultural principals (see website)

Ons Groene Huis Woongemeenschap in Haarlem i.o.

Heussensstraat 108 | 2023 JT Haarlem +31-(0)235 402 019
info@onsgroenehuis.net onsgroenehuis.net

2010 14

Dutch, g, e, f

all, by supported decision-making ('gedragen besluitvorming')

individual financing property of all

sustainability

ecological, self-managed, non-violent, inner growth, environmental protection, organic gardening, alternative energy, new technologies, ecological construction, no cell phones/wireless, non-smoking

Ons Groene Huis ("Our Green House") wants to realize a sustainably built or renovated cohousing project in Haarlem or its direct surroundings. The future inhabitants consciously choose for sustainability and commitment to each other and the neighbourhood.
The project is based on three main values: sustainability, commitment and self-management.

Norway

Bergen Ecovillage

Pinnelien 9 | 5053 Bergen ☎+47-555 908 10
info@bergenokologiskelandsby.no bergenokologiskelandsby.no

2014 3.9 ha (near Bergen) 10 % ♀4 ♂2 0 6

e, g, Norwegian, Danish, Dutch and Icelandic 0

a board and a project group (consensus) individual financing

permaculture - and in 2020 an agricultural cooperative

permaculture, agriculture, green courses - and in the future: cultural activities, yoga and meditation

ecological, self-managed, non-violent, inner growth, meditation, yoga, organic diet, adult education, pedagogical, environmental protection, self-sufficiency, permaculture, organic agriculture, organic gardening, organic farming, alternative energy, ecological construction, accessible to the disabled, self-sufficiency, fermentation, Bokashi

☺ ()

Bergen Ecovillage is building an ecovillage near Bruvik on Osterøy in western Norway, close to Bergen. Our goal is to build 39 houses and a community house with shared facilities. The houses are built in small clusters, and there will be a variety of small family houses, terrace houses and apartments. Some might also be tiny houses. Our agricultural and ideological foundation is permaculture. Our aim is to become partly self-sufficient with food, by planting and growing a large number of perennials. We wish to create local jobs and an agricultural cooperation. Please contact us to get more information. See photos and information here:
www.instagram.com/bergenokologiskelandsby/
www.opaform.no/bergen-okologiske-landsby/

→See also description of network "**Norsk Økosamfunnsforening/GEN Norge**"

☺ membership, financial obligations and shared ideological foundation - as stated in vision, goals and articles of association

Portugal

Dragon's Place Quinta do Tapado

Fiais da Beira, Ervedal - Oliveira do Hospital | 3405-077 Fiais da Beira
+351-966-728 293
scorpio.irondragon@gmail.com dragonsplace.webs.com

2003 (near Viseu, Coimbra) 60 % 2

Dutch, e, (Portuguese) all

depends on the responsibility (natural hierarchy and responsibility)

shared economy property (10 ha) of Dirk and Christine

Celtic

farming, technical assistance, sustainable energy

ecological, self-managed, anarchistic, hierarchical, spiritual, tantra, sexuality, environmental protection, self-sufficiency, permaculture, organic agriculture, organic gardening, organic farming, forestry, alternative energy, new technologies, ecological construction, humour, non-smoking, music

Dragon's Place is an old Celtic place in a beautiful natural area in mid Portugal - where everybody is welcome, especially all those who wish to learn about the new Celtic view, and put into practice the harmonious integration of landscape and people providing their own food, energy, shelter, and other material and non-material needs in a sustainable way.
On our quinta, we grow our own food together with local traditional and some biological farmers in order to learn from each other. There are fruit trees, a terraced vineyard for making our own wine and large areas of woodland with goats who help to keep it clean. When we arrived in 2003 there was only a wild area, a jungle. For over 30 years no one had done anything. There are several sources for water, and ingenious distribution networks deliver the water to where it is required. On some terraces we cultivate vegetables and we demonstrate special water saving irrigation systems. Of the three stone buildings on the property, we have restored one already, and one of the others is planned. In the firestorm disaster of October 2017 we have lost a large stock of materials that we use for building.
I am Dirk, Tech freak but also a spiritual (Celtic) master (mentor). Christine is a former nurse, who practiced and studied alternative methods for healthcare. She makes progress in agriculture and cultivates herbs. We want to build up a community with 3 or 4 interns and lots of external helpers.

☺ 60 % compatible

Russian Federation

Community in Kinseluk

Lesnaya 7 | 662923 Imisskoye
boris.prem@gmx.de wwoofindependents.org/user/500285

2010 | 4 ha (near Kuragino) | 50 % | ♀2 ♂1 4 7

g, e, Russian — all, all
all who are concerned (consensus) — common purse
informally used by us
critical of the system, resilience
ecological, self-managed, non-violent, organic diet, self-sufficiency, organic gardening, organic farming, animal friendly husbandry, alternative energy, construction

We, a family with six children between the ages of twenty and seven, left Germany eight years ago. Since that time we live in the Siberian taiga, since four years on a little farm in a secluded location at the brook Kinseluk in a wide, west-opened valley, where in the summer a large herd of horses grazes and in the winter lumberjacks work. The next settlement on a paved road (no shop, but with a bus stop) is located about nine kilometers away on the river Kasyr. To provide us with food, we keep our milk cows and chickens, cultivate a large vegetable garden and bake bread. We built one of our two huts ourselves. We teach those children who are required to be taught at home. In all areas of our work we learn year after year.
During the last winters, our eldest son traveled to Germany for six months to earn some money. Here we had to supply the animals and in addition to the lessons in the school subjects we were dealing with languages (Vedic, English), literature (religious writings in the Vedic language, Homer, German literature of various eras ...), sports (long distance running, gymnastics, floor gymnastics, horizontal bar) and music (folk songs, instrumental music). In 2019 there will be work again in the barn and in the vegetable garden, which has to be fenced again. We'll have to work at the hay harvest for two cows and a calf and to prepare the supplies for the winter (firewood, cheese, butter, sauerkraut, cucumbers, tomatoes, berries, herbs ...).
Our construction project for 2019 is a hut in wood-sawdust-loam technique, which can be used as a bathhouse and summer kitchen on the ground floor and where two or three people can sleep on the first floor.And: We are thinking of moving to a location about 40 kilometers away. If this becomes necessary and possible already in 2019, paths and bridges have to be built first.

So there is a lot to do and a lot to learn. Can you imagine working for some time - or sharing your life with us?

☺ none

Staroe Kubaevo

Village Staroe Kubaevo, Sudogda district | Vladimir region

☎+7-(0)917-541-6888

maharishi542@gmail.com vk.com/staroe_kubaevo

near Wladimir (40 km), Moskau (300 km)

Russian, (e) ☺

Staroe Kubaevo ("Old Kubaevo") is a settlement project in an old abandoned village in the forest of the taiga. Only one original inhabitant has stayed there, but the lonely place is currently being revived by a settler and two small settler families and new houses are being built.
The construction is based on old, natural construction methods with clay, wood, stones, lime, cortwood or straw. The economy within the settlement is currently being supplemented by permaculture cultivation and beehives. Currently, all local people decide on the next steps for the further development of the village. In and around the village there is still plenty of land for new settlers.

Spain

Cultivemonos Cultivando

C/u 14 | 22438 Panillo (Huesca) ☎+34-690 738 921 / +44-(0)772 967 18 21

fabioalonsoortiz@gmail.com www.cultivemonos-cultivando.org.uk

2008 (near Barbastro) ♂1

e, sp

ecological, self-managed, non-violent, inner growth, permaculture, organic gardening, animal friendly husbandry, ecological construction, humour

☺ ()

The name of the project means: "We want to cultivate our being by cultivating the land". We act consciously to develop more noble attitudes and qualities in this way. The constant improvement of our own character is our primary motivation for gardening together. We intend to establish a foundation so that our land will no longer be private property and are looking for people with the necessary resources to build their own home (stone or wood). There is enough space to build houses at an altitude of 750 m, south of the Pyrenees of Huesca. On our homepage you can find more information in English language as well as many pictures. (Editor's note: In March 2019 neither one nor the other could be found on the website.)

Ecovila Amat

Sant Privat den Bas | 17178 Catalunya ☎+34-9-691 136 169

ecovilamat@riseup.net ecovilamat.org

2015 70 ha (near Olot) ♂1 1

e, f, sp, Portuguese, Catalan 1

1 by now (assembly of the house cooperativists)
individual financing to start, soon collective financing and working
property of one owner, willing to share it with the new partners of the housing coop
eco-libertarian, freedom for small nations without state
agro forestry, accommodation, seminars
ecological, self-managed, anarchistic, left-wing political understanding, feminist, non-violent, ecumenical, vegetarian diet, adult education, pedagogical, seminar-house, peace-work, animal protection, environmental protection, self-sufficiency, permaculture, organic agriculture, organic gardening, organic farming, forestry, alternative energy, new technologies, ecological construction, humour, woodwork, adventure sports

☺ (✋✋)

The project is just beginning. It is now a private property of an activist after 25 years of visiting and creating communities and libertarian networks of social economy. It will become a community when 4 more families become partners of the housing cooperative, which implies sharing the cost of the land and the reconstruction of the three existing buildings.

Fair Coop, Catalan Integral Cooperative, Bank of the Commons
☺ ideological, financial and good attitude

La Rosina

C/ Camino Arroyos 3 | 10500 Valencia de Alcantara ☎+34-640 805 065
larosinadelashuertas@gmail.com www.larosina.jimdo.com

2002 (near Portalegre) 50 % ♀1 ♂1 2 4
Portuguese, sp, e, f, i ♻2 2
all who are concerned (consensus process)
limited common financing, shared economy rented, property of all
ecumenical and matriarcal
arts
ecological, self-managed, anarchistic, feminist, non-violent, spiritual, ecumenical, anthroposophic, inner growth, meditation, yoga, tantra, organic diet, vegetarian diet, macrobiotic diet, adult education, pedagogical, seminar-house, environmental protection, self-sufficiency, permaculture, organic agriculture, organic gardening, new technologies, ecological construction, construction, humour, accessible to the disabled, non-smoking

☺ ✋✋

The Rosina is named after grandmother Rosa, mother of José, born and raised on the farm and symbol of female resilience. Our family is currently a co-housing project under construction, we organize intensive courses related to permaculture, art, personal development and ecological housing.
Our farm is located near the Sierra de S. Pedro and S. Mamede between Extremadura and Alentejo. There are many signs of a forgotten and ignored culture that reveres the sanctity of the feminine. It is ideal to rest, recuperate, learn to be good about oneself and the sur-

roundings. Art, holistic healing, food, gardening, communication with nature, low impact building. In a word: Permaculture as an experience.

☺ ideological

Q Center of Experimentation

Plaça Mayor, 1 | 44556 Molinos ☎+34-634 320 010
info@qcentro.org www.qcentro.org/eng/index.html

2011 8 ha (near Alcañiz) 10 % 2

- sp, f, g, e, Catalan, Sanskrit
- shared economy respecting the individual economy
- owned by the two founding members of the project
- Non dual spiritual traditions, ecological and degrowth culture
- garden, firewood, maintenance, office, organization of spiritual retreats and seminars
- ecological, self-managed, non-violent, spiritual, ecumenical, inner growth, meditation, yoga, tantra, organic diet, vegetarian diet, seminar-house, peace-work, environmental protection, self-sufficiency, permaculture, organic agriculture, organic gardening, alternative energy, non-smoking, degrowth culture, ascetic life

☺

Q is a retreat space surrounded by nature, where we offer the possibility of experiencing an austere, simple and conscious life, in harmony with nature. Our modus vivendi integrates the ecological culture, the degrowth culture, and the ascetic culture, being an example of its symbiosis. Q is based on 3 main daily practical principles:
1) Respect to Resources: Use conscientiously the natural and industrial resources.
2) Respect to Space: Be responsible for your actions: clean up what you made dirty, pick up what you threw away.
3) Respect to Solar Time: The daily life is governed according to the cycles of sunlight and temperature.

Q plans to be carried by a small community of 6 like-minded people who are able to develop an agile and effective organization without leader. This requires in each person full capacity for self-determination, initiative and performance; a lot of awareness, responsibility and transparency.
This is why we are looking for people between 42 and 63 years old with a clear and firm decision to retire in nature; with serious experience in internal work, with full consciousness of the ego; who enjoy austerity and asceticism; with polyvalency to work in the different tasks of the farm, with a communitarian spirit (while recognizing the importance of individuality); and with economic solvency or a minimum of savings.
Time will tell if the propitious people appear to direct and develop Q in its splendor!

→See also description of network **"GEN Europe"**

☺ ideological and financial

Serene Light Gardens

Carretera General de Bajamar 100A | 38250 Bajamar ☎+34-654 579 155
escualconcienciacreativa@gmail.com www.serenelightgardens.org

2016 1 ha 20 % becoming more ♀1 ♂1 0 2

e, f, sp, g, Portuguese 10

all in weekly meetings (consensus) limited common financing

all

yoga, meditation, devotional singing

farming, ecobuilding, receiving volunteers and guests

ecological, self-managed, non-violent, spiritual, meditation, yoga, tantra, organic diet, vegetarian diet, vegan diet, animal protection, environmental protection, self-sufficiency, permaculture, organic agriculture, organic gardening, organic farming, forestry, alternative energy, new technologies, ecological construction, construction, humour, non-smoking

☺ ( )

We are a little non-profit organization called "La Escuela de la Conciencia Creativa" ("The School of Creative Consciousness") registered in Spain, and we have been tirelessly dedicating our lives to nourish an eco-village project for the past 10 years. After a long process of learning and an experiment in a rented farm, we finally found a land in Tenerife (Spring 2016) that suits perfectly the creation of a therapeutic and educational eco-center which will serve as a model and living reference for an integrative and symbiotic way of life with our natural environment, as well as a space of research where we can teach, learn and share with each other all about sustainable living.

Even though the land is small for a proper eco-village, we aim to inspire from there either the creation of eco-villages and diverse eco-projects, as the advancement of society as a whole towards a more harmonious, conscious and responsible way of living.

☺

vegetarian, drugs and smoke free.
Financially strong.

Sweden

Vedapark A Scandinavian Ayurveda Village

Korsorvagen 23 B | 21747 Malmö ☎+45-(0)212 305 83
trigunnar@gmail.com www.vedapark.com

2018 40 ha (near Lund) 75 %

e, Scandinavian languages 0 2 + volunteers on and off

those who are involved with the issue in question (consensus)
individual financing and limited common financing
individual or collective property
the Veda - regarded as a Science of Consciousness
ecological, non-violent, spiritual, inner growth, meditation, yoga, organic diet, vegetarian diet, adult education, pedagogical, seminar-house, peace-work, animal protection, environmental protection, self-sufficiency, permaculture, organic agriculture, organic gardening, animal friendly husbandry, forestry, alternative energy, new technologies, ecological construction, humour, non-smoking, Veda, Vedic Science, Vastu houses, Ayurveda, retreat facilities

☺

We will welcome an international social environment in Vedapark, a holistic village with an interest in Ayurveda and Vastu architecture.
It is for all open minded people, including artists, business, intellectuals and retired, but also for children and students - in short, for all peace-loving individuals who want a balanced life in nurturing contact with Nature and themselves.
A key factor of this project is to create healthy homes that will improve our life, well-being and outer success.

→See also description of network "**GEN Europe**"

Villa Billebro

Höglidsv. 28A | 13550 Tyresö ☎+46-(0)8-7127850
kent@billebro.se www.billebro.se

1995 1000 m² (near Ljusdal) ♀2 ♂3 1 6

e, Swedish 1-2
varies (majority vote) individual financing
one member
love, joy, spirituality, transformation, meditation, dance, fun and games
art, dance, meditation
ecological, self-managed, non-violent, spiritual, Buddhist, Taoist, anthroposophic, inner growth, meditation, yoga, tantra, sexuality, free love, organic diet, vegetarian diet, macrobiotic diet, seminar-house, permaculture, forestry, alternative energy, new technologies, ecological construction, humour, accessible to the disabled, non-smoking, voice dialogue, sports, art

☺

Very much is changing now.
Big empasis on body soul health, healing, expansion, access to consciousness, love, joy, your own inner greatness and sexuality, orgasmic living & loving, creating a lot also in physical world and money also.
We have a lot of property now in 4 places around Sweden and a lot of machinery, cars, vehicles, bicycles and boats.
Kent Billebro, founder and owner, also has a company which sells and rents out sound and light equipment and works as a disc jockey. Kent likes to dance and sometimes we stage therapeutic dance parties in our house.

Switzerland

Lebensraum Belmont

Oberdorfweg 4 | 3812 Wilderswil ☎ +41-(0)796 549 170
b.stoffel@bluewin.ch www.lebensraumbelmont.ch

2013 4000 m² (near Interlaken, Thun, Bern)

g

property of the cooperative

☺

In 2013, the cooperative LEBENSRAUM Belmont bought three parcels of land with a hotel in order to live in this beautiful place with the imposing mountain scenery of Jungfrau, Mönch and Eiger, once in a natural coexistence of young and old. Since then, the cooperative has been in the process of being established.
There are no communal projects yet - most of the members of the cooperative do not live on site (yet!). In 2019, we are not connected by specific spiritual or ecological literatures, but by curiosity about what will happen when everyone unfolds their uniqueness in the greatest possible freedom and at the same time feels connected to one another and to a bigger whole.
In 2019, two families with a small child live in the newly built 5-family house and two members of the cooperative each live in the other three apartments. A large permaculture garden is growing. The conversion of the former hotel Belmont into residential units is planned for 2019-2020. In the final stage, 30-40 people will live in the Belmont habitat. At the beginning of 2019, we are 14 adult members of the cooperative and two toddlers; 6-8 new members will join in 2019.

SoWo - Solidary Living

Unterdorfstr. 2 | 3322 Urtenen-Schönbühl
info@so-wo.ch unserhausprojekt.ch, so-wo.ch

2018 4000 m² (near Bern) 10 % ♀8 ♂7 4 19

g, f, e, sp, i, Hungarian

all inhabitants (consensus) partially shared economy

collective property

no

construction, gardening, childcare

ecological, self-managed, anarchistic, left-wing political understanding, feminist, non-violent, organic diet, vegetarian diet, vegan diet, adult education, pedagogical, environmental protection, permaculture, organic agriculture, organic gardening, alternative energy, new technologies, ecological construction, construction, humour, intergenerational

☺

We are planning our housing project and we started the reconstruction of an old farm in spring 2019. We are very happy to host guests who are willing to help us with the construction work. We are still looking for people who want to join our project in the long term. We will create space for about 23 adults and 10 kids. For more information please feel free to contact us via e-mail!

United Kingdom

Coastal Community

B5

Portsmouth, Hampshire UK | Hampshire
+44-(0)778 711 95 66
info@coastalcommunity.org.uk
www.coastalcommunity.org.uk

2016 (near Exeter, Plymouth)

e

the collective (consensus)

individual financing

planned: shared ownership

vegan, eco-friendly, organic

sustainable living, food growing, education centre

ecological, self-managed, non-violent, inner growth, meditation, yoga, organic diet, vegan diet, adult education, peace-work, one-world development, animal protection, environmental protection, self-sufficiency, permaculture, organic agriculture, organic gardening, organic farming, forestry, alternative energy, new technologies, ecological construction, healthy living

☺

We are a small group of vegans looking to form a cohousing eco community in South Devon, England. We are currently chatting about possibilities with an eye on the property market. Ideally we are looking for a small farm with outbuildings that could be converted into separate residential units. We are looking for one or two more people or families to invest in 1-3 bedroom units. Based on permaculture and cohousing principles, we would like to forge a healthier way of life: grow most of our own food and fuel; generate our own energy; share skills; provide a place for courses, activities, holidays, retreats; and create an abundant biodiverse forest garden to share. For more details please visit our website or contact us!

Cohousing Hafan Las

B4

Pant y Maen, Llangolman, Clunderwen | Pembrokeshire SA66 7Xj
+44-(0)1437-532 575
info@cohousinghafanlas.org.uk
www.cohousinghafanlas.org.uk

e, Welsh

the members (consensus decision making)

individual financing

none

ecological, self-managed, left-wing political understanding, feminist, non-violent, vegetarian diet, permaculture, organic gardening, accessible to the disabled, non violent communication

☺

We are aiming to develop an eco-cohousing community in West Wales, consisting of low-impact, affordable homes, a good mix of people at different stages of life, and with all ages having an active involvement in the creation and management of the community; to have at least 1/3 of people over 50, helping older people to live actively into old age.

We want to create a culture where attention is given to emotional, physical, psychological and spiritual well-being, as well as care for the earth, we will be implementing an open and transparent communication system, using the Nonviolent Communication (NVC) model, and consensus decision-making methods. All members will undertake training in both areas. We want to adopt sustainable lifestyles, i.e. recycling and sharing resources, and looking at positive ways of dealing with waste, both human and commercial, also a legal structure that allows for purchase or rental of dwellings.
There will be 20 or more individual household units as well as a large, shared common house. Each person or family will have a simple, private living space, with one to three bedrooms, and share a large common house. Members will be involved in the design and construction, with the aim of low-impact buildings that will enable ecological sustainability in every aspect of daily life.

- UK Cohousing Network, Permaculture Wales
- ☺ agreement with our vision, mission and aims; financial, legal and group-related determinations

Cohousing Harrow

D5

26 Dukes Avenue | Harrow HA1 1XW ☎ +44-(0)790 613 73 03
cew.acc.ukcn@woodbridge.f9.co.uk

(near London)

- e, f, (sp)
- all of us (consensus)
- limited common financing
- ecological, self-managed, sustainable, intergenerational, inclusive

☺

Cohousing Harrow is still at the stage of forming a big enough group of local people, who will then actively work to develop cohousing projects in the Harrow area. We are currently putting on information events and gradually being joined by interested people. Our intention is to build developments of between 20 and 30 dwellings, which will be affordable in perpetuity. Our main values are Community, Sustainability and Inclusivity. We are part of Cohousing London, which is a London-wide group coordinating Cohousing developments across London.

- Cohousing London and UK Cohousing Network

Hope House Ko-Thriving

C5

Kompany of Intentional Lifestyle Designers | Bristol
asha100hope@gmail.com https://autopia.co/a/hopehouse/apply
https://www.facebook.com/groups/hopehousekothriving

2019

- cultivate the joy of peak physical, mental, emotional, spiritual, social and financial health and vitality, and embark on and/or progress and fulfill our defined personal mission and ambition
- inner growth, meditation, organic diet, new technologies, nutrition, ketogenic diet, fitness, self-development, health, automation, multi-generational, hacking, adopting change, lifestyle redesign

☺ (✋✋)

As "Hope House Ko-Thriving", we are good kompany living together to inspire and support one another by co-creatively designing & practicing intentional lifestyles that are informed by cutting-edge understandings of our HOSs*.
We do this specifically to work on and achieve the following for ourselves personally and for each other: (a) cultivate the joy of peak physical, mental, emotional, spiritual, social and financial health and vitality, and
(b) embark on and/or progress the adventure to fulfill whatever constitutes our personal, deepest or wildest mission and ambition.
Join us if you:

- have defined your personal health, mission and ambition goals and are ready to design a lifestyle that will enable you to achieve them
- wish to live together with inspiring, motivated people and will inspire others in turn
- value and will contribute to living together in a group and enjoying its benefits: financially, personally & socially
- will work positively with the challenges of co-living together, appreciating this as an opportunity for learning
- will offer your influence to shape our co-creative journey
- are comfortable with organisation, structures, procedures and values as invaluable to create lasting change in our lifestyle and to create functioning ko-thriving
- will use digital communication & automation technologies designed specifically to enable lifestyle change and functioning ko-thriving

* 'Human Operating System': the term we have invented for an analytical conceptualisation of how our bodies and minds function in a measurable way that we can hack (i.e., influence, change, adapt for our purposes)

☺ application process

✎ KindaVillage

B4

Glanrhyd y Pysgod | Pencader SA399LX ☎+44-(0)1559-395 253
admin@kindavillage.org kindavillage.org

2017 32 ha (near Carmarthen, Swansea) 3

e 6

limited common financing — company limited by guarantee

kindness

social impact, organic gardening, retreats and wellbeing

self-managed, non-violent, inner growth, vegetarian diet, vegan diet, peace-work, animal protection, organic agriculture, organic gardening, organic farming, alternative energy, ecological construction, construction, humour, non-smoking, social work, crafts, therapy, arts, media

☺ ✋✋

KindaVillage is a visionary project supporting the values of 'The Missing Kind' charity (tMK), a development of our first sustainable kindness community. With well-being and sufficiency at its core, KindaVillage is an innovative social and environmental impact centre supporting the local community, creating national initiatives and inspiring global change.
We are currently seeking focused, hard-working, passionate and dedicated pioneers, volun-

teers, and social entrepreneur partners to put this together; those in search of a kinder and more balanced world in which to live. We also seek sponsors and supporters with a kind and giving heart to help us realise our vision.
Our call goes out to those who have a social or environmental project, a passion and a commitment to kindness. Those who may be in need of resource and community to support the realisation of their project and those who would like to support a project that is creating a more positive and sustainable living and working model for us all.

The Missing Kind
volunteering first

Rubha Phoil

Armadale Pier Road | Isle of Skye, IV45 8RS
+44-(0)7393-830 403
info@earth-ways.co.uk
www.earth-ways.co.uk

2016 | 6 ha | 25 % | 8

- long term/permanent members (sociocracy)
- Food, accomodation, community expenses are paid via a communally run business.
- nature-based permaculture
- Running our tourist business, growing food and natural building.
- vegan diet, meditation, ecological , self-managed, inner growth, environmental protection, self-sufficiency, permaculture, Animal friendly husbandry, Forestry, ecological construction, alcohol/ drugs free, pet free, art of mentoring, 8 shields, regenerative

After many years of searching for a suitable location to start a small permaculture community, we are pleased to announce the birth of a new community in a paradise on the Isle of Skye: Rubha Phoil. Rubha Phoil is a nature-based permaculture community with a regenerative culture, where a small group of people of all ages can create and demonstrate an abundant and thriving existence on and from the land, while enhancing the ecosystem they live in. Rubha Phoil has ALL the potential to become a beautiful abundant nature-based permaculture community. There is a lot of development; check our website often:
https://www.earth-ways.co.uk/rubha/

Permaculture, Art of Mentoring, 8 Shields
successful application process

Czech Republic

Spolek Zemesouznení

+42-(0)732 426 226
info@zemesouzneni.cz http://zemesouzneni.cz

2017 20

- sociocratic approach
- cooperative, associations
- we are inspiration for a conscious life in harmony with nature
- permaculture gardening, natural buildings, alternative energies, traditional crafts, alternative currency, homeschooling, courses of personal development and spirituality, cooperative entrepreneurship, solidarity economy, closed regional production-consumption circle of basic food, water, energy, buildings

☺

We plan to build the first full-featured ecovillage in Czechia with integrated social, ecological and economic sustainable holistic solutions. Our goal is to organically grow up to 100-150 tribal members till 5 years. We don't have land yet. We would like to gain min. 24 ha of land ideally in the Central Bohemia region about 1 hour from Prague with water and building permission. Our Vision: "We develop a regenerative culture in Czechia, being conscious participants of the ecological transformation of Planet Earth, collaborating to live in harmony with nature and creative energy." Our Goals:

- Zemesouzneni Community: A growing number of people in Czechia discover, learn and co-create ways for sustainable, regenerative and holistic living inspired by real ecovillage projects and business.
- Zemesouzneni Ecovillage: 150 people form a tribal community and have a place to live and work sustainably, following a holistic lifestyle.
- Holistic Business: Integrated production-consumption systems for bio-products and services with alternative local currency in the region.

☺ holistic worldview, shared vision and values, commitment to contribute based on capacity and possibilities, agreement to code of conduct and Big Circle rules, basic compassionate communication skills

Hungary

Biofalu Máriahalom

Pataksor 12 | 2527 Máriahalom +36-(0)630-942-8670
geza@biofalu.hu www.biofalu.hu

50 ha (near Piliscsaba) ♀1 ♂2 3

Hungarian, e 1 2

the founder

property of one member

retreat center, natural building experiments

ecological, spiritual, inner growth, adult education, seminar-house, environmental protection, permaculture, alternative energy, construction, small, simple, natural/re-cycled material houses, stove experiments

☺ ( )

I (Géza) started the project with the intention to build/experiment a new way of community and simple living close to Budapest. I'm still on this path, but people join only temporarily (days, weeks) as volunteers... We have a vegetarian retreat center with 70 beds and four common rooms, which is managed by volunteers. We help people to get first hand experiences about simple living techniques and conscious living. The basic values are: freedom, healing, self awareness, creativity.
We have 50 ha of land on which up to 70 houses may be built, but I would like to start this process only when at least five families arrive with similar intentions: Families who would like to be each other's neigbours and make small steps together on creating something new. In the meantime, trees are growing on the land (40.000).
We welcome volunteers in the retreat center and on small construction projects. We are on Facebook (in Hungarian): Biofalu Máriahalom - with pictures!

→See also description of network "**GEN Europe**"

GEN Hungary

Turkey

Bayramiç Yeniköy Yeniköy Çiftligi

Yenimahalle/Muratlar Köyü | 17720 Çanakkale/Bayramiç

+90-(0)555 898 05 23

mustafaalperulgen@hotmail.com www.bayramicyenikoy.com

2009 40 ha (near Çanakkale) 60 % ♀2 ♂2 4

e 4 4

all who are concerned (consensus) individual financing

property of a community

production of traditional wheat in ecological ways

ecological, left-wing political understanding, non-violent, organic diet, seminar-house, peace-work, self-sufficiency, organic agriculture, organic gardening, organic farming, animal friendly husbandry, ecological construction

☺

The Bayramiçyeniköy group tries to build a self-sufficient village by applying the philosophy of permaculture, using local seeds, applying natural agricultural practices and passing on what we have learned and produced.

with food communities

Addresses from the 2014 edition of the eurotopia Communities Directory

...from communities that are not listed with complete details and self-description in this edition, but were listed like that in the 2014 book (back copies available). We contacted them for this edition but did not get confirmation or new infos.
The following data is surely a valuable supplement to the entries on the previous pages. For the following reasons, however, we ask you to handle it with care:

- Some communities might not have responded on purpose. We do not list addresses of communities that have responded and asked us to not publish them anymore, of course - but still, some of the following groups might not want to have visitors or inquiries.
- It's possible that some of the addresses in the following list have not even been contacted by us because we have decided that they do not match our community definitions.

ZAP:
This symbol stands for a ceased community. We use this rather light-hearted symbol to remind us to look on the bright side: Not all dreams come true. That doen't mean that we should stop dreaming. There's plenty of examples in this book to justify hope for people being able to live together in peace. However: Some addresses marked as 'ceased' might be existing after all. Or again. That does not have to mean they're zombie communities, more likely it was one of those mistakes that happen when humans interact with databases...

Austria

Arts & Living ZAP

Friedrichshof Wohnungsgenossenschaft ZAP

Lebens- und Innovationsraum Schönwasser ZAP

Liv - Leben individuell, verbunden ZAP

Naturhof Pramtal - Nahopra

Nachhaltig leben lernen in Gemeinschaft | Reischenbach 8
4755 Zell an der Pram | ☎+43-(0)680-2377949 www.nahopra.at

Wohnprojekt Wien

Verein für nachhaltiges Leben | Krakauer Straße 19 | 1020 Wien
presse@wp-wien.at, office@wp-wien.at www.wohnprojekt-wien.at

Belarus

Windsnest

☎+375-(0)297664683

Belgium

Clos Convivial Geder
Francis Busigny | Place Madelin 3a | 5020 Temploux | ☎+32-(0)496-586755

Cohousing Vinderhoute
Federico Bisschop | Bergstraat 14 | 9921 Vinderhoute | ☎+32-(0)9-3359199
✉info@cohousingvinderhoute.be 💻www.cohousingvinderhoute.be

De Regenboog
Norbert Gillelaan 20 | 1070 Bruxelles
☎+32-(0)2-5206586 ✉marc.vantente@skynet.be

Kasteel Nieuwenhoven
Sint-Truiden | ✉info@kasteelnieuwenhoven.be

Croatia

Recycled Estate
Reciklirano imanje | Vukomeric 23/4 | 10418 Dubranec ☎+385-(0)1-6267-223
☎+385-(0)95-843-3493 | ✉volontiraj@zmag.hr 💻www.zmag.hr

Cyprus

Buyukkonuk Eco-Village, North Cyprus
71 Ataturk Blv., Buyukkonuk, Iskele Kibris | 99870 MERSIN-10, Turkey
✉delcraft2000@hotmail.com 💻www.ecotourismcyprus.com

Denmark

Den Selvforsynende Landsby
Højgårdsvej 27 | 5762 Vester Skerninge
☎+45-62243232 ✉kontakt@selvforsyning.dk 💻www.selvforsyning.dk

Enggården
Tåningvej 44 | 8660 Skanderborg
☎+45-61473016 ✉om@omri.dk 💻www.facebook.com/enggaarden

Fri & Fro
Under Himlen 8 | 4500 Nykøbing Sjælland
☎+45-22929877 💻www.friogfro.dk

Rumlepotten
Pottemagertoften 184 | 8270 Højbjerg
☎+45-86298126 ✉rumling@gmail.com 💻www.rumlepotnet.dk

Svanholm
Svanholm Allé 2 | 4050 Skibby
eberhard@svanholm.dk, kirstenh@svanholm.dk www.svanholm.dk

Finland

Metsäkylä
Organic Village Gaija
Kaijantie 283 | 63700 Ähtäri | +358-(0)45-1139787 | www.gaija.fi

France

Communauté du Pain de Vie
c/o Bauchet Bruno | 16, Avenue E, Martelange | 84000 Avignon
+33-(0)490866799 smaf64@laposte.net

Eco Chateau
Chateau de Cautine | Saint-Privat | 19220 Correze
+33-(0)977217606, +44-(0)20-3287-9212
ruth.eco.chateau@gmail.com www.eco-chateau.com

Ecolieu - Ermitage Fondale
Fondale | 20270 Tallone
+33-(0)777065685, -(0)623743970, -(0)612557915 s-rot@web.de,
sylvainfondale@yahoo.fr www.ecovillagefondalecorsica.fr

Hameau de la Brousse
Michel Gayout et LN Le Chevillier | La Brousse | 16410 Sers | +33-(0)545249572
michelgayout@orange.fr www.hameaudelabrousse.com

La Goutte d'Eau
Le Camp du Geai
Le Pré Aux Yourtes
Ecoasis de la Pinsole | 07690 Vanosc
+33-(0)678346652 www.yourtes.supersite.fr

Les Amarantes
Les Jardins d'Espaladous
Le Crouzet | 43260 Saint Pierre Eynac
contact@espaladous.com www.espaladous.com

Oasis Pen Ann Hoat
Penthoatvian | 22110 Kergrist-Moëlou | +33-(0)625556695
oasispenanhoat@yahoo.fr www.agroecologie-bretagne.com

Plum Village

Village des Pruniers | Le Pey | 24240 Thénac
☎+33-(0)553584858 💻www.plumvillage.org, www.villagedespruniers.net

Germany

Allmende Wulfsdorf

Bornkampsweg 36 | 22926 Ahrensburg | ☎+49-(0)4102-458230
✉info@allmende-wulfsdorf.de 💻www.allmende-wulfsdorf.de

Am Lappach Wohnen

avatary-land

Wald bei Cosul | 02692 Cosul | ☎+49-(0)15151890602
✉solarusplexus@freenet.de, schenker_oeff@yahoo.de 💻www.global-love.eu

Beginengemeinschaft Malgarten

Kloster Malgarten 2b | 49565 Bramsche | 💻www.beginen-in-malgarten.de

Beginenhof Bielefeld

Bielefelder Beginenhöfe e.V. | Karl-Oldewurtel-Str. 31 | 33659 Bielefeld
💻www.bielefelderbeginenhoefe.de

Beginenhof Bochum

Beginen Heute e.V. Essen | Kornharpener Str. 176 | 44791 Bochum-Kornharpen

Beginenhof Essen

Beginenhof Essen e.V. | Goethestraße 63 - 65 | 45130 Essen
💻www.beginenhof-essen.de

Beginenhof Gelsenkirchen

Beginenhof Köln

Beginenhof Köln eG | Unter Linden 119 | 50859 Köln-Widdersdorf
✉leben@beginenhof-koeln.de 💻www.beginenhof-koeln.de

Beginenhof Meschede (geplant)

Meschede
✉inwie@gmx.de

Beginenhof Münster (geplant)

Beginenhof Münster e.V. |
✉beginen@muenster.de 💻www.dachverband-der-beginen.de/muenster.php

Beginenhof Unna

Förderverein Beginenkultur Unna e.V. | Märkische Str. 4 | 59423 Unna
✉beginenhof.unna@gmx.de

Beginenhöfe Schwerte
Förderverein Schwerter Beginenhöfe e.V. | Kiebitzweg 2 | 58239 Schwerte
beginenhoefe-schwerte@gmx.de
www.schwerte.de/schwerte-fuer-schwerter/frauen/beginenhoefe/

Beginenhöfe Tübingen
Torbogenweg 6 | 72070 Tübingen-Hagelloch | www.beginenstiftung.de

Begineninitiative an Ems und Wieken (geplant)

Bewegung und Wandel in Pankow-Rosenthal

Brückendorf

Christliche Gemeinde in See e.V.
See | 91230 Happurg
+49-(0)9157-9292600 kontakt@cg-see.de www.cg-see.de

Emmaus Gemeinschaft in Köln
Geestemünder Straße 42 | 50735 Köln
+49-(0)221-9711731 info@emmaus-koeln.de www.emmaus-koeln.de

Emmaus Sonsbeck
Alpener Str. 40 | 47665 Sonsbeck
+49-(0)2838-96446 info@emmaus-sonsbeck.de www.emmaus-sonsbeck.de

Forum Kreuzberg
Förderverein e.V. | Eisenbahnstr. 21 | 10997 Berlin | +49-(0)30-6127275
foerder@forumkreuzberg.de www.forumkreuzberg.de

Gemeinschaft am Gieberg
Am Gieberg 6 | 29490 Glienitz/Neu Darchau
+49-(0)5858-971767 giebergwerk@gmx.de

Gemeinschaft Grützdorf

Gemeinschaft in Fresdorf

Gemeinschaft MutterLand
z.Zt. vertreten durch RaKuna | Ottermühle 1 | 07368 Remptendorf/Liebschütz
+49-(0)36640-139944 info@kraft-der-kreise.de www.kraft-der-kreise.de,
buendnis-fuer-das-leben.de, frauen-kreise-weltweit.be2web.de

Gemeinschaftskeimling Leipzig

Grundstein Neukirchen e.V.

Hammer Kommune

Haus LichtlanD Oberbrunn

Haus 'Weg der Genesung und der Gesundheit'
Fam. Westphal-Born | Kahlweg 2 | 78355 Hohenfels
+49-(0)7557-9285894 info@licht-born.de www.licht-born.de

Haus Yoga Vidya Westerwald
Gut Hoffnungstal 1 | 57641 Oberlahr | ☎+49-(0)2685-8002-0
✉westerwald@yoga-vidya.de 💻www.westerwald.yoga-vidya.de

Heiloase-Südtürkei

Hofgemeinschaft Lübnitz
Dorfstr. 5 | 14806 Lübnitz
☎+49-(0)33841-34270 ✉An-Yue@web.de 💻www.hofgemeinschaft.net

Hofgemeinschaft Uranushof
Arbacher Str. 3 | 56769 Retterath
☎+49-(0)2657-633 ✉uranushof@web.de

Johannishöhe - Umweltbildungshaus
Natürlich leben und lernen e.V. | Dresdner Str.13a | 01737 Tharandt
☎+49-(0)35203-37181 ✉info@johannishoehe.de 💻www.johannishoehe.de

Kaiserbacher Mühle
Kaiserbacher Mühle | 76889 Klingenmünster
☎+49-(0)6349-928414 ✉info@stadt-und-natur.de, info@kaiserbacher-muehle.de
💻www.kaiserbacher-muehle.de

KlosterGut Schlehdorf
Kirchstraße 15 | 82444 Schlehdorf
☎+49-(0)8851-9291980 ✉kontakt@klostergut-schlehdorf.de

Kombinat Gatschow

Kommune im KuBiZ

Land.Leben.Kunst.Werk.e.V.
Geschwister-Scholl-Str. 25 | 06780 Quetzdölsdorf | ☎+49-(0)34604-92069
✉info@landlebenkunstwerk.de 💻www.landlebenkunstwerk.de

Lebensgemeinschaft "anders leben"

Lebensgemeinschaft am Höllbachhof

Lebensgemeinschaft Happurg

Lebensgut Pommritz
Pommritz Nr. 1 | 02627 Hochkirch
☎+49-(0)35939-81385, -81884 ✉lebensgut.pommritz@yahoo.de 💻www.lebensgut.de

LebensQuell Ammertal eG
... für ein Leben in Gemeinschaft. | Reschstraße 6 | 82418 Murnau
☎+49-(0)176-82182001 ✉info@LebensQuell-Ammertal.de 💻www.LebensQuell-Ammertal.de

Lebenswiese

Ökodorf Erlenweide

Projekt Gaia
Projekt Lebensdorf
ProWoKultA

Projekte für Wohnen, Kultur und Aktion e.V. | Am Ulmenrück 7-11 | 60433 Frankfurt
vorstand@prowokulta.de www.prowokulta.de

Raum für Gewahrsein und Entwicklung e.V.
Regenbogenstamm

Rathausstr. 8 | 56357 Reitzenhain | +49-(0)179-3158466
regenbogenstamm@yahoo.de www.regenbogenstamm.de

Rhöner Gemeinschafts-Projekte e. V.

Spielgasse 14 | 97654 Bastheim | +49-(0)179-72487922

Sandburg
Selbsthilfeprojekt Freundeskreis Lychen

c/o Roy Rempt | Fürstenberger Str. 3 | 17279 Lychen | +49-(0)39888-3076
roy.ly@web.de transition-initiativen.de/standorte/transition-town-lychen

Sonnental

Anke Schubert | Hennersbach 9 | 01825 Bad Gottleuba
+49-(0)35025-57598 anke_bluemli@gmx.de zweisonnen.de

Tenne-Steinfurth

Steinfurther Hauptstr. 18-20 | 61231 Bad Nauheim
+49-(0)6032-868719 www.gemeinschaft-tenne-steinfurth.weebly.com

Tollense Lebenspark

(worth reading, in German: #6018 at eurotopia.directory/booklinks)

Torhaus Trebitz

Hauptstr. 15 | 14822 Brück im Hohen Fläming
+49-(0)33844-51034 torhaus@riseup.net www.torhaus.blogsport.de

Waldhaus

Dorfstraße 35 | 17391 Krusenfelde | +49-(0)39723-27912, +49-(0)178 9202176
info@allmende-ev.de allmende-ev.de

Weda Elysia

Ein Gärtnerhof-Siedlungsprojekt für Familien-Stammes-Landsitze
Ginsterkopf 6 | 38889 Wienrode
+49-(0)3944-366388 info@weda-elysia.de www.weda-elysia.de

Wohnfühldorf Bayern

c/o Uschi Ausfelder | Albert-Schweitzer-Straße 10 | 84539 Ampfing
+49-(0)8636-695832 info@wohnfuehldorf.de www.wohnfuehldorf.de

Wohnprojekt Layenhof e.V.

Am Finther Wald 42 | 55126 Mainz | +49-(0)6131-6226723
vorstand@wohnprojekt-layenhof.de www.wohnprojekt-layenhof.de

Wohnprojekt Schellingstraße

Schellingstr. 6 | 72072 Tübingen | ☎+49-(0)7071-792136, +49-(0)7071-792136
info@schellingstrasse.de www.schellingstrasse.de

Zwölf Stämme Gemeinschaft in Klosterzimmern

Zwölf Stämme Gemeinschaft in Wörnitz

Greece

Kalikalos

Xenonas Martiou | Kissos (Tsagarada) | 37012 Magnesia
☎+30-(0)24260-32039 info@kalikalos.com www.kalikalos.com

Meltemi

Rafina | Nh@power.ece.ntua.gr

Pelion Holistic Education Centre

Spiti ton Kentavron | Spiti ton Kentavron | 37001 Anilio
☎+30-(0)24260-32176 swhic@hotmail.com

Hungary

Galgafarm - Ecovillage Initiative

Gaia Foundation | Fõ 337, út 83 | 2193 Galgahévíz
gaia@zpok.hu, gaboreszter@gaiaalapitvany.hu www.gaiaalapitvany.hu

Nagyszékelyi KÖRTE

Conscious Cooperation Towards Communal Self-Sufficiency | Nagyszékely
kortemag@googlegroups.com

Iceland

Sólheimar vistvænt byggðarhverfi - Sólheimar Eco village

Selfoss | gap@solheimar.is

Ireland

Tigroney House/Ballyman House

Israel

Kibbutz Tamuz

46 Hanasi St. | 99037 Beit Shemesh | info@tamuz.org.il www.tamuz.org.il

Italy

Comunita Mariana "Oasi della Pace"

Casa Generalizia | Strada Grotte di torri, 54 | 02032 Passo Corese
☎+39-0765-488993 ✉oasispacis_pc@yahoo.it

Consolida

Le Mogli Farm

Italy Ecovillage | Via Mogli 8, Abruzzi National Park | 03030 Pescosolido
☎+39-340-2517941 ✉italyecovillage@gmail.com 💻www.italyecovillage.com

Nomadelfia

C.P.103 | 58100 Grosseto (GR)
☎+39-0564-338243 ✉info@nomadelfia.it 💻www.nomadelfia.it

Panta Rei

CEA Panta Rei | Località le pierle | 06065 Passignano sul trasimeno (PG)
☎+39-075-8296164 ✉segreteriapantarei@gmail.com 💻www.pantarei-cea.it

Rainbow Eco Peace Village

Villaggio Upacchi

Località Upacchi | 52031 Anghiari (AR)
☎+39-0575-749323 ✉evalotz@yahoo.it 💻www.eccoupacchi.eu

Latvia

Bruknas manor house

☎+371-(0)29633291

Kristian David School

Barkavas pag. Madonas nov. LV 48344834 | ☎+371-(0)64828320, -(0)22059737
✉k.davida.skola@tl.lv 💻www.kdskola.lv/Grostona/Sveicinati.html

Lithuania

IzReaL.eu - It's Really you

Nosenai g. 9 | 33292 Nosenai kaimas | ✉info@izreal.eu 💻www.izreal.eu

Melkys ecovillage

☎+370-616-13553 ✉zina.gineitiene@gmail.com

Moldova

Schastlivoe
107/1 Tarinei Steet Apartment 7 | 2069 Chisinau
✉schastlivoe@gmail.com, aline_shnay@hotmail.com 💻schastlivoe.com

Netherlands

Academie Tokudu Ecodorp Bergen
Jaap Weijandweg 20 | 1862XL Bergen
✉mailfredjan@gmail.com 💻www.ecodorpbergen.nl

Casa De Pauw
Oude Velperweg 36 - 38 | 6824 HE Arnhem
✉info@casadepauw.nl 💻www.casadepauw.nl

De Bondgenoot
De Bondgenoot 22 | 3772 BX Barneveld
☎+31-(0)342-492090 ✉info@bondgenoot.nl 💻www.bondgenoot.nl

De Hobbitstee
Van Zijlweg 3 | 8351 H.W. Wapserveen/Drente
☎+31-(0)521-321324 💻www.dehobbitstee.org

De Wonne
Noorderhagen 25 | 7511 EK Enschede
☎+31-(0)53-4318787 ✉enschede@wonne.nl 💻www.wonne.nl

Ecodorp Gaasterland

Ecodorp Groenlingen

Emmaus Feniks
Ulingsheide 3 | 5932 NA Tegelen
☎+31-(0)77-3747500 💻www.emmaus-feniks.nl

Franciscaans Milieuproject

Katholieke Werker

Meergeneratie Woonproject Nijmegen
Prinses Beatrixstraat 20 | Ooij
☎+31-(0)6-15467966 ✉woonproject.go@gmail.com 💻www.meergeneratie-wonen.nu

Nikola-Kommuniteit
Maliesingel 63 | 3581 BS Utrecht
☎+31-(0)30-2314184 ✉nikkom@online.nl 💻www.nikola-kommuniteit.nl

Project Anna's Hoeve
Zeevaertweg 67 | 1212 EE Hilversum
+31-(0)6-49602515 annashoeve@hilversum.nl www.annas-hoeve.nl

Vereniging Aardehuis
Rietgors 2a | 8121 JZ Olst
+31-(0)6-20415480 email@aardehuis.nl www.aardehuis.nl

Norway

Hurdal Ecovillage
Hurdalsjøen Økologiske Landsby SA | Landsbyvegen 3 | 2090 Hurdal
+47-(0) 400 877 85 tonekarif@hotmail.com www.hurdalecovillage.no

Poland

Bardo-Siedlung

Bushcraft Berlin & Biber Camp Bronków

L.A.S. - Local Alternative Society
Pryszczowa Góra 44 | 21-025 Niemce | +48-(0)692325054
auramm@wp.pl www.permakultura.com.pl, www.dlaziemi.org

Przysiólek Macze
Macze 2 | 19-300 Elk | +48-(0)602-713114 sameisky@elk.com.pl

Portugal

Terramada
Tanoeiro Cx P 506 M | 8950-277 Tanoeiro, Castro Marim
+351-964-525536 terramada@gmail.com

Romania

Copii Soarelui
Casuta postala nr. 3 | 335200 Brad, jud. Hunedoara
contact@meditatie.ro www.meditatie.ro

Kászonszék
Ökofalu | Kászonimpér 157 | 537238 com. Plaiesii de Jos, jud. Harghita
+40-(0)740-545374 okokaszon@gmail.com

Stanciova Village

Russian Federation

Big Stone Ecovillage
☎+7-89213674351 ✉masha-kulyasova@yandex.ru,Ivan-kulyasov@yandex.ru

Bolshoj Kamen (Big Stone)
"Net" Ecovillage | 13, 36 Furshtadskaya | 199000 St.Petersburg
☎+7-(0)8122732844 ✉masha-kulyasova@yandex.ru, laysanm@yandex.ru

Edinstvo Ecovillage
c/o Alexander Roumega | PO Box 5132 | 344015 Rostov-on-Don
☎+7-90-34-34-37-84 ✉no confirmed email address available

Medvinka

Slovakia

Alter Nativa o.z.
Brdárka 11 | 04934 Brdárka
☎+421-(0)910-170790 ✉alternativa@alter-nativa.sk 💻www.alter-nativa.sk

Slovenia

Skupnost Barka
Arche Slovenia | Zbilje 66 | 1215 Medvode
☎+386-(0)1361-6459 ✉barka.skup@guest.arnes.si

Spain

8th Life
EcoVillage Project | ☎+34-697590185, +34-636498904, +34-922-400133
✉8thlife@8thlife.org 💻www.8thLife.net

Aldea de Pano
Kurt Fridez | única sin número | 22438 Pano (Huesca)
☎+34-(0)974-347000 ✉panofridez@yahoo.es

Argayall
Place of Light | Valle Gran Rey | 38870 La Gomera/Canary Islands
☎+34-922-697008 ✉info@argayall.com 💻www.argayall.com

BIO-SOL
c/ El cerro Nr. 7, Lújar | 18614 Granada (Andalucia)
☎+34-(0)670658027/-617726951 after 21 h ✉iluminati16@hotmail.com

Cal Cases
Masia Cal Cases | 08273 Santa Maria d'Oló
☎+34-938384027 calcases@calcases.info www.calcases.info

Els Isards
Saleta del Mas | 17403 St. Hilary Sacalm, Girona, Catalonia
☎+34-660065957

Kanawen Ecovillage

La Base

Sweden

Ängsbacka
☎+46-(0)553-10035 booking@angsbacka.se

Brf Ekobyn Mjölnartorpet
Skaldens gata 65 | 656 38 Karlstad
brfekobyn@gmail.com www.mjolnartorpet.se

Kampetorp ekoby
Ällelien Kampetorp | 45750 Bullaren
☎+46-(0)525-53006 Kampetorp@kampetorp.se www.kampetorp.se

Kursgarden Lindsberg
Lindsberg 10, (Pl. 4795) | 79191 Falun
☎+46-(0)23-43030 Lindsberg10@gmail.com lindsberg.org

Tullstugan Collective housing unit

Switzerland

Association Epidaure
Epiquerez

Clos du Doubs
Ecovillage Clos du Doubs | 2887 Soubey
☎+41-(0)32-9551704 ffjaco@yahoo.fr

Hof allne Wind
Bemerberg | 8903 Birmensdorf
☎+41-(0)44-7371547 info@hofallnewind.ch www.hofallnewind.ch

Objectif Gaia
nicolasbriet@yahoo.co.uk

Turkey

Marmariç Permakültür

Marmariç, Dernekli Köyü, Bayindir | 35840 Izmir
✉info@marmaric.com 💻marmaric.org

Ukraine

Baranivka

House in Baranivka | 20-a Vul. Tsentralna, Baranivka village, Shyshaky district
38021 Baranivka | ☎+380-(0)95-547-4866 ✉info@baranivka.org

Dolyna Dzherel

Kiev | ✉Pilatova@dow.com

United Kingdom

Argyle Street Housing Co-op

3 Fletchers Terrace | CB1 3LU Cambridge
☎+44-(0)1223-411615 ✉office@ash.coop 💻www.ash.coop

Beech Grove Community

Forest Drive, Nonington | Kent CT15 4FB
☎+44-(0)1304-842980 ✉beechgrove@bruderhof.com 💻www.bruderhof.com/

Beech Hill Community

Beech Hill, Morchard Bishop, Devon | EX17 6RF Exeter
☎+44-(0)1363-877228 ✉beechhill@gmail.com

Bhaktivedanta Manor

International Society for Krishna Consciousness (ISKCON)
Aldenham, Near Watford | WD25 8EZ Herts
☎+44-(0)1923-851000 ✉info@krishnatemple.com 💻www.krishnatemple.com

Cambridge K1 Cohousing

129 Ditton Walk | CB5 8QD Cambridge
☎+44-(0)1223-740326 ✉robert@enlinca.org.uk 💻www.cambridge-k1.co.uk

Camphill School Aberdeen

Central Office, Murtle House | Bieldside | AB15 9EP Aberdeen
☎+44-(0)1224-867935 ✉b.porter@crss.org.uk 💻www.camphillschools.org.uk

Culdees

Boreland Farm | Fearnan by Aberfeldy | PH15 2PG Perthshire
☎+44-(0)1887-830519 / -(0)7904-954116 (mobile)
✉contact@culdees-ecovillage.co.uk, contact@culdeesbunkhouse.co.uk
💻www.culdees-ecovillage.co.uk

Earthworm Housing Co-op

Leintwardine, nr. Craven Arms | SY7 0LH Shrops | ☎+44-(0)1547-540461
✉earthwormcooperative@yahoo.co.uk
💻www.earthwormhousingcooperative.org.uk

Ecoforest

Enlinca

129 Ditton Walk | CB5 8QD Cambridge
✉robert@enlinca.org.uk 💻www.enlinca.org.uk

Faslane Peace Camp

A814, Shandon | Nr Helensburgh | G84 8NT Dumbartonshire
☎+44-(0)1436-820901, +44-(0)7511793227 ✉faslane30@gmail.com
💻www.faslanepeacecamp.wordpress.com

Glasgow Catholic Worker

26 Rose Street | G3 6RE Glasgow
✉glw@catholicworker.org.uk 💻www.catholicworker.org.uk

Heartwood Housing Co-operative

Blaen y Wern, Llangyndeyern | SA17 5ES Carmarthenshire
☎+44-(0)1267 275960 💻#4192 at eurotopia.directory/booklinks

Hoathly Hill Community

26 Hoathly Hill | RH19 4SJ East Grinstead
☎+44-(0)1342-810744 ✉chloe.anthony@gmail.com

Hockerton Housing Project

The Watershed, Gables Drive | NG25 0QU Southwell
☎+44-(0)1636-816902 ✉contact@hockertonhousingproject.org.uk
💻www.hockertonhousingproject.org.uk

Holy Isle Project

Centre for World Peace and Health | Lamlash Bay | KA27 8GB Isle of Arran
Scotland | ☎+44-(0)1770-601100 ✉reception@holyisle.org 💻www.holyisle.org

Islington Park Street Community

38-44 Islington Park Street | N1 1PX Islington, London
✉44@ips.eclipse.co.uk, 44figtree@gmail.com 💻islingtonparkstreet.org/

L'Arche Bognor Regis

Bradbury House | 51 Aldwick Road | PO21 2NJ Bognor Regis
☎+44-(0)1243-863426 ✉enquiries@larchebognor.org.uk
💻www.larchebognor.org.uk

Lighthouse

214 Allerton Road | Mossley Hill | L18 6JN Liverpool
☎+44-(0)151-7241376 ✉liverpool@jesus.org.uk

Ned

73 Beech Avenue | Forest fields | NG7 7LR Nottingham
☎+44-(0)845-4589595 ✉nedshousingcoop@riseup.net

Oakleigh

☎+44-(0)114-2420344 ✉susannah.diamond@gmail.com

Old Chapel Farm

Tylwych | Llanidloes, PowysSY18 6JR
☎+44-(0)1686-412744 ✉franblockley@yahoo.co.uk 💻www.oldchapelfarm.org

OWCH - New Ground Cohousing

Older Women's Cohousing Company | 74 Inglethorpe Street | SW6 6NX London
☎+44-(0)7385-6239 ✉owch.org@googlemail.com 💻www.owch.org.uk

Postlip Hall

Winchcombe | GL54 5AQ Cheltenham
✉PHA.cohousing@gmail.com 💻postliphall.org.uk

Saint Fransis House

227 Crowley Road | Oxford OXU 1XG
☎+44-(0)1865-248288 ✉susanclarkson@hotmail.com
💻www.eurocatholicworker.org/index.php?id=10&L=2

Springfield House

2 Springfield Road | Stoneygate | LE2 3BA Leicester
☎+44-(0)116-2705425 ✉leicester@jesus.org.uk

Tipi Village

Tipi Valley | Wernlas Cwmdu, Talley, Llandeilo
SA19 7EE Carmarthenshire, Wales
☎+44-(0)1558-685186 ✉revrmayes@hotmail.co.uk
💻www.diggersanddreamers.org.uk

Woodgate

Community Networks

Here we publish addresses and descriptions of community networks that we know of, that we wrote to and which have responded. Of networks that have not responded we still publish web addresses, since these are freely available to the public, especially as those networks might still be mentioned by communities that have responded to our call for submission.

The network descriptions have been composed by the networks themselves. The lists under the network descriptions show members of those networks: Some networks could allow us to publish addresses of even those communities that have not answered our individual call for submission - so these are additional existing communities which you can often check out via their websites. With the European Camphill office we agreed on publishing all addresses and only very few individual descriptions, but a long article about the movement (page 44ff). They are just too many, and much is similar in every Camphill community.

An arrow in front of a community or organisation name means that it can be found with a detailed description in the main part of the book or with the "useful addresses" section (starting on page 542ff)(you can also refer to the index in the end to find it). Bold addresses are official contact addresses of a network, e.g. the national branches of GEN Europe. Sometimes they are communities, sometimes only an office.

Please refer to page 607 (next to last) for an explanation of the country initials.

L'Arche (Jean Vanier)/Federation Internationale des Communautes de L'Arche

A L'Arche community is...
... mutual relationships: At the heart of L'Arche communities are relationships between people with and without intellectual disabilities. A respectful relationship between people who treat each other as of equal value provides security, allowing for growth, personal development and freedom to become more fully the people we want to be. Most importantly, mutual relationships foster the acceptance of each person as a unique and valuable individual, whatever his or her abilities or disabilities.

... spirituality: From its first community, founded in France in the Roman Catholic tradition, many other communities have been established in various cultural and religious traditions. Today, faith life continues to be an essential element of all L'Arche communities, which may be rooted in a particular faith tradition or have an ecumenical or an inter-faith character. Communities are respectful of the particular faith traditions of people who come to the community and seek to support members to deepen in their own faith experience and values, whether they situate themselves within or outside a religious affiliation.

... know-how: Almost 50 years of experience and cooperative work with doctors, psychologists, social workers and other professionals have enabled L'Arche communities to develop strong competencies and a high standard of quality in the field of care-giving. In several countries, L'Arche is called upon by governments for advice on policy formation.

... community life: Living with others emphasises what a person can accomplish, give and learn. Just as in a family, all members of the community are responsible for the group's well-being, be it at home, in a workshop or a day programme, through cooking or washing up, through a helping hand, a smile or a joke.
(Description from 2013)

Federation Internationale des Communautes de L'Arche

F L'Arche Internationale | 25, rue Rosenwald | 75015 Paris
☎+33-(0)1 53 68 08 00 ✉international@larche.org 💻www.larche.org

→L'Arche Communauté d'Aywaille (B)
→Arche im Nauen (CH)
→Goshen (F)

L'Arche de Lanza del Vasto

The Ark Community has members in France, Germany, Italy, Switzerland, Belgium, Spain, Brazil, Argentina, Mexico and Ecuador. It goes back to an inspiration that writer, artist and philosopher Lanza del Vasto had in 1937 during a meeting with Gandhi in India. The Ark is made up of individuals and groups who bring the spirit and means of non-violence to their respective places in order to contribute to solving the problems of our time. Ark members live together in communities in four places: In La Fleyssiere and St. Antoine (F), in Chambrelien (CH) and at the Friedenshof (D). Thanks to its roots in non-violence and to the intuitive foresight of the founding group around Lanza del Vasto, the Ark has been a pioneer of themes that have proven to be vital for our society:

- Relationships in the spirit of non-violence, in which people recognize in each other the sister/brother with whom they can create something together.
- The search for coherence between our thoughts and our actions.
- Respect for the otherness of the other, especially in the area of spirituality and religion.
- Culture of community, to find a way out of the destructive forms of individualism.
- Simplification of life, as a sign of our responsibility in an economy and society based on destructive consumption.
- Commitment to justice and peace, in the certainty that social change begins with personal transformation and becomes concrete through active participation in political action.
- Freedom of thought and opinion, which is preserved by the practice of consensus decision; consensus strengthens personal responsibility and protects against subordination.

The members of the Ark strive to bring their lives more and more into harmony with the deepest human desires. They know that this requires a living correlation between spiritual life, personal ethics, and social commitment.

L'Arche de Lanza del Vasto

WWW ✉info@archegemeinschaft.de 💻www.archecom.org

→Biotopia (D)
→Friedenshof-Kommunität (D)
→Haus der Gastfreundschaft (HdG) (D)
→Lilitopia (D)
→L' Arche de Saint Antoine (F)
→La Borie Noble (F)
→La Fleyssière (F)

Bruderhof Communities

We are a movement of intentional Christian communities with some 2700 people living in over twenty locations on four continents. We are a covenant of families and singles who want to follow Jesus without compromise in the spirit of the first church in Jerusalem. We renounce private property and share everything with one another.
We try to practice the commandment of Jesus to love God and our neighbour in everyday life; the early church in Jerusalem, as described in the New Testament, serves as our inspiration and model. We recognize the authority of the Bible in all areas of life and confess the rule of faith expressed in the Apostolic Creed. Historically, we are rooted in the Anabaptist tradition of the Radical Reformation, but feel connected to all who take following Jesus seriously. Further, we believe that Jesus can work in all people, regardless of their background, their professed faith, or their circumstances.

Bruderhof Community/ Bruderhofgemeinschaft

D Talweg 18 | 07639 Bad Klosterlausnitz
☎+49-(0)36601-922 988
holzland@bruderhof.com
www.bruderhof.com

→Holzlandgemeinschaft (D)
→Sannerzgemeinschaft (D)
→Darvell Bruderhof (UK)

Camphill

See our article about Camphill on page 44ff and the map on eurotopia.directory/camphill

Camphill - Middle Europe

D Dorfgemeinschaft Hermannsberg | 88693 Heiligenberg
Ch.Heemann@hermannsberg.de

Camphill Liebenfels

A Marienhöhe, Lebmach 15 | 9556 Liebenfels-Kärnten
☎+43-(0)4215-2355 wertsch@camphill.at www.camphill.at

→Humanushaus Beitenwil (CH)

Camphill na soutoku, z.s.

CZ Camphill Ceské Kopisty | Ceské Kopisty 6 | 41201 Terezín
camphill@camphill.cz www.camphill.cz

Camphill Alt-Schönow

D Alt-Schönow 5 | 14165 Berlin (Zehlendorf)
☎+49-(0)30-845 71890 info@camphill-alt-schoenow.de
www.camphill-alt-schoenow.de

Camphill Dorfgemeinschaft Hausenhof

D Hausenhof 7 | 91463 Dietersheim
+49-(0)9164-9984-0 info@Hausenhof.de www.Hausenhof.de

Camphill Dorfgemeinschaft Hermannsberg

D 88633 Heiligenberg-Hattenweiler
+49-(0)7552-260 10 K.Kwiek@hermannsberg.de,
ch.mueller@hermannsberg.de, Lichthof@hermannsberg.de
www.hermannsberg.de

Camphill Dorfgemeinschaft Sellen e.V.

D Sellen 101 / 98 | 48565 Steinfurt-Burgsteinfurt | +49-(0)2551-936 60
info@camphill-steinfurt.de www.camphill-steinfurt.de

Camphill Lebensgemeinschaft Königsmühle

D Schöntalstr. 11 | 67434 Neustadt a.d. Weinstraße | +49-(0)6321-7289
koenigsmuehle@t-online.de www.camphill-rheinland-pfalz.de

Camphill Schulgemeinschaft Bruckfelden

D Adalbert Stifter Weg 3 | 88699 Frickingen-Bruckfelden
+49-(0)7554-982 511 info@camphill-schulgemeinschaften.de

Camphill Schulgemeinschaft Föhrenbühl

D Föhrenbühlweg 5 | 88633 Heiligenberg-Steigen
foehrenbuehl@camphill-schulgemeinschaften.de
camphill-schulgemeinschaften.de/foehrenbuehl/

Camphill Schulgemeinschaft und Hof Brachenreuthe

D 88662 Überlingen | +49-(0)7551-800 70
info@brachenreuthe.de www.brachenreuthe.de

Markus Gemeinschaft e.V.

D Camphill-Initiative für Mensch und Umwelt | Hauptstraße 1
06577 Hauteroda | +49-(0)34673-736 923
info@markus-gemeinschaft.de www.markus-gemeinschaft.de

SKID gGmbH

D SozialKulturelle IntegrationsDienste | Kronengasse 1 | 88662 Überlingen
info@skid-ggmbh.de skid-ggmbh.de

Thomas-Haus Berlin

D für Heilpädagogik und Sprachtherapie | Peter-Lenné-Straße 42
14195 Berlin (Dahlem)
thomas-haus@gmx.de www.thomas-haus-berlin.de

Pahkla Camphilli Küla

EE Kohila vald, Rapla Maakond | 79702 Eesti
+372-48-98300/-97535 pahklack@hot.ee www.pahklack.org

Kaupunkikylä

SF yksilöllistä elämää yhteisössä | Kyläkatu 140 | 15700 Lahti
www.kaupunkikyla.fi

Myllylähde-yhteisö
SF Aimo Kuusisto | Aholantie 133, Palomaa | 16800 Koski HL
☎+358-(0)3-7641944

Sylvia-Koti
SF Kyläkatu 140 | 15700 Lahti
☎+358-(0)3-883130 ✉info@sylvia-koti.fi 💻www.sylvia-koti.fi

Tapolan Kyläyhteisö
SF Raikkolantie 40 | 16350 Niinikosi
☎+358-(0)18-882450 💻www.tapola-camphill.fi

Ballymoney Community
IRL Sea Road | Gorey, Co. Wexford
✉info@camphillballymoney.ie 💻www.camphill.ie/ballymoney/index

Callan Community
IRL The Granary | Bridge St | Callan, County Kilkenny | ✉callan@camphill.ie

Camphill Ballytobin Community
IRL Callan | Co. Kilkenny
☎+353-(0)56-25114 ✉ballytobin@camphill.ie 💻www.camphill.ie

Camphill Community Ballybay
IRL Corraskea | Robb Farm | Ballybay, Co. Monaghan
☎+44-(0)42-9741939 ✉ballybay@camphill.ie 💻www.camphill.ie

Camphill Community Dingle
IRL Doonsheane | Dingle Co. Kerry
☎+353-(0)45-481143 ✉dingle@camphill.ie 💻www.camphill.ie/Dingle

Camphill Community Dunshane
IRL Dunshane House | Brannockstown, Near Naas | Co. Kildare
☎+353-(0)45-483628 ✉dunshane@camphill.ie 💻www.camphill.ie

Camphill Community Greenacres
IRL 1a Farmhill Park | Goatstown | 14 Dublin
☎+353-(0)1-2987618 ✉greenacres@camphill.ie 💻www.camphill.ie

Camphill Community Kyle
IRL Coolagh, Callan | Co. Kilkenny
☎+353-(0)56-25737 ✉kyle@camphill.ie 💻www.camphill.ie

Camphill Community Mountshannon
IRL Clai House | Millpool Road | Mountshannon, Co. Clare
✉camphillmountshannon@gmail.com
💻www.camphill.ie/mountshannon/index

Camphill Community Thomastown
IRL The Watergarden | Ladywell Street | Thomastown, Co. Kilkenny
☎+353-(0)56-24844, -54132 ✉jerpoint@camphill.ie 💻www.camphill.ie

Camphill Village Community Duffcarrig
IRL Duffcarig, Gorey | Co. Wexford
☎+353-(0)55-25911 ✉duffcarrig@camphill.ie 💻www.camphill.ie

Carrick-on-Suir Camphill Community
IRL Castle Street | Carrick-on-Suir, Co. Tipperary
☎+353-(0)51-645080 ✉carrick@camphill.ie 💻www.camphill.ie

Grangebeg Camphill Community
IRL Dunlavin | Co. Kildare
✉grangebeg@camphill.ie 💻www.camphill.ie/grangebeg/index

Grangemockler Camphill Community
IRL Templemichael | Carrick-on-Suir, Co. Tipperary | ☎+353-(0)51-647202
✉grangemockler@camphill.ie 💻www.camphill.ie/grangemockler

Jerpoint Camphill Community
IRL Thomastown | Co. Kilkenny
✉jerpoint@camphill.ie 💻www.camphill.ie/jerpoint/index

KCAT Camphill Community
IRL Mill Lane | Callan, Co. Kilkenny
✉info@kcat.ie 💻www.camphill.ie/kcat/index

The Bridge Camphill Community
IRL Main Street | Kilcullen, Co. Kildare
☎+353-(0)45-481597 ✉thebridge@camphill.ie 💻www.camphill.ie

Camphill Village Rozkalni
LV Rencenu pagasts | 4232 Burtnieku novads
✉camphillrozkalni@gmail.com 💻camphillrozkalni.lv

Camphill Community Rotvoll
N Hans Collins vei 5 | 7053 Ranheim | ☎+47-(0)73-826850
✉applications@camphill-rotvoll.no 💻rotvoll.camphill.no/hjem/

Camphill landsbystiftelse i Norge
N Camphill Villages in Norway | Reidar Jenssensgate 10 | 7550 Hommelvik
☎+47-73-978460 ✉sekretariat@camphill.no 💻www.camphill.no

→Camphill Rotvoll (N)

→Camphill Solborg (N)

Hogganvik Landsby
N Hogganvik | 5583 Vikedal | ☎+47-(0)5276-0111
✉hogganvik@camphill.no 💻www.camphill.no/Hogganvik/

Jøssåsen Landsby
N Jøssåsen Landsby | 7550 Hommelvik
☎+47-(0)73979-180,-950 ✉application@jossasen.no

Vallersund Gård

N Camphill Landsbystiftelse i Norge | Vallersund Gård
7167 Valsøyveien, Vallersund | ☎+47-(0)7252-7080 💻vallersund.camphill.no

Vidaraasen Village (Camphill)

N Vidaråsen Landsby | Landsbyveien 34 | 3158 Andebu | ☎+47-3344-4100
✉office@vidaraasen.no 💻vidarasen.camphill.no, www.vidaraasen.no

Camphill Gemeenschap Christophorus

NL Duinweg 35 | 3735 LC Bosch en Duin
💻www.amerpoort.nl/locaties/wonen/christophorus.html

Camphillgemeenschap Maartenhuis

NL Ruyslaan 81 | 1796 AZ De Koog (Texel) | ☎+31-(0)22-231 76 82
✉secretariaat@maartenhuis.nl 💻www.maartenhuis.nl

Stowarzyszenie 'Wspólnota' w Wójtówce

PL Camphill Community 💻https://tinyurl.com/eurotopia8884

→Camphill Village Svetlana (RUS)

Tourmaline

RUS official Camphill-partner | Borisa Galushkina st. 26 | 129626 Moscow
✉turmaline.nko@gmail.com 💻turmaline.ru, eng.turmaline.ru

Camphill Häggatorp

S Häggatorp 1 | 53461 Vedum
✉camphillhaggatorp@telia.com 💻www.camphillhaggatorp.se

Foreningen Staffansgården

S Camphill | Furugatan 1 | 82060 Delsbo | ☎+46-(0)653-168 50
✉info@staffansgarden.com 💻www.staffansgarden.com

Mickelsgården

S Camphill | Lund 2905 | 82060 Delsbo | ☎+46-(0)653-169 82/-109 17
✉info@staffansgarden.com, matti@staffansgarden.com,
jens@staffansgarden.com 💻www.staffansgarden.com

→Beannachar Camphill Community (UK)

Botton Village Community

UK Danby, Whitby | YO21 2NJ North Yorkshire
☎+44-(0)1287-660 871 💻www.cvt.org.uk/communities/botton-village

Camphill Blair Drummond

UK Blair Drummond House | Cuthil Brae | By Stirling, Perthshire FK9 4UT
☎+44-(0)1786-841 573 ✉office@camphillblairdrummond.org.uk
💻camphillblairdrummond.org.uk

Camphill Coleg Elidyr (incl. Victoria House)

UK Rhandirmwyn, Llandovery | Carmarthenshire SA20 0NL
☎+44-(0)1550-760 400 1 ✉admin@colegelidyr.com 💻colegelidyr.com/

Camphill Communities East Anglia

UK Thornage Hall | Thornage, Holt | Norfolk, Scotland NR25 7QH
ccea@btconnect.com www.camphill.org.uk

Camphill Community Glencraig

UK Craigavad | BT18 0DB Holywood, Co. Down | +44-(0)1232-423 396
office@glencraig.org.uk, reception@glencraig.org.uk
www.glencraig.org.uk

Camphill Community Holywood

UK 8 Shore Road | Holywood, BT18 9HX
info@camphillholywood.co.uk www.camphillholywood.co.uk

Camphill Community Mourne Grange

UK 169 Newry Road, Kilkeel | BT34 4EX Co. Down, Northern Ireland
+44-(0)28-417 601 28 info@mournegrange.org
www.mournegrange.org

Camphill Devon Community

UK Hapstead Village, Buckfastleigh | Devon TQ11 0JN | +44-(0)1364-642 631
admin@camphilldevon.org.uk www.camphilldevon.org.uk

Camphill Milton Keynes Community

UK Japonica Lane, Willen Park South | MK15 9JY Milton Keynes
+44-(0)908-235 000 office@camphill-miltonkeynes.co.uk
www.camphillmk.co.uk

Camphill St. Albans

UK 50 Carlisle Avenue, St Albans | Herts AL3 5LX | +44-(0)1727-811 228
www.camphillvillagetrust.org.uk/locations/st-albans/

Camphill: Sturts Farm

UK Three Cross Road | West Moors | Ferndown, Dorset BH22 0NF
+44-(0)20-287 05 72 office@sturtsfarm.com www.sturtsfarm.com

Camphill: The Hatch

UK St. John's House | Kington Lane, Thornbury | Bristol, Avon BS12 1NA
+44-(0)1454-413 010 admin@hatchcommunity.org.uk
hatchcommunity.org.uk

Cherry Orchards Camphill Community

UK Canford Lane, Westbury-on-Trym | BS9 3PE Bristol | +44-(0)117-950 31 83
cherryorchards@camphill.org.uk www.cherryorchards.co.uk

Clanabogan Camphill Community

UK 15 Drudgeon Rd, Clanabogan | BT78 1TJ Omagh, County Tyrone
+44-(0)28-822 561 11 info@camphillclanabogan.com
www.camphillclanabogan.com

Corbenic Camphill Community

UK Trochry, Near Dunkeld | PH8 0DY Perthshire | +44-(0)1350-723 206
betty@corbeniccamphill.co.uk www.corbeniccamphill.co.uk

Delrow Camphill Community

UK Hilfield Lane, Aldenham | WD2 8DJ Watford, Herts | ☎+44-(0)1923-856 006
✉delrowhouse@hotmail.co.uk 💻tinyurl.com/eurotopia3642

Esk Valley Camphill Community

UK Stormy Hall | YO21 2NN Danby Head
✉info@eskvalleycamphill.org 💻www.eskvalleycamphill.org

Gannicox Camphill Community

UK 57 Cainscross Road | GL5 4EX Stroud, Gloucestershire
✉gannicox@camphill.org.uk 💻www.camphill.org.uk/~gannicox

Glasallt Fawr Camphill Centre

UK Llangadog | Carmarthenshire SA19 9AS
✉info@glasallt-fawr.com 💻www.glasallt-fawr.com

Grange Camphill Village

UK Littledean Road | GL14 1HJ Newnham | ☎+44-(0)1594-516 246
✉grange@cvt.org.uk 💻www.cvt.org.uk/communities/grange-village

Larchfield Camphill Community

UK Stokesley Road, Hemlington | Middlesbrough, Cleveland TS8 9DY
☎+44-(0)642-593 688 ✉larchfield.enquiries@cvt.org.uk
💻www.cvt.org.uk/communities/larchfield-community

Loch Arthur Camphill Community

UK Loch End Farm | Beeswing | Dumfries DG2 8JQ
☎+44-(0)1387-760 687 ✉admin@locharthur.org.uk 💻locharthur.org.uk

Milltown Camphill Community

UK Arbuthnott, Laurencekirk | AB30 1PB Kincardineshire
☎+44-(0)561-361 349 ✉milltownoffice@btconnect.com,
✉milltownworkshop@yahoo.co.uk 💻www.milltowncommunity.org.uk

Newton Dee Village Community

UK Camphill Community | Bieldside | AB15 9DX Aberdeen
☎+44-(0)1224-868 701 ✉info@newtondee.org.uk
💻www.newtondee.co.uk

Oaklands Park

UK Newnham | GL14 1EF Gloucestershire | ☎+44-(0)1594-516 551
✉oaklands@cvt.org.uk 💻www.cvt.org.uk/communities/oaklands-park

Ochil Tower Camphill School

UK 140 High Street | PH3 1AD Auchterarder | ☎+44-(0)1764-62416
✉office@ochiltowerschool.org 💻www.ochiltowerschool.org.uk

Orchard Leigh Farm Community

UK Bath House, Eastington, Stonehouse | GL10 3AY Gloucestersire GL10 3AY
☎+44-(0)1453-823 811 ✉orchard.leigh@virgin.net
💻www.camphill.org.uk/~orchardleigh

Pennine Camphill Community
UK Wood Lane, Chapelthorpe, Wakefield | WF4 3JL West Yorkshire
☎+44-(0)1924-255 281 ✉office@pennine.ac.uk, volunteering@pennine.ac.uk
💻pennine.ac.uk

Ringwood Camphill Communities
UK The Sheiling School | Horton Road, Ashley | nr. Ringwood, Hampshire, BH24 2EB
☎+44-(0)1425-477 488 ✉enquiries@thesheilingringwood.co.uk
💻www.sheilingschool.co.uk

Simeon Care for the Elderly
UK Cairnlee Road, Bieldside | Aberdeen AB15 9BN | ☎+44-(0)224-862 415
✉info@simeoncare.org.uk 💻www.simeoncare.org.uk

The Croft Community
UK Highfield Road, Old Malton | YO17 7DB North Yorkshire
☎+44-(0)1653-694 391 💻croftcommunity.com

The Lantern Camphill Community
UK Folly Farm Lane | BH24 2NN Ringwood, Hampshire
✉info@lanterncommunity.org.uk 💻www.lanterncommunity.org.uk

The Mount Camphill Community
UK Faircrouch Lane, Wadhurst | East Sussex TN5 6PT | ☎+44-(0)892-782 025
✉office@mountcamphill.org 💻www.mountcamphill.org

Thornbury Camphill Community
UK Thornbury Park | Park Road | Bristol BS35 1HP | ☎+44-(0)1454-412 194
✉mail@sheilingschool.org.uk 💻www.sheilingschool.org.uk

Tigh A' Chomainn Camphill
UK 4 Craigton Crescent, Peterculter | Aberdeenshire AB14 0SB
☎+44-(0)1224-732 656 💻camphilltac.org.uk

Tiphereth Camphill Edinburgh
UK Torphin Holding, Torphin Road 55 | Colinton, Edinburgh EH13 0PQ
☎+44-(0)131-441 20 55 ✉Sandy.Walton@tiphereth.org.uk
💻www.tiphereth.org.uk

William Morris Camphill Community
UK William Morris House | Eastington, Stonehouse | Gloucestershire GL10 3SH
☎+44-(0)453-824 025 ✉info@wmcc.ac.uk 💻www.learning-disability.org.uk

Church and Peace

Church and Peace is the European ecumenical network of more than 100 Christian communities, congregations, peace organizations, peace service agencies, churches and individuals from 14 European countries and from a wide variety of Christian traditions and denominations. Their common conviction is that the peace witness and nonviolence of the Gospel are essential characteristics of the Church and lead to nonviolent service for peace and reconciliation. Its manifold commitment, especially in war-torn countries, is rooted in the combination of peace theology, spirituality and practice of nonviolence. This means: striving for reconciliation, peace education, nonviolent conflict management, commitment to justice in the distribution of the worlds resources, advocacy for the integrity of creation.
Church and Peace has its origins in the ecumenical discourses that have been held since 1949 by the Historical Peace Churches (Quakers, Mennonites and Church of the Brethren) and the International Fellowship of Reconciliation with the churches of the World Council of Churches on the significance of nonviolence in the discipleship of Jesus.
Church and Peace invites the churches to become peace churches and offers space to meet, connect with and strengthen each other. International and regional conferences bring together up to 250 participants each year from over 15 countries to share their experiences and encourage each other. People in the Church and Peace network engage in concrete projects of ecumenical and interfaith cooperation, locally or internationally, in joint activities with other civil society groups and in the Conference of European Churches (CEC). Church and Peace is an organisation in partnership of the CEC, and a member of the Network of Christian Peace Organizations (UK). Ecumenical dialogue, especially the development of a common peace ethic, is a high priority for Church and Peace.
It is also a member of regional umbrella groups such as the Network of Christian Peace Organisations (Britain and Ireland) and the Action Committee Service for Peace (Germany).
Find our member groups: https://www.church-and-peace.org/en/network/members/

Church & Peace e.V.

D International Office | Mittelstr. 4 | 34474 Diemelstadt-Wethen
☎+49-(0)5694-990 55 06 ✉intloffice@church-and-peace.org
💻www.church-and-peace.org

→Basisgemeinde Wulfshagenerhütten (D)
→Brot & Rosen - Diakonische Basisgemeinschaft (D)
→Hofgemeinschaft Guggenhausen e.V. (D)

church and peace

Colibris

Colibris did not respond to our contact attempts, but many French communities say that they are part of this network. We also suspect that the network "Mouvement des Oasis en Tous Lieux", printed in the last volume, has merged into Colibris.

Colibris

F www.colibris-lemouvement.org

→Coogite (F)
→Eco-hameau d'Andral (F)
→Ecohameau de Verfeil (F)
→Ecovillage "Ecohameau Les Trois sources" (F)
→Goshen (F)
→Hameau du Ruisseau (F)
→La ferme de Chenèvre (F)
→La Lune Nette (F)
→Le Broglio (F)
→Le collectif du Moulinage de Chirols (F)
→Le Hameau des Buis (F)
→Oasis de Belledonne (F)
→Oasis de Lentiourel (F)
→Solenvie (F)
→Terra Flor (F)

Common Life Network

The Common Life Network is a network dedicated to the founding and encouragement of Christian communities in Switzerland. We maintain a website with a search platform, hints to resources, a list of communities etc.

Common Life Network

CH Hutti 1 | 3250 Lyss
info@we-share-it.ch www.we-share-it.ch/commonlife

Emmaus

Emmaus has not responded to our call, but it is a well known and important network and we describe several communities in this book that are part of the Emmaus-Network.

Secretariat General d'Emmaus International

F www.emmaus-international.org

→Communauté de la Poudrière (B)

→Haus der Gastfreundschaft (HdG) (D)
→Village Emmaüs Lescar-Pau (F)
→Emmaus Haarzuilens (NL)
→Emmaus in Langeweg and Breda (NL)

European Catholic Worker

The Catholic Worker is a movement of radical Christians and has a pacifist and anarchist orientation. Our goal is to connect the Gospel with political action and practical aid to the poorest. In this way we want to give a sign of the Kingdom of God on earth. The Catholic Worker Movement was founded in the US by Dorothy Day and Peter Maurin during the depths of the Great Depression in the 1930s.
Today over 200 Catholic Worker communities remain committed to nonviolence, voluntary poverty, prayer, and hospitality for the homeless, exiled, hungry and forsaken. Catholic Workers continue to protest injustice, war, racism and violence of all forms.
Inspired by the ideas and power of the Catholic Workers we brought the movement to Europe and started houses and communities in the Netherlands, Great Britain, Belgium, Germany, France and Sweden.

The European Catholic Worker
NL info@eurocatholicworker.org www.eurocatholicworker.org

→Biotopia (D)
→Brot & Rosen - Diakonische Basisgemeinschaft (D)
→Haus der Gastfreundschaft (HdG) (D)
→Kana (D)
→Lilitopia (D)
→London Catholic Worker (UK)
→The Catholic Worker Farm (UK)
→Amsterdam Catholic Worker (NL)
→The Fig Tree –
A Jona House Resistance and Catholic Worker Community (S)

Foreningen Bofællesskab.dk

Bofællesskab.dk is a digital platform and meeting place for all residential communities in Denmark. This means that the website is the place on the web where everyone who either lives in, wants to live in, or just has an interest in community living, is meeting and networking, sharing good ideas, selling homes in communities and much more.
In January 2018, Bofællesskab.dk became an association, which, in addition to providing the framework for the operation of the website, is also the organizational link focusing on many diverse activities. The association is led by a board of five persons who are rooted in the community of co-housing in Denmark. The website is run by an administrator who,

together with an operating group and the board, ensures that the portal is kept updated and developed so that it is always a source of inspiration for all members and users.
Developing the association is an ongoing process. Ad hoc working groups are set up, each of which works with projects and events under the auspices of the association and as a contribution to the extensive movement that is underway in both Denmark and the rest of the world. The association actively strives to participate in and organize events and to prepare material that disseminates knowledge of community housing and other forms of living together. Bofællesskab.dk will also help new initiatives in the process of creating housing communities. Therefore, we will shortly prepare a start-up package that aims to make the creation of a common residence more accessible and easier to get through.

Foreningen Bofællesskab.dk
DK Sædekildegårdvej 1 | 5250 Odense SV
info@bofaellesskab.dk bofaellesskab.dk

→Bofællesskabet Bondebjerget (DK)
→Bofællesskabet Fladengrund (DK)
→Bofællesskabet Lille Grundet (DK)
→Holdepunktet (DK)

GEN - Global Ecovillage Network

The Global Ecovillage Network connects thousands of communities dedicated to developing and demonstrating four dimensions of ecological, social, cultural and economic sustainability in whole systems-designed projects around the world. GEN ecovillages include both intentional communities and traditional ecovillages in rural, urban and nomadic settings. GEN is organised in 5 regional networks: CASA Latina in Latin America, the GENNA Alliance in North America, GEN Europe in Europe, GEN Africa on the African continent and GENOA in Oceania and Asia (see http://ecovillage.org/regions).
Since 1995, GEN facilitates exchange and provides information, tools and examples to inspire sustainable thinking, living and acting. In addition to providing education, consultancy, and incubation to support ecovillage development, since 2000 GEN holds consultative status at the UN-Economic and Social Council (ECOSOC) commission, and is represented at regular briefing sessions at UN Headquarters. GEN's mission is "to innovate, catalyze, educate and advocate in global partnership with ecovillages and all those dedicated to the shift to a regenerative world." Its goals are:

- To advance the education of individuals from all walks of life by sharing the experience and best practices gained from the networks of ecovillages and sustainable communities worldwide.
- To advance human rights, conflict resolution, and reconciliation by empowering local communities to interact globally, while promoting a culture of mutual acceptance and respect, effective communications, and cross-cultural outreach.
- To advance environmental protection globally by serving as a think tank, incubator, international partner organization, and catalyst for projects that expedite the shift to sustainable and resilient lifestyles.
- To advance citizen and community participation in local decision-making, influencing policy-makers, and educating the public, to accelerate the transition to sustainable living.

Global Ecovillage Network

UK The Park, Findhorn | IV36 3TZ Forres, Scotland | ☎+44-(0)1309-692194
welcome@ecovillage.org www.ecovillage.org

Global Ecovillage Network of Europe

E Calle Abajo 1 | 31489 Artieda
secretariat@gen-europe.org www.gen-europe.org

GEN Austria

Our vision is to inspire and encourage people to find and form their way of living together. We facilitate exchange between community projects, transfer knowledge and experience and act as a role model for a biosphere friendly and communal culture.
Specifically, we organize excursions to existing community projects or invite to lectures. We regularly take part in network events in Austria and Europe focusing on community, sustainable lifestyles and regional cooperation.
We are happy about people who want to participate actively!

GEN Austria

A Die österreichische Netzwerkgruppe für gemeinschaftliches Leben
Bandgasse 27 | 1070 Wien
☎+43-(0)664 862 43 94 mail@gen-austria.at gen-austria.at

→Wohnprojekt Hasendorf (A)

GEN Deutschland (GEN Germany)

The association GEN Germany was founded in 2015. Currently, GEN Germany consists of 15 communities. What unites the people in this network is the desire to try out "new things" and find ways to lead a good and sustainable life - ecologically, economically and socially. Nobody can answer the social questions we face alone. Ecovillages provide research spaces for experimental living in peace and in harmony with nature. At the same time, GEN Germany promotes the formation of communities and solidarity as core elements of a more resilient society. Part of the vision is the creation of a diverse and shared pool of knowledge for a sustainable lifestyle.
GEN Germany is set up as a sociocratic organisation. Formally it consists of the extended board, the working groups on the dimensions of sustainability (economy, ecology, social affairs and culture) and the members. Many communities of the association GEN Germany have an impact on society. Through public relations work, cooperation with communities, universities and civil society, we want to make an experimental, non-material and practical contribution to showing the wider public that there is a future suitable for the grandchildren.
GEN Germany offers the following membership options:
1. full membership for communities that want to actively participate in the community network as a member.
2. sponsoring membership for communities, institutions and individuals who would like to support the work of our network.

Are you interested? Feel free to visit our website at www.gen-deutschland.de or contact us at info@gen-deutschland.de

GEN Deutschland

D Sieben Linden 1 | 38489 Beetzendorf
info@gen-deutschland.de gen-deutschland.de

→Akademie für angewandtes gutes Leben (D)
→gASTWERKe e.V. (D)
→Gemeinschaft Clan B (D)
→Gemeinschaft Haslachhof e.V. (D)
→Gemeinschaft Lebensbogen (D)
→Gemeinschaft Schloss Blumenthal (D)
→Gemeinschaft Sulzbrunn (D)
→Kunesto Cobstädt eG (D)
→Lebensgarten Steyerberg (D)
→Lebensgemeinschaft auf Schloss Tonndorf eG (D)
→Lebenslernort Am Windberg (D)
→Lebenstraum Gemeinschaft Jahnishausen (D)
→Mittendrin leben eG (D)
→Nature Community eG (D)
→Rawaule (D)
→Sieben Linden (D)
→Villa Locomuna (D)
→ZEGG - Zentrum für experimentelle Gesellschaftsgestaltung (D)

GEN Europe

GEN Europe is the network for ecovillages and sustainable communities in Europe. It is the regional branch of the Global Ecovillage Network, and connects over 100 member ecovillages and over 20 national ecovillage networks in Europe. Our network connects communities to promote learning and knowledge exchange; builds capacity for the creation of new ecovillages; and promotes and advocates for ecovillage lifestyles for the transition to a resilient society. We warmly invite all those curious about ecovillages to our annual gathering, the European Ecovillage Conference, which takes place in a different European ecovillage each year.

Global Ecovillage Network of Europe

E Calle Abajo 1 | 31489 Artieda
secretariat@gen-europe.org www.gen-europe.org

→Cambium (A)

→ Cohousing Pomali (A)
→ Garten der Generationen (A)
→ GEN Austria (A)
→ Ces - Fondazione per la rinascita di Chiesso (CH) ❋
→ Herzfeld Sennrüti (CH)
→ Schloss Glarisegg (CH)
→ Smala Ecovillages (CH) ❋
→ Yamagishi Verein (CH)
→ Zentrum der Einheit Schweibenalp (CH)
→ Freie Feldlage - Gemeinschaft in der Heilstätte Harzgerode (D) ❋
→ Friedenshof-Kommunität (D)
→ Gemeinschaft Lebensbogen (D)
→ Gemeinschaft Sonnenwald Schernbach (D) ❋
→ Gemeinschaft Sulzbrunn (D) ❋
→ Lebensgarten Steyerberg (D) ❋
→ Lebensgemeinschaft auf Schloss Tonndorf eG (D) ❋
→ Lebenstraum Gemeinschaft Jahnishausen (D) ❋
→ Puls der Erde (D)
→ Schloss Tempelhof eG (D) ❋
→ Schöpferisches Zentrum OASE (D)
→ Sieben Linden (D) ❋
→ Stamm der Likatier (D)
→ Villa Locomuna (D)
→ ZEGG - Zentrum für experimentelle Gesellschaftsgestaltung (D) ❋
→ Bjergager Solby (DK) ❋
→ Dyssekilde Ecovillage (DK) ❋
→ Freetown Christiania (DK)
→ Gaia Trust (DK)
→ Hertha Levefællesskab (DK) ❋

LØS (Landsforeningen for Økosamfund)

DK The Danish Association of Sustainable Communities

☎ +45-229 522 92 ✉ info@okosamfund.dk 💻 www.okosamfund.dk

→ Amalurra Artzentales (E)
→ Arterra Bizimodu (E) ❋
→ El Calabacino (E)

→ Los Portales (E)
→ Possibilica (E)
→ Q (E)

Red Ibérica de Ecoaldeas (RIE)

E Calle Abajo 1 | 31480 Artieda (Navarra)
☎ +34-(9)645 541 333 info@ecohabitar.org rie.ecovillage.org

→ Sunseed Desert Technology (E)
→ Lilleoru (EE)
→ ECOlonie - Centre Ecologique International (F)
→ Eotopia (F)
→ La Borie Noble (F)
→ La Lune Nette (F)

GEN Finland

SF Suomen Kestävän Elämäntavan Yhteisöt ry | c/o Kallio, Postintie 3
09770 Valittula
☎ +358-(0)407 773 869 skey.pj@gmail.com www.skey.fi

→ Keuruun Ekokylä (SF)
→ The Findhorn Foundation (UK)
→ Kalamos Island Eco-community (GR)
→ Biofalu Máriahalom (H)
→ Gyürüfü Ecovillage (Gyurufu) (H)
→ Comunità di Etica Vivente (I)
→ Damanhur (I)
→ Ecovillaggio Torri Superiore (I)
→ Giardino della Gioia (I)
→ La Città della Luce (I)
→ La Comune di Bagnaia (I)

Rete Italiana dei Villagi Ecologici - RIVE

I Strada Bagnaia 37 | 53018 Sovicille (SI)
segreteria@ecovillaggi.it www.ecovillaggi.it

→ Zonca (I)
→ Cloughjordan Ecovillage (IRL)
→ Kibbuz Lotan (IL)
→ Gaia Eco Village (N)

Norsk Økosamfunnsforening/GEN Norge

N Norwegian Ecovillage Association (NEA)
leder@okosamfunn.no www.okosamfunn.no

Vallersund Gård

N Camphill Landsbystiftelse i Norge | Vallersund Gård
7167 Valsøyveien, Vallersund
+47-(0)7252-7080 vallersund.camphill.no

GEN Nederland

NL Ecodorpen Netwerk Nederland | p/a Waalbandijk 8 | 4064 CB Varik
info@ecodorpennetwerk.nl ecodorpennetwerk.nl

→ Ecoaldea Vegetariana Espiral (P)

→ Tamera (P)

→ Global Campus Palestine (PS)

→ Ecovillage Grishino (RUS)

GEN Russia (REEN)

RUS Russian Ecovillage and Ecoinitiatives Network | ul. Zhygulyovskaya, 18-3
109117 Moscow | +7-909-156-03-36, -926-323-04-60
irina.sardarova@gmail.com; dmitry.solovyev@list.ru; genrussia@gmail.com
www.gen-russia.ru, Mitglieder auf tinyurl.com/eurotopia8751

→ Kind Land (Dobraya Zemlya) (RUS)

→ Suderbyn Permaculture Ecovillage (S)

→ Vedapark (S)

→ Sunny Hill (SLO)

→ Zaježka (SK)

→ Imece Evi (TR)

Knidia Eco Farm

TR Ali Somer | Yazi koy degirmen buku Datça | 48900 Mugla
+90-(0)532-211 38 58 ali@datca.com www.knidia.com

GEN Finland/Suomen kestävän elämäntavan yhteisöt ry

GEN Finland is a national network of ecovillages and individuals interested in communal and sustainable lifestyle. The NGO was founded in 2003. In 2019 it has 12 intentional communities as members (to be found at http://genfinland.weebly.com/yhteisoumlt.html).
Our aims are:

- To promote and develop a sustainable lifestyle and communality
- To support the development of new ecovillages and to strengthen the collaboration of existing ones
- To reinforce the common significance of eco-communities in society

Our main activities for 2019:

- Sustainable lifestyle trainings and gatherings for all ages
- Various Erasmus+ partnership trainings in Europe
- Website, FB, email list, info events
- International connections with GEN Europe, Baltic Ecovillage Network etc.

GEN Finland

SF Suomen Kestävän Elämäntavan Yhteisöt ry | c/o Kallio, Postintie 3
09770 Valittula
☎+358-(0)407 773 869 ✉skey.pj@gmail.com 💻www.skey.fi

→Keuruun Ekokylä (SF)
→Väinölä (SF)

GEN Netherlands/Ecodorpennetwerk/ Vereniging Ecodorp Nederland

The Dutch Ecovillage Network aims to share the joint knowledge and experiences from different projects and initiatives in the Netherlands and to join forces in common eco-village interests. You can also contact the network if you want to know more about the various eco villages in the Netherlands.
Keep up-to-date with all developments by subscribing to the free newsletter which comes out minimum twice a year or visit the yearly festival (celebration) or one of the, more substantive, network weekends.

GEN Nederland

NL Ecodorpen Netwerk Nederland | p/a Waalbandijk 8 | 4064 CB Varik
✉info@ecodorpennetwerk.nl 💻ecodorpennetwerk.nl

→De Vlierhof (D)
→C.W. Lismortel (NL)
→Coöperatieve Vereniging Ecodorpen Gelderland (NL)
→De Bongerd (NL)
→Ecodorp Bergen (NL)
→Emmaus Haarzuilens (NL)
→GWL-terrein (NL)

German Association of Beguines

We today call ourselves Beguine, modeled after thousands of women who lived and worked independently and charitably since the early 13th century. At that time they freed themselves from the influence of the church, the estates, the patriarchal dictates. They founded large and small communities, earned their own money with crafts, nursing and terminal care. They were teachers, administrators, counselors. They were also called “good people”, “field nuns”, “forest sisters”, “sisters of the free spirit”.

Since the 1990s, women have been bringing this spirit back to life. Whether Christian-faithful to the church, ecumenical-cosmopolitan, feminist-spiritual, or political-social - we feel connected to the tradition of the beginning thought: together women move a lot. Women´s power is special. Women do good in society.
There are now 16 beguinages in Germany, where more than 200 women live and work. Another 300 or so Beguine live as single in the sense of our tradition.
Since 2004 we have joined together under the umbrella organization of the Beguines. We regularly organize trips to the old and modern places of women´s culture. Every year on the last weekend of August, we organize a nationwide get-together with a lecture, information, singing, playing and dancing. And a lot of fun!
Detailed information about all our beguinages, meetings and travel can be found on the website www.dachverband-der-beginen.de.

Dachverband der Beginen e.V.

D Geschäftsstelle Essen Beginenhof | Goethestraße 63-65 | 45130 Essen
☎+49-(0)201-316 69 23 kontakt@dachverband-der-beginen.de
www.dachverband-der-beginen.de

→Beginenhof Dortmund (D)

Givers-Movement

The 'GIVERS-/DONATORS- or GIFT-MOVEMENT' is the "movement of people, who only want to exchange gifts in (global) love", i.e. non-violent sharing, in order to live in holistic responsibility. We believe that one can only be truly happy ("heart-joyous", "soul-fulfilled") by following the conscience and leading a holistically responsible life (like a healthy cell in the organism of the world), and that a "reason-instead-of-'fists'-culture" ('fist' meaning all non-argumentative manipulation) is necessary in sufficient measure amongst people. We offer this as "MOST-POSSIBLE "UNIDEOLOGICAL" RESP. "OPEN TO CONSENSUS" "(INITIAL-) AGREEMENT-DENOMINATOR FOR GLOBAL RESPONSIBILITY": "AS 'GLOBAL CONSENSUS-BROTHERS AND SISTERS' LET US PLAN, WHAT IS NECESSARY – IN REALLY OPEN AND FREE ARGUMENTATION-CULTURE..." "FREE EARTH FOR FREE SHARING PEOPLE!" When individuals do not align themselves according to the wellbeing of the whole, and build corresponding responsibility-community-moulds, then it clearly cannot produce decisions, which are responsible for all. If, however, they are prepared to do so and step into a "Global thinking, local acting!"-mindset, then one should not preset for them other fixed ("ideological") content, but put one's trust in their free consensus search. We try to form "consensus-communities of global sharing", as a path to "global consensus democracy (demogracy)".
For instance, we work in the following forms of "life-culture":

1. "(Truth-)pilgrims", without money or identity-papers of the state, on the street with banners, as "collectors for the communities of the future" (today in part combined with internet-work in communes).
2. "Houses of Hospitality", or "families-of-the-poor" for as much as possible unconditional sharing with those in need (our core group, the "givers", merely "own" houses and land according to "moral usage rights", i.e. through "organic integration" into the 'biotopic' relations of life, without purchase, enforceable contracts etc. They live without own commercial electricity, water etc, just from leftovers, gifts and self-sufficiency).
3. Self-sufficient "village families" as a "survival model" for humanity, for instance as a

seedling our little "Friedensgarten" (Peace-Garden) community.

Our movement is – in the sense of searching for the "smallest necessary denominator for global responsibility" – open to people of diverse religious backgrounds and world views, and has "spiritually non-violent" references to Mahavira, Buddha, Lao-Tse, Diogenes, Jesus, Tolstoi, Gandhi, Maurin, Gräser, Abbe Pierre, "Peace Pilgrim", Gröning etc. We regard people already as so called "giver-allies" as soon as they believe, that our core-ideal is theoretically correct, in a self-determined application of practical consequences. From the smallest steps (participating in thought exchange [=decisive root!] or in charitable associations etc.) up to radical steps, everything is possible, so that nobody should be overstrained.

Givers-(or Gift-)Movement/Schenker-(oder Schenk-)Bewegung

D Lilitopia-Projekt, Eulenweg 16 | 35260 Stadtallendorf
☎+49-(0)157 850 656 76 ✉schenker.oeff@holistic-love.net
💻www.global-love.eu, www.holistic-love.net

Andreas Paradies

D c/o Öff Öff, Eulenweg 16 | 35260 Stadtallendorf
☎+49-(0)6428-605 87 73, +49-(0)35939-806 02
✉schenker.oeff@holistic-love.net 💻www.global-love.eu

→Biotopia (D)
→Friedensgarten (D)
→Haus der Gastfreundschaft (HdG) (D)
→Lilitopia (D)

ISKCON

We have not received an updated description of the International Society for Krishna Consciousness

International Society for Krishna Consciousness

WWW 💻iskcon.org

→ISKCON Köln e.V. (D)
→La Nouvelle Mayapour (F)
→Krisna Völgy (H)
→Villa Vrindavana (I)

Kollektivhus NU

The Swedish Association Kollektivhus NU is a national association working for collaborative housing and other alternative ways of living. We have appr. 45 cohouses all over Sweden in our network and that they can all be found under "Hus" on our website.

Kollektivhus NU

S ✉info@kollektivhus.se 💻www.kollektivhus.se

BoAktiv Landgången
S Sadelmakarebyn 6C | 21849 Bunkeflostrand
boaktiv@gmail.com www.boaktiv.se

→ Bogemenskapsföreningen Tunnan (S)

Dunderbacken
S Hägerstensvägen 237 C | 129 35 Hägersten
info@dunderbacken.net dunderbacken.se

→ Färdknäppen Kollektiv (S)

→ Kollektivhus Angsviksgarden (Ängsviksgården) (S)

→ Kollektivhuset Södra Station (S)

→ Tersen (S)

KommuJa

KommuJa is a German network of communes that consider themselves politically active. With life in a commune we try to come closer to our (political) utopias, however we do not see community as an end in itself. The groups are libertarian, emancipatory, they function in solidarity and are inspired by ecological ideas and thought processes. The communities within the network share the following fundamental ideas:

- We want equitable cooperation and we object to power structures. Therefore our communes exhibit structural equality. Discriminatory structures must be uncovered and changed. We endeavour to create comprehensibility, transparency and traceability in all communal processes. We make decisions according to the principle of consensus, in order to find solutions that can be supported by all. It is important to us that the people in the commune can further their personal development and that they show a fundamental willingness to debate, to participate in and learn about group activities. The fundamental idea of our communal living incorporates direct communication, we don't want to solve conflicts and differences via structures and want to treat each other mindfully. We try to provide children with a space in which they can develop according to their tendencies and speed, and where they can independently choose their co-parental figures. People of different sexual orientations and in a variety of different relationship structures live in the communities and manifest an alternative to the societal normalisation of the nuclear family, of single person households and heterosexual couple relationships.
- We want to free ourselves from the prevailing models of accountancy and ownership. Therefore our economics are based on solidarity, e.g. in that we share a communal kitty for everyone, from which we take according to personal needs. In doing so we question our consumer habits and make time and space for interpersonal encounters. Since private ownership, ranging from capital to real estate and means of production, leads to power inequalities and dependencies, our communes are based on communal ownership. We do not want to define people simply as a function of their performance. Therefore we strive to disengage from externally determined work. In many communes work is collective and self-determined. The communes within our network support one another through interpersonal exchanges of information, goods and work.

- The political communes want to chane gsociety's standards and conditions. We already contribute to this to some extent by our alternative lifestyles. Distant goals are egalitarian (free and equal) structures that enable an emancipated, solidary and ecological life. We therefore make a stand for human togetherness and sharing, as well as against the increasing individualisation within a society that is strongly focused on consumption, purchase and the exploitation of nature. The individuals within the communes are also active in a variety of political fields, such as anti-fascism, anti-atomic energy, BUKO, media freedom, international solidarity, preservation of cultivated plants and anti-GM.... Further information can be found on our website (a list of associated projects is available under the heading and/or by keyword search "communes" ("Kommunen").

KommuJa!

D Netzwerk politischer Kommunen www.kommuja.de

→BlackSwan (D)
→Finkenburg (D)
→Galions e.V. (D)
→gASTWERKe e.V. (D)
→Gemeinschaft Lebensbogen (D)
→Holunderhof e.V. (D)
→Kommune Buchhagen (D)
→Kommune Hof Rossee (D)
→Kommune Karmitz (D)
→Kommune Lossehof (D)
→Kommune Niederkaufungen (D)
→Kommune Olgashof (D)
→Kommune Uthlede (D)
→Kommune Volzendorf - AufBruchLandung (D)
→Kommune Waltershausen (D)
→Krumme Eiche e.V. (D)
→Laakenhof (D)
→Mühle - Wohn- und Baugenossenschaft e.G. (D)
→ÖkoLeA (D)
→SSM - Sozialistische Selbsthilfe Köln-Mülheim (D)
→Vereinigte Varben WaWaVox (D)
→Villa Locomuna (D)
→Vitopia (D)
→Wendland-Hof e.G. (D)

Lebenshäuser ("Life Houses")

Our social system is increasingly overstrained and cannot provide the necessary feeling of security for many people. Therefore the association "Life House – ecumenical community for social integration" ("Lebenshaus - ökumenische Gemeinschaft für soziale Integration e.V.") was founded in Trossingen in 1985. The founding members were nearly all active in peace circles, various churches and/or the International Fellowship for Reconciliation. The goal is to provide a framework for people who are looking for a protective community, for social or mental health reasons. During the temporally restricted stay of each "guest", we support her/him in her/his efforts to overcome her/his "life crisis" and help him search for other possibilities in life. In 1987 we were able to move into a renovated listed building, a former factory in Trossingen. In the meantime, further "life houses" have arisen in Southern Germany, which fulfil essentially the same goals as the Lebenshaus in Trossingen and are connected with one another, but not part of the same "organisation". We meet once a year for an exchange of experiences.

Lebenshäuser

D **Kirchstr. 23 | 78647 Trossingen | ☎+49-(0)7425-1446**
lebenshaus@t-online.de www.lebenshaus-trossingen.de

→Lebenshaus Heitersheim e.V. (D)
→Lebenshaus im Chiemgau (D)
→Lebenshaus Trossingen (D)

Longo Mai

At the request of the parties involved, we do not publish any descriptions or community addresses of the Longo Mai Network, only the information published on the website.

Pro Longo Mai

CH **St. Johanns-Vorstadt 13, Postfach 1848 | 4001 Basel**
☎+41-(0)61 262 01 11 info@prolongomai.ch www.prolongomai.ch

LØS - Landforeningen Økologisk Samfund/The Danish Association of Sustainable Communities

A preliminary meeting of socially, ecologically and spiritually orientated communities took place at Fjordvang farm in Thy, Denmark, January 1993. This meeting led to the formation of a network of sustainable communities. The purpose is: To further and enhance a way of life that makes possible wholesome cycles between human beings and nature, in a caring and respectful way towards all sentient beings and our natural resources. LØS is dedicated to establish living environments that are ecologically and socially sustainable in rural, as well as in urban areas. Our visions:

- Co-operation based on ecological, social and spiritual awareness
- Ethical and ecological use of energy resources - mineral, biological, human and cultural
- Autonomy, intimacy and accountability in governance and economy

- Openness to multiplicity of views of life and ways of living
- Promoting the growing awareness of global interconnectedness
- Founding and co-operation of ecological and socially sustainable communities in rural and urban areas
- Continuously letting the vision evolve through seminars, exchange, vision quests and work forums.

A list of our members can be found at www.okosamfund.dk/medlemmer

LØS (Landsforeningen for Økosamfund)

DK The Danish Association of Sustainable Communities

☎+45-229 522 92 info@okosamfund.dk www.okosamfund.dk

→A-70 Toustrup Mark (DK)
→Bjergager Solby (DK)
→Dyssekilde Ecovillage (DK)
→Freetown Christiania (DK)
→Hertha Levefællesskab (DK)
→Kilian Water Ltd. (DK)
→Nordic Folkecenter for Renewable Energies (DK)
→Økosamfundet Soleng (DK)
→Tranehøj (DK)
→Udgården - Lading Andelsboligforening (DK)

New Creation Christian Community

The New Creation Christian Community is part of the Jesus Fellowship Church. The Fellowship is an evangelical Christian church with a charismatic emphasis. The Community consists of around 350 residents in over 30 houses around the U.K. Each community house contains anything between six and sixty people, who live as a large "family". Each house has a "common purse" arrangement, with members pooling their income to meet all personal and household expenses. All members are on an equal footing, with no privileges or extra financial incentives being accorded to anyone, including any of the leaders. The New Creation Christian Community has pioneered co-operative Christian businesses which provide paid employment for many who were previously unemployed. We receive visitors on a temporary basis for up to twelve months. If your stay is longer than a week, you pay a board and lodging charge, but the remainder of your income is entirely at your disposal.

New Creation Christian Community

UK Nether Heyford | NN7 3LB Northampton

☎+44-(0)1327-344500 info@jesus.org.uk christiancommunity.org.uk

→Jesus Fellowship (UK)

GEN Norge/Norsk Økosamfunnsforening

The purpose of the association is to develop and facilitate sustainable development by supporting the establishment of ecovillages, ecoclusters and ecotours or other measures that can promote a more ecological or holistic lifestyle and values that are in line with such a goal. NEA/GEN Norway will also support work that communicates knowledge about ecology, sustainable development and a more holistic life philosophy in society, in order to contribute to a positive development for the whole world. NEA/GEN Norway has over 390 members on their mailing list and 1095 members on the Facebook group. We send out between 4 and 6 newsletters each year and have a number of open meetings in Oslo every spring and autumn. There is often a good mix with people, and many different projects from home and abroad are presented. We also arrange seminars, study circles and courses. In addition, we have some activity on facebook where we spread useful information about trips, projects, courses and other activities that may be relevant. We also allow members to post items. Meetings are held in Oslo and you will find more information by visiting facebook and our website.
Every year we publish an updated "yearbook" that lists established projects, projects in planning, and partner associations. At the beginning of 2018 we had 11 established ecovillages, 21 planned projects and 8 partners associations. There are certain to be more than this, but not all projects are known to us, and others do not answer our e-mails or telephone calls. We are also in touch with a number of projects around the world, in the Middle East and Africa, that have active members or supporters living in Norway. Our current contact person for overseas relations is Ingunn Grande (integralion@hotmail.com).

GEN Norge/Norsk Økosamfunnsforening

N **✉ leder@okosamfunn.no, integralion@hotmail.com 💻 www.okosamfunn.no**

→Bergen Ecovillage (N)

Radical Routes

Radical Routes is a network of radical co-ops whose members are committed to working for positive social change. The network is made up mainly of housing co-ops of various sizes (none with more than 17 members), a few workers co-ops and a couple of social centres.

Radical Routes (General Enquiries)

UK **16 Sholebroke Avenue, Leeds, | Leeds, West Yorkshire, LS7 3HB**
☎ +44-(0)845 330 4510, -(0)1422 842 327 💻 www.radicalroutes.org.uk/

→Brambles Housing Co-op (UK)
→Cornerstone Housing Co-op (UK)
→Coventry Peace House (UK)
→Fireside Housing Co-op (UK)
→Keveral Farm (UK)
→Lifespan Community Collective (UK)

Red Ibérica de Ecoaldeas

Iberian Ecovillage Network (RIE) is the network that relates, connects and promotes:

- the different Ecovillages wich are working in Spain and Portugal
- the different projects that are wanting to create Ecovillages
- Some different associations with similar objectives
- People who are living, searching, finding and expressing themselves, to be fully present in this world: respecting the planet, respecting people.

In November 2001, the Founding Assembly of the Iberian Ecovillage Network was held in Madrid. From then until now our main objectives are:

- sharing information and resources among network members, other individuals or groups.
- dissemination of the concept of ecovillage and sustainable settlement
- facilitate rapprochement of people from towns and cities who want to engage in these life forms.

Red Ibérica de Ecoaldeas (RIE)

E Calle Abajo 1 | 31480 Artieda (Navarra)
☎+34-(9)645 541 333 ✉info@ecohabitar.org
rie.ecovillage.org

RIE
RED IBERICA DE ECOALDEAS

→Amalurra Artzentales (E)
→Amalurra Granada (E)
→El Calabacino (E)
→Lakabe (E)
→Los Portales (E)
→Matavenero y Poibueno (E)
→Sunseed Desert Technology (E)
→Ecoaldea Vegetariana Espiral (P)

REEN (Russian Ecovillage and Ecoinitiative Network)

The Russian Ecovillage and Ecoinitiative Network (GEN Russia) has been existing informally since 2005, and was formalized in 2011. It unites those eco-villages and eco-initiatives (both living in communities and not) which would like to spread their ideas of sustainable living in wider society, as well as build up a larger and tighter community of ecovillages and intentional communities. Today, in 2019, we have 15 full members and several associate members. Our full members include ecovillages and ecoinitiatives from the first «wave» (created in the 1990s) and from the second «wave» of new villages which have grown up from the popular idea of 1-hectar kin domains (also: Anastasia-Villages) and the third «wave» (post-Anastasia villages and new direction projects).

Participants of the Network regardless of their confession believe that living sustainably on our planet depends on the realization of the following principles:

- Environmental care, preservation and recovery of natural biodiversity as absolute priority of any human activity on Earth
- Principle of conscious minimalism coming up to replace the philosophy of unrestrained, constantly growing consumerism
- Aspiration to the personal development and spiritual growth instead of striving to material accumulation and power
- Peace as a universal principle concerning all sides of daily activity and interaction in the society.

REEN has several on-going projects:
1. It conducts activities to marry the ecovillage movement with old traditional villages to share experience and empower the villages
2. Since 2012 it has been conducting the Festival of eco-technologies with 40 master-classes conducted in 4 days and 300 participants hosted
3. It conducts events for students and the general public to spread the ideas of ecovillages
4. It is developing new approaches to education in rural areas

GEN Russia (REEN)

RUS **Russian Ecovillage and Ecoinitiatives Network | ul. Zhygulyovskaya, 18-3 | 109117 Moscow**

☎ +7-909-156-03-36, -926-323-04-60 ✉ irina.sardarova@gmail.com; dmitry.solovyev@list.ru; genrussia@gmail.com 💻 www.gen-russia.ru, Mitglieder auf tinyurl.com/eurotopia8751

→ Ecovillage Grishino (RUS)
→ EcoVillage Yasnoe (RUS)
→ Kind Land (Dobraya Zemlya) (RUS)

Réseau Français des Eco-Villages (RFEV)

We are working on a Label "French Ecovillage Network" which will be open to all ecoplaces that share respect of human being, respect of animal, respect of nature, respect of everything that is living. And also respect of sociocraty and horizontal decision making.
We offer training courses, workshops and support for project leaders and ecovillage candidates. We have a lot to do in France and we try to communicate with the town halls and local officials and hope laws will change to facilitate the growing of real ecovillages.

Réseau Français des Eco-Villages (RFEV)

F **💻 ecovillagesfrance.wordpress.com**
💻 www.facebook.com/EcovillagesFrance

RIVE - Rete Italiana dei Villaggi Ecologici

Rete Italiana Villaggi ecologici – RIVE (Italian ecovillage network), founded in 1996, is the umbrella organization for ecovillages, intentional communities and several Italian transition projects and associations. Since its conception RIVE promoted and developed cooperation among its members and it encouraged and supported successful transition projects.
RIVE's members are spread out on the entire Italian territory and they present different approaches concerning their organizational structures, economical sustainability and ecological standards. RIVE's members are active citizens working in the educational field, promoting seminars and practical workshops for the identification and dissemination of "best practices" of sustainability through a responsible attitude towards the territory, its resources and all its inhabitants.
RIVE is aware of the importance of educating, involving and promoting the active participation of young people through youth exchange projects, education of youth leaders, making education and international exchanges available to as many of its members as possible, in order to facilitate youth participation in this moment of global transition towards a more sustainable life style. Formal and informal trainings for youth leaders and for young people are put in place by most of RIVE's members both locally and internationally, wishing to turn the frustration of youth regarding their future and their ability to adapt to global changes into empowerment for constructive change implying the active participation of young people in the development of a "new society".
One of the main objectives of RIVE is to find and implement innovative techniques and ideas in the social, ecological, technological, economical and world view dimensions for the promotion of sustainable development. RIVE is partner of several international organization such as ECOLISE, Global Ecovillage Network (GEN); together with its partners RIVE constitutes groupes for the direct interaction with external actors (institutional and private level) for the dissemination of best practices in the fields of sustainability and socio-ecology.; thanks to their knowledge in sustainable and ecological practices ecovillages and intentional communities are often able to start up a new economy respectful of the environment, the local culture as well as future generations. With time they also showed to be very effective in the reconstruction of a social network in the rural area where they are located and they often become bioregional hubs as well as bridges to the international scene thanks to their strong transnational network.

Rete Italiana dei Villagi Ecologici - RIVE

I Strada Bagnaia 37 | 53018 Sovicille (SI)

segreteria@ecovillaggi.it www.ecovillaggi.it

→Comunità di Etica Vivente (I)
→Damanhur (I)
→Ecovillaggio Torri Superiore (I)
→Fattoria solidale Brugine (I)
→Giardino della Gioia (I)
→La Città della Luce (I)
→La Comune di Bagnaia (I)

→La torre di mezzo (I)
→Progetto Gaia Terra (I)
→Reggioli (I)
→Roncadella (I)
→Tempo di vivere (I)
→Utopiaggia (I)

Soshona Freundeskreis

We, Soshona, are an open circle of friends which currently comprises about 100 people from the German Rhein-Main area. We strive to provide a trustworthy framework for exchanges and experiences regarding, amongst others, the following subjects: Love, friendship, community, spirituality, healing, children and personal development. Approximately 20-25 people come to our group meetings, which have occurred regularly since 1997. A small and variable team organises our meetings every 4-6 weeks. Alongside group meetings there are regular working groups who meet separately, for instance women's and men's groups. Two communities have evolved from our midst (see below). In the summer we organise our own summer camp and at New Year's Eve we organise communal time with friends from our regional network. On two further long weekends in the year we meet for communal time. We celebrate seasonal festivities with the help of our fire pit and rituals. It's worth being amongst us! Interested people are welcome!

Soshona
D **Schlegelstr. 12 | 63128 Dietzenbach**
☎+49-(0)6074-481 1218 ✉soshonien@yahoo.de 💻www.soshona.de

→Lebensgemeinschaft Berstadt aka Rainbow Community Wetterau (D)
→Maibacher Schweiz (D)

Twelve Tribes

Our life in the Communities of the Twelve Tribes can be compared to a beehive: Bees have no concern for themselves but have a common identity. The workers are responsible to collect nectar and guard the hive. The young bees keep the hive in good condition, feed the larvae, and support in other household chores. There is never a dull moment in the busy life of a little bee. This is much like the life that we have. We love to do all together. Daily we gather to thank our Master for His salvation, and to hear Him speak to us through one another. This gathering keeps alive a genuine love and care for each other. As we work, we take advantage of the daily situations, guarding ourselves from the selfishness and pride that would come in to separate us and take away our love. Our children are a vital part of our life. We not only educate them, but we work with them to accomplish the simple tasks necessary to maintain a family life. Our life is not a dull routine of chores, but is full of the warmth that comes from the sweet fellowship of friends speaking their hearts to one another, celebrating the Sabbath every week, and participating in weddings and festivals.

We welcome anyone to experience our life with us. Please come and see what it is like to be part of a beehive of people expressing the warmth and love of our Creator.

Editorial note: We publish the description of the Twelve Tribes Communities Network and the addresses of the associated communities here, with the explicit note that we did not invite the Twelve Tribes Communities into the main part with the communities' self-descriptions. Too different are the eurotopia editorial team's and the Twelve Tribe communities' world-views of how to deal with their children. There, physical punishment is part of education, and because this is forbidden in Germany, the Twelve Tribes Community of Klosterzimmern (Bavaria) has moved to the Czech Republic. We wrote more about this in the last edition of eurotopia (the article can also be found online on eurotopia.directory/articles2014.html). However, in many ways the Twelve Tribes are also an inspired community movement. We thank both our readers and the members of the Twelve Tribes communities for accepting the compromise to present the Twelve Tribes (only) here as a network.

Communities of the Twelve Tribes

CZ Preserved Seed Farm | Msecke Zehrovice 150 | 27064 Msecke Zehrovice
☎+42-(0)721-305 558 ✉kontakt@zwoelf-staemme.de
💻www.zwoelf-staemme.de

Komunita ve Skalné

CZ Ceská 481 | 351 34 Skalná | ☎+420-910 117 288
✉kontakt@zwoelf-staemme.de 💻www.zwoelf-staemme.de

The Preserved Seed Farm

CZ Farma Uchované Semínko z.s. Mšecké Žehrovice c. 150
27064 Mšecké Žehrovice | ☎+420-721 305 558
✉kontakt@zwoelf-staemme.de 💻www.zwoelf-staemme.de

Communidad de Irún

E Caserio Barracas 88 | 20305 Irún, Guipuzcoa
☎+34-943 632 316 ✉barracas@docetribus.com 💻www.docetribus.org

Comunidad en Corella

E Avenida de Navarra 20 | 31591 Corella Navarra
☎+34-948-78-26-92 ✉contacto@docetribus.com 💻www.docetribus.org/

Comunidad en San Sebastian

E Paseo Ulía 375 | 20013 San Sebastian, Guipúzcoa
☎+34-943-327 983 💻www.docetribus.com

Communauté des douze tribus de Sus

F ☎+33-(0)559 661 428 ✉contact@douzetribus.fr 💻www.douzetribus.fr

The Community at Stentwood Farm

UK c/o Shebet (Philip Parish) | Dunkeswell, Honiton | Devon, EX14 4RW
☎+44-(0)1823-681 155 ✉thecommunity@stentwoodfarm.co.uk
💻www.stentwoodfarm.co.uk

Vereniging Gemeenschappelijk Wonen/ Landelijke Vereniging Centraal Wonen (LVCW)

Vereniging Gemeenschappelijk Wonen (formerly known as LVCW) is a Dutch network of intentional communities and co-housing projects, founded in 1977.
Today, the 71 communities consist of a total of 1700 households, or 3000-4000 adults and children. Find addresses at gemeenschappelijkwonen.nl/adressen-lidprojecten

Vereniging Gemeenschappelijk Wonen

NL **Postbus 19163 | 3501 DD Utrecht | ☎+31-(0)6-130 158 69**
info@gemeenschappelijkwonen.nl www.gemeenschappelijkwonen.nl

WWOOF (World Wide Opportunities on Organic Farms)

WWOOF is a worldwide movement linking volunteers with organic farmers and growers to promote cultural and educational experiences based on trust and non monetary exchanges, thereby helping to build a sustainable global community. We are primarily a network of farmers and growers - but many of our farms are run by communities.
WWOOF is organised always when possible at a national level to provide a better service to our members. Please join or contact directly the WWOOF organisation responsible for the hosts on the country where you want to volunteer and they will be pleased to assist you. On this page you can find a complete list of the official websites and contacts: wwoof.net/#destination

WWOOF

UK **World Wide Opportunities on Organic Farms**
Green Fish Resource Centre, 46-50 Oldham Street | M4 1LE Manchester
contact@wwoof.net www.wwoof.net

→Co-Housing HuisElf (B)
→Dorfgemeinschaft Klein Hundorf (D)
→Eine Spinnerei (D)
→Gemeinschaft Sulzbrunn (D)
→Hofgemeinschaft Kotthausen (D)
→Kastanienhof Marth (D)
→WWOOF Deutschland (D)
→Zar Nekla die Zweite (D)
→El Pardal (E)
→Los Portales (E)
→Tanquián (E)
→La ferme de Chenèvre (F)
→La Ferme du Collet (F)
→Ramounat (F)

→ Keveral Farm (UK)
→ Old Hall Community (UK)
→ WWOOF United Kingdom (UK)
→ WWOOF Greece (GR)
→ Fattoria solidale Brugine (I)
→ LaCasaRotta (I)

WWOOF Italia

I Via Casavecchia 109 | 57022 Castagneto Carducci LI
☎ +39-0565-765001 ✉ info@wwoof.it 💻 www.wwoof.it

WWOOF Moldova

MOL WWOOF Moldova | Moscova 21 of. 405 | 2068 Chisinau
☎ +373-78705222 ✉ info@wwoof.org.md 💻 www.wwoof.org.md

→ Atlantis Exchange (N)

WWOOF Netherlands

NL Lizzy Ansinghstraat 65 | 1072 RE Amsterdam
☎ +31-(0)6-42154177 ✉ info@wwoofnetherlands.org 💻 wwoofnetherlands.org

→ La Belle Verte (P)

WWOOF Portugal

P ✉ info@wwoof.pt 💻 www.wwoof.pt

→ WWOOF Serbia (SRB)
→ Imece Evi (TR)

Useful Addresses

eurotopia is a communities directory. Still, besides the networks of communities there's a lot of other initiatives and organisations that help to make community life possible and/or work towards a sustainable life, which is a base as well as a consequence of community life. Also, there's many addresses that will be precious resources for community seekers and dwellers. Therefore we are happy to publish, in this chapter, many useful addresses. Since we concentrate our research efforts on intentional communities, the following section is not complete in any way - it's just a ot of good people, initiatives and organisations that you might be interested in.

The addresses are in alphabetical order; find online resources at the end of the chapter.

Photo: Fejø Permakultur og Omstillingsinitiativ, Denmark

Austria

Permakultur Austria

Akademie | Schiffmühlenstr. 53/15 | 1220 Wien

☎+43-(0)664-862 43 94 ✉mail@permakultur-austria-akademie.at

💻www.permakultur-austria-akademie.at

Networking - we are the oldest permaculture association in Austria and always strive to combine different initiatives in the field of sustainability. Over the years we have had many different collaborations (e.g., SOL, VISIpedia).

We see communities as important laboratories for practiced regenerative strategies on social, water and energy issues, etc. In our certification courses we regularly visit communities or invite residents to us.

projektentfaltung

... für erfolgreiche Projekte! | ☎+43-(0)720-978 837

✉r.wytek@projektentfaltung.at 💻www.projektentfaltung.at

We live in exciting times full of change. And we are lucky to be able to co-create these changes in a responsible and joyful manner.

Our organisation supports community founders and existing communities in doing so, and offers workshops and consultation for successful project development.

One focus of our work is the financing of community and residential projects.

Belgium

ECOLISE

European Network for Community-Led Initiatives
on Climate Change & Sustainability | Rue d'Edimbourg 26 | 1050 Brussels
✉info@ecolise.eu 💻www.ecolise.eu/

ECOLISE, the European network for community-led initiatives on climate change and sustainability, is a coalition of national and international networks of community-led initiatives on sustainability and climate change, as well as organisations that support a community-led transition to a resilient Europe.

ECOLISE aims to celebrate, stimulate and increase the positive impact of community-led action on sustainability and climate change, by:

- Raising the profile and highlighting the potential of what is already happening across Europe and beyond
- Sharing and co-creating knowledge and catalysing effective co-operation among member networks and other stakeholders
- Influencing European and national policy development and delivery to empower, enable and build upon the benefits of community-led action.

The members of ECOLISE include international networks of community-based initiatives such as the Transition Network (representing over 1200 Transition initiatives), the Global Ecovillage Network (15,000 ecovillages), the Permaculture movement (3 million practitioners globally) and ICLEI, the association of local governments for sustainability; national and regional networks; and other specialist bodies engaged in European-level research, training and communications to support community-led action on climate change and sustainability.

By bringing these organisations together, ECOLISE seeks to establish a common, Europe-wide agenda and a platform for collective action

JNM vzw

Jeugbond voor Natuurstudie en Milieubescherming
Kortrijksepoortstraat 192 | 9000 Gent
☎+32-(0)496-902069 ✉info@jnm.be 💻www.jnm.be

JNM vzw is a youth movement in Flanders and Brussels for everyone who's got an interest in nature and cares for the environment.

JE BEESTIGSTE JEUGDBEWEGING

Samenhuizen

Vooruitgangstraat 323 / 11 | 1030 Schaarbeek
☎+32-(0)48-992 64 64 ✉info@samenhuizen.be
💻www.samenhuizen.be

Samenhuizen is a Flemish non-profit organisation that promotes and supports all types of collaborative/intentional housing.

SamenHuizen

Croatia

Gea Viva - eco island retreat

Prodolina 864 | 21405 Milna

☎+385-(0)997 493 711 ✉info@geaviva.net 💻www.geaviva.net

Gea Viva is a transient community on the Croatian island Brac, an eco retreat camp, inviting you to re-establish your connection with nature, take time out, feel part of all life around you, awaken and transform. Our 1,35 ha permaculture site is off-grid, and we offer courses, events, Yoga and meditation.

Denmark

Fejø Permakultur og Omstillingsinitiativ

Stormosevej 36 | 4944 Fejø

☎+45-239 048 00 ✉trickynicy@gmx.de 💻fejoepermakultur.wordpress.com

Fejø Permaculture and Transition Initiativ is a association located on the island Fejø. We are a group of people living in different constellations but working together in various activities. Our perspective is to contribute to the transition of our island into a resilient and regenerative community.

Fornyet Energi/Renewable Energy

Øbrovej 9 | 4295 Stenlille

✉fornyetenergi@fornyetenergi.dk 💻www.fornyetenergi.dk

Fornyet Energi (English: Renewable Energy) is an architecture- and building firm. We help with projecting work, consulting, practical building and workshops. Our designs include COB, passive solar, mass- and rocket stoves, light, "no waste of space", clay in many ways, and smart sanitation. Building with consciousness & care - for humans & environment.

Gaia Trust

Stavnsholt Gydevej 52 | 3460 Birkerod

☎+45-(0)405 875 01 ✉rossjackson@gaia.org 💻www.gaia.org

Gaia Trust, Denmark, is a charitable entity supporting sustainability projects around the world since 1987. Two NGOs especially have been supported for many years and continue to be supported, namely (1) Global Ecovillage Network and (2) Gaia Education. Many other smaller projects receive supporting grants from time to time.

Kilian Water Ltd.

Natural wastewatertreatment, reuse of water and vitalizing drinkingwater

Torupvej 4, Vrads | 8654 Bryrup

☎+45-757 579 01 ✉info@kilianwater.com 💻www.kilianwater.com

We are a firm delivering natural systems for water treating, reuse, and vitalizing water to especially eco-villages. We have made systems for following eco-villages in Denmark: Himmerlandsbyen (2006), Friland Fase 2 (2007/2008), Friland Fase 3 (2010/2011), Den Selvforsynende Landsby (2016), Økosamfund Dyssekilde (2018). We have also contact with eco-villages in planning.

Nordic Folkecenter for Renewable Energies

Kammersgårdsvej 16, Sdr. Ydby | 7760 Hurup Thy

☎+45-979 566 00 ✉info@folkecenter.dk 💻www.folkecenter.eu

Nordic Folkecenter is a non-governmental organisation situated in the north of Denmark. The main tasks and aims of the institution are development, testing and demonstration. There are four topics NFC is active in:

1. Working with micro SME (small and middle scale enterprises) which have 2-20 employees: NFC helps the enterprises to find money, build a prototype and to test it.
2. Testing small wind turbines with a power up to 25kW.
3. Education:
   a) Visitors can book guided tours from two to six hours
   b) Schools, universities and interested people of the civil society can participate in specific trainings (e.g. already people from Turkey and Norway visited the center)
   c) Engineers, architects, physicists and students in mainly these fields can participate in the education/trainee program. The duration can vary from one to twelve months.
4. Working in developing countries: Two Folkecenters in Africa (Mali and Uganda) were already built and another one is planned in Chile.

Finland

Vegan Society of Finland

Suvilahdenkatu 4 | 00500 Helsinki

✉info@vegaaniliitto.fi 💻www.vegaaniliitto.fi

We are the Finnish Vegan society and promote and support a vegan lifestyle. We attend exhibitions and festivals as well as give presentations to a variety of organisations as part of our outreach work. We aim to make veganism an easily adopted and widely recognised approach to reducing animal and human suffering and environmental damage by means of meaningful, peaceful and factual dialogue with individuals and organisations.

France

Eco-Quartier Strasbourg

1 a place des Orphelins | 67000 Strasbourg | ☎+33-(0)776 693 978

✉association@ecoquartier-strasbourg.fr 💻www.ecoquartier-strasbourg.fr

We are an NGO who has experimented the first co-housing project in Strasbourg. Than, we took the leadership of the co-housing projects network in Strasbourg. Today, we federate more than 20 co-housings, for more than 160 inhabitants. We want now to develop the affordable co-housing projects. We are part of the french national network for co-housing projects "Coordinaction de l' habitat participatif".

F.E.V.E.

Formation et Expérimentation au Vivre Ensemble | L'Arche, 34 la basse cour
38160 Saint Antoine l'Abbaye
☎+33-(0)476 364 825 feve@feve-nv.com www.feve-nv.com

Every year since 2010, the community l'Arche de St Antoine holds a year-long training called FEVE (stands for "training and experimenting : living together") . The training includes sharing the daily life of the 30-years-experience community house (at least 3 months), and workshops about group dynamics, non-violent communication, conflict management, sociocracy... It's also possible to attend each workshop without following the whole cycle (one year includes 7 workshops, each one being between 2 and 7 days). The community welcomes many foreigners but the workshops are in French.

Les jardins de Nipponia

La Scierie | Chemin de Craste Neuve | 33680 Le Porge | ☎+33-(0)652 148 741
julesmoya@hotmail.com www.facebook.com/lesjardinsdenipponia/

We are developing a network of self-suficient places (food and energy) around the world. Each places is open and respects human and nature and gather people with similar passions (arts, crafts, sciences). There are gates in local area with information about the network to guide people from the main society to this alternative society.

Germany

AGBeratung (AGB)

Projekte beraten Projekte | Richardstr. 99 | 12043 Berlin
agb@agberatung-berlin.org www.agberatung-berlin.org

Collectively organised projects are faced with a variety of tasks: to meet ones own demands in terms of equal rights and self-organisation and at the same time to satisfy the formal demands of a capitalistically organised society. This is a path that is not very familiar and on which there are many stumbling blocks, for instance in the development of a constructive conflict culture, appropriate decision-making structures or common goal setting. The questioning view from the outside, advice on overcoming obstacles or guidance in parts can be helpful, as can an overview of already existing beaten paths.

The focus is always on the groups own concerns. To take these seriously and to support them in their implementation, in the sense of helping people to help themselves and in spite of the adversities of a legal society hostile to collectives, the AGBeratung sees its task.

Akademie für Suffizienz

Groß Pankower Weg 7 | 16928 Reckenthin | ☎+49-(0)176 2036 2434
corinna.vosse@posteo.de www.akademie-suffizienz.de

The Akademie für Suffizienz ("Academy for Sufficiency") offers opportunities to discover a richness that comes from acting together and from what the region has to offer.

In this experience, we see a force for transformation, for the desire and the readiness for a different everyday life. The academy provides access to means of production, practical skills and knowledge for community care. There are accommodations, group rooms, a self-catering kitchen, various workshops, agricultural trial fields, vegetable, fruit and berry gardens, a kitchen for food processing and preservation. The use of these is open to all interested parties and is based on self-selected (monetary or other) contributions.

Allmende e.V. - Waldhaus

Verein für Zusammenkunft und Gemeinschaft | Dorfstraße 35 | 17391 Krusenfelde

☎ +49-(0)39723 279 12 ✉ info@allmende-ev.de 💻 www.allmende-ev.de

The association supports communities in their founding phase by (among other things) providing them with land which it has bought out. The association's members pay a monthly fee of 5 to 15 Euros, which the Allmende people think everyone can afford to help improve the world. This money and other donations finance the buyout. The association is also active in the field of permaculture.

AllmendeLand-Projekt

Kölnische Str. 183 | 34119 Kassel

✉ info@allmendeland.de 💻 www.allmendeland.de

The AllmendeLand-Projekt is not a comunity, but was developed in the surroundings of the community-movement. It is attached to the common property movement, in which ground is bought out of the market economy and then used for organic agriculture. More information on the website.

Auroville International (AVI) Deutschland e.V.

Solmsstr. 6 | 10961 Berlin

☎ +49-(0)30-428 031 50 ✉ info@auroville.de 💻 www.auroville.de

AVI-Deutschland is a regional contact point of the world-wide network around the large, international, eco-spiritual Auroville settlement in India. The association gives information about Auroville, it's ideals and the latest developments, creating a network of people interested in Auroville. Further contact info for Europe at www.auroville-international.org

Bildungs-und Begegnungsstätte für gewaltfreie Aktion e.V.

KURVE Wustrow | Kirchstr. 14 | 29462 Wustrow

☎ +49-(0)5843-987 10 ✉ info@kurvewustrow.org 💻 www.kurvewustrow.org

The Centre for Training and Networking in Nonviolent Action – KURVE Wustrow was founded in 1980 with the aim of turning concerns about violent conflict, environmental degradation and social injustice into conscious nonviolent action. This concern finds expression in the mission statement "spreading nonviolence".

KURVE Wustrow conducts trainings, workshops and events in the field of civil, nonviolent conflict transformation at home and abroad. We send young volunteers as part of the "weltwärts"-programme as well as international peace workers as part of the "Civil Peace Service" to selected crisis regions.

For this purpose, KURVE Wustrow is closely cooperating with partner organisations in the Balkans, in the Middle East and in South Asia. With our training work in Germany we primarily empower activists in social movements and work together with them against nuclear power, racism and right-wing extremism.

Bundesverband der Regionalbewegung e.V.

Hindenburgstraße 11 | 91555 Feuchtwangen

☎+49-(0)9852-1381 ✉info@regionalbewegung.de 💻www.regionalbewegung.de

The Federation of German Regional Movements ("Bundesverband der Regionalbewegung e.V.") comprises, supports and connects diverse activities in the field of regional sustainable development. The federation considers regionally based economic policy to play an important role in securing and creating work, in preserving regional identity and improving quality of life.

Deutsche Arbeitsgemeinschaft Soziale Landwirtschaft (DASoL)

Hofsuche-Datenbank

✉Thomas.vanElsen@petrarca.info 💻www.soziale-landwirtschaft.de

The aim of DASoL is an exchange forum for the diversity of social agriculture and its further development. On our website there is a database of farms, whose activities range from the integration of people with physical, mental or spiritual impairments, the inclusion of socially disadvantaged people, delinquent or learning disabled young people, drug addicts, the long-term unemployed and active senior citizens to educational initiatives such as school and kindergarten farms.

Dorfschmiede Deckenbach

Bau- und Energiegenossenschaft Deckenbach eG | Lindengasse 2

35315 Homberg (Ohm) | ☎+49-(0)6633-643 75 33

✉post@bauenergie-deckenbach.de 💻www.bauenergie-deckenbach.de

The cooperative and its members implement projects through the acquisition, development and use of land and buildings in the village of Deckenbach with the aim of achieving holistic sustainability.

With the "village smithy" Deckenbach a meeting point for the inhabitants of the village and the surrounding area is created.

Doris De Cruz

Service-Büro (Aufbruch-Bremen-Gruppe) | Wilseder-Berg-Str. | 28329 Bremen

☎+49-(0)4219-440 20 97 ✉decruzdoris@gmail.com 💻www.aufbruch-bremen.de

An initiative for people,

- who want to change their normal everyday life,
- who want to live more holistically and collectively,
- who want to take personal responsibility for the dream of another, better world.

Ecobasa.org

The sustainable gift-economy network | Sonnenallee 28 | 12047 Berlin

☎+49-(0)163-631 06 22 ✉info@ecobasa.org 💻ecobasa.org

ecobasa is a directory of sustainable communities, ecovillages and individuals looking for alternative ways of living. The online platform facilitates communication for the exchange of experiences, knowledge, services, products and other resources.

Travelers and communities can find each other to volunteer, join a community, to form a new one or travel together in Community-Tours. When people want to visit a community, they can offer to help on wishes of the communities profiles in return for food and accommodation.

Resources are located and coordinated in Community-Tours on the platform and the travelers naturally form a logistic network. All members offer their gifts without expectations for them to be returned directly. They may recieve what they whish for from someone else. Positive references build trust into the network and its participants. Responsible relationships are emerging where people get to know their needs and qualities. We dont want to create just another platform, but an open source technology that other networks can use in their own systems. We need you to make it work, so join the network!

Ecotopia Biketour

✉info@ecotopiabiketour.net 💻www.ecotopiabiketour.net/

Ecotopia Biketour is a self-organized, international community that has been organizing a yearly bicycle tour in different regions of Europe every summer since 1990. During the tour we visit environmental and social projects and practice forms of activism and sustainable living. We cook communal vegan food, practice consensus decision making, and share skills by doing workshops. One of our core values is to create a non-discriminatory environment. Ecotopia Biketour is for anyone interested in travelling by bike, community life, DIY, an ecological lifestyle, and/or learning by experience.

FARN - Fachstelle Radikalisierungsprävention und Engagement im Naturschutz ("Department for Radicalization Prevention and Commitment in Nature Conservation")

c/o NaturFreunde Deutschlands | Warschauer Str. 58a/59a | 10243 Berlin

☎+49-(0)30-297 732-60 ✉info@nf-farn.de 💻www.nf-farn.de

FARN was founded in October 2017. FARN...

- examines the historical and current links between nature conservation and environmental protection with extreme right-wing and nationalist tendencies.
- makes biological and racist continuities in nature and environmental protection visible.
- identifies extreme-right and inhumane ideologies and thought patterns in nature and environmental protection and draws up human-positive and democracy-promoting drafts in natural protection.
- provides information, advice and qualification for actors in nature and environmental protection, child and youth welfare as well as for young people and young adults.

FARN is funded by the German Federal Ministry for Family Affairs, Senior Citizens, Women and Youth within the federal program "Demokratie leben!".

Forum Gemeinschaftliches Wohnen e.V.

Bundesvereinigung (FGW) Hildesheimer Str. 15 | 30169 Hannover

☎+49-(0)511-165 910-0 info@fgw-ev.de fgw-ev.de

The FORUM is a nationwide consortium of interested parties, experts, house and settlement communities, housing companies and municipalities to promote new forms of housing and residential care as well as committed neighbourhoods for older people.

The FORUM advises, informs and networks interested persons and other actors and multiplicators who want to promote new forms of living, new forms of residential care as well as binding neighbourhood structures with a focus on older people.

Gemeinschaftsbildung nach Scott Peck

Götz Brase und Team | Schloss Oberbrunn, Seeonerstr 4

83132 Oberbrunn/Pittenhart | ☎+49-(0)8624-879 59 62, +49-(0)40-22691720

gemeinschaftsbildung@t-online.de www.gemeinschaftsbildung.com

In recent years, Scott Peck's Foundation for Community Encouragement (FCE) has become increasingly popular in German-speaking areas. It includes not only personal disarmament and non-violent communication, but also self-government of groups and individual contributions towards finding a path that sustains the communal heart. We offer short and weekend workshops and are concerned with the exploration of intensive community.

Gemeinschaftskompass (Eva Stützel)

c/o Ökodorf Sieben Linden | 38489 Beetzendorf

☎+49-(0)39000-512 37 (Eva)

eva.stuetzel@gemeinschaftskompass.de

www.gemeinschaftskompass.de

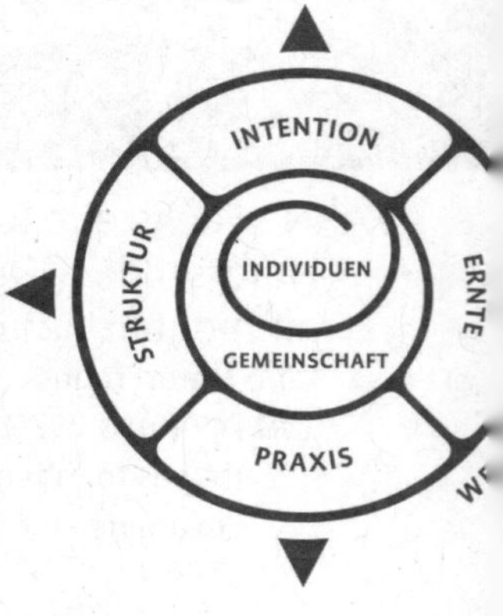

The Community Compass is the essence of the experience of building Sieben Linden Ecovillage. It gives an overview about what is important to start a community project. I offer process assistance for community initiatives as well as seminars on the Community Compass.

Helga Mestemacher Immobilien GmbH

Eibengrund 1 | 49152 Bad Essen

☎+49-(0)5472-3535

Real estate brokerage for residential groups and cohabitation groups

ijgd - Internationale Jugendgemeinschaftsdienste e. V. -

Bundesverein | Kasernenstraße 48 | 53111 Bonn

☎+49-(0)228-228 00 14 workcamp.leader@ijgd.de www.ijgd.de

We organise international 3-week workcamps in Germany and send German volunteers to workcamps of our partner organisations abroad. The camps are made up of youth groups of 15 volunteers (16-28 years) with the aim of international understanding through encounters in everyday situations and community work.

ISKCON Heidelberg Rhein Neckar Odenwald e.V.

Bhakti-Yoga Zentrum Nava Navadvip | Zuzenhäuser Straße 14 | 74909 Meckesheim

☎+49-(0)6226-953 07 41 ✉info@iskcon-heidelberg.de

💻www.iskcon-heidelberg.de

We follow the mission of Srila Prabhupada and His ISKCON movement. At the moment our center is only open on Sundays.

Kana

Dortmunder Suppenküche e.V. | Mallinckrodtstraße 114 | 44145 Dortmund

☎+49-(0)231-839 853 ✉info@kana-suppenkueche.de

💻www.kana-suppenkueche.de

A Christian grassroots community – members live independently, but commit themselves for the common project. Five times a week we welcome poor and homeless people and share soup and coffee together.

We are part of the Catholic Worker movement. We also have round table discussions, worship services and speak out publically and politically for peace and justice issues.

Komm-Rat

Beratungsgruppe in der Kommune Niederkaufungen | Kirchweg 1 | 34260 Kaufungen

☎+49-(0)5605-800 70 ✉info@kommune-niederkaufungen.de

We offer project consultation for beginning or already existing groups. We can draw on our long-term community experience. The focal points of our consultation are: Legal forms, finances, building, work, organising everyday life, pension schemes, group dynamics, non-violent communication. Depending on needs and issues, we provide support through supervision, moderation and future-workshops.

When contacting us by e-mail, please note "Komm-Rat" as subject.

Lebensform Gemeinschaft

Marita Bergmann | Pappelweg 3 | 19217 Carlow | ☎+49-(0)38873-209 69

✉marita@bergmann-seminare.de

💻www.marita-bergmann.de, www.pappelweg-drei.de

Provides identification, consultancy, guidance and training for persons seeking communities, for groups in the founding process, and for existing communities in the form of seminars and supervision.

Liebeskunst-Netzwerk Berlin

Living and Loving in Community | 10551 Berlin

☎+49-(0)30-390 370 00 ✉news@berlin007.de 💻www.liebeskunst-netzwerk.de

The "Loving Arts Network Berlin" brings people in contact who are interested in "living and loving in community". Our origin lies in the ZEGG community (page 276).

We turn to people who are interested in personality development in the field of love, lust and community, also for polyamory and tantra.

The contacts are made through the "Loving Arts Network Newsletter".

The newsletter contains many events and news from the scene, also in english.

You can subscribe to the newsletter on request. Mail to: news@berlin007.de

Netzkraftbewegung

Marsstr. 70 | 46509 Xanten

☎+49-(0)281-3106 ✉info@netzkraftbewegung.de 💻www.netzkraft.net

International network

- of organisations and groups who commit themselves for a better world.

Mutual support

- based on the willingness to meet other net participants and to support them if possible.

Common objectives:

- international body for peace, human rights and environment within the United Nations
- sustainable ecological-social economic development
- decentralization of political power.

Netzwerk Frankfurt für gemeinschaftliches Wohnen e.V.

Adickesallee 67/69 | 60322 Frankfurt am Main

✉info@gemeinschaftliches-wohnen.de 💻www.gemeinschaftliches-wohnen.de

The network is a non-profit organization that has a professional coordination and advice center since 2009. It sees itself as a forum for around 70 Frankfurt housing initiatives and advises and supports them in implementing their living concepts. In addition, it is a contact point for those who are new interested and works for a better framework for community living in the city.

Netzwerk Selbsthilfe Berlin - der politische Förderfonds

im MehringHof | Gneisenaustraße 2a | 10961 Berlin | ☎+49-(0)30-691 30 72

✉mail@netzwerk-selbsthilfe.de 💻netzwerk-selbsthilfe.de

Netzwerk Selbsthilfe Berlin is the founder of an independent solidarity fund supported by its members. The idea of Netzwerk Selbsthilfe is implemented in three ways: Support - Network - Promote. In addition to direct financial support, projects from Berlin and Brandenburg are also personally advised on all questions of self-organization. As a working aid for projects, Netzwerk regularly publishes the brochure "Fördertöpfe für Vereine, selbstorganisierte Projekte und politische Initiativen"; and the brochure for promotional prizes "Schwein gehabt".

oekojobs.de - Ecologic Jobs in Europe

✉info@oekojobs.de 💻www.eco-jobs.info

With eco-jobs.info you can engage for a better world. Here you can find possibilities for volunteer places, internships, jobs, workcamps and community work in environment and nature protection in Europe. Employers from organisations, ecological businesses or social entrepreneurs can add or edit new job offers in the database.The project is open for new ideas and is happy about help of any kind.

ÖKODORF-Institut für Wohnprojekte

Beratung bei Gemeinschaftssuche/-gründung | Alpenblickstr.12

79737 Herrischried | ☎+49-(0)7764-933 999

✉oekodorf@gemeinschaften.de 💻www.gemeinschaften.de

Providing support for community research/foundation since 1980: address exchange (over 30.000 contacts), interspiritual Ecovillage Festivals (encounters with communities, founding initiatives & interested persons), email newsletters to some 6,000 persons interested in the community movement, media (mail-order books, videos, music, games...), real estate information. Consultations and seminars are also offered in our Black-Forest-Community, also possible while prana-walking.

Permakultur Institut e.V.

Im Garten 11 | 51503 Hoffnungsthal

☎+49-(0)2205-919 83 83 ✉info@permakultur.de 💻www.permakultur.de

Permakultur Institut e.V. (PKI) is a non-profit association which spreads permacultural design approaches and practical experience through public relations work, education and networking. We support projects, initiatives and people which feel connected to the ethics of permaculture.

We develop permaculture as an applied design approach for sustainable solutions through holistic observation, connected thinking, creative design and responsible action.

pro... gemeinsam bauen und leben

Wohngenossenschaft eG | Hornbergstr. 82 | 70188 Stuttgart

☎+49-(0)711-234 81 62 ✉info@pro-wohngenossenschaft.de

💻www.pro-wohngenossenschaft.de

The cooperative "pro... gemeinsam bauen und leben eG" (eng: living and building together) was founded in 1999 to support groups who wish to create communal multi-generational housing projects – through advice and concept development, property searches and negotiation with communities, and during the building stages in close cooperation with the project group.

Projektwerkstatt Saasen

Villa Kunterbunt des kreativen Widerstands | Ludwigstr. 11 | 35447 Reiskirchen

☎+49-(0)6401-903 283 ✉saasen@projektwerkstatt.de

💻www.projektwerkstatt.de/saasen

Die Projektwerkstatt (the Project Workshop) is above all a political centre with everything that entails: libraries, archives, group rooms, even a conference house, plus technical equipment ranging from computers and layout archives all the way to fax machines, photocopiers etc. Nearly 40 beds, kitchens, etc. create an atmosphere to work on political resistance projects for a long time. Thats exactly what the house is for!

Simplified world explorers, people looking for a substitute for mommy/social welfare office, etc., esos, party soldiers/club meers, etc., must count on being invited to self-organization and self-reflection when they stay in this open platform for action. If you want to fall into a comfortable nest, you should look elsewhere. This also applies to those who want to follow the pre-punched channels of life (for example, wage labor) or stay on the way to this (education, study)!

Events: www.projektwerkstatt.de/termine

Materials: www.aktionsversand.tk

Training and workshop offers: www.vortragsangebote.tk

SCI (Service Civil International)

Deutscher Zweig e.V. | Blücherstr. 14 | 53115 Bonn

☎+49-(0)228-212086 info@sci-d.de www.sci-d.de

Service Civil International (SCI) is an international non-profit organization that works for peace, non-violence, social justice, sustainable development and intercultural exchange through volunteering. Some workcamps take place in intentional communities.

Stadtleben mit Landliebe trifft Landleben mit Stadtliebe

Lindengasse 2 | 35315 Homberg (Ohm) | ☎+49-(0)6633-6437533

post@stadtleben-marburg.de www.stadtleben-marburg.de

In the present, we decide who will steer our economy in the future. In three steps we want to participate again in the production of our food and also in the local barter trade with these fair manufactured products. In the course of this it is necessary to fill Marburg with new local economic life and to support this through increased cultivation and production of fruit, vegetables and other foodstuffs in our region.

Stadleben Marburg eG was founded in 2015. The cooperative sees itself as an umbrella organisation for residential and living communities in Marburg an der Lahn and its surroundings. We support the organisation of people and initiatives who are interested in the development and establishment of alternative living and working communities in Marburg.

Stiftung FreiRäume

c/o Projektwerkstatt in Saasen | Ludwigstraße 11 | 34447 Reiskirchen

☎+49-(0)6401-903283 kontakt@stiftung-freiraeume.de

www.stiftung-freiraeume.de

The FreiRäume Foundation wants to protect open spaces or to build or develop them. They should always be - at least partially - a platform for acts directed at removing power relations, exploitation, environmental destruction, etc. The aim is to rdisplace (formal) hierarchical structures and to promote horizontal, equitable action and cooperation of people. The Foundation aims to achieve this by providing user groups with buildings, areas, spaces, carts and other facilities under the conditions of openness and horizontality, negotiating cooperation agreements setting out such goals with the users, but forgoing further influence.

Stiftung trias

Gemeinnützige Stiftung für Boden, Ökologie und Wohnen | Martin-Luther-Str. 1

45525 Hattingen | ☎+49-(0)2324-9022213

info@stiftung-trias.de www.stiftung-trias.de

The Stiftung trias (e: trust trias) is a non-profit foundation supporting cooperative and ecological housing projects as well as reforms concerning ownership of ground and property. Trias studies, finances and networks for collaborative housing across Germany, focusing on self-organized projects including ecological building methods.

The Foundation has been working closely since 2002, both locally and nationally, to not only gain attention for community-orientated housing but support housing projects in a general way. Activities include a wide range of events and seminars, relevant publications and an online database documenting such housing projects across the country. See www.wohnprojekte-portal.de.

Tiefenökologie Netzwerk

Ökodorf Sieben Linden | 38489 Beetzendorf

☎ +49-(0)39000-908 62 ✉ info@tiefenoekologie.de 💻 www.tiefenoekologie.de

Deep ecology is about the insight that the world is a living system, in which everything is connected and interdependent. It is open to the dynamics of self-organisation of living and developing systems and sees us humans as a strand in the web of life, interwoven into the natural connections from which we were created. Deep ecology combines cognitive, emotional and spiritual elements and levels of tangible societal and political actions. We offer presentations, seminars and trainings.

UnaVision Initiative

Herzershof 10 | 15328 Kuestriner Vorland

☎ +49-(0)172-862 89 47 ✉ info@unavision.eu 💻 www.unavision.eu

Our first location is in Oderbruch, 90 km east of Berlin. Our network is made up of all active players from UnaVision, UnaVersity and the UnaVillage regions who work on the implementation of the vision, the UnaVersity idea and the prototypes of the UnaVillages (learning locations for transformative learning) or support this work with their resources (finances, knowledge, time, network). UnaVersity coordinates research activities and transformative learning for the UnaVision network and the UnaVillage regions. The UnaVillage Region is not a "community" in the sense of ecovillage communities or other spiritual, political or social groups, but a learning and working community for achieving the UnaVision vision and building prototypes of UnaVillages.

Werkstatt für Gewaltfreie Aktion, Baden

Vaubanallee 20 | 79100 Freiburg

☎ +49-(0)761-432 84 ✉ buero.freiburg@wfga.de 💻 www.wfga.de

The Workshop for Non-Violent Action in Baden was founded in 1984 as an independent educational and training facility. We support groups and individuals in their quest to become active against violence, to accomplish non-violent resistance, to resolve conflicts in a non-violent manner and to develop non-violent alternatives. We offer extensive training in alternatives to military and war, non-violent action, campaigning, civic courage, mediation, consensus decision-making, non-violent communication and conflict resolution. Our workshop offers a learning space which allows us to experience that the world can be changed without resorting to violence.

WohnBund-Beratung NRW GmbH

Humboldtstraße 42 | 44787 Bochum

☎ +49-(0)234-90440-0 ✉ kontakt@wbb-nrw.de 💻 www.wbb-nrw.de

WohnBund-Beratung NRW works as a consultancy in the field of Co-Housing. We support initiatives by organising open meetings, workshops, moderation of the group-process, project development, financing, legal form etc.

Wohnprojekte-Portal

c/o Stiftung trias, see above portal@stiftung-trias.de
www.wohnprojekte-portal.de

The www.wohnprojekte-portal.de is a directory for co-housing projects. You will find people looking for a project, groups, projects under development as well as established co-housing projects. Furthermore there are addresses of consultants, downloads, literature and other interesting information.

WohnStrategen e.V.

Gemeinsam Planen & Bauen in Thüringen | Jakobstraße 10 | 99423 Weimar
+49-(0)3643-773 6496 info@wohnstrategen.de www.wohnstrategen.de

Since 2005 we have focused our voluntary efforts on information and education regarding communal living as a possible future-proof way of life in Thuringia. We also strive to make connections between projects and to provide initial consultation for new project groups or people in search of projects.

WWOOF Deutschland

Hauptstr. 30 | 99752 Bleicherode
+49-(0)36338-589 375 kontakt@wwoof.de www.wwoof.de

WWOOF is a worldwide network of volunteers (WWOOFers) and organically managed farms. WWOOF volunteers are involved in everyday life at those farms, volunteer to help, and thereby gain first-hand knowledge of organic agriculture, horticulture and self-sufficiency in the countryside and gain their own experience. WWOOF-Germany publishes a list of participating farms in Germany (online and printed), which is used by WWOOFers in order to independently arrange stays on farms.

There are family farms, farm communities and individuals among the farms, and always the quality of a WWOOF stay increases with the quality of the relationship of the people involved, which is essentially related to the way in which they communicate with each other. WWOOF is thus an exciting field of learning in terms of communication of own needs and limits, the mutual appreciation and common conflict management.

WWOOF Germany is recognized as charitable due to the promotion of environmental protection and landscape management. It has been shown that the association achieves this goal in particular by raising awareness among the participants: People learn how to produce healthy food by doing it themselves and by getting to know different natural/alternative/sustainable life concepts.

Greece

WWOOF Greece

Linking volunteers with organic growers and sustainability projects
Paralia Markopoulou Oropou Avenue | 19015 Nea Palatia Oropou
info@wwoof.gr www.wwoof.gr

We support any intentional community in Greece that cultivates land organically or truly promotes sustainability. Every such community is welcome to join the WWOOF family and gain access to a worldwide network of volunteers and inspiration seekers.

Israel

International Communes Desk

Kibbutz Volunteers Program Center (KPC) | 1 Yasmin Street, Seminar Efal
52960 Ramat Efal | ☎+972-(0)525 348 550 ✉anton@communa.org.il
💻www.communa.org.il

The Intentional Communities Desk, renamed from the International Communes Desk in 2015, is a contact body between the communes of the world - where the term "commune" covers the wide spectrum of intentional communities, kibbutzim and other types of communal living. The Desk was started in 1976 at the initiative of the late Mordechai Bentov, a kibbutz member and cabinet minister. Our modest headquarters are situated at the Yad Tabenkin Institute, on the campus of Seminar Efal, on the outskirts of Tel Aviv, Israel.

Our major regular activity has been the twice-a-year publication, in English, C.A.L.L. (Communities At Large Letter), published in order to promote the aims of the desk. We now publish Kol, our hebrew version of C.A.L.L., biannually too

Most members of the Desk - our committee - are kibbutzniks, from all types of kibbutzim. Almost all of our work is voluntary and the few who are paid receive very little for their labour. Our very modest budget is provided by the kibbutz movements, a generous gesture of support, considering their financial problems.

The lCD occasionally initiates meetings with members of communes and other alternative communities from Israel and abroad.

The ICD has access to the Library of Yad Tabenkin, with its comprehensive collection of publications and books about communes and other alternative communities. This material is at the disposal of researchers, students and anyone who is interested.

We are a Registered NGO, so we are now legally permitted to receive tax-free donations (even a modest sum will be very welcome!). Do feel free to contact us!

Italy

Cohousing.it

Viale Toscana 13/B | 20136 Milano | ✉info@cohousing.it 💻www.cohousing.it

We have a huge community of people who live in intentional communities (cohousing), manage lots of cohousing project and help people who are interested in cohousing (from the start of the project until the real life inside): We manage all the aspects (architectural, building, participation process, starting of community...)

Fiera delle Utopie Concrete

Via Marconi 8 | 06012 Città di Castello | ☎+39-075-855 43 21
✉coordinatore@utopieconcrete.it 💻www.utopieconcrete.it

The Fiera delle Utopie Concrete (e: Fair of Practical Utopias) presents since 1988 annually in Città di Castello, Central Italy, expriences and solutions of an ecological conversion of economy and society in seminars, conferences, performances and exhibitions. With its pragmatic bottom-up approach of bringing together persons of good ecological will, the initiative considers itself close to the movement of intentional communities.

Luxembourg

LIFE

ecoCreActiv Plattform | 22, rue Baltzing | 3413 Dudelange
☎+352-518168 ✉housesharing@lifeproject.lu 💻www.lifeproject.lu
We are an independent and self-organized youth platform. We want to give the youth a forum to participate at life, politics and society and take action on different topics concerning sustainability, friendship and creativity.

Netherlands

Omslag

Werkplaats voor Duurzame Ontwikkeling | Hoogstraat 301a | 5654 NB Eindhoven
☎+31-(0)40-2910295 ✉omslag@omslag.nl 💻www.omslag.nl/wonen

In 2005 Omslag started the Dutch Information Service on Different Ways of Housing and Living. Our service consists of an extended website and a library with books, magazines, DVDs. Besides a two-monthly printed magazine (ZOZ), we publish a free monthly electronic newsletter (in Dutch) and organise meetings on specific related subjects. Our library is open to the public each friday from 14:00-17:00.

Purusha

Station 24 | 6063 NP Vlodrop | ☎+31-(0)475-539531 ✉info@purusha.de, ✉info@purusha.eu 💻www.purusha.eu, www.purusha.de, www.purusha.org (USA)
Our intention: to raise the level of consciousness (intelligence, health, creativity etc.) of each member of our Purusha community and at the same time create a strong influence of coherence for the whole society. Scientific research confirms the benefits and positive results on both levels - individual: better health, more creativity, increased self confidence etc, and sociological: less violence, fewer accidents, fewer cases of illness, economical improvements and more.

This is accomplished by practising the TM- and TM-Sidhi-Programme in a group as part of Maharishi Vedic Science - according to the teachings and instructions by His Holiness Maharishi Mahesh Yogi.

Applicants should check for their nearest place for learning Transcendental Meditation on: www.meditation.de (or: www.tm.org).

Norway

Atlantis Exchange

Atlantis Utveksling | Skippergata 9 | 0152 Oslo

☎+47-(0)22 47 71 70 ✉atlantis@atlantis.no 💻www.atlantis.no

Atlantis Utveksling (Atlantis Exchange in English) organizes cultural exchange programs in Norway for young foreigners who would like to work as working guests in agriculture or tourism, university trainees in agriculture and au pairs.

Atlantis Exchange is a foundation established by the Norwegian Youth Council in 1987. Our goal is to promote respect and understanding between cultures through youth exchange. We have specialized in work experience programs for people between 18 and 30 years old.

All programs are based on the idea of working for free accommodation, board and pocket money. The participant lives with a Norwegian host family, sometimes with other participants as well.

Palestine

Global Campus Palestine

Farkha Ecovillage | c/o Tamera, Monte Cerro | 7630 Colos | Portugal

☎+351-(2)-960 305 607 (Address and phone number are at Tamera Ecovillage)

✉aida.shibli@gmail.com 💻www.globalcampuspalestine.weebly.com

Global Campus Palestine (GCP) provides Palestinians with the opportunity to study regenerative ecological practices, as a first step towards establishing decentralized, autonomous models of how human beings can live sustainably and in regenerative lifestyles.

GCP is today located in the village Farka, a small village of 1500 people who has a good amount of projects and groups that want to make the move from a traditional Palestinian village towards becoming the first Ecovillage in Palestine.

The village coined the term of voluntary work in Palestine and every year, we are holding an international voluntary festival bringing together 200 young people from all around the world to share ten days of political activism that puts life in the center of the work.

Portugal

Haliotis

Place for Ecology, Permaculture & Health | Hindenburgstraße 12
76470 Ötigheim | Germany | ☎+49-(0)1590 849 8445 bghaliotis@gmail.com
www.haliotisport.wordpress.com, www.youtube.com/berndgerken

Haliotis near Sao Bartholomeu de Messines, Ourique is not an eco-village or an "eco-city", but a tangible contribution to the revival of the rural regions by family and friend groups (1-25 people of all ages) by scattered settlement. We encourage interested people to visit this landscape and raise a self-sufficient life there! Many abandoned places with good land around are inviting.

Institute for Global Peace Work (IGP) Tamera

Tamera | Monte do Cerro | 7630-392 Reliquias
☎+351-283 635 484 igf@tamera.org www.tamera.org

The IGP is working to weave an international peace network to establish a global force for peace. The main goal is to develop a global cooperation for the establishment of Peace Research Villages in order to create alternatives for the global system of violence.

Serbia

WWOOF Serbia

Visocka Rzana bb | 18306 Pirot | info@wwoofserbia.org www.wwoofserbia.org

WorldWide Opportunities on Organic Farms, Serbia (WWOOF Serbia) is part of a worldwide movement linking visitors with organic farmers and growers to promote cultural and educational experiences based on trust and non-monetary exchange thereby helping to build a sustainable global community.

Many of our hosts are intentional communities, so we believe that WWOOF Serbia and its wwoofers actively support the development of intentional communities in Serbia. Therefore, our slogan is "...living, learning & sharing organic lifestyles".

Spain

AUTarcaMatricultura - Autonomous Academy for Permaculture

Camino La Ermita 26 | 38780 Tijarafe
☎+34-922 490 215 autarca@matricultura.org www.matricultura.org

AUTarcaMatricultura advises on how to build Life-Biotopes: The patriarchal economic, finance and society systems are neither to be reformed nor to be controlled. Therefore everybody should urgently start the construction of the most vivid and self-responsible Life-Biotopes. The building of a Life-Biotope can be started at the most different places and under most different circumstances. E.g. in a city apartment of one single person, in a multiple family dwelling, in a community, on a farm, in a school, in a village, a city, on a market place...

Cambium Permacultura en Formación

c/Niça 13-15, 3° 2 | 08024 Barcelona

☎+34-666 794 796 ✉info@permacultura-es.org 💻www.permacultura-es.org

Cambium is a website and an NGO for the dissemination of permaculture, especially among the Spanish speaking community worldwide. We offer introduction to PC, and PDC courses, as well as public talks. We also offer consulting and design services.

Ecohabitar Visiones Sostenibles

Abajo 1 | 31480 Artieda (Navarra)

☎+34-948 577 384 ✉info@ecohabitar.org 💻www.ecohabitar.org

Editor of the quarterly journal EcoHabitar and of books on bio-construction, renewable energies, permaculture and ecovillages. The head office is located in the municipality of Arterra Bizimodu (Navarre, Spain). In the magazine EcoHabitar, a page with the latest news from GEN Europe is published in addition to the topics mentioned above.

Permacultura Montsant

Lista de Correos | 43360 Cornudella, Tarragona | ☎+34-71821197, +34 977 262 017

✉info@permacultura-montsant.org 💻www.permacultura-montsant.org

We are a permaculture institute that offers long term residential apprenticenship and training programs.

Possibilica

...a nomadic archearchal nanonation delivering healing and transformational initiations | Paseo Illetas 31, Apt 3b | 07181 Illetas-Calvia

☎+49-(0)162-910 82 08 ✉clinton@nextculture.org 💻www.possibilica.org

Since 2008, Possibilica has facilitated adulthood and archetypal healing and initiatory processes in GEN ecovillage gameworlds. We specialize in deepening and unfolding the context and thoughtware of conscious communities. We have served ZEGG, Schloss Glarissegg, Niederkaufungen, Findhorn, New Culture, Tempelhof, Tamera, Lebensbogen, Boom Festival, Sieben Linden, etc.

Sweden

Alternativ.nu

☎+46-(0)73-032 71 28 ✉petter@alternativ.nu 💻www.alternativ.nu

Swedish website about self sufficiency and country living. Forum, wiki and magazine about small scale farming, eco building and living. More that 300,000 visitors per month.

Ecotopia På Österlen

Bondrum 1334 | 27395 Skåne Tranås

☎+46-(703)330 814 ✉info@ecotopia.se 💻www.ecotopia.se

Holistically oriented educational institution in southern Sweden. Education in ecological building, ecological cultivation/permaculture, alternative medicine, circular economy, vegan or wild plant kitchen as well as initiatives to help shape a sustainable future. As a private non-profit organisation, we base our activities on the ethical economic model of the Common Welfare Economy for the well-being of humans and the environment as the primary goal of economic activity.

Sambruket i Sösdalabygden

Svensköpsvägen 8 | 298 94 Linderöd

☎+46-(0)706 525 677 ✉info@sambruket.se 💻www.sambruket.se

Sambrukets aim is to extend eco village ideas to the large number of people who cannot or do not want to join something as tight and definitively located as a formal community. It is instead a network in which the members arrange their own living and family after preference, but work together on a joint project for the future. In our case, we try to organise a sustainable living pattern trough small-scale farming, education, discussions, building renovation, process and equipment innovation etc. Sambruket is in this respect quite unusual, but there is a strong need for practice, training, introduction and education before one dives into the full-time community experience. *Photo on page 15*

Switzerland

CoHabitat

Sentier des Saules 3 | 1205 Genève

☎+41-(0)22-320 86 06 ✉contact@co-habitat.net 💻co-habitat.net

CoHabitat is a global geographical database of collaborative housing projects. It is multilingual and just like community-led housing initiatives, it is open and collaborative. It offers the opportunity for existing projects to share their experiences with the world, establish direct links with their peers and for start-up initiatives to access valuable information to begin their journeys.

COMMENDE - Kommunitäres Geld

c/o Alfred von Euw | Oberfeldstr. 16 | 9437 Marbach

☎+41-(0)718 880 344 ✉insula@bluewin.ch

We regard ourselves as a free, "anarchist" study and working group with the goals of developing an appropriate understanding of money and to explore ideational, economic and social conditions for the creation of a stable community, as well as the significance of work in the biography of humankind.

HOLON

Netzwerk für integrale Entwicklung | 3003 Bern | ☎+41-(0)81-413 73 09

✉info@holon-net.net 💻www.holon-net.net, www.integrale-politik.org

HOLON connects people and groups who strive for a more humane, solidary and ecologically sustainable society, who question material culture and the thinking behind growth and who wish to contribute to a more sustainable culture. HOLON takes a holistic view of humankind.

Our summer conference as well as our congress for integral politics are tangible examples of the implementation of our thinking, where societo-political questions are addressed. The annual "Vernetzungswanderung" (= connection hike) offers a platform for bringing people together and is organised by a different set of people each time.

Stiftung Edith Maryon

zur Förderung sozialer Wohn- und Arbeitsstätten | Gerbergasse 30 | 4001 Basel
☎+41-(0)61-337 78 78 💻maryon.ch
With the help of gifts, donations and bequests, the Foundation Edith Maryon takes over and acquires land and real estate of all kinds, withdraws them from speculation and thus preserves and enables socially innovative housing projects, places of culture, education and training, biodynamically managed farms and other places of work. In many cases, the Foundation limits itself to the role of the landowner and leaves the land to the residents and users in the heritable building right.

urbaMonde

sustainable cities by and for the inhabitants | Sentier des Saules 3 | 1205 Genève
☎+41-(0)22-320 86 06 ✉contact@urbamonde.org 💻urbamonde.org
UrbaMondes mission is to support groups of inhabitants who are committed to the transformation of the city towards sustainability and inclusion. Our NGO, based in Switzerland and France, consists of a team of enthusiasts from various backgrounds who share the vision of a sustainable city "by and for the inhabitants".

Turkey

Bugday Association for Supporting Ecological Living

WWOOF Turkey & Camtepe Ecological Living Center
Osmanaga Mh, Yogurtçu Sükrü Sk, A.H. Uzunal Is Merkezi, 48/6 | 34714 Istanbul
☎+90-216-349 99 33 ✉info@bugday.org
💻www.bugday.org/english, www.tatuta.org, www.wwoofturkey.org
Bugday ecological movement has been tirelessly working to support, create and promote fair and sustainable production-consumption patterns in Turkey and beyond.
The main working areas of Bugday can be summarized as organic agriculture, farmers markets, heirloom seeds, urban gardening, community supported agriculture, participatory guarantee systems, and coordination of WWOOF Turkey.

United Kingdom

Active Distribution

72-73 Old Market | BS2 0EJ Bristol
✉info@activedistribution.org 💻www.activedistribution.org
Active Distribution is a not for profit DIY mailorder distribution of "all things anarchist" and a few other radical things too! We also do wholesale, book and pamphlet publishing, badgemaking and more.

Local Futures

The Economics of Happiness | PO Box 239 | TQ9 9DP Totnes

☎+44-(0)1392-581175 ✉info@localfutures.org 💻www.localfutures.org

A pioneer of the new economy movement, Local Futures has been raising awareness for four decades about the need to shift direction – away from dependence on global monopolies, and towards decentralized, regional economies.

We produce books, films, and other 'education for action' tools, as well as organizing activist-oriented conferences and workshops worldwide.

Permaculture Association

BCM Permaculture Association | WC1N 3XX London

☎+44-(0)113-230 74 61, +44-(0)845-4581805 ✉office@permaculture.org.uk 💻www.permaculture.org.uk

The Permaculture Association is working to radically and positively change the way we live in the UK and actively supports a worldwide movement. Helping educate, network, support and redesign all aspects of our daily lives to transform ourselves, our communities and our landscapes according to the core ethics of Earth Care, People Care and Fair Shares.

Permaculture Magazine & Permanent Publications

The Sustainability Centre | East Meon | GU32 1HR Hampshire

☎+44-(0)1730-823 311 ✉enquiries@permaculture.co.uk, maddy@permaculture.co.uk 💻permaculture.co.uk

We have published Permaculture Magazine International since 1992 in print and digital formats. Every subscriber can read our back issue archive free of charge. We have also published over 80 practical and visionary books via Permanent Publications' to encourage people to live more healthy, self-reliant, community focused and ecologically sound ways of life. See http://permanentpublications.co.uk. We have a busy website packed with information at www.permaculture.co.uk and a popular YouTube channel https://www.youtube.com/PermacultureMedia. We run the £25,000 Permaculture Magazine Prize annually. Follow us on Twitter @PermacultureMag, Instagram @permaculturemagazine and Facebook @PermacultureMag

Totnes Eco Education Community (TEECom) Cooperative

Totnes Devon | TQ9 5RL Totnes Devon

☎+44-(0)7842-573 922 sustainable.ecoplanning@yahoo.com teecom.org

We are a small group of people, mostly local to Totnes, involved in permaculture planning and design of ecologically and sociologically sustaining housing, as well as in eco-building and organic gardening.

Our aims are to share our collective skills with others in very practical learning-by-doing sessions and workshops.

We are planning to build a permaculture community of four houses.

WWOOF United Kingdom

PO Box 2207 | MK18 9BW Buckingham

info@wwoof.org.uk www.wwoof.org.uk

Get hands-on experience with WWOOF UK on organic farms, gardens and small-holdings – all offering food, accommodation and learning in exchange for practical help on the land. Our listing includes communities - but also individuals willing to open their doors to you and share a different way of life.

United States

Gaia University International

P.O. Box 24 | 95679 Rumsey, California

☎+1-805-610 08 99, +1-805-400 44 09, +1-831-917 5780

info@gaiauniversity.org gaiauniversity.org

Gaia University is an international learning community collaborating for ecosocial regeneration.

Gaia University provides a unique approach to higher learning by offering students (called Associates) access to Certificates, Diplomas, and Bachelors and Masters Degrees; whilst they are actively engaged in self and planetary transformation. As a holistic education collective, we engage the heart (emotional) and hands (practical) as well as the head (mind) aspects of whole-self learning.

Our action learning methodology enables you to study locally, at work or on a project, in your own language, supported by Gaia Universitys worldwide network of learning providers, expert advisors, tutors and mentors. We also have a robust alumni network and offer free webinars from graduates and elders in the permaculture and regenerative fields to our associates and the public.

Internet

Ecovillage Research Guide (by GEN)

research-wg@ecovillage.org ecovillage.org/en/page/research-ecovillages

Aim and services of the research group are

- making the existing research on ecovillages more visible to approaching researchers, academia, ecovillages, and the society, (with data base and bibliography resource on website)
- to better connect the single research projects

Mission for future research projects:

- Control the scientific quality of research on, for and from ecovillages
- Inform and connect initiatives of research including searching for funding

Ethical vision:

- Emphasizing the value of ecovillages as holistic and transdisciplinary experiments and innovative models
- Visualizing the value of ecovillages to the outside world and approaching researcher by e.g. connecting the results and what has been proved by these research projects

Why? Ecovillages are increasingly approached by students and other researchers who discover their value as holistic "laboratories" of the future in almost all areas of life. Members in ecovillages often invest their time in research projects. Further they have their own research questions they would like to follow and it could help to network resources to start projects.

For whom? This research group is helpful for

- For students, researchers, academia who want to examine ecovillages
- For GEN to have another bridge to academia, media, society and for themselves
- For ecovillage members, who
  - receive research request by students, universities etc.
  - want to research in their life or understand their way of life more deeply

NuMundo

numundo.org | ☎+1-510-541-1107 ✉contact@numundo.org 💻numundo.org

NuMundo connects "eco-nomads" around the globe. We are a database and a community of "impact centers" and meaningful travel experiences facilitating personal transformation and accelerating the development of regenerative living centers. A network of over 700 locations for intentional community, regenerative living, and reconnection with nature, our online platform answers the growing demand for purpose and physical participation in new ways of living. Through the NuMundo network, more people are connecting to their purpose, and joining a global revolution toward regenerative lifestyle design.

Oasen des Lebens

✉info@oasendeslebens.com 💻www.oasendeslebens.com

Holidays in oases all over the world - do some environmental protection on holiday and actively plant trees and harvest what has been sown there next year!

Open Hearts Hiking

✉info@openheartshiking.de 💻openheartshiking.de

Hiking is wonderful - preferably for several days. There are plenty of long-distance hiking trails. After just a few hours, everyday life is already forgotten and you are a traveller, curious, with time, thoughts wandering around and your heart opening... Those who also like exciting encounters while hiking will wander where the Open Hearts are on their way: Open Hearts Hiking is the idea of a network of people who want to go hiking. Open Hearts Hiking takes place at different times on different long-distance hiking trails. There you might have the pleasure to meet other Open Hearts Hikers. It remains a matter of luck, but you know that you are not alone on the trail...

wagendorf.de

💻www.wagendorf.de

Resources for trailer (community) living, see article on page 47f.

www.oeko.siedlungen.eu

Web Portal for Sustainable Settlements

✉wolpensinger@oekosiedlungen.de 💻www.oekosiedlungen.de

www.oekosiedlungen.de presents more than 250 eco-settlements including more than 60,000 appartments created in Germany since 1980. In many eco-settlements, communal aspects play a vital role, but the extent of communal work and interdependence vary enormously.

Bibliography

This bibliography gives some idea about community related literature. Surely there's more out there, especially if you take into account all that has been written about sustainable life and collaboration, about communication and ecological agriculture etc. - in the end, community living touches almost all aspects of life. We never tire to ask communities about recommendations, but the following list is far from a comprehensive library (we know some more German books that are not included in this bibliography). What's especially missing is good films about community life. There are very few. This is why eurotopia is planning to produce one in 2020. Do sign up to our newsletter at eurotopia.directory to stay informed. Don't worry about being bothered, we send it out next to never.

Some of these books are out of print, but you might find them second-hand.

DIRECTORIES OF INTENTIONAL COMMUNITIES & ECOVILLAGES

- **eurotopia - Verzeichnis von Gemeinschaften und Ökodörfern in Europa** (almost identical to the book your looking at, but in German) ISBN 978-3-9816860-3-6, available at www.eurotopia.directory (Shop)
- **Diggers & Dreamers** (mostly UK communities) ISBN 978-0954575755 (2015 edition), available at www.edgeoftime.co.uk
- **Communities Directory** (mostly US communitues and some in the rest of the world, only few in Europe) ISBN 978-0-9718264-9-6, available at www.ic.org → select "Directory Book" in the "Communities Map" menu
- **Comuni, Comunità, Ecovillaggi** Italian Communities Directory. ISBN 978-0-9565620-0-5
- Guidotti, Francesca: **Ecovillaggi e cohousing** Italian Directory from 2013. ISBN 88-66810-09-4
- **Anglican Religious Life 2018-19** A yearbook of religious orders and communities in the Anglican Communion and tertiaries, oblates, associates and companions. ISBN 9781848259621, available at Church House Bookshop: www.chbookshop.co.uk

INTENTIONAL COMMUNITIES & ECOVILLAGES PAST & PRESENT

- Bang, Jan Martin: **Ecovillages**. A Practical Guide to Sustainable Communities, ISBN 978-0865715387
- Christian, Diana Leafe: **Finding Community: How to Join an Ecovillage or Intentional Community**. ISBN 978-0865715783
- Christian, Diana Leafe und Adams, Patch: **Creating a Life Together: Practical Tools to Grow Ecovillages and Intentional Communities**. ISBN 978-0865714717
- Coates, Chris: **Communes Britannica**, A History of Communal Living in Britain, ISBN 9780951494592, available at www.edgeoftime.co.uk
- **Cohousing in Britain**, ISBN 978-0954575731, available at www.edgeoftime.co.uk
- Dregger, Leila: **Ecovillage: 1001 Ways to Heal the Planet** ISBN 978-1909470750, 2015, available at ecovillage.org (Resources→Market→Books)
- Fellowship for Intentional Communities: **Communities Magazine**. Extremely valuable resource for all interested in community living, published quarterly in print and as an e-paper. *At the time of printing, the publishing scheme of Communities Magazine is being evaluated*

- Global Ecovilla Network: **Ecovillage Living: Restoring the Earth and Her People** available at ecovillage.org (Resources→Market→Books)
- Global Ecovilla Network: **Ecovillages Around the World** available at ecovillage.org (Resources→Market→Books)
- Global Ecovilla Network: **various publications** available at ecovillage.org (Resources→Market→Books)
- Litfin, Karen T., **Ecovillages. Lessons for Sustainable Community**. ISBN 978-0745679501
- McCamant, Kathryn and Durerett, Charles: **Creating Cohousing. Building Sustainable Communities**, 2011. ISBN 978-0-86571-672-8
- Metcalf, Bill: **From Utopian Dreaming to communal Reality**: Cooperative Lifestyles in Australia. University of new South Wales Press, Sydney 1995
- Metcalf, Bill: **Shared Visions, Shared Lives**: Communal Living around the Globe 16 Stories by elders from communities around the globe, having lived at an average of 30 years in communities. Bill Metcalf is an Australian historian and sociologist, 1996. ISBN 1-899171-01-0
- Metcalf, Bill: **The Findhorn Book of Community Living**. Both an introduction to the different types of intentional community and a guided tour of communities around the globe. Findhorn Press, ISBN 1-844090529
- Vestbro, Dick Urban (editor): **Living together – Cohousing Ideas and Realities Around the World**. Proceedings from the international collaborative housing conference in Stockholm 5-9 May 2010, 2010, ISBN 978-91-7415-738-3
- Wain, David (director): **Wanderlust**, DVD. Community life in this movie might be full clichés, but mainstream life is even crazier.
  USA 2012, available at eurotopia.directory (Shop)
- Wolf, Stefan (director): **A New We**, DVD, available at eurotopia.directory (Shop)

SPECIFIC COMMUNITIES

- Bruderhof: numerous free e-books and some printed books, available at www.plough.com, i.e. Arnold, Emmy: **A Joyful Pilgrimage** (ISBN 978-0874869569)
- Findhorn: **The Findhorn Garden: Pioneering a New Vision of Humanity and Nature in Co-operation**, Findhorn Press Ltd (2003), ISBN 978-1844090181
- Findhorn: Caddy, Eileen: **Flight Into Freedom and Beyond**, Deep Books (2002), ISBN 978-1899171644
- Findhorn: Caddy, Eileen: **Opening Doors Within**, Findhorn Press (2007), ISBN 978-1844091089
- Findhorn: Caddy, Peter: **In Perfect Timing**, Findhorn Pubn, ISBN 978-1899171262
- Findhorn: Dorothy MacLean, **Memoirs of an Ordinary Mystic**, Lorian Press (2010), ISBN 978-0936878317
- Findhorn: Dorothy MacLean, **To Hear the Angels Sing**, Lorian Press (2008), ISBN 978-0936878010
- Findhorn: Meltzer, Graham, **Findhorn Reflections: A very personal take on life inside the famous spiritual community and ecovillage**. 2015, ISBN 978-1512006513

- Sieben Linden Ecovillage: Würfel, Michael (director): **Life under Palm Trees**, DVD, 59 min, available at eurotopia.directory (Shop)
- Tamera: Dregger, Leila: **Tamera: A Model for the Future**, Verlag Meiga, 2010

SUSTAINABILITY & ECOLOGY

- Dawson, Jonathan: **Ecovillages**. New frontiers for sustainability. Chelsea Green Publishing, 2006
- Martina Widmer, Sylvie Seguin (directors), Longo Maï Co-operatives: **From Seed to Seed - Educational films on seed production**, available at www.seedfilmd.org
- The Grace Foundation: **Defend the Sacred**. This book features women and men from all over the world who, despite everything, do not give up.They know that we can only master the challenges of our times when we unite. It's only by standing together that we'll be able to defend what lives, what is sacred, both inside and outside ourselves. 2019, ISBN 978-3927266667
- Holloway, John: **Crack Capitalism**. 2010, ISBN 978-0745330082
- Holmgren, David: **Permaculture: Principles and Pathways Beyond Sustainability**. ISBN 978-0646418445
- Hopkins, Rob: **The Transition Handbook: From Oil Dependency to Local Resilience** The book that inspired the Transition Town Movement - might just inspire you, too... 2008, ISBN 978-1900322188
- Kennedy and Kennedy: **Designing Ecological Settlements**. Ecological Planning and Building: Experiences in new housing and in the renewal of existing housing quarters in European countries. ISBN 3-496-02630-8
- Korten, David: **The Post-Corporate World**. ISBN 978-1887208031
- Langford, Andrew: **EcoSocial Design**. An exciting exploration of the tools and philosophies needed to redesign our human cultures as regenerative systems for the Earth, Ecosocial Design is also the basis of the first certificate at Gaia U. (page 565). Available at leanpub.com/ESD
- Mayur & Hernandez: **Learning for the New Millennium**: Challenges of education for the 21st century. Provides new approaches to learning. Essays by a.o. Bertrand Russell, Paolo Freire, E.F. Schumacher, Ivan Illich, Theodore Roszak. International Institute for Sustainable Future and Bgaratiya Vidya Bhavan, Mumbai, India 1997
- McLaughlin and Davidson: **Builders of the Dawn**: Community Lifestyles in a Changing World. Presents an overview of community lifestyles based on interviews with many community founders. It offers workable guidelines for those interested in building communities. 1990, ISBN 0-913990-68-X
- Mollison, Bill: **Permaculture: A Designers' Manual**. ISBN 978-0908228010, 1997
- Norberg-Hodge, Helena: **Ancient Futures**: Learning from Ladakh. This book raises important questions about the whole notion of progress and explores the root causes of the malaise of industrial society. At the same time, the story of Ladakh serves as a source of inspiration for our own future, 1991. ISBN 0-87156-559-5
- Pedagogy of the Earth: **Education for a sustainable future**. Ideas and information by some of the finest ecologists, developmentalists, thinkers, scientists, poets and even philosophers from around the world, 1999. ISBN 81-87404-00-01

- Whitefield, Patrick: **The Earth Care Manual: A Permaculture Handbook for Britain and Other Temperate Climates**. ISBN 978-1856232791, 2016

ALTERNATIVE & COMMUNAL LIFE IN GENERAL

- Adams, Patch/Mylander: **Gesundheit!: Bringing Good Health to You, the Medical System, and Society through Physician Service, Complementary Therapies, Humor, and Joy**. 1998, ISBN 978-0892817818
- Callenbach, Ernest: **Ecotopia: The Notebooks and Reports of William Weston**, 1975, various editions available
- Field, Martin: **Thinking About CoHousing**. This book explores the whys and wherefores of the UK CoHousing scene. It seeks both to contribute ideas to that task and tools which will be of assistance to groups trying to develop CoHousing Schemes. ISBN: 0951494570.
- Gaia Education (Hrsg.): **The Four Keys**, four comprehensive books compiled by Gaia Education associates, offering an overview of cutting-edge thinking on design for sustainability at local, regional and global scales. Available for free download at www.gaiaeducation.org or as individual books:
  - Kosha Anja Joubert (ed.), Robin Alfred (ed.): **Beyond you and me - Inspirations and Wisdom for Building Community**. ISBN 9781856230384
  - E. Christopher Mare & Max O. Lindegger: **Designing Ecological Habitats - Creating a Sense of Place**, 9781856230612
  - Jonathan Dawson (ed.), Ross Jackson (ed.) and Helena Norberg-Hodge (ed.): **Gaian Economics – Living Well Within Planetary Limits**, ISBN 9781856230568
  - Maddy Harland & William Keepin: **The Song of the Earth – A Synthesis of the Scientific & Spiritual Worldviews**, ISBN 9781856230957
- Landua, Gregory and Ethan Roland (Gaia U associates): **Regenerative Enterprise: Optimizing for Multi-Capital Abundance** ISBN 978-1-3290652-2-2, 2015
- McCamant & Durett: **Cohousing, A Contemporary Approach to Housing Ourselves** ISBN: 978-0898155396, 1994
- Peck, M. Scott: **The Different Drum**. The process of community building, ISBN 978-0684848587, 1998

TOOLS FOR COMMUNITY LIVING

- Jackson, Hildur: Creating Harmony: **Conflict Resolution in Communities**. An anthology with 25 interesting case studies from ecovillages and GEN members around the world. ISBN 1-85623-014-7, 1999
- Laloux, Frederic: **Reinventing Organisations**. Has inspired hundreds of organizations throughout the world to take a radical leap and adopt a whole different set of management principles and practices. ISBN 978-2960133509, 2014
- Starhawk: **The Empowerment Manual: A Guide for Collaborative Groups** ISBN 978-0865716971, 2011.

Check out https://eurotopia.directory and our shop for further inspiration

Empty Space

What is this, you may ask. Twelve empty pages in a Directory that's too heavy already and printed in small fonts with small pictures? Why not leave out those pages? Or make the pictures bigger?
This is the story: When the German directory (632 pages) was printed in the summer of 2019, the English cover was printed at the same time. We knew that the English book would be a little thinner, but we didn't expect it to be this much thinner. English is just so much leaner than German. This book, now, has 608 pages, and the cover just about wraps around them. That's still 12 pages more than the book would need to have, but if we'd lose any more pages, the cover wouldn't fit anymore (and it would look like it has to wear it's brother's shirts which are much too large). The layout is finished by now, so no enlarging of photos is possible, and we want the index to be in the back. So here we are, filling space inside the book. It would be so great to have a comic strip to insert here (and I'm sure I'll find one as soon as it's too late to insert it), but I write this two hours before the whole book file is sent to the printer's at a Friday at the end of November 2019. I considered copying jokes from the internet, but there's copyright issues and some of you might find our tone too silly already. So, here's two things that we'll do: We'll print "Notes" in the page headers to make it look like empty pages for notes is a really useful feature in a book. Additionally, we will insert a grey square on each empty page. If you liked cutting out the cover page*, you can cut along the square, which will get you a super cool secret hiding compartment that no thief will ever find, because what thief would read a communities directory?

I'll go to work with these valuable extra features now, goodbye, it was great to compile this book for you. Make sure to visit our website eurotopia.directory, it should have a newsletter option, so we can stay in touch!

Micha Würfel - for the eurotopia team.

* we hope you didn't because the rest of the cover page breaks so much easier if the window is cut or ripped out. On the other hand - if you really like it, continue to cut through this book as much as you like. We'll gladly sell you more.

Index

Here you can find names and page numbers of all communities and organisations of this book that are presented

- with detailed descriptions, existing or planned (**bold page numbers**),
- in the chapter "addresses from the last edition of eurotopia",
- in the chapter "community networks" or
- in the chapter of "useful addresses".

When communities might be known with different names, we list them separately with each name and lead you to their "main name" with an arrow - so if you have heard of a Dutch community called "Buitenplaats Sion", you look that up and find the entry "Buitenplaats Sion → Geworteld wonen (NL)". Under "Geworteld wonen (NL)" you will then find that entry's page numbers in this book.

The letters in (parenthesis) are country initials, explained on page 607. Special characters have their own list segment (i.e. "Ä" comes below all names starting with "A")

C

D

H

I

J

X

Y

Z

Country Initials (found with the addresses or in the index):

A	Austria
B	Belgium
BH	Bosnia and Herzegovina
BOL	Bolivia
CH	Switzerland
CZ	Czech Republic
D	Germany
DK	Denmark
E	Spain
EE	Estlonia
F	France
GR	Greece
H	Hungary
HR	Croatia
I	Italy
IL	Israel
IRL	Ireland
IS	Iceland
LT	Lithuania
LV	Latvia
N	Norway
NL	The Netherlands
NZ	New Zealand
P	Portugal
PL	Poland
RO	Romania
RUS	Russian Federation
S	Sweden
SF	Finland
SK	Slovakia
SLO	Slovenia
TR	Turkey
UE	Ukraine
UK	United Kingdom
USA	USA
ZP	Cyprus
WWW	*Internet*

:

Is eurotopia a recognised non-profit organisation? No. Once all costs for research, production, printing, advertising, administration and distribution of this directory have been paid, we hope to make a profit from this undertaking. This profit remains with the association „Blühende Landschaften e.V.", which publishes the book. The association consists of four adults with a shared economy, who raise three children together as co-parents in the ecovillage Sieben Linden (Christmas 2019, the children are 4, 4 and 7 years old). The group also took over an experimental straw bale building („Villa Strohbunt") from a group that broke up (and was the former publisher of this directory until 2009) and built a new straw bale house (Einhorn), both with (bank) loans. Therefore, if a profit remains from the publication of the eurotopia Directory, debts for ecological houses in Sieben Linden will be paid off. The Blühende Landschaften e.V. also plans to produce a film on community life and to expand the eurotopia Directory into digital and even more international spheres.
Decisions on the publication of this directory are be made primarily by Michael Würfel, whereby the others will be informed about major decisions and can stop him if necessary.

Legend

	planned community
	this project was not described in the last edition of eurotopia (2014)
	co-housing (with emphasis on community living)
	settlement (with emphasis on community living)
	ecovillage (self-assessment)
A1	coordinate on the respective map (countries with many communities)
	telephone number, email address, website
	year of founding
	urban/small town or suburban/in the country (behind: size of premises))
(...)	behind that, if applicable: name of bigger cities close by.
♀/♂/	number of women/men/children (if this information is given)
	number of members (inconsistencies between this and the answers above have simply been adopted)
	degree of self-sufficiency with food
	languages spoken in the community (other than language of that country) e=English, f=French, g=German, i=Italian, sp=Spanish
	how many of you are working on the community site? (i.e. how many of you are usually on location?)
	how many of you are working within community-owned enterprises?
	who decides important issues? (in parenthesis: how are these decisions made?
	handling of finances
	ownership of premises and buildings
	ideological, spiritual or religious focus
	main fields of work
	keywords
	open to new members - behind: requirements for new members
	working guests welcome after prior arrangement (free board and lodging)
( )	in parenthesis: as above, though not necessarily with free board and lodging
	member of the following community networks